PATTERN FORMATION AND DYNAMICS IN NONEQUILIBRIUM SYSTEMS

Many exciting frontiers of science and engineering require understanding of the spatiotemporal properties of sustained nonequilibrium systems such as fluids, plasmas, reacting and diffusing chemicals, crystals solidifying from a melt, heart muscle, and networks of excitable neurons in brains.

This introductory textbook for graduate students in biology, chemistry, engineering, mathematics, and physics provides a systematic account of the basic science common to these diverse areas. This book provides a careful pedagogical motivation of key concepts, discusses why diverse nonequilibrium systems often show similar patterns and dynamics, and gives a balanced discussion of the role of experiments, simulation, and analytics. It contains numerous illustrative worked examples, and over 150 exercises.

This book will also interest scientists who want to learn about the experiments, simulations, and theory that explain how complex patterns form in sustained nonequilibrium systems.

MICHAEL CROSS is a Professor of Theoretical Physics at the California Institute of Technology, USA. His research interests are in nonequilibrium and nonlinear physics including pattern formation, chaos theory, nanomechanical systems, and condensed matter physics, particularly the theory of liquid and solid helium.

HENRY GREENSIDE is a Professor in the Department of Physics at Duke University, USA. He has carried out research in condensed matter physics, plasma physics, nonequilibrium pattern formation, and theoretical neurobiology. He is also involved with outreach programs to stimulate interest in science and physics at junior high school and high school levels.

"This book by Cross and Greenside presents a comprehensive introduction to an important area of natural science, and assembles in one volume the essential conceptual, theoretical, and experimental tools a serious student will need to obtain a modern understanding of pattern formation outside of equilibrium. The masterful 50-page Introduction lays out the essential questions and provides motivation to the reader to explore the subsequent chapters, beginning with simple ideas and growing progressively in mathematical sophistication and physical depth. Careful attention is paid to the relationship between the theoretical methods and controlled laboratory experiments or numerical simulations. I can highly recommend this book to any student or researcher interested in a deepened understanding of nonequilibrium spatiotemporal patterns."

Pierre Hohenberg, New York University

"This book gives an excellent didactic introduction to pattern formation in spatially extended systems. It can serve both as the basis for an advanced undergraduate or graduate course as well as a reference. It is one of those books that will never outlive its usefulness. It is a must for anyone interested in nonlinear, nonequilibrium physics."

Eberhard Bodenschatz, MPI for Dynamics and Self-Organization, University of Goettingen, Cornell University

"This book fills a long-standing need, and is certain to be an instant classic. The physics of pattern forming systems is diverse but the theoretical core of the subject, along with many of the most important applications, can be learned from this splendid book. It is bound to be a key text for courses, as well as a much cited reference."

Stephen Morris, University of Toronto

PATTERN FORMATION AND DYNAMICS IN NONEQUILIBRIUM SYSTEMS

MICHAEL CROSS
California Institute of Technology

HENRY GREENSIDE
Duke University

CAMBRIDGE
UNIVERSITY PRESS

CAMBRIDGE
UNIVERSITY PRESS

University Printing House, Cambridge CB2 8BS, United Kingdom

Cambridge University Press is part of the University of Cambridge.

It furthers the University's mission by disseminating knowledge in the pursuit of education, learning and research at the highest international levels of excellence.

www.cambridge.org
Information on this title: www.cambridge.org/9780521770507

© M. Cross and H. Greenside 2009

This publication is in copyright. Subject to statutory exception and to the provisions of relevant collective licensing agreements, no reproduction of any part may take place without the written permission of Cambridge University Press.

First published 2009

A catalogue record for this publication is available from the British Library

ISBN 978-0-521-77050-7 Hardback

Cambridge University Press has no responsibility for the persistence or accuracy of URLs for external or third-party internet websites referred to in this publication, and does not guarantee that any content on such websites is, or will remain, accurate or appropriate.

To our families

Katy, Colin and Lynn
Peyton, Arthur and Noel

Contents

Preface		*page* xiii
1	**Introduction**	**1**
	1.1 The big picture: why is the Universe not boring?	2
	1.2 Convection: a first example of a nonequilibrium system	3
	1.3 Examples of nonequilibrium patterns and dynamics	10
	1.3.1 Natural patterns	10
	1.3.2 Prepared patterns	20
	1.3.3 What are the interesting questions?	35
	1.4 New features of pattern-forming systems	38
	1.4.1 Conceptual differences	38
	1.4.2 New properties	43
	1.5 A strategy for studying pattern-forming nonequilibrium systems	44
	1.6 Nonequilibrium systems not discussed in this book	48
	1.7 Conclusion	49
	1.8 Further reading	50
2	**Linear instability: basics**	**56**
	2.1 Conceptual framework for a linear stability analysis	57
	2.2 Linear stability analysis of a pattern-forming system	63
	2.2.1 One-dimensional Swift–Hohenberg equation	63
	2.2.2 Linear stability analysis	64
	2.2.3 Growth rates and instability diagram	67
	2.3 Key steps of a linear stability analysis	69
	2.4 Experimental investigations of linear stability	70
	2.4.1 General remarks	70
	2.4.2 Taylor–Couette instability	74

	2.5	Classification for linear instabilities of a uniform state	75
		2.5.1 Type-I instability	77
		2.5.2 Type-II instability	79
		2.5.3 Type-III instability	80
	2.6	Role of symmetry in a linear stability analysis	81
		2.6.1 Rotationally invariant systems	82
		2.6.2 Uniaxial systems	84
		2.6.3 Anisotropic systems	86
		2.6.4 Formal discussion	86
	2.7	Conclusions	88
	2.8	Further reading	88
3	**Linear instability: applications**		**96**
	3.1	Turing instability	96
		3.1.1 Reaction–diffusion equations	97
		3.1.2 Linear stability analysis	99
		3.1.3 Oscillatory instability	108
	3.2	Realistic chemical systems	109
		3.2.1 Experimental apparatus	109
		3.2.2 Evolution equations	110
		3.2.3 Experimental results	116
	3.3	Conclusions	119
	3.4	Further reading	120
4	**Nonlinear states**		**126**
	4.1	Nonlinear saturation	129
		4.1.1 Complex amplitude	130
		4.1.2 Bifurcation theory	134
		4.1.3 Nonlinear stripe state of the Swift–Hohenberg equation	137
	4.2	Stability balloons	139
		4.2.1 General discussion	139
		4.2.2 Busse balloon for Rayleigh–Bénard convection	147
	4.3	Two-dimensional lattice states	152
	4.4	Non-ideal states	158
		4.4.1 Realistic patterns	158
		4.4.2 Topological defects	160
		4.4.3 Dynamics of defects	164
	4.5	Conclusions	165
	4.6	Further reading	166

5	**Models**		**173**
	5.1	Swift–Hohenberg model	175
		5.1.1 Heuristic derivation	176
		5.1.2 Properties	179
		5.1.3 Numerical simulations	183
		5.1.4 Comparison with experimental systems	185
	5.2	Generalized Swift–Hohenberg models	187
		5.2.1 Non-symmetric model	187
		5.2.2 Nonpotential models	188
		5.2.3 Models with mean flow	188
		5.2.4 Model for rotating convection	190
		5.2.5 Model for quasicrystalline patterns	192
	5.3	Order-parameter equations	192
	5.4	Complex Ginzburg–Landau equation	196
	5.5	Kuramoto–Sivashinsky equation	197
	5.6	Reaction–diffusion models	199
	5.7	Models that are discrete in space, time, or value	201
	5.8	Conclusions	201
	5.9	Further reading	202
6	**One-dimensional amplitude equation**		**208**
	6.1	Origin and meaning of the amplitude	211
	6.2	Derivation of the amplitude equation	214
		6.2.1 Phenomenological derivation	214
		6.2.2 Deduction of the amplitude-equation parameters	217
		6.2.3 Method of multiple scales	218
		6.2.4 Boundary conditions for the amplitude equation	219
	6.3	Properties of the amplitude equation	221
		6.3.1 Universality and scales	221
		6.3.2 Potential dynamics	224
	6.4	Applications of the amplitude equation	226
		6.4.1 Lateral boundaries	226
		6.4.2 Eckhaus instability	230
		6.4.3 Phase dynamics	234
	6.5	Limitations of the amplitude-equation formalism	237
	6.6	Conclusions	238
	6.7	Further reading	239
7	**Amplitude equations for two-dimensional patterns**		**244**
	7.1	Stripes in rotationally invariant systems	246
		7.1.1 Amplitude equation	246
		7.1.2 Boundary conditions	248

		7.1.3	Potential	249
		7.1.4	Stability balloon	250
		7.1.5	Phase dynamics	252
	7.2	Stripes in anisotropic systems		253
		7.2.1	Amplitude equation	253
		7.2.2	Stability balloon	254
		7.2.3	Phase dynamics	255
	7.3	Superimposed stripes		255
		7.3.1	Amplitude equations	256
		7.3.2	Competition between stripes and lattices	261
		7.3.3	Hexagons in the absence of field-inversion symmetry	264
		7.3.4	Spatial variations	269
		7.3.5	Cross-stripe instability	270
	7.4	Conclusions		272
	7.5	Further reading		273
8	**Defects and fronts**			**279**
	8.1	Dislocations		281
		8.1.1	Stationary dislocation	283
		8.1.2	Dislocation dynamics	285
		8.1.3	Interaction of dislocations	289
	8.2	Grain boundaries		290
	8.3	Fronts		296
		8.3.1	Existence of front solutions	296
		8.3.2	Front selection	303
		8.3.3	Wave-number selection	307
	8.4	Conclusions		309
	8.5	Further reading		309
9	**Patterns far from threshold**			**315**
	9.1	Stripe and lattice states		317
		9.1.1	Goldstone modes and phase dynamics	318
		9.1.2	Phase diffusion equation	320
		9.1.3	Beyond the phase equation	327
		9.1.4	Wave-number selection	331
	9.2	Novel patterns		337
		9.2.1	Pinning and disorder	338
		9.2.2	Localized structures	340
		9.2.3	Patterns based on front properties	342
		9.2.4	Spatiotemporal chaos	345
	9.3	Conclusions		352
	9.4	Further reading		353

10	**Oscillatory patterns**	**358**
10.1	Convective and absolute instability	360
10.2	States arising from a type-III-o instability	363
	10.2.1 Phenomenology	363
	10.2.2 Amplitude equation	365
	10.2.3 Phase equation	368
	10.2.4 Stability balloon	370
	10.2.5 Defects: sources, sinks, shocks, and spirals	372
10.3	Unidirectional waves in a type-I-o system	379
	10.3.1 Amplitude equation	380
	10.3.2 Criterion for absolute instability	382
	10.3.3 Absorbing boundaries	383
	10.3.4 Noise-sustained structures	384
	10.3.5 Local modes	386
10.4	Bidirectional waves in a type-I-o system	388
	10.4.1 Traveling and standing waves	389
	10.4.2 Onset in finite geometries	390
	10.4.3 Nonlinear waves with reflecting boundaries	392
10.5	Waves in a two-dimensional type-I-o system	393
10.6	Conclusions	395
10.7	Further reading	396
11	**Excitable media**	**401**
11.1	Nerve fibers and heart muscle	404
	11.1.1 Hodgkin–Huxley model of action potentials	404
	11.1.2 Models of electrical signaling in the heart	411
	11.1.3 FitzHugh–Nagumo model	413
11.2	Oscillatory or excitable	416
	11.2.1 Relaxation oscillations	419
	11.2.2 Excitable dynamics	420
11.3	Front propagation	421
11.4	Pulses	424
11.5	Waves	426
11.6	Spirals	430
	11.6.1 Structure	430
	11.6.2 Formation	436
	11.6.3 Instabilities	437
	11.6.4 Three dimensions	439
	11.6.5 Application to heart arrhythmias	439
11.7	Further reading	441

12 Numerical methods — 445
- 12.1 Introduction — 445
- 12.2 Discretization of fields and equations — 447
  - 12.2.1 Finitely many operations on a finite amount of data — 447
  - 12.2.2 The discretization of continuous fields — 449
  - 12.2.3 The discretization of equations — 451
- 12.3 Time integration methods for pattern-forming systems — 457
  - 12.3.1 Overview — 457
  - 12.3.2 Explicit methods — 460
  - 12.3.3 Implicit methods — 465
  - 12.3.4 Operator splitting — 470
  - 12.3.5 How to choose the spatial and temporal resolutions — 473
- 12.4 Stationary states of a pattern-forming system — 475
  - 12.4.1 Iterative methods — 476
  - 12.4.2 Newton's method — 477
- 12.5 Conclusion — 482
- 12.6 Further reading — 485

Appendix 1 Elementary bifurcation theory — 496
Appendix 2 Multiple-scales perturbation theory — 503
Glossary — 520
References — 526
Index — 531

Preface

This book is an introduction to the patterns and dynamics of sustained nonequilibrium systems at a level appropriate for graduate students in biology, chemistry, engineering, mathematics, physics, and other fields. Our intent is for the book to serve as a second course that continues from a first introductory course in nonlinear dynamics. While a first exposure to nonlinear dynamics traditionally emphasizes how systems evolve in time, this book addresses new questions about the spatiotemporal structure of nonequilibrium systems. Students and researchers who succeed in understanding most of the material presented here will have a good understanding of many recent achievements and will be prepared to carry out original research on related topics.

We can suggest three reasons why nonequilibrium systems are worthy of study. First, observation tells us that most of the Universe consists of nonequilibrium systems and that these systems possess an extraordinarily rich and visually fascinating variety of spatiotemporal structure. So one answer is sheer basic curiosity: where does this rich structure come from and can we understand it? Experiments and simulations further tell us that many of these systems – whether they be fluids, granular media, reacting chemicals, lasers, plasmas, or biological tissues – often have *similar* dynamical properties. This then is the central scientific puzzle and challenge: to identify and to explain the similarities of different nonequilibrium systems, to discover unifying themes, and, if possible, to develop a quantitative understanding of experiments and simulations.

A second reason for studying nonequilibrium phenomena is their importance to technology. Although the many observed spatiotemporal patterns are often interesting in their own right, an understanding of such patterns – e.g. being able to predict when a pattern will go unstable or knowing how to select a pattern that maximizes some property like heat transport – is often important technologically. Representative examples are growing pure crystals, designing a high-power coherent laser, improving yield and selectivity in chemical synthesis, and inventing new electrical control techniques to prevent epilepsy or a heart attack. In these and other cases such as forecasting the weather or predicting earthquakes, improvements in the design, control, and prediction of nonequilibrium systems are often limited by our incomplete understanding of sustained nonequilibrium dynamics.

Finally, a third reason for learning the material in this book is to develop specific conceptual, mathematical, and numerical skills for understanding complex phenomena. Many nonequilibrium systems involve continuous media whose quantitative description is given in terms of nonlinear partial differential equations. The solutions of such equations can be difficult to understand (e.g. because they may evolve nonperiodically in time and be simultaneously disordered in space), and questions such as "Is the output from this computer simulation correct?," "Is this simulation producing the same results as my experimental data?" or "Is experimental noise relevant here?" may not be easily answered. As an example, one broadly useful mathematical technique that we discuss and use several times throughout the book is multiscale perturbation theory, which leads to so-called "amplitude equations" that provide a quantitatively useful reduction of complex dynamics. We also discuss the role of numerical simulation, which has some advantages and disadvantages compared to analytical theory and experimental investigation.

To help the reader master the various conceptual, mathematical, and numerical skills, the book has numerous worked examples that we call etudes. By analogy to a musical etude, which is a composition that helps a music student master a particular technique while also learning a piece of artistic value, our etudes are one- to two-page long worked examples that illustrate a particular idea and that also try to provide a non-trivial application of the idea.

Although this book is intended for an interdisciplinary audience, it is really a physics book in the following sense. Many of the nonequilibrium systems in the Universe, for example a germ or a star, are simply too complex to analyze directly and so are ill-suited for discovering fundamental properties upon which a general quantitative understanding can be developed. In much of this book, we follow a physics tradition of trying to identify and study simple idealized experimental systems that also have some of the interesting properties observed in more complex systems.

Thus instead of studying the exceedingly complex dynamics of the Earth's weather, which would require in turn understanding the effects of clouds, the solar wind, the coupling to oceans and ice caps, the topography of mountains and forest, and the effects of human industry, we instead focus our experimental and theoretical attention on enormously simplified laboratory systems. One example is Rayleigh–Bénard convection, which is a fluid experiment consisting of a thin horizontal layer of a pure fluid that is driven out of equilibrium by a vertical temperature difference that is constant in time and uniform in space. Another is a mixture of reacting and diffusing chemicals in a thin layer of gel, with reservoirs of chemicals to sustain the reaction. The bet is then that to understand aspects of what is going on in the weather or in an epileptic brain, it will be useful to explore some basic questions first for convection and other well-controlled laboratory systems. Similarly, as we discuss later in the book, there are conceptual, mathematical, and computational advantages if one studies simplified and reduced mathematical models such as the Swift–Hohenberg and complex Ginzburg–Landau equations when trying to understand the much more difficult partial differential equations that describe physical systems quantitatively. The experiments, models, simulations, and theory discussed in this book – especially the numerous comparisons of theory and simulation with experiment – will give the reader valuable insights and confidence about how to think about the more complex systems that are closer to their interests.

As background, readers of this book should know the equivalent of an introductory nonlinear dynamics course at the level of Strogatz's book [99]. Readers should feel comfortable with concepts such as phase space, dissipation, attractors (fixed points, limit cycles, tori, and strange attractors), basins of attractors, the basic bifurcations (super- and subcritical, saddle-node, pitchfork, transcritical, Hopf), linear stability analysis of fixed points, Lyapunov exponents, and fractal dimensions. A previous exposure to thermodynamics and to fluid dynamics at an undergraduate level will be helpful but is not essential and can be reviewed as needed. The reader will need to be competent in using multivariate calculus, linear algebra, and Fourier analysis at a junior undergraduate level. Several appendices in this book provide concise reviews of some of this prerequisite material, but only on those parts that are important for understanding the text.

There is too much material in this book for a single semester class so we give here some suggestions of what material could be covered, based on several scenarios of how the book might be used.

The first six chapters present the basic core material and should be covered in most classes for which this book is a main text. By the end of Chapter 6, most of the main ideas have been introduced, at least qualitatively. The successive chapters present more advanced material that can be discussed selectively. For example, those particularly interested in the systematic treatment of stationary patterns may choose to complete the semester by studying all or parts of Chapters 7 and 8, which provide quantitative discussions of two-dimensional patterns and localized structures, and Chapter 9 which is a more qualitative discussion of stationary patterns far from onset. For a less mathematical approach, it is possible to leave out the more technical Chapters 7 and 8 and move straight to Chapter 9 although we recommend including the first three subsections of Section 7.3 on the central question of the competition between stripes, two-dimensional lattices, and quasiperiodic patterns (these sections can be read independently of the remainder of the chapter). If the interest is more in dynamical phenomena, such as oscillations, propagating pulses, and waves (which may be the case if applying the ideas to signalling phenomena in biology is a goal), the class may choose to skip Chapters 7–9, pausing briefly to study Section 8.3 on fronts, and move immediately to Chapters 10 and 11 on oscillatory patterns and excitable media. Numerical simulations are vital to many aspects of the study of pattern formation, and to nonlinear dynamics in general, and so any of the above suggestions may include all or parts of Chapter 12.

In learning about nonequilibrium physics and in writing this book, the authors have benefited from discussions with many colleagues and students. We would like to thank Philip Bayly, Bob Behringer, Eshel Ben-Jacob, Eberhard Bodenschatz, Helmut Brand, Hugues Chaté, Peilong Chen, Elizabeth Cherry, Keng-Hwee Chiam, Bill Coughran Jr., Peter Daniels, David Egolf, Bogdan Epureanu, Paul Fischer, Jerry Gollub, Roman Grigoriev, James Gunton, Craig Henriquez, Alain Karma, Kihong Kim, Paul Kolodner, Lorenz Kramer, Andrew Krystal, Eugenia Kuo, Ming-Chih Lai, Herbert Levine, Ron Lifshitz, Manfred Lücke, Paul Manneville, Dan Meiron, Steve Morris, Alan Newell, Corey O'Hern, Mark Paul, Werner Pesch, Joel Reisman, Hermann Riecke, Sam Safran, Janet Scheel, Berk Sensoy, Boris Shraiman, Eric Siggia, Matt Strain, Cliff Surko, Harry Swinney, Shigeyuki Tajima, Gerry Tesauro, Yuhai Tu, Wim van Saarloos, and Scott Zoldi. We would like to express our appreciation to John Bechhoefer, Roman Grigoriev, Pierre Hohenberg, Steven Morris, and Wim Van Saarloos for helpful comments on early drafts of this book. And we would like especially to thank Guenter Ahlers and Pierre Hohenberg for many enjoyable and inspiring discussions over the years.

We would also like to thank the Department of Energy, the National Science Foundation, and the National Institute of Health for supporting our research over many years. We also thank the Aspen Center for Physics and the Institute of Theoretical Physics, Santa Barbara, where the authors were able to work productively on parts of this book.

Henry Greenside would like to thank his wife, Noel Greis, for her patience, support, and good cheer while he was working on this book.

Readers can find supplementary material on the book website at

`http://mcc.caltech.edu/pattern-formation-book/`,

or

`http://www.phy.duke.edu/~hsg/pattern-formation-book/`.

A list of errata will also be available on these websites.

We would be grateful to readers if they could forward to us comments they might have about the material and its presentation. These comments can be e-mailed to us at `mcc@caltech.edu` or `hsg@phy.duke.edu`.

1
Introduction

In this opening chapter, we give an informal and qualitative overview – a pep talk – to help you appreciate why sustained nonequilibrium systems are so interesting and worthy of study.

We begin in Section 1.1 by discussing the big picture of how the Universe is filled with nonequilibrium systems of many different kinds, a consequence of the fact that the Universe had a beginning and has not yet stopped evolving. A profound and important question is then to understand how the observed richness of structure in the Universe arises from the property of not being in thermodynamic equilibrium. In Section 1.2, a particularly well studied nonequilibrium system, Rayleigh–Bénard convection, is introduced to establish some vocabulary and insight regarding what is a nonequilibrium system. Next, in Section 1.3, we extend our discussion to representative examples of nonequilibrium patterns in nature and in the laboratory, to illustrate the great diversity of such patterns and to provide some concrete examples to think about. These examples serve to motivate some of the central questions that are discussed throughout the book, e.g. spatially dependent instabilities, wave number selection, pattern formation, and spatiotemporal chaos. The humble desktop-sized experiments discussed in this section, together with theory and simulations relating to them, can also be regarded as the real current battleground for understanding nonequilibrium systems since there is a chance to compare theory with experiment quantitatively.

Next, Section 1.4 discusses some of the ways that pattern-forming nonequilibrium systems differ from the low-dimensional dynamical systems that you may have seen in an introductory nonlinear dynamics course. Some guidelines are also given to determine qualitatively when low-dimensional nonlinear dynamics may not suffice to analyze a particular nonequilibrium system. In Section 1.5, a strategy is given and explained for exploring nonequilibrium systems. We explain why fluid dynamics experiments have some advantages over other possible experimental systems and why certain fluid experiments such as Rayleigh–Bénard convection are especially attractive. Finally, Section 1.6 mentions some of the topics that we will *not* address in this book for lack of time or expertise.

1.1 The big picture: why is the Universe not boring?

When people look at the world around them or peer through telescopes at outer space, a question that sometime arises is: why is there something rather than nothing? Why does our Universe consist of matter and light rather than being an empty void? While this question remains unanswered scientifically and is intensely pursued by researchers in particle physics and cosmology, in this book we discuss a second related question that is also interesting and fundamental: why does the existing matter and light have an interesting structure? Or more bluntly: why is the Universe not boring?

For it turns out that it is not clear how the existence of matter and light, together with the equations that determine their behavior, produce the extraordinary complexity of the observed Universe. Instead of all matter in the Universe being clumped together in a single black hole, or spread out in a featureless cloud, we see with our telescopes a stunning variety of galaxies of different shapes and sizes. The galaxies are not randomly distributed throughout space like molecules in a gas but are organized in clusters, the clusters are organized in super-clusters, and these super-clusters themselves are organized in voids and walls. Our Sun, a fairly typical star in a fairly typical galaxy, is not a boring spherical static ball of gas but a complex evolving tangled medium of plasma and magnetic fields that produces structure in the form of convection cells, sunspots, and solar flares. Our Earth is not a boring homogeneous static ball of matter but consists of an atmosphere, ocean, and rocky mantle that each evolve in time in an endless never-repeating dynamics of weather, water currents, and tectonic motion. Further, some of the atoms on the surface of our Earth have organized themselves into a biosphere of life forms, which we as humans particularly appreciate as a source of rich and interesting structure that evolves dynamically. Even at the level of a biological organism such as a mammal, there is further complex structure and dynamics, e.g. in the electrical patterns of the brain and in the beating of the heart.

So again we can ask: why does the matter and light that exist have such interesting structure? As scientists, we can ask further: is it possible to explain the origin of this rich structure and how it evolves in time? In fact, how should we define or quantify such informal and qualitative concepts such as "structure" or "patterns" or "complexity" or "interesting?" On what details does this complexity depend and how does this complexity change as various parameters that characterize a system are varied?

While this book will explain some of what is known about these questions, especially at the laboratory level which allows controlled reproducible experiments, we can say at a hand-waving level why the Universe is interesting rather than boring: the Universe was born in a cosmological Big Bang and is still young when measured in units of the lifetime of a star. Thus the Universe has not yet lasted long enough to come to thermodynamic equilibrium: *the Universe as a whole is a nonequilibrium system*. Because stars are young and have not yet reached thermodynamic equilibrium, the nuclear fuel in their core has not yet been consumed. The flux of energy from this core through the surface of the star and out into space drives the complex dynamics of the star's plasma and magnetic field. Similarly, because the Earth is still geologically young, its interior has not yet cooled down and the flux of heat from its hot core out through its surface, together with heat received from the Sun, drives the dynamics of the atmosphere, ocean, and mantle. And it is this same flux of energy from the Earth and Sun that sustains Earth's intricate biosphere.

This hand-waving explanation of the origin of nonequilibrium structure is unsatisfactory since it does not lead to the quantitative testing of predictions by experiment. To make progress, scientists have found it useful to turn to desktop experimental systems that can be readily manipulated and studied, and that are also easier to analyze mathematically and to simulate with a computer. The experiments and theory described in this book summarize some of the systematic experimental and theoretical efforts of the last thirty years to understand how to predict and to analyze such desktop nonequilibrium phenomena. However, you should appreciate that much interesting research remains to be carried out if our desktop insights are to be related to the more complex systems found in the world around us. We hope that this book will encourage you to become an active participant in this challenging endeavor.

1.2 Convection: a first example of a nonequilibrium system

Before surveying some examples that illustrate the diversity of patterns and dynamics in natural and controlled nonequilibrium systems, we first discuss a particular yet representative nonequilibrium system, a fluid dynamics experiment known as Rayleigh–Bénard convection. Our discussion here is qualitative since we wish to impart quickly some basic vocabulary and a sense of the interesting issues before turning to the examples discussed in Section 1.3 below. We will return to convection many times throughout the book, since it is one of the most thoroughly studied of all sustained nonequilibrium systems, and has repeatedly yielded valuable experimental and theoretical insights.

A Rayleigh–Bénard convection experiment consists of a layer of fluid, e.g. air or water, between two horizontal plates such that the bottom plate is warm and

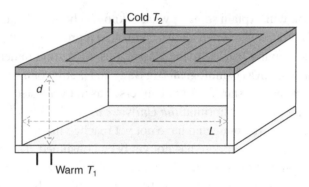

Fig. 1.1 Rayleigh–Bénard convection of a fluid layer between two horizontal plates is one of the simplest sustained nonequilibrium systems. The drawing shows a featureless square room of lateral width L and height d with copper-covered floor and ceiling, and supporting walls made of wood. By appropriate plumbing and control circuits, the floor and ceiling are maintained at constant temperatures of T_1 and T_2 respectively. When the temperature difference $\Delta T = T_1 - T_2$ is sufficiently large, the warm less-dense air near the floor and the cold more-dense air near the ceiling spontaneously start to move, i.e. convection sets in. The rising and falling regions of air eventually forms cellular structures known as convection rolls. The characteristic roll size is about the depth d of the air.

the upper plate is cool. As an example to visualize (but a bit impractical for actual experimentation as you will discover in Exercise 1.5), consider a square room whose lateral width L is larger than its height d, and in which all furniture, doors, windows, and fixtures have been removed so that there is only a smooth flat horizontal floor, a smooth flat horizontal ceiling, and smooth flat vertical walls (see Fig. 1.1). The floor and ceiling are then coated with a layer of copper, and just beneath the floor and just above the ceiling some water-carrying pipes and electronic circuits connected to water heaters are arranged so that the floor is maintained at a constant temperature T_1 and the ceiling is maintained at a constant temperature T_2.[1] Because copper conducts heat so well, any temperature variations within the floor or within the ceiling quickly become negligible so that the floor and ceiling can be considered as time-independent constant-temperature surfaces. The supporting sidewalls are made of some material that conducts heat poorly such as wood or Plexiglas.

A typical nonequilibrium experiment for the room in Fig. 1.1 would then be simply to fix the temperature difference $\Delta T = T_1 - T_2$ at some value and then to observe what happens to the air. "Observe what happens" can mean several

---

[1] Uniformly warming the floor and cooling the ceiling is not the usual way that a room is heated. Instead, a convector – a localized heat source with a large surface area – is placed somewhere in the room, and heat is lost through the windows instead of through the ceiling. (What we call a convector everyone else calls a radiator but this is poorly named since the air is heated mainly by convection, not by radiation.) But this nonuniform geometry is more complicated, and so less well suited, than our idealized room for experiment and analysis.

things depending on the questions of interest. By introducing some smoke into the room, the pattern of air currents could be visualized. A more quantitative observation might involve recording as a function of time t some local quantity such as the temperature $T(\mathbf{x}_0, t)$ or the x-component of the air's velocity $v_x(\mathbf{x}_0, t)$ at a particular fixed position $\mathbf{x}_0 = (x_0, y_0, z_0)$ inside the room. Alternatively, an experimentalist might choose to record some global quantity such as the total heat $H(t)$ transported from the floor to the ceiling, a quantity of possible interest to mechanical engineers and architects. These measurements of some quantity at successive moments of time constitute a time series that can be stored, plotted, and analyzed. A more ambitious and difficult observation might consist of measuring multivariate time series, e.g. measuring the temperature field $T(\mathbf{x}, t)$ and the components of the velocity field $\mathbf{v}(\mathbf{x}, t)$ simultaneously at many different spatial points, at successive instants of time. These data could then be made into movies or analyzed statistically. All of these observations are carried out for a particular fixed choice of the temperature difference ΔT and over some long time interval (long enough that any transient behavior will decay sufficiently). Other experiments might involve repeating the same measurements but for several successive values of ΔT, with each value again held constant during a given experiment. In this way, the spatiotemporal dynamical properties of the air in the room can be mapped out as a function of the parameter ΔT, and various dynamical states and transitions between them can be identified.

The temperature difference ΔT is a particularly important parameter in a convection experiment because it determines whether or not the fluid is in thermodynamic equilibrium. (It is precisely the fact that the nonequilibrium properties of the entire room can be described by a single parameter ΔT that constitutes the idealization of this experiment, and that motivated the extra experimental work of coating the floor and ceiling with copper.) If $\Delta T = 0$ so that the ceiling and floor have the same common temperature $T = T_1 = T_2$, then after some transient time, the air will be in thermodynamic equilibrium with zero velocity and the same uniform temperature T throughout. There is typically a transient time associated with approaching thermodynamic equilibrium because the air itself is rarely in such equilibrium without taking special precautions. For example, there might be a small breeze in the air when the door to the experimental room is closed or some part of the air may be a bit warmer than some other part because someone walked through the room. But as long as the room is sealed and the floor and ceiling have the same temperature, all macroscopic motion in the air will die out and the air will attain the same temperature everywhere.

As soon as the temperature difference ΔT becomes nonzero (with either sign), the air can no longer be in thermodynamic equilibrium since the temperature is spatially nonuniform. One says that the air is driven out of equilibrium by the temperature difference since the nonequilibrium state is maintained as long as there is a temperature difference. For the case $\Delta T > 0$ of a warm floor and cool ceiling, as ΔT becomes larger and larger (but again held constant throughout any particular experiment), more and more energy flows through the air from the warm floor to the cooler ceiling, the system is driven further from equilibrium, and more and more complicated spatiotemporal dynamical states are observed. A temperature difference is not the only way to drive a system out of equilibrium as we will discuss in other parts of the book. Other possibilities include inducing relative motion (e.g. pushing water through a pipe which creates a shear flow), varying some parameter in a time-dependent fashion (e.g. shaking a cup of water up and down), applying an electrical current across an electrical circuit, maintaining one or more chemical gradients, or creating a deviation from a Maxwellian velocity distribution of particles in a fusion plasma.

For any particular mechanism such as a temperature difference that drives a system out of equilibrium, there are dissipative (friction-like) mechanisms that oppose this driving and act in such a way so as to restore the system to equilibrium. For the air convecting inside our room, there are two dissipative mechanisms that restore the air to a state of thermodynamic equilibrium if ΔT is set to zero. One is the viscosity of the fluid, which acts to decrease any spatial variation of the velocity field. Since it is known from fluid dynamics that the velocity of a fluid is zero at a material surface,[2] the only possible long-term behavior for a fluid approaching equilibrium in the presence of static walls is that the velocity field everywhere decays to zero. A second dissipative mechanism is heat conduction through the air. The warm regions of air lose heat to the cooler regions of air by molecular diffusion, and eventually the temperature becomes constant and uniform everywhere inside the room. These dissipative mechanisms of viscosity and heat conduction are always present, even when $\Delta T \neq 0$, and so one often talks about a sustained nonequilibrium system as a driven-dissipative system.

Rayleigh–Bénard convection is sometimes called buoyancy-induced convection for reasons that illustrate a bit further the driving and dissipative mechanisms competing in a nonequilibrium system. Let us consider an experiment in which the air in the room has reached thermal equilibrium with $\Delta T = 0$ and then the temperature difference ΔT is increased to some positive value. Small parcels of air near the floor will expand and so decrease in density as they absorb heat from the floor, while small parcels of air near the ceiling will contract in volume and increase in density as they lose heat to the ceiling. As illustrated in Fig. 1.2, buoyancy forces then appear that accelerate the lighter warmer fluid upwards and the heavier colder

---

[2] More precisely, the fluid velocity at a wall is zero in a frame of reference moving with the surface. Exercise 1.9 suggests a simple experiment using an electric fan to explore this point.

1.2 Convection: a first example

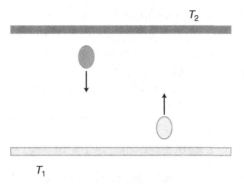

Fig. 1.2 Illustration of the driving and dissipative forces acting on small parcels of air near the floor and ceiling of the experimental room in Fig. 1.1 whose floor is warmer than its ceiling. The parcels are assumed to be small enough that their temperatures are approximately constant over their interiors. The acceleration of the parcels by buoyancy forces is opposed by a friction arising from the fluid viscosity and also by the diffusion of heat between warmer and cooler regions of the fluid. Only when the temperature difference $\Delta T = T_1 - T_2$ exceeds a finite critical value $\Delta T_c > 0$ can the buoyancy forces overcome the dissipation and convection currents form.

fluid downwards, in accord with the truism that "hot air rises" and "cold air falls." These buoyancy forces constitute the physical mechanism by which the temperature difference ΔT "drives" the air out of equilibrium. As a warm parcel moves upward, it has to push its way through the surrounding fluid and this motion is opposed by the dissipative friction force associated with fluid viscosity. Also, as the parcel rises, it loses heat by thermal conduction to the now cooler surrounding air, becomes more dense, and the buoyancy force is diminished. Similar dissipative effects act on a cool descending parcel.

From this microscopic picture, we can understand the experimental fact that making the temperature difference ΔT positive is a necessary but not sufficient condition for the air to start moving since the buoyancy forces may not be strong enough to overcome the dissipative effects of viscosity and conduction. Indeed, experiment and theory show that only when the temperature difference exceeds a threshold, a critical value we denote as ΔT_c, will the buoyancy forces be sufficiently large that the air will spontaneously start to move and a persistent spatiotemporal structure will appear in the form of convection currents. If the room's width L is large compared to its depth d so that the influence of the walls on the bulk fluid can be ignored, a precise criterion for the onset of convection can be stated in the form

$$R > R_c. \tag{1.1}$$

Table 1.1. *The isobaric coefficient of thermal expansion α, the kinematic viscosity ν, and the thermal diffusivity κ for air, water, and mercury at room temperature* T = 293 K *and at atmospheric pressure. These parameters vary weakly with temperature.*

Fluid	α (K^{-1})	ν (m^2/s)	κ (m^2/s)
Air	3×10^{-3}	2×10^{-5}	2×10^{-5}
Mercury	2×10^{-4}	1×10^{-7}	3×10^{-6}
Water	2×10^{-4}	1×10^{-6}	2×10^{-7}

The parameter R is defined in terms of various physical parameters

$$R = \frac{\alpha g d^3 \Delta T}{\nu \kappa}, \tag{1.2}$$

and the critical value of R has the approximate value

$$R_c \approx 1708. \tag{1.3}$$

The parameters in Eq. (1.2) have the following meaning: g is the gravitational acceleration, about 9.8 m/s^2 over much of the Earth's surface; $\alpha = -(1/\rho)(\partial \rho/\partial T)|_p$ is the fluid's coefficient of thermal expansion at constant pressure, and measures the relative change in density ρ as the temperature is varied; d is the uniform depth of the fluid; ΔT is the uniform temperature difference across the fluid layer; ν is the fluid's kinematic viscosity; and κ is the fluid's thermal diffusivity. Approximate values of the parameters α, ν, and κ for air, water, and mercury at room temperature ($T = 293$ K) and at atmospheric pressure are given in Table 1.1.

The combination of physical parameters in Eq. (1.2) is dimensionless and so has the same value no matter what physical units are used in any given experiment, e.g. System Internationale (SI), Centimeter-Gram-Seconds (CGS), or British. This combination is denoted by the symbol "R" and is called the Rayleigh number in honor of the physicist and applied mathematician Lord Rayleigh who, in 1916, was the first to identify its significance for determining the onset of convection. The pure number R_c is called the critical Rayleigh number R_c since it denotes the threshold that R must exceed for convection to commence. The value R_c can be calculated directly from the equations that govern the time evolution of a convecting fluid (the Boussinesq equations) as the criterion when the motionless conducting state of the fluid first becomes linearly unstable. The general method of this linear stability analysis is described in Chapter 2.

Despite its dependence on six parameters, you should think of the Rayleigh number R as simply being proportional to the temperature difference ΔT. The reason is that all the parameters in Eq. (1.2) except ΔT are approximately constant in a typical series of convection experiments. Thus the parameters α, ν, and κ in Eq. (1.1) depend weakly on temperature and are effectively fixed once a particular fluid is chosen. The acceleration g is fixed once a particular geographical location is selected for the experiment and the depth of the fluid d is typically fixed once the convection cell has been designed and is difficult to vary as an experimental parameter. Only the temperature difference ΔT is easily changed substantially and so this naturally becomes the experimental control parameter.

You should also note that the numerator $\alpha g d^3 \Delta T$ in Eq. (1.2) is related to quantities that determine the buoyancy force, while the denominator $\nu\kappa$ involves quantities related to the two dissipative mechanisms so Eq. (1.1) indeed states that instability will not occur until the driving is sufficiently strong compared to the dissipation. Most nonequilibrium systems have one or more such dimensionless parameters associated with them and these parameters are key quantities to identify and to measure when studying a nonequilibrium system.

What kind of dynamics can we expect for the air if the Rayleigh number R is held constant at some value larger than the critical value R_c? From Fig. 1.2, we expect the warm fluid near the floor to rise and the cool fluid near the ceiling to descend but the entire layer of ascending fluid near the floor cannot pass through the entire layer of descending fluid near the ceiling because the fluid is approximately incompressible. What is observed experimentally is pattern formation: the fluid spontaneously achieves a compromise such that some regions of fluid rise and neighboring regions descend, leading to the formation of a cellular convection "pattern" in the temperature, velocity, and pressure fields. The distance between adjacent rising and falling regions turns out to be about the depth of the air. Once the air begins to convect, the dynamics becomes too complicated to understand by casual arguments applied to small parcels of air and we need to turn to experiments to observe what happens and to a deeper mathematical analysis to understand the experimental results (see Figs. 1.14 and 1.15 below in Section 1.3.2). However, one last observation can be made. The motion of the fluid parcels inside the experimental system transport heat and thereby modify the temperature gradient that is felt in a particular location inside the system. Thus the motion of the medium changes the balance of driving and dissipation in different parts of the medium, and this is the reason why the dynamics is nonlinear and often difficult to understand.

The general points we learn from the above discussion about Rayleigh–Bénard convection are the following. There are mechanisms that can drive a system out of thermodynamic equilibrium, such as a flux of energy, momentum or matter through the system. This driving is opposed by one or more dissipative mechanisms such as viscous friction, heat conduction, or electrical resistance that restore the system to thermal equilibrium. The relative strength of the driving and dissipative mechanisms can often be summarized in the form of one or more dimensionless parameters, e.g. the Rayleigh number R in the case of convection. Nonequilibrium systems often become unstable and develop an interesting spatiotemporal pattern when the dimensionless parameter exceeds some threshold, which we call the critical value of that parameter. What happens to a system when driven above this threshold is a complex and fascinating question which we look at visually in the next section and then discuss in much greater detail throughout the rest of the book. However, the origin of the complexity can be understood qualitatively from the fact that transport of energy and matter by different parts of the pattern locally modifies the balance of driving and dissipation, which in turn may change the pattern and the associated transport.

1.3 Examples of nonequilibrium patterns and dynamics

1.3.1 Natural patterns

In this section we discuss examples of pattern-forming nonequilibrium systems as found in nature while in the next section we look at prepared laboratory systems, such that a nonequilibrium system can be carefully prepared and controlled. These examples help to demonstrate the great variety of dynamics observed in pattern-forming nonequilibrium systems and provide concrete examples to keep in mind as we try to identify the interesting questions to ask.

We begin with phenomena at some of the largest length and time scales of the Universe and then descend to human length and time scales. An example of an interesting pattern on the grandest scales of the Universe is the recently measured organization of galaxies into sheets and voids shown in Fig. 1.3. Observation has shown that our Universe is everywhere expanding, with all faraway galaxies moving away from each other and from the Earth, and with the galaxies that are furthest away moving the fastest. The light from a galaxy that is moving away from Earth is Doppler-shifted to a longer wavelength (becomes more red) compared to the light coming from an identical but stationary galaxy. By measuring the extent to which known spectral lines are red-shifted, astronomers can estimate the recessional speed v of a galaxy and convert this speed to a distance d by using the so-called Hubble law $v = H_0 d$, where the Hubble constant H_0 has the approximate value $65 \, \text{km s}^{-1} \, \text{Mpc}^{-1}$ (and a megaparsec Mpc is about 3×10^{19} km or about 3×10^6 light years).

Figure 1.3 summarizes such distance measurements for about 100 000 galaxies out to the rather extraordinary distance of about four billion light years which is

1.3 Examples of nonequilibrium patterns and dynamics 11

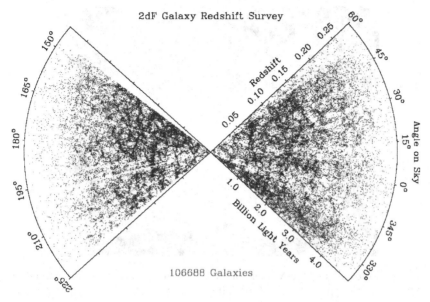

Fig. 1.3 Spatial distribution of 106 688 galaxies as measured in the 2dF (Two-degree Field) Galaxy Redshift survey out to a depth of over 4 billion light years from Earth. The left and right halves represent investigations over two separate arcs of the sky; the angles indicate astronomical declination, which is the angular latitude of a celestial object north or south of the celestial equator. Each point represents a galaxy whose distance from Earth is indicated in billions of light years or equivalently in terms of its redshift $z = \Delta\lambda/\lambda = v/c$ of the galaxy's light spectrum, where v is the velocity of recession from Earth. The distribution of galaxies is a nonuniform fractal-like structure with huge voids and walls.

comparable to the size of the Universe itself. Rather surprisingly, the galaxies do not fill space uniformly like molecules in a gas but instead are clustered in sheets and walls with large voids (relatively empty regions of space) between them. Here the pattern is not a geometric structure (e.g. a lattice) but a statistical deviation from randomly and uniformly distributed points that is difficult for the human visual system to quantify. Perhaps the closest earthly analogy would be a foam of bubbles in which the galaxies are concentrated on the surfaces of the bubbles. The reason for this galactic structure is not known at this time but is presumably a consequence of the details of the Big Bang (when matter first formed), the expansion of the Uuniverse, the effects of gravity, and the effects of the mysterious dark matter that makes up most of the mass of the Universe but which has not yet been directly observed or identified.

A second example of grand pattern formation is the M74 galaxy shown in Fig. 1.4. Now a galaxy consists of a huge number of about 10^{10} stars and has a net angular momentum from the way it was born by the condensation of a large hydrogen cloud.

Fig. 1.4 Photograph of the M74 spiral galaxy, a gravitationally bound island of 100 billion stars, approximately 100 000 light years wide, that lies about 35 000 000 light years from Earth in the Pisces constellation. Why galaxies form in the first place and why they appear in spiral, elliptical, and irregular forms remains incompletely understood. (Gemini Observatory, GMOS team.)

From just these facts, you might expect galaxies to be rotating featureless blobs of stars with a mass density that varies monotonically as a function of radius from the center. Such blobs do in fact exist and are known as elliptical galaxies. However many galaxies do have a nonuniform mass density in the form of two or more spiral arms as shown in Fig. 1.4. Our own galaxy, the Milky Way, is such a spiral galaxy and our Solar System lives in one of its high-density spiral arms.

Why galaxies evolve to form spiral arms is poorly understood and is an important open question in current astrophysical research. As we will see in the next section and in Chapter 11, laboratory experiments show that spiral formation is common for nonequilibrium media that have a tendency to oscillate in time or that support wave propagation. Further, experiments show that a tendency to form spirals is insensitive to details of the medium supporting the spiral. So a galactic spiral may not be too surprising since there are mechanisms in galaxies that can produce wave propagation. For example, some researchers have proposed that the spiral arms are detonation waves of star formation that propagate through the galaxy, somewhat analogous to the excitation waves observed in the Belousov–Zhabotinsky reaction–diffusion system shown below in Fig. 1.18 and discussed later in Chapter 11. Some interesting questions to ask about Fig. 1.4 are what determines the frequency of rotation of the spiral arms (which is not the same as the orbital rotation rate of the matter within a spiral arm) and what determines the spiral pitch (how tightly the spiral is wound)?

1.3 Examples of nonequilibrium patterns and dynamics

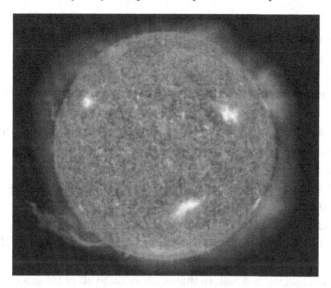

Fig. 1.5 Photograph of the Sun's surface in ultraviolet light, showing a complex time-dependent granular structure. The small bright regions are granules approximately 1000 km across (the Sun itself is about 100 Earth diameters in size) and correspond to hot plasma rising from the interior while the darker borders of the granules correspond to cooler plasma descending back to the interior. The filamentary structure exuding from the surface are plasma filaments following magnetic field lines.

Figure 1.5 descends from the scale of a galaxy to that of a star and shows a snapshot of the ultraviolet light emitted from the highly turbulent plasma in the so-called photosphere of the Sun. Heat diffuses by collisions from the Sun's small dense and extremely hot core (20 million degrees Kelvin) out to about two-thirds of the radius of the Sun, at which point the heat is transported to the cooler surface (about 6000 K) by convective motion of the Sun's plasma. The small bright dots in Fig. 1.5 are 1000 km-sized features called "granules" and correspond to the top of convection cells, the darker boundaries are where the cooler plasma descends back into the interior. The Rayleigh number R in Eq. (1.1) can be estimated for this convecting plasma and turns out to have the huge value of 10^{12} so Fig. 1.5 represents a very strongly driven nonequilibrium system indeed.

Figure 1.5 and related movies of the Sun's surface suggest many interesting questions related to pattern formation, many of which are not yet answered. One question is that of what determines the distributions of the sizes and lifetimes of the granules. Another question is that of how any organized structure persists at all since, at any given point, the plasma is varying rapidly and chaotically. Other solar images show that the smallest granules are found to cluster together to form convective structures called super-granules which may be 30 times larger on average. Why does this happen and what determines this new length scale? And what is the role of the magnetic field in all of this? Unlike the convecting air in Fig. 1.1, the Sun's plasma is a highly conducting electrical medium and its motion is influenced by the Sun's magnetic field (by a Lorentz force acting on currents in the plasma) and the magnetic field in turn is modified by the motion of the plasma (by Ampère's law, since currents generate a magnetic field). The magnetic field is known to be especially important for understanding the occurrence of sunspots, whose number varies approximately periodically with a 22-year cycle. There is evidence that the Earth's climate is partly influenced by the average number of sunspots and so a full understanding of the weather may require a deeper understanding of the Sun's spatiotemporal dynamics.

Our next example of pattern formation should be familiar to readers who have followed the observations of the planet Jupiter by the Voyager spacecraft and by the Hubble Space Telescope. Figure 1.6 shows a photograph of Jupiter in which one can see a nonequilibrium striped pattern that is common to all of the gas giants (Jupiter, Saturn, Neptune, and Uranus). Careful observation of the bands and of their dynamics shows that they are highly turbulent time-dependent flows of the outer portion of Jupiter's atmosphere, with adjacent bands flowing in opposite directions with respect to Jupiter's axis of rotation. Again numerous questions suggest themselves such as why do the bands form, what determines their wavelength of approximate

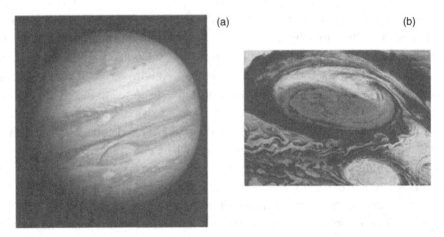

Fig. 1.6 (a) Photograph of the planet Jupiter (about 11 Earth diameters in size), showing a colored banded nonequilibrium structure. Such bands and spots are common to all the outer gaseous planets and arise from convection together with shear flow driven by the planet's rotation. (b) Blow-up of the famous great Red Spot, which is about the same size as the Earth. The persistence over many centuries of this turbulent spot within the surrounding turbulent atmosphere remains an intriguing mystery.

1.3 Examples of nonequilibrium patterns and dynamics

periodicity (their spacing is fairly uniform across the planet), and how do such bands survive in such a highly fluctuating fluid medium? The Red Spot and other similar spots pose a problem of their own, namely how do such large localized features (which are themselves time-dependent and strongly turbulent) survive in the middle of such a strongly fluctuating time-dependent fluid? The Red Spot was first observed by Galileo in the 1500s and so has been a persistent feature for at least 500 years.

The mechanisms that drive these nonequilibrium stripes and spots are not hard to identify. Jupiter's core is known to be hot and the transport of heat from the core out through its atmosphere causes convection in the outermost layer, just as in Fig. 1.1. However, the convection is substantially modified by Jupiter's rapid rotation around its axis, about once every 10 hours. As warm and cold parcels of fluid rise and descend, they are pushed to the side by large Coriolis forces and so follow a spiraling path.

We next turn to terrestrial examples of natural pattern formation and dynamics. While visiting a beach or desert, you have likely seen nonequilibrium pattern formation in the form of approximately periodic ripples found in sand dunes or sand bars, an example of which is shown in Fig. 1.7(a). The driven-dissipative nature of sand ripples is readily understood although the particular details of this pattern formation are not. The driving comes from wind (or water) flowing over the sand. When moving fast enough, the wind lifts sand grains into the air, transferring translational and rotational energy to them. These grains eventually fall back to earth and dissipate their energy into heat by friction as they roll and rub against other sand grains. The formation of nearly regular stripes is understood in rough outline, both from laboratory experiments and from computer simulations that can track the motion of tens of thousands of mathematical grains that collide according to specified rules. One surprise that came out of studying the stripes in sand dunes is that there is not a well-defined average wavelength as is the case for a convection pattern, for which the average wavelength is determined simply by the depth of the fluid. Instead, the average wavelength grows slowly with time, and can achieve kilometer length scales as shown in the Martian sand dunes of Fig. 1.7(b).

Another familiar and famous terrestrial example of pattern formation is a snowflake (see Fig. 1.8). This is a nonequilibrium system rather different than any described so far in that the pattern is formed by crystalline dendrites (these are the needle-like branches of a snowflake) that grow into the surrounding air. Unlike a convecting fluid in a fixed geometric box, the dendrite's shape itself changes as the system evolves. The nonequilibrium driving for snowflake formation is the presence of air that is supersaturated with water vapor. (In contrast, an equilibrium state would involve a static ice crystal in contact with saturated water vapor.) Each tip of the snowflake grows by adsorbing water molecules onto its surface from the

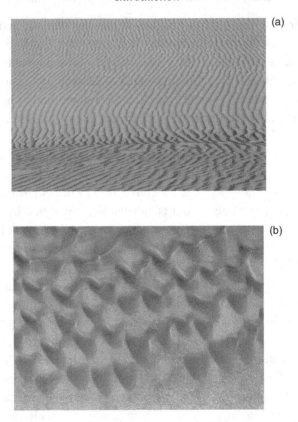

Fig. 1.7 (a) Pattern formation in wind-swept sand at the Mesquite Flat Sand Dunes in Death Valley, California. The ripple spacing is about 10 cm. The foreground of the picture is the top of a dune, and the remainder shows ripples on the valley floor. (Photo by M. C. Cross.) (b) Sand dunes in the Proctor Crater on Mars, as taken by the Mars Global Surveyor spacecraft in September of 2000 (Malin Space Science Systems). The average distance between dune peaks is about 500 m.

surrounding air, and the rate at which the tip grows and its shape are determined in a complex way by how rapidly water molecules in the surrounding air can diffuse to the crystalline tip, and by how rapidly the heat released by adsorption can be dissipated by diffusion within the air.

There are many fascinating questions associated with how snowflakes form. For example, what determines the propagation speed of the tip of a dendrite and is there a unique speed for fixed external conditions? Is there a unique shape to the tip of a dendrite and on what details does this shape depend? Why are the arms of a snowflake approximately the same length and have approximately the same intricate shape but are not exactly identical? And what causes the formation of the side-branches, whose rich spatial structure is such that no two snowflakes are presumably ever alike? Scientists have made progress over the last twenty years in

1.3 Examples of nonequilibrium patterns and dynamics

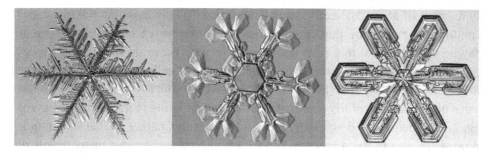

Fig. 1.8 Photographs by Kenneth Libbrecht of three snowflakes. The overall hexagonal symmetry reflects the underlying crystalline structure of water molecules while the intricate structure of the similar but not identical branches is a consequence of the instability of the dendritic tips, which propagate at an approximately constant speed into the surrounding supersaturated water vapor. (Photos courtesy of Kenneth Libbrecht.)

answering many of these questions. As a result of this progress, snowflakes rank among the best understood of all nonequilibrium systems.

Beyond their aesthetic beauty, you should appreciate that snowflakes belong to a technologically valuable class of nonequilibrium phenomena involving the synthesis of crystals and alloys. An example is the creation of meter-sized ultra-pure single crystal boules of silicon from which computer chips are made. One way to create such an ultra-pure boule is to pull a crystal slowly out of a rotating and convecting liquid silicon melt. In such a process, scientists and engineers have found that they need to understand the instabilities and dynamics of the solid–liquid interface (called a solidification front) since the extent to which undesired impurities can be prevented from diffusing into and contaminating the crystal depends delicately on the dynamics of the front. The metals that are the fabric of our modern world are also usually formed by solidification from the melt. Their strength, flexibility, and ductility are largely determined by the size and intermingling of small crystalline grains rather than by the properties of the ideal crystal lattice. This microstructure depends sensitively on the nonequilibrium growth process, for example how the solidification fronts propagate from the many nucleation sites. The tip of a snowflake dendrite is also a solidification front (although now between a solid and gas) and basic research on snowflakes has provided valuable insights for these other more difficult technological problems.

We turn finally to two examples of biological natural patterns and dynamics. A heroine of biological pattern formation is the slime mold *Dictyostelium discoideum*, which is a colony of tiny amoeba-like creatures – each about 10 microns in size – that live on forest floors. These cells spend most of their lives as solitary creatures foraging for food but when food becomes scarce, the cells secrete an attractant (cyclic AMP) into their environment that triggers pattern formation and the eventual aggregation of about 10^5 cells into a central mass. (This mass later evolves into a multicellular structure that can distribute cells to new regions where resources may be available, but that is another story.) Of interest to us is the spontaneous spatiotemporal pattern of propagating spiral waves of cells that is observed in the early stages of aggregation (Fig. 1.9). This pattern turns out to be remarkably similar to that observed in carefully prepared reacting and diffusing inorganic reagents (see Fig. 1.18(a) below), and you will indeed learn later in Chapter 11 that such multi-spiral states are observed in many nonequilibrium media and that many details of such states are understood theoretically.

Experiment and theory have shown that, in rough outline, the slime-mold pattern arises from a nonlinear dynamics in which cells secrete an attractant, cells move toward higher concentrations of the attractant (a process known as chemotaxis), and attractant is destroyed by secretion of an appropriate enzyme. The slime-mold dynamics is nonequilibrium because there are sustained chemical gradients; temperature and velocity gradients are not important here as they were for a convecting flow. Figure 1.9 suggests some quantitative questions similar to those suggested by Fig. 1.4, namely what determines the frequency and velocity of the waves in the spirals and how do these quantities vary with parameters? And biologists would like to know why slime molds use this particular spatiotemporal pattern to self-organize into a new multicellular structure.

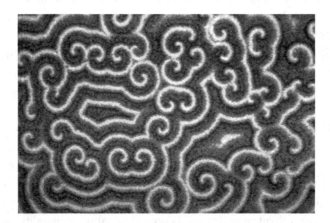

Fig. 1.9 Photograph of a starving slime-mold colony in the early stages of aggregation. The cells were placed on an 8-cm-wide caffeine-laced agar dish with an average density of 106 cells/cm^2. The field of view covers 4 cm. The light regions correspond to elongated cells that are moving with a speed of about 10 microns/minute by chemotaxis toward higher secretant concentrations. The dark regions correspond to flattened cells that are stationary. The spiral waves rotate with a period of about 5 minutes. This early aggregation stage persists for about four hours after which the pattern and cell behavior changes substantially, forming thread-like streams. (Figure courtesy of Dr. Florian Siegert.)

1.3 Examples of nonequilibrium patterns and dynamics

Our second biological example of a natural pattern formation is the important medical problem of ventricular fibrillation. This occurs when the thick muscle tissue surrounding the left ventricle (the largest of the four heart chambers) enters into an irregular spatiotemporal electrical state that is no longer under the control of the heart's pacemaker (the sinoatrial node). In this fibrillating state, the ventricular muscle cannot contract coherently to pump blood, and the heart and the rest of the body start to die from lack of oxygen. A common but not always successful treatment is to apply a massive electrical current to the heart (via a defibrillator) that somehow eliminates the irregular electrical waves in the left ventricle and that resets the heart tissue so that the ventricle can respond once again to the sinoatrial node.

Why the dynamics of the left ventricular muscle sometimes changes from periodic coherent contractions to a higher-frequency nonperiodic incoherent quivering is still poorly understood. One intriguing observation is that ventricular fibrillation is observed primarily in mammals whose hearts are sufficiently large or thick. Thus mice, shrews, and guinea pigs do not easily suffer ventricular fibrillation while pigs, dogs, horses, and humans do. Experiments, theory, and simulations have begun to provide valuable insights about the spatiotemporal dynamics of ventricular fibrillation and how it depends on a heart's size and shape, as well as on its electrical, chemical, and mechanical properties. An example is Fig. 1.10, taken

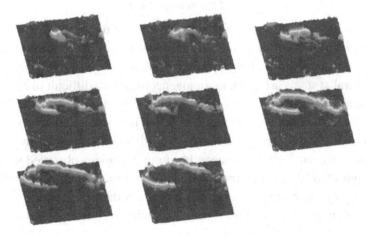

Fig. 1.10 Visualization of the surface voltage potential of an isolated blood-perfused dog heart in a fibrillating state. The surface of the left ventricle was painted with a dye whose fluorescent properties are sensitive to the local transmembrane voltage (which is of order 80 millivolts), and then the fluorescence under a strong external light source was recorded as a function of time. The waves propagate at a speed of about 20–40 cm/s (heart tissue is anisotropic and the speed varies with the direction of propagation), a range that is about the same for most mammalian hearts. (From Witkowski *et al.* [114].)

from a pioneering experiment that visualized the time-dependent voltage pattern on the surface of a fibrillating dog heart. This electrical pattern is complex and consists of spiral-like waves that move around, and that sometimes terminate or are created through collisions with other waves. Since muscle tissue contracts shortly after an electrical wave front passes through it, the irregular geometric shape of the waves in Fig. 1.10 explains directly why the heart is not contracting coherently and so has difficulty pumping blood. The similarity of the dynamics to that observed in slime molds and in reacting and diffusing chemical solutions is likely misleading. Left ventricular muscle is a rather thick three-dimensional nonequilibrium medium and recent theoretical research suggests that the surface patterns in Fig. 1.10 are likely intersections by the surface of more intricate three-dimensional electrical waves inside the heart wall that experimentalists have not yet been able to observe directly.

Given that ventricular fibrillation kills over 200 000 people in the United States each year and is a leading cause of death in industrial countries worldwide, understanding the onset and properties of ventricular fibrillation and finding ways to prevent it remain major medical and scientific goals. Chapter 11 will discuss heart dynamics in more detail since it turns out to be one of the more exciting current frontiers of nonequilibrium pattern formation and illustrates well many of the concepts discussed in earlier chapters.

1.3.2 Prepared patterns

The previous section surveyed some of the patterns and dynamics that are observed in natural nonequilibrium systems. For the most part, these natural systems are difficult to study and to understand. Unlike the idealized room of convecting air in Fig. 1.1, natural systems are often inhomogeneous and so difficult to characterize, they are subject to many different and simultaneous mechanisms of driving and dissipation (some of which are not known or are not well understood), and some systems are simply too remote or too big for direct experimental investigation. In this section, we survey some nonequilibrium phenomena observed in idealized carefully controlled laboratory experiments and reach the important conclusion that even such highly simplified systems can produce a dazzling variety of complex patterns and dynamics, often with properties similar to those observed in natural systems.

Figure 1.11 shows several patterns and dynamical states in a Taylor–Couette fluid dynamics experiment. The experiment is named after the French scientist Maurice Couette who, in the late 1800s, was one of the first to use this apparatus to study the shearing of a fluid, and after the British scientist Geoffrey Taylor who used this system in the early 1920s to make the first quantitative comparison in fluid dynamics of a linear stability analysis with experiment (see Figs. 2.5 and 2.6 in Section 2.4).

1.3 Examples of nonequilibrium patterns and dynamics

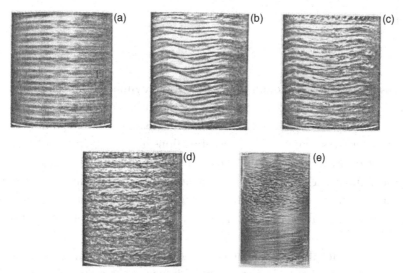

Fig. 1.11 Five examples of pattern formation in a Taylor–Couette fluid dynamics experiment, in which a fluid filling the thin annular gap between two concentric cylinders is sheared by rotating the inner and outer cylinders at constant but different speeds. The flow pattern was visualized by shining light through the transparent outer cylinder and scattering the light off a low concentration of shiny impurities such as aluminum flakes. In panels (a)–(d), the outer cylinder is at rest, while the inner cylinder is rotating at different angular frequencies corresponding respectively to inner Reynolds numbers of $\mathcal{R}_i/\mathcal{R}_c = 1.1, 6.0, 16.0$, and 26.5, where $\mathcal{R}_c$ is the critical value at which laminar flow becomes unstable to Taylor cells. The fluid is water at $T = 27.5\,°C$ with kinematic viscosity $\nu = 8.5 \times 10^{-7} m^2/s$. The height of the two glass cylinders is $H = 6.3$ cm while the outer and inner radii are respectively $r_2 = 2.54$ cm and $r_1 = 2.22$ cm. Only pattern (a) is time-independent. (e) A so-called stripe-turbulent state found in a different Couette experiment with parameters $r_1 = 5.3$ cm, $r_2 = 5.95$ cm, and $H = 20.9$ cm. The inner and outer Reynolds numbers are $\mathcal{R}_i = 943$ and $\mathcal{R}_o = -3000$. (Panels (a)–(d) from Fenstermacher et al. [35]; panel (e) from Andereck et al. [3].)

Taylor–Couette flow has some similarities to Rayleigh–Bénard convection in that a fluid like water or air is placed between two walls, here the inner and outer boundaries of two concentric cylinders (with the outer cylinder usually made of glass to facilitate visualization). But instead of being driven out of equilibrium by a temperature gradient, the fluid is driven out of equilibrium by a velocity gradient that is sustained by using motors and gears to rotate the inner and outer cylinders at constant angular frequencies ω_i and ω_o respectively (not necessarily with the same sign). The fluid temperature is constant throughout.

As was the case for convection, a dimensionless combination of system parameters can be identified as the "stress" parameter that measures the strength of driving compared to dissipation. When both cylinders are spinning, there are two such parameters and they are traditionally called the inner and outer Reynolds numbers.

They have the form

$$\mathcal{R}_i = \frac{\omega_i r_i (r_o - r_i)}{\nu}, \quad \mathcal{R}_o = \frac{\omega_o r_o (r_o - r_i)}{\nu}, \quad (1.4)$$

where r_i is the radius of the outer wall of the inner cylinder, r_o is the radius of the inner wall of the outer cylinder, and ν is again the fluid's kinematic viscosity. Note that there are many other valid ways to define dimensionless stress parameters here, e.g. for an inner stress parameter the combinations $\omega_i r_i^2/\nu$, $\omega_i(r_o^2 - r_i^2)/\nu$, or even $(1/2)\omega_i^2(r_o+r_i)(r_o-r_i)^3/\nu^2$ could be used instead. (This last combination is called the Taylor number and is basically the square of a Reynolds number.) Often the appropriate choice of a stress parameter is suggested by a linear stability analysis of the dynamical equations but in some cases the choice is simply set by historical precedent.

Let us consider first the situation of a fixed outer cylinder so that $\omega_o = 0$ and $\mathcal{R}_o = 0$. Then experiments show – in agreement with theory – that the velocity field **v** of the fluid is time-independent and featureless[3] until the inner Reynolds number $\mathcal{R}_i$ exceeds a critical value $\mathcal{R}_c \approx 100$. (The specific value of $\mathcal{R}_c$ depends on the ratio of radii r_o/r_i and on the height of the cylinders.) For $\mathcal{R} > \mathcal{R}_c$, interesting patterns appear and these can be visualized by doping the fluid with a small concentration of metallic or plastic flakes that reflect external light. Unlike the natural systems described in the previous section, a Taylor–Couette cell can be accurately controlled with the temperature, inner and outer radii, and rotational velocities all determined to a relative accuracy of 1% or better. The results of such experiments are highly reproducible and so the response of the system to small changes in the parameters can be carefully and thoroughly mapped out.

Figure 1.11 shows examples of the patterns observed in particular Taylor–Couette cells of fixed fluid, height, and inner and outer radii, for different constant values of the Reynolds numbers $\mathcal{R}_i$ and $\mathcal{R}_o$. For $\mathcal{R}_o = 0$ and $\mathcal{R}_i > \mathcal{R}_c$ just larger than the critical value $\mathcal{R}_c$ at which a uniform state becomes unstable, Fig. 1.11(a) shows that the fluid spontaneously forms a time-independent pattern of uniformly spaced azimuthally symmetric donut-like cells called Taylor cells. Given the static one-dimensional nature of this pattern, there is really only one interesting question to ask, which is the question of wave number selection: what determines the wavelength of the Taylor cells and is this wavelength unique for fixed external parameters? This is a basic question in pattern formation, and we will return to it a number of times in this book.

As the inner Reynolds number $\mathcal{R}_i$ is increased further, the static Taylor cells become unstable to a time-periodic state consisting of waves that propagate around

---

[3] "Featureless" here means that the velocity field **v** is independent of the azimuthal and axial coordinates and has a simple monotonic dependence on the radial coordinate.

each Taylor cell. Such a transition in a dynamical system is often called a bifurcation.[4] This regime is called the wavy vortex state, a snapshot of which is shown in Fig. 1.11(b). The angular frequency of the waves is somewhat less than the rotational frequency of the inner cylinder and is known experimentally to depend on the values of the parameters r_o, r_i, and v. For still larger $\mathcal{R}_i$, a second Hopf bifurcation takes place, leading to time-quasiperiodic dynamics and a more complex spatial motion known as the modulated wavy vortex state (see Fig. 1.11(c)). For still larger values of $\mathcal{R}_i$ (see Fig. 1.11(d)), the fluid becomes turbulent in that the time dependence is everywhere nonperiodic, there is no longer any identifiable wave motion, and the spatial structure is disordered. Note how one can perceive the ghostly remains of the Taylor cells in this turbulent regime, raising again a question similar to the one we asked about the Red Spot of Jupiter, namely how can some kind of average structure persist in the presence of strong local fluctuations? (A theoretical understanding of this strongly driven regime has not yet been developed.) A practical engineering question to answer would be to predict the average torque on the inner cylinder as a function of the Reynolds number $\mathcal{R}$. How does the complex fluid motion modify the resistance felt by the motor, which is turning the inner cylinder at constant speed?

Figure 1.12 summarizes many of the dynamical states that have been discovered experimentally in Taylor–Couette flows for different values of the inner and outer Reynolds numbers.[5] This figure raises many interesting questions. Where do all these different states come from and are these the only ones that can occur? Is the transition from one state to another, say from Couette flow to Taylor vortex flow or from spiral turbulence to featureless turbulence, similar to an equilibrium phase transition corresponding to the melting of a crystal to form a liquid or the evaporation of a liquid to form a gas? Little is known about most of these states and their transitions. One of the few theoretical successes is the heavy black line, which Taylor predicted in 1923 as the boundary separating the featureless laminar regime of Couette flow from various patterned states.

While quite interesting, the patterns in Taylor–Couette flows tend to have mainly a one-dimensional cellular structure and so we turn next to laboratory experiments of Rayleigh–Bénard convection that show two- and three-dimensional pattern formation and dynamics. As you learn by answering Exercise 1.5, a room like that described in Fig. 1.1 is impractical for convection experiments because the onset of convection is reached for a tiny difficult-to-achieve temperature difference, and the

---

[4] A bifurcation of a dynamical system that introduces an intrinsic temporal oscillation is called a Hopf bifurcation, so the transition of Taylor cells to the wavy vortex state in Fig. 1.11(b) is a spatiotemporal example of a Hopf bifurcation.
[5] This diagram is not complete since some regimes are hysteretic. One can then observe different states for the same values of $\mathcal{R}_i$ and $\mathcal{R}_o$, depending on the history of the experiment.

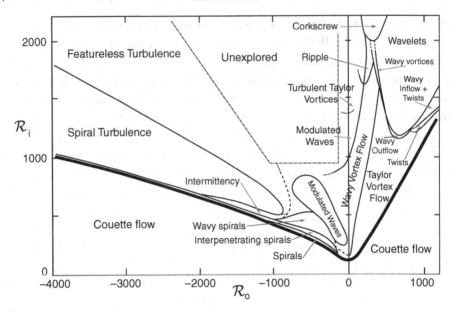

Fig. 1.12 Phase diagram of patterns observed in Taylor–Couette flow as a function of the inner Reynolds number $\mathcal{R}_i$ and the outer Reynolds number $\mathcal{R}_o$. The heavy line denotes the boundary between featureless flow below the line and patterned states above the line. (Redrawn from Andereck et al. [3].)

time scales for observation are uncomfortably long. Instead, experimentalists use tiny convection cells that are Swiss watches of high precision, with a fluid depth of perhaps $d \approx 1$ mm and a width $L \approx 5$ cm. The bottom and top plates of such apparatuses are machined and polished to be flat to better than one micron and then aligned to be parallel to better than one part in 10^4. The bottom plate may be made of gold-plated copper which has a thermal conductivity about 1000 times higher than water or air. The upper plate is often made of a thin, wide (and expensive!) sapphire plate, which has the nice properties of being optically transparent (allowing visualization of the flow) and of being an excellent thermal conductor. The mean temperature of the fluid and the temperature difference ΔT across the plates can be controlled to better than 1 milliKelvin (again about one part in 10^4) for more than a month of observation at a time. A Rayleigh–Bénard convection experiment has a significant advantage over Taylor–Couette and other fluid experiments in having no moving parts in contact with the fluid. Thus motors or pumps that oscillate and vibrate can be avoided and the observed dynamics of the convecting fluid is intrinsic since the fluid is bounded by time-independent spatially homogeneous boundaries.

The patterns of a convecting flow are usually visualized by a method called shadowgraphy (Fig. 1.13). The index of refraction of a fluid is weakly dependent

1.3 Examples of nonequilibrium patterns and dynamics

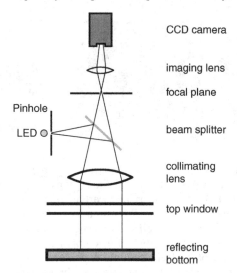

Fig. 1.13 Schematic drawing of the shadowgraphy method for visualizing a Rayleigh–Bénard convection pattern. Monochromatic light from a light emitting diode (LED) is reflected by a beam splitter through the transparent top plate (often made of sapphire) of the convection experiment. The light beams are refracted by the warm and cold regions of the fluid which act as diverging and converging lenses respectively. The light is then reflected off the mirror bottom plate, is refracted once more through the convection rolls, back through the beam splitter, and is then analyzed by a CCD (charge-coupled-device) videocamera. (From deBruyn *et al.* [30].)

on temperature so that the warm rising plumes of fluid act as a diverging lens and the cold descending plumes of fluid act as a converging lens. A parallel beam of light passing through the convecting fluid will be refracted by the convection rolls, and focused toward the regions of higher refractive index. The convection rolls act as an array of lenses producing, at the imaging plane, a pattern of alternating bright and dark regions, with the bright regions corresponding to the cold down-flow and the dark regions corresponding to the warm up-flow. These images are often sufficient to identify the interesting patterns and their dynamics, and can be recorded by a video camera and stored in digital form for later analysis.

Given this background, you can now appreciate the experimental data of Fig. 1.14, which shows three convection patterns in large cylindrical geometries. As we will see, the typical size of the structures in the convecting flow is set by the depth of the layer of fluid. The important parameter describing the "size" of the experimental system is therefore the aspect ratio defined as the lateral extent of the convecting fluid (e.g. the radius of a cylindrical cell) divided by the fluid depth. We will use the symbol Γ for the aspect ratio. The convecting

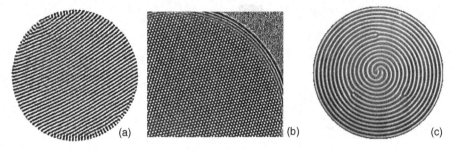

Fig. 1.14 Visualization of convecting fluid patterns in large cylindrical cells by the shadowgraphy method of Fig. 1.13. The white regions correspond to descending colder fluid and the dark stripes to rising warmer fluid. (a) A stationary stripe pattern just above onset for $R = 1.04R_c$, with a local wavelength that is close to twice the depth of the fluid. This cell has an aspect ratio of $\Gamma = r/d = 41$. (b) A remarkably uniform stationary lattice of hexagonal convection cells is found in gaseous CO_2 for $R = 1.06R_c$. Here the fluid is descending in the middle of each hexagonal cell and rising along its boundaries. Only a portion of an aspect-ratio $\Gamma = 86$ cell is shown. With a small increase in the Rayleigh number to $R = 1.15R_c$, the hexagonal cells in (b) change into a slowly-rotating spiral. A three-armed spiral pattern is shown in (c). (Panels (a) and (b) from deBruyn et al. [30]; (c) from the website of Eberhard Bodenschatz.)

fluid in the experiments shown in Fig. 1.14 is compressed carbon dioxide at room temperature.

Figure 1.14(a) shows a most remarkable fact. After some transient dynamics not shown, the rising and falling plumes of fluid self-organize into a time-independent periodic lattice of straight lines, often called "stripes." The surprise is that the circular geometry of the surrounding walls has little effect on this final geometric pattern; one might have expected instead the formation of axisymmetric (circular) convection rolls with the same symmetry as that of the lateral walls.

Under slightly different conditions Fig. 1.14(b) shows that a nearly perfect time-independent lattice of hexagonal convection cells forms. In each hexagon, warm fluid rises through its center and descends at its six sides, and the diameter of each hexagon is about the depth of the fluid. As was the case for the stripe pattern, the cylindrical shape of the lateral walls seems to have little effect on the pattern formation within the fluid except for the few cells directly adjacent to the lateral wall.

In Fig. 1.11(b) and in Figs. 1.14(a) and (b), we seem to be observing an intrinsic ordering of the convection cells which you might guess is analogous to the formation of a crystalline lattice of atoms as some liquid is slowly cooled. However, as we discuss in the next chapter and later in the book, the mechanism for formation of stationary nonequilibrium periodic lattices is fundamentally different than the mechanism by which periodic equilibrium crystalline lattices form, e.g. the cubic

1.3 Examples of nonequilibrium patterns and dynamics

lattice of sodium and chlorine atoms in table salt. In the latter case, atoms attract each other at long distances and repel each other at short distances so that the lattice spacing is determined by the unique energy minimum for which the repulsive and attractive forces balance. In contrast, the lattice spacing in a nonequilibrium system is determined by dynamic mechanisms that have nothing to do with repulsion or attraction and with which no energy-like quantity can generally be associated. A consequence is that nonequilibrium lattices may lack a unique lattice spacing for specified experimental conditions.

For the same fluid and geometry of Fig. 1.14(b), if the Rayleigh number is increased just a tiny bit more to the value $R = 1.15R_c$, the hexagonal lattice disappears and is replaced by a large slowly rotating spiral. A similar spiral pattern in a smaller aspect ratio cell is shown in Fig. 1.14(c). If you look carefully, you will see that the spiral terminates before reaching the lateral wall by merging with three topological defects called dislocations. The three convection patterns of Fig. 1.14 raise obvious interesting questions about pattern formation in nonequilibrium systems. Why do we see stripes in one case and hexagons in another? What determines the lattice spacing? Why do the hexagons disappear with a small increase in R, to be replaced by a large rotating spiral? And what determines the angular frequency of the spiral's rotation? We will be able to answer some of these questions in Chapters 6 through 9 later in the book.

The patterns in Fig. 1.14 are time-independent or weakly time-dependent. In contrast, Fig. 1.15 shows snapshots from two different time-dependent states that have been observed in a convecting fluid close to the onset of convection. Figure 1.15(a) shows a most remarkable dynamical state called spiral defect chaos. Spirals and striped regions evolve in an exceedingly complex way, with spirals migrating through the system, rotating (with either sense) as they move, sometimes annihilating with other spirals, and sometimes giving birth to spirals and other defects. It seems almost inconceivable that rising and falling air can spontaneously develop such a complicated dynamical dance, especially under conditions such that the air is constrained by time-independent and spatially homogeneous boundaries. Experiments and numerical simulations have further shown that, under identical experimental conditions (although in a large rectangular convection cell), one can see spiral defect chaos or a time-independent lattice of stripes similar to Fig. 1.14(a), i.e. there are two dynamical attractors and which one is observed depends on the initial conditions of the experiment. So the same fluid can convect in two very different ways under the same external conditions. Another interesting feature of spiral defect chaos is that it is found only in convection systems that are sufficiently big. For geometries with aspect ratio Γ smaller than about 20, one finds other less-disorganized patterns. This dependence on size is not understood theoretically.

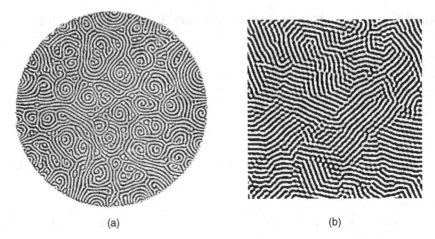

Fig. 1.15 Two spatiotemporal chaotic examples of Rayleigh–Bénard convection. (a) A snapshot of spiral defect chaos in a large cylindrical cell of radius $r = 44$ mm, and depth $d = 0.6$ mm ($\Gamma = 73$), for Rayleigh number $R = 1.894\, R_c$. The fluid is gaseous carbon dioxide at a pressure of 33 bar. (From Morris *et al.* [75]) (b) A snapshot of domain chaos, which occurs in a Rayleigh–Bénard convection cell that is rotating with constant angular frequency about the vertical axis. The figure actually shows results from numerical simulations of equations to be introduced in Chapter 5 but the experimental pictures are similar. Each domain of rolls is unstable to the growth of a new domain with rolls oriented at about 60° with respect to the old angle, and the pattern remains dynamic. (From Cross *et al.* [27].)

If a convection apparatus is rotated at a constant angular frequency ω about its center, a different chaotic dynamics is observed called domain chaos (Fig. 1.15(b)). The rotation rate can be expressed in dimensionless form using a rotational Reynolds number

$$\Omega = \frac{\omega d^2}{\nu}, \qquad (1.5)$$

where we use an uppercase Greek omega, Ω, to distinguish this quantity from the Reynolds numbers $\mathcal{R}$ defined in Eq. (1.4) for Couette flow. Provided that Ω exceeds a critical value Ω_c (which has the value $\Omega_c \simeq 12$ for gaseous CO_2), experiments show that the domain chaos persists arbitrarily close to the onset of convection. As the Rayleigh number for convection R approaches R_c from above, the size of the domains and the time for one domain to change into another domain of a different orientation both appear to diverge. The experimental discovery of domain chaos was quite exciting for theorists since, in the regime arbitrarily close to onset, there is a good chance of understanding the dynamics by developing a perturbation theory in the small quantity $\varepsilon = (R - R_c)/R_c \ll 1$.

These two kinds of spatiotemporal chaos raise some of the most difficult conceptual questions concerning sustained nonequilibrium states. How do we understand

1.3 Examples of nonequilibrium patterns and dynamics

such disordered states, as well as the transitions into and out of such states? Are such states analogous to the liquid and gas phases of some equilibrium system and are the transitions between spatiotemporal chaotic states possibly similar to thermodynamic phase transitions? As one example to consider, as the Rayleigh number R is increased with zero rotation rate, a pattern consisting mainly of stripes evolves into the spiral defect chaos state. What then is the effect of rotation on this spiral defect chaos, and how does this state change into domain chaos with increasing Ω?

We now turn away from convection to discuss patterns found in other controlled laboratory experiments. Figure 1.16 shows three new kinds of nonequilibrium

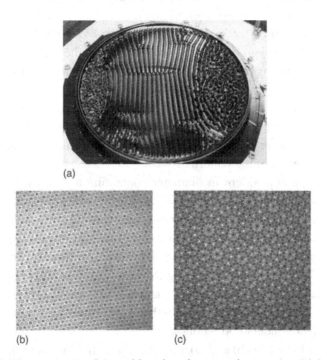

Fig. 1.16 Three patterns observed in crispation experiments, in which a shallow horizontal layer of fluid (here a silicon oil with viscosity $\nu = 1 \text{ cm}^2/\text{s}$) is shaken up and down with a specified acceleration $a(t)$ Eq. (1.6). Bright areas correspond to flat regions of the fluid surface (peaks or troughs) that reflect incoming normal light back toward an imaging device. (a) A mixed pattern of stripe and chaotic regions. The fluid is being driven sinusoidally with parameters $a_1 = 0$ and $a_2 \approx 8g$ (about $1.45a_c$) and $f_1 = 45 \text{ Hz}$ in Eq. (1.6). (From Kudrolli and Gollub [58].) (b) A time-periodic superlattice pattern consisting of two superimposed hexagonal lattices with different lattice constants (ratio $\sqrt{3}$) and rotated with respect to one another by $30°$. This figure was obtained by averaging over two drive periods for parameter values $m = 4, n = 5, a_4 \approx 4.4g, a_5 \approx 7.9g, f = 22 \text{ Hz}$, and $\phi = 60°$ in Eq. (1.6). (c) For the same parameters as those of panel (b) but with a relative driving phase ϕ set to $16°$, a spatially quasiperiodic pattern is observed that has 12-fold symmetry around various points in the pattern. ((b) and (c) from Kudrolli et al. [59].)

patterns that are observed in a so-called crispation or Faraday experiment, named after the British scientist Michael Faraday who was the first to report some observations of such a system in the year 1831. In these particular crispation experiments, a dish containing a fluid layer was shaken up and down with a specified acceleration of the form

$$a(t) = a_n \cos(2\pi nft) + a_m \cos(2\pi mft + \phi). \tag{1.6}$$

Here the acceleration amplitudes a_n and a_m are measured in units of the Earth's acceleration g, the basic frequency f is varied over the range 10–200 Hz, m and n are integers, and ϕ is a specified phase.[6] For the case of sinusoidal driving with $a_n = 0$ and $m = 1$, when the acceleration amplitude a_m or shaking frequency f exceeds some threshold, the fluid's flat surface becomes unstable to the formation of capillary waves (short-wavelength surface waves for which the surface tension of the fluid is a stronger restoring force than gravity) and the nonlinear interaction of these waves leads to intricate patterns, including lattices of stripes, squares, or hexagons, and spatiotemporal chaos.[7] An advantage of crispation experiments over convection and Taylor–Couette experiments is that the effective system size of the system can be easily increased by simply increasing the driving frequency f, which decreases the average wavelength of the patterns.

In Fig. 1.16(a), a cup 32 cm in diameter containing a 3 mm layer of viscous silicon oil was shaken sinusoidally ($a_n = 0$, $m = 1$) with frequency $f = 45\,\text{Hz}$ and acceleration $a_m \approx 8g$. For these parameters, the fluid surface spontaneously evolves to a novel mixed state for which part of the fluid is evolving chaotically and part is an approximately stationary stripe pattern. Unlike the stripe-turbulent state of Fig. 1.11(e), the fronts separating the chaotic and laminar regimes do not propagate. Increasing the amplitude a_1 further causes the chaotic regions to grow in size at the expense of the stripe region until the stripe region disappears completely. These results are not understood theoretically.

The patterns in Figs. 1.16(b) and (c) are obtained for the case of periodic external driving with two frequencies such that the ratio of the driving frequencies is a rational number m/n. If one frequency $f_m = 4f$ is an even multiple and the second frequency $f_n = 5f$ is an odd multiple of a base frequency $f = 22\,\text{Hz}$, there is a regime of parameters such that the fluid surface spontaneously forms a new kind of structure called a superlattice that can be understood as the superposition of two different lattices. Superlattices are found in other nonequilibrium systems as well

---

[6] As a simple experiment, you can try placing a loud speaker face up to the ceiling, put a small board on the speaker cone, and then put a cup of water on the board. Playing various tones at different volumes through the speaker will then shake the cup up and down with a prescribed amplitude and frequency and you should be able to see some interesting patterns.

[7] Crispation patterns are all time dependent and vary subharmonically with the driving frequency f. A stationary pattern is then one that looks the same after two driving periods.

1.3 Examples of nonequilibrium patterns and dynamics

as in equilibrium structures. If the relative phase of the driving is decreased from $\phi = 60°$ to $16°$, an intricate time-independent pattern is now observed (Fig. 1.16(c)) that is called quasicrystalline since it is spatially nonperiodic yet highly ordered in that its wave number spectrum $P(\mathbf{k})$ has sharp discrete peaks.

Quasicrystalline states are rather extraordinary. Since the development of X-ray crystallography and associated theory in the early twentieth century, scientists had believed that sharp peaks in a power spectrum (corresponding to discrete points in a X-ray film) could arise only from a periodic arrangement of the objects scattering the X-rays. This orthodoxy was proved wrong in 1984 when experimentalists announced the synthesis of the first quasicrystal, an Al-Mn alloy whose X-ray diffraction pattern had the seemingly impossible properties of a 5-fold symmetry (not possible for a space-filling periodic lattice) and sharp peaks (indicating the absence of disorder). Figure 1.16(c) is a nonequilibrium example of a quasicrystalline pattern with a 12-fold symmetry, and experiments show that the pattern is intrinsic since it is not sensitive to the shape or size of the container. Why is panel (c) quasicrystalline rather than striped or hexagonal as we saw for Rayleigh–Bénard convection in Figs. 1.14(a) and (b)? As various parameters are varied, what kinds of transitions into and out of this state exist?

Pattern formation in a rather different kind of crispation experiment is shown in Fig. 1.17, which involves the vertical shaking of a granular medium consisting of thousands of tiny brass balls.[8] This pattern formation is not related to the capillary waves of a Faraday experiment since a granular medium does not possess a surface tension (the brass balls do not attract each other as do the molecules in a fluid). When the dimensionless amplitude of shaking is sufficiently large, the granular layer is actually thrown into the air, somewhat like a pancake from a frying pan, and then the layer starts to spread out vertically since the brass balls are not all moving with identical velocities. It is then possible for the bottom of the container to be moving upwards at the same time as the bottom of the granular layer is descending, causing some balls to strike the bottom (changing their direction) while other balls remain suspended in the air and continue to fall. This can lead to an alternating pattern in which peaks and valleys of balls formed at one cycle become respectively valleys and peaks at the next cycle or every four cycles, and so on.

In Figs. 1.17(a)–(e), we see stripe and hexagonal regions somewhat similar to those observed in convection near onset (Fig. 1.14) but the regions appear in new and unusual combinations, e.g. two kinds of hexagonal regions separated by a front (b), two flat regions (c), a region of locally square cells abutting a stripe region (d), a pattern consisting of three kinds of hexagonal regions (e), and spatiotemporal chaos

---

[8] This experiment was originally carried out not to study crispation dynamics but to explore the properties of granular media, a major research area of current nonequilibrium science.

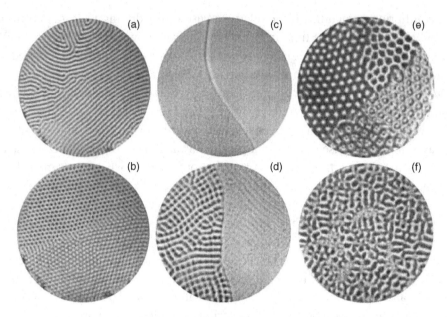

Fig. 1.17 Six patterns observed in a granular crispation experiment. A shallow layer (depth = 1.2 mm) of tiny brass balls (diameters ranging from 0.15–0.18 mm) was vertically shaken up and down in an evacuated cylinder of diameter 127 mm at a constant angular frequency $\omega = 421\,\mathrm{s}^{-1}$ with a varying vertical amplitude A. Each pattern is characterized by the dimensionless acceleration parameter $\Gamma = \omega^2 A/g$ where g is the gravitational acceleration. (a) $\Gamma = 3.3$, a disordered stripe state that is found just above the onset of the instability of a flat uniform state; (b) $\Gamma = 4.0$, a state consisting of two different kinds of locally hexagonal structures; (c) $\Gamma = 5.8$, two flat regions divided by a kink; (d) $\Gamma = 6.0$, a phase of locally square-symmetry states coexisting with a phase of stripes; (e) $\Gamma = 7.4$, different kinds of coexisting hexagonal phases; (f) $\Gamma = 8.5$, a spatiotemporal chaotic state. (From Melo *et al.* [73].)

consisting of short stripe-like domains (f). Unlike convection, Taylor–Couette flow, or crispation experiments with a fluid, it is not clear what sets the length scales of these cellular patterns. The similarities of these patterns to those observed in fluids and in other systems such as lasers is intriguing and puzzling. Is there an underlying continuum description of these brass balls, similar to the Navier–Stokes equations of fluid dynamics? If so, what is that description and how is it derived? What are the properties of the granular media that determine these different spatiotemporal phases?

For our last example of controlled nonequilibrium patterns and dynamics, we discuss experiments involving chemical solutions. Figure 1.18(a) shows a snapshot of a time-dependent two-dimensional chemical reaction known as the Belousov–Zhabotinsky reaction, named after two Russians who, in the 1950s and 1960s,

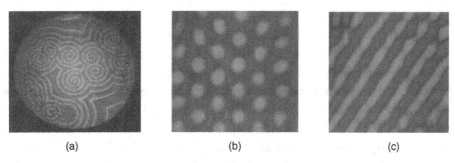

Fig. 1.18 Representative patterns in reaction–diffusion systems. (a) Time-dependent many-spiral state observed in a Belousov–Zhabotinsky excitable reaction consisting of chemical reagents in a shallow layer of fluid in a Petri-like dish. Since the system is closed, the pattern becomes time-independent after a long-lived complicated transient state. (From Winfree and Strogatz [112].) (b) Nearly time-independent hexagonal pattern of spots observed in a chlorite-iodide-malonic-acid (CIMA) system of chemicals that are reacting in a thin cylindrical polyacrylamide gel of diameter 25.4 mm and thickness 2.00 mm. Unlike (a), this is a sustained nonequilibrium system since reagents are fed to and reaction products removed from the gel. The gel suppresses fluid motion and provides a way to visualize the iodide concentration field since the iodide binds with starch embedded in the gel to produce a blue color. The spacing between dots is about 0.2 mm, substantially smaller than the thickness of the gel. (c) For slightly different external conditions, a nearly time-independent stripe pattern is observed instead of comparable wavelength. (From Ouyang and Swinney [84].)

established the remarkable fact that chemical systems could approach equilibrium with a non-monotonic dynamics, e.g. by oscillating in time or by propagating waves in space. When first announced, the experimental discovery was greeted with disbelief and ridicule since most scientists at that time believed incorrectly that the second law of thermodynamics (that the entropy of a system can only increase monotonically toward a maximum value corresponding to thermodynamic equilibrium) implied a monotonic evolution of chemical concentrations toward their asymptotic equilibrium values. With the hindsight of several decades of nonlinear dynamics research that has established convincingly the existence of periodic, quasiperiodic, and chaotic attractors in many experimental systems as well as the occurrence of complicated transients leading to these attractors, it is difficult for contemporary scientists to appreciate this initial disbelief.

Current interest in the Belousov–Zhabotinsky reaction lies primarily in its value as an experimental metaphor for studying more complicated continuous media such as lasers and heart tissue that have the property of being excitable. An excitable medium is such that a local weak perturbation decays while a perturbation whose strength exceeds some threshold grows rapidly in magnitude and then decays. An example is a field of dry grass for which a local increase in temperature causes no change until the temperature exceeds the kindling point. The temperature then rapidly increases as the grass combusts, a wall of flame propagates through the grass, and then the temperature decays to the ambient temperature once the grass has been consumed. Excitable media such as the Belousov–Zhabotinsky reaction, heart tissue, and dry grass fields show similar spatiotemporal patterns and so investigations of the disordered rotating and propagating spiral waves in Fig. 1.18(a), of target patterns of concentric circular propagating waves (not shown and believed to be induced by impurities such as a piece of dust), and their generalization to three-dimensional chemical media in the form of scroll waves give insight simultaneously to many different systems. The questions of interest are ones that we have discussed earlier in the context of a galaxy's spiral arms and of the aggregation of slime mold. What determines the speed of a front and its frequency of rotation and how do these quantities vary with parameters? For many-spiral states, what happens when one spiral interacts with another spiral or with a boundary? In a three-dimensional medium, what are the possible wave forms and how do their properties vary with parameters?

The reaction–diffusion patterns of Figs. 1.18(b) and (c) involve different chemicals and are qualitatively different in that the medium is not excitable and there are no propagating waves. Also, these figures represent true sustained nonequilibrium states since porous reservoirs in physical contact with opposing circular surfaces of a thin cylindrical gel (see Fig. 3.3 on page 110) feed chemical reagents into the interior of the gel where the pattern formation occurs, and also withdraw reaction products. The small pores of the gel suppress fluid motion which greatly simplifies the theoretical analysis since the spatiotemporal dynamics then arises only from the reaction and diffusion of chemicals within the gel. The patterns were visualized by using the fact that one of the reacting chemicals (iodide) binds to starch that is immobilized in the gel, causing a color change that reflects the local iodide concentration.

A typical experiment involves holding the temperature and reservoir chemical concentrations constant except for one chemical concentration which becomes the control parameter. For one choice of this concentration (Fig. 1.18(b)), the chemicals spontaneously form a locally hexagonal pattern similar to the convection pattern Fig. 1.14(b) and granular crispation patterns in Fig. 1.17. Other parameter values lead to a stripe state, superlattices, and spatiotemporal chaos. These chemical patterns are cellular just like the convection patterns that we discussed in Figs. 1.14 and 1.15 but here the length scale is determined dynamically by a balance of diffusion and chemical reaction rates rather than by the geometry of the container.

When the experiments in Figs. 1.18(b) and (c) and others were first reported around 1990, there was great scientific excitement, not because they were the first examples of nonequilibrium pattern formation but because they were the first to

1.3 Examples of nonequilibrium patterns and dynamics

confirm a remarkable insight of Alan Turing. In 1952, Turing observed that diffusion, which by itself tends to make chemical concentrations spatially uniform, could lead to the formation of cellular patterns if chemical reactions were also to occur. (We will discuss some details of Turing's insight in Chapter 3, when we discuss the linear stability of a uniform state.) Turing then went on to speculate that patterns generated by reaction and diffusion might suffice to explain biological morphogenesis, the formation of structure during the growth of a biological organism. Examples Turing had in mind included the formation of stripes or spots on animal surfaces (tigers, zebras, cheetahs, giraffes, fish, seashells), the formation of symmetrically arranged buds that grow into leaves or tentacles, and the question of how a presumably spherically symmetric fertilized egg (zygote) could start the process of dividing and differentiating into the many different kinds of cells found in an adult organism. Ironically, while Turing's insight of pattern formation by reaction and diffusion was finally confirmed forty years later by nonbiological experiments, pattern formation in biological systems has turned out to be more complicated than originally conceived by Turing, and a picture as simple as the one he proposed has not yet emerged.

1.3.3 What are the interesting questions?

To summarize the many nonequilibrium states discussed in Sections 1.3.1 and 1.3.2, let us list here the scientific questions raised by our discussion:

Basic length and time scales: Many of the patterns that we discussed have a cellular structure, consisting locally of stripes, squares, or hexagons of a certain typical size, or of waves or spirals that evolve with a certain frequency and velocity. An obvious question is what determines the basic length and time scales of such patterns? In some cases such as Taylor–Couette flow and Rayleigh–Bénard convection, the cellular size is determined by the experimental geometry (e.g. the thickness of the fluid layer) but this is not always the case. For example in the limit that the thermal conductivity of a convecting fluid becomes large compared to that of the floor and ceiling in Fig. 1.1 (liquid mercury between two glass plates would be an example), the cellular length scale can become much larger than the depth of the fluid. For reaction–diffusion chemical systems and related media such as the heart or a slime mold, the length and time scales are determined dynamically by diffusion constants and reaction rates.

Wave number selection: For some parameter ranges, stationary spatially periodic lattices are observed that can be characterized by a single number, the lattice spacing. In these cases, we can ask the question of wave number selection: is a unique lattice spacing observed for specified parameters and boundary conditions and what determines its value? If there are multiple spacings, what determines their values?

Related questions arise in other systems. Thus the spirals in spiral galaxies (Fig. 1.4), in slime-mold aggregation (Fig. 1.9), in spiral defect chaos (Fig. 1.15), and on the surface of a fibrillating heart (Fig. 1.10) all raise the question of what determines the frequency of rotation and the velocity of the arms in the observed spirals. In snowflakes (Fig. 1.8) and in the domain chaos of a convecting fluid (Fig. 1.15), there is propagation of a tip or front and one can ask if the propagation occurs with a unique velocity and what determines that velocity.

You should appreciate that the question of length and time scales and the question of wave number selection are distinct. The question of scales corresponds to knowing what unit of measurement is appropriate (say meters versus millimeters or days versus seconds). The question of wave number selection then corresponds to making precise measurements on this scale. For example, the lattice spacing of hexagonal rolls may be about one millimeter in order of magnitude (the length scale) but in actual experiments we would be interested to know if the precise value is 0.94 or 1.02 mm and whether these values repeat from experiment to experiment.

Pattern selection: For patterns that form in two- and three-dimensional domains, multiple patterns are often observed for the same fixed external conditions. The question of pattern selection is why one or often only a few patterns may be observed or why, in other cases, certain patterns are not observed? An example we discussed above was the occurrence of a spiral defect chaos state (Fig. 1.15(a)) or of a stripe-pattern state under identical conditions in a large square domain. On the other hand, a quasiperiodic pattern like Fig. 1.16(b) has never been observed in a convecting flow. Why not?

Transitions between states: A given nonequilibrium state will often change into some other state as parameters are varied and so we can ask: what are the possible transitions between nonequilibrium states? Of special interest are supercritical transitions – in which a new state grows continuously from a previous state – since analytical progress is often possible near the onset of such a transition.

A related question is whether supercritical nonequilibrium transitions have interesting critical exponents associated with the transition. For second-order thermodynamic phase transitions and for nonequilibrium supercritical transitions, a quantity Q that characterizes the system may converge to zero or diverge to infinity at the transition point as a power law of the form

$$Q \propto |p - p_c|^\alpha, \qquad \text{as } p \to p_c, \tag{1.7}$$

where p is the parameter that is being varied with all others held fixed, p_c is the critical value of p at which the supercritical transition occurs, and the quantity α is the critical exponent which determines the rate of convergence or divergence of Q.[9] Some of the great advances in twentieth-century theoretical and experimental science concerned the discovery and explanation for "universal" values of these critical exponents. In equilibrium systems, their values turn out to depend remarkably only on the symmetry

---

[9] An example of a second-order equilibrium phase transition is the loss of magnetism of pure iron as its temperature T is increased to its Curie temperature $T_c \approx 1043$ K. The magnetization M of the iron decreases to zero according to Eq. (1.7) with an exponent $\alpha \approx 0.3$. The onset of convection in Fig. 1.1 is a supercritical nonequilibrium transition for which the maximum magnitude of the velocity field $\max_\mathbf{x} \|\mathbf{v}\|$ vanishes according to Eq. (1.7) with an exponent $\alpha \approx 1/2$.

1.3 Examples of nonequilibrium patterns and dynamics

and dimensionality of the system undergoing the phase transition, but not on the possibly complicated details of its atomic structure. So we can ask: do universal exponents occur in spatially extended nonequilibrium systems? If so, on what details of the system do they depend?

Stability: In addition to classifying the possible transitions between different nonequilibrium states, can we predict when transitions will occur for particular states as particular parameters are varied? For many experiments, the underlying dynamical equations are known and a linear stability analysis of a known state can be attempted numerically (more rarely, analytically). In other cases such as granular flow or neural tissue, the underlying equations are not known (or might not exist or might not be practical to work with mathematically or computationally) and one might instead try to identify empirical features in experimental data that could suggest when a transition is about to occur. Two examples are attempts to predict an economic crash from stock market time series, and efforts to predict the onset of an epileptic seizure from 19-electrode multivariate EEG time series.

Boundaries: Even in experimental systems that are large compared to some basic cellular length scale, the lateral boundaries confining the medium (e.g. the walls in Fig. 1.1) can strongly influence the observed patterns and dynamics. How do the shape, size, and properties associated with lateral boundaries influence the dynamics? One example we discussed in Section 1.3.2 was the fact that spiral defect chaos (Fig. 1.15(a)) is not observed until a convection system is sufficiently wide. Somehow the lateral boundaries suppress this state unless the boundaries are sufficiently far from each other.

Transients: For spatially extended nonequilibrium systems, it can be difficult to determine how long one must wait for a transient to end or even if an observed state is transient or not. Mathematical models of spatially extended systems suggest that the average time for transients to decay toward a fixed point can sometimes grow exponentially rapidly with the system size and so be unobservably long even for systems of moderate size. What determines the time scale for a transient spatiotemporal pattern to decay? Is it possible to distinguish long-lived transient states from statistically stationary states?

Spatiotemporal chaos: Many systems become chaotic when driven sufficiently away from equilibrium, i.e. their nontransient dynamics are bounded, are neither stationary, periodic, nor quasiperiodic in time, and small perturbations grow exponentially rapidly on average. Chaotic pattern forming systems in addition may develop a nonperiodic spatial structure that is called spatiotemporal chaos. Several examples discussed above include spiral defect chaos and domain chaos in Fig. 1.15 and a chaotic pattern in a granular crispation experiment, Fig. 1.17(f). Spatiotemporal chaos raises difficult conceptual questions about how to characterize the spatiotemporal disorder and how its properties depend on parameters.

Transport: For engineers and applied scientists, an important question is how does the transport of energy and matter through a spatiotemporal nonequilibrium system depend on its parameters? For example, computer chips generate heat and a mechanical engineer may need to design the geometry of a computer board so that convection can remove the heat efficiently. Similarly, the synthesis of an ultrapure crystal from a molten substrate is a nonequilibrium problem for which a chemical engineer needs to know how the transport of impurities into the crystal depends on parameters so that this transport can be minimized.

Control: For many applied science problems, it is not sufficient to observe a nonequilibrium system passively, one needs to control a system actively by applying an external perturbation. An example is the dynamics of left ventricular muscle (Fig. 1.10), for which one might hope to use gentle electrical perturbations to prevent the onset of fibrillation when an arrhythmia appears. Similarly, an electrical engineer may need to apply an external perturbation to a laser to stabilize a regime of high-power coherent emission, or a plasma physicist may want to confine a hot thermonuclear plasma for long times by modulating some external magnetic field, and one can speculate about a futuristic technology that perturbs the atmosphere to prevent the formation of a tornado or hurricane. These goals raise many unsolved questions regarding how a nonequilibrium system responds to external perturbations and how to choose such perturbations to achieve a particular goal.

1.4 New features of pattern-forming systems

The variety of nonequilibrium patterns discussed in the previous sections and the many scientific questions suggested by these patterns are possibly overwhelming if you are learning about these for the first time. To give you some sense about what features of these systems are significant, we discuss in this section some ways that pattern-forming nonequilibrium systems differ from those that you may have encountered in introductory courses on nonlinear dynamics and on thermodynamics. We first discuss some conceptual differences and then some specific new properties.

1.4.1 Conceptual differences

An important new feature of pattern-forming systems and a direct consequence of their nonequilibrium nature is that *the patterns must be understood within a dynamical framework*. This is the case even if we are interested in just time-independent patterns such as panels (a) and (b) of Fig. 1.14. In strong contrast, the geometry and spacing of a spatial structure in thermodynamic equilibrium can be understood as the minimum of the system's energy (or, more precisely, as the minimum of the system's free energy for systems at a finite temperature T). A familiar equilibrium example is a periodic lattice of atoms or molecules in a crystal. The positions of the atoms can be determined directly from the energy of their mutual interaction, independently of the dynamics of the atoms, since the minimum energy is achieved for zero velocity of each atom and so depends only on the positions of the atoms.[10] This is not the case for nonequilibrium systems *since generally there is no free-energy-like quantity whose extremum corresponds to the static nonequilibrium pattern.* (We will discuss this point in more detail later in the book but note for now that the absence of a free energy is partly a consequence of the fact that nonequilibrium systems are open systems subjected to imposed external fluxes and so a system's energy, mass, and momentum are often not conserved.) Furthermore, we are often interested in the breakdown of stationary patterns to new patterns that remain dynamic indefinitely, a phenomenon that obviously requires a dynamical formulation and that has no thermodynamic analogy.

Introductory nonlinear dynamics courses discuss systems that are well described with just a few variables, e.g. the logistic map, the Hénon map, the standard map, the driven Duffing equation, and the Lorenz model. A pattern-forming system differs fundamentally in that many variables are needed to describe the dynamics (the phase space is high-dimensional). New concepts and methods are then needed to study pattern formation, and indeed many basic ideas that you may have learned in an introductory nonlinear dynamics course will not be applicable in this book. For example, the strategy of studying a continuous-time dynamical system via its associated discrete-time Poincaré map is no longer useful, nor is it productive to analyze an experimental time series by embedding it into some low-dimensional phase space. As mentioned above, a high-dimensional phase space is needed even to describe a fixed point (an attractor with zero fractal dimension) such as the static pattern of convection rolls in Fig. 1.14(a) since the transient orbit meanders through the high-dimensional space as it approaches the fixed point.

Figure 1.19 gives a physical insight into why pattern-forming systems have a high-dimensional phase space by contrasting two convection systems. The thermosyphon shown in Fig. 1.19(a) is a thin closed circular pipe filled with a fluid that is heated over its bottom half and cooled over its top half. From our discussion of convection in Section 1.2, you will not be surprised to learn that if the circular pipe is placed vertically in a gravitational field, then for a temperature difference between the top and bottom halves that exceeds some critical value, the fluid begins to circulate around the tube, forming a simple convection "roll." (Which way does the fluid begin to circulate? This is not determined *a priori*, and the circulation will be clockwise for some experimental runs, anticlockwise for others – a good example of what is known as a broken symmetry.) If the temperature difference in

---

[10] We are thinking about the atoms classically here, a good approximation except for the lightest atoms such as hydrogen and helium. In a full quantum mechanical description the atoms are no longer at rest in the crystal even at zero temperature due to zero point motion. However, the lattice structure is still obtained as the minimum energy state, albeit involving a more complicated calculation of the energy.

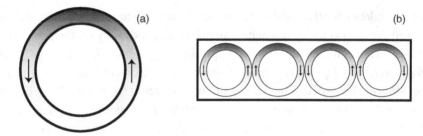

Fig. 1.19 Two convection systems: (a) a thermosyphon and (b) Rayleigh–Bénard convection with many rolls. The light and dark regions denote respectively warm and cool fluid regions. The thermosyphon is well-described by attractors in a three-dimensional phase space whereas the patterns in Rayleigh–Bénard convection require a high-dimensional dynamical description even to describe its static states.

this simple system is set to some still larger value, the direction of flow shows spontaneous chaotic reversals at what appear to be random time intervals. And indeed experimentalists have shown that these reversal events are described well by the three-variable Lorenz equations, one of the most famous systems in the study of low-dimensional chaos.[11] We can easily understand why a three-variable description might be adequate for Fig. 1.19(a). Although there are other dynamical degrees of freedom of the fluid such as variations of the flow transverse to the axis of the pipe, these turn out to be rapidly damped to constant values by the fluid viscosity since the walls are close together and the fluid velocity is zero at the wall. Thus these transverse degrees of freedom do not enter into the thermosyphon dynamics.

This thermosyphon should be contrasted with the pattern-forming convection system in Fig. 1.19(b). Roughly, we might consider each convection roll to be analogous to a separate thermosyphon loop so that the dimension of the phase space needed to describe the convection system will be proportional to the number of convection rolls (or alternatively proportional to the length of the convection experiment). In fact, if we drive this system at a strength corresponding to the onset of chaos in the thermosyphon, we would find that we not only have to include for each roll three Lorenz-type variables of circulation velocity, temperature perturbation and heat flow, but also new variables associated with distortions of each roll caused by the coupling of each roll to other rolls. A pattern like Fig. 1.14(a) with about 40 rolls may then well involve a phase space of dimension at least 150, huge compared to any dynamical system described in introductory nonlinear dynamics courses. We will indeed have to develop new concepts and methods to work with such high-dimensional phase spaces.

---

[11] The Lorenz variables X, Y, and Z are now interpreted as the fluid circulation velocity, the asymmetry of temperature between the right and left halves of the loop, and the heat transported.

1.4 New features of pattern-forming systems

By the way, Fig. 1.19 illuminates another (although somewhat technical) point, that the essential difference between the two convection systems is not that the thermosyphon is described by a few coupled ordinary differential equations while the convection system is described by a few coupled partial differential equations (abbreviated throughout this book as pdes). This might appear to be the case since dynamical systems described by partial differential equations in principle have infinite-dimensional phase spaces with the dynamical variables specified at a continuum of points labeled by their position in space. In fact, both systems in Fig. 1.19 are described by the partial differential equations of fluid dynamics and heat transport. However for the thermosyphon, the dynamics of interest can be understood within a truncated approximation of a few important dynamical degrees of freedom (because of the strong damping by nearby lateral walls), whereas the interesting dynamics in the pattern-forming system, such as the approach to a stationary state or the transition of a stationary state to persistent dynamics, cannot. In Chapter 6, we will see that even reduced descriptions of pattern-forming systems are often described by partial differential equations (amplitude equations) and so remain infinite dimensional.

We have argued that an important difference between the dynamical systems studied in an introductory nonlinear dynamics course and the pattern-forming systems discussed in this book is that a description of the latter requires a high-dimensional phase space. This is all well and good but you may ask, is there some easy way to determine the phase-space dimension of an experiment or of a simulation? Fortunately, there are two informal ways to estimate whether some system has a high-dimensional phase space. One is simply visual inspection. If a nonequilibrium system is large compared to some basic characteristic length (e.g. the width of a convection roll), then a high-dimensional phase space is likely needed to describe the dynamics. We will call such systems spatially extended. A second way to identify a high-dimensional phase space is to simulate the system on a computer. If many numerical degrees of freedom are needed (e.g. many spatial grid points or many modes in a Fourier expansion) to reproduce known attractors and their bifurcations to some reasonable accuracy, then again the phase space is high-dimensional.

Using visual inspection to estimate whether the phase space of some system is high-dimensional assumes that we somehow know some characteristic length scale with which we can compare the size of an experimental system. Identifying such a length can be subtle but is often clear from the context of the system being studied. As we have seen in Section 1.3, many pattern-forming systems involve cellular structures (e.g. stripes and hexagons) or propagating waves in the form of spirals or scroll waves which all have a well-defined wavelength. In these cases, a system is spatially extended with a high-dimensional phase space if its geometric size is large compared to the size of this wavelength. For more strongly driven nonequilibrium systems, there can be structure over a range of length scales[12] and often no single length can be identified as being special (for example, in Fig. 1.11(d), there is structure on length scales substantially smaller than the width of a Taylor cell). But then the fact that there is structure over a range of lengths itself indicates the need for a high-dimensional phase space. For strongly driven systems we might need to introduce other length scales called correlation lengths that quantify over what distance one part of the system remains correlated with another part. A system is spatially extended and high dimensional if it is geometrically large compared to any one of these correlation lengths.

The high-dimensional phase space of pattern-forming systems has the unfortunate implication that data analysis is challenging, both technically and conceptually. Current desktop spatiotemporal experiments and simulations may require storing and analyzing hundreds of gigabytes of data as compared to a few tens of megabytes for low-dimensional dynamical systems. Several satellite-based observational projects investigating the dynamics of Earth's ecology, geology, and meteorology are approaching hundreds of terabytes in storage. By comparison, the total printed contents of the Library of Congress constitute about 10 terabytes and several independent estimates suggest that the amount of information stored in the human brain over a lifetime is perhaps 1–10 gigabytes of compressed data. There is then a great need for theoretical insight that can suggest ways to reduce such vast quantities of data to manageable amounts and to identify questions that can be answered. We will touch on some of these data analysis issues several times throughout the book but you should be aware that these are difficult and unsolved questions.

Closely related to the challenge of analyzing large amounts of spatiotemporal data is the challenge of simulating spatially extended dynamical systems. In an introductory nonlinear dynamics course, no sophistication is needed to iterate a map with a few variables or to integrate the three-variable Lorenz equations for a long period of time, one simply invokes a few appropriate lines in a computer mathematics program like Maple or Mathematica. But to integrate numerically in a large box the partial differential equations that describe Rayleigh–Bénard convection (the Boussinesq equations) for the long times indicated by experiments is much more challenging. One can rarely look up and just use an appropriate algorithm because there are numerous subtleties concerning how to discretize three-dimensional time-dependent nonlinear partial differential equations and how to solve the related linear algebra problems efficiently (which may require solving hundreds of millions of

---

[12] A "range of length scales" can be made more quantitative and objective by Fourier analyzing some observable $u(\mathbf{x}, t)$ associated with the system and then by calculating the time-averaged wave-number spectrum $P(k)$. The range of length scales then corresponds to the range of wave numbers $[k_1, k_2]$ such that $P(k)$ differs significantly from zero.

1.4 New features of pattern-forming systems

simultaneous linear equations at each successive time step). Writing, validating, and optimizing an appropriate code can take several years, even for someone with a Ph.D. with special training in computational science. Further, even if a validated code were instantly available, powerful parallel computers are needed to simulate such equations in large domains over long time scales and such computers are still not widely available or easy to use.

1.4.2 New properties

In addition to a high-dimensional phase space, spatially extended sustained nonequilibrium systems have some genuinely new features when compared to dynamical systems that evolve only in time. One such feature concerns how a fixed point becomes unstable. For a system that evolves only in time (e.g. a driven nonlinear pendulum), an infinitesimal perturbation of an unstable fixed point simply grows exponentially in magnitude (or perhaps exponentially with oscillations if the imaginary part of the growth rate is nonzero). But for pattern-forming systems, an infinitesimal perturbation of an unstable fixed point can grow spatially as well as temporally. Further, there are two distinct kinds of spatial growth. One kind is an absolute instability in which a perturbation that is localized over some region of space grows at a fixed position. The second kind is a convective instability in which the instability propagates as it grows. For this second kind of instability, there is exponential growth only in a moving frame of reference. At any observation point fixed in space, there is growth and then asymptotic decay of the instability as the propagating disturbance moves beyond the observation point.

The linear instability of the uniform motionless state of air to convection rolls in the convection experiment Fig. 1.1 is an example of an absolute instability. The instability of snowflake dendritic tips in Fig. 1.8 is an example of a propagating convective instability. This convective instability explains why no two snowflakes are ever alike since it has the remarkable property of magnifying noise arising from the molecular collisions near the tip of the growing dendrite. Perturbations from this noise later influence the formation of the dendrite's side branches.

Another new property is that a large pattern-forming system can consist of different spatial regions that each, by itself, could be a pattern for the entire system. Further, such a quilt of different states can persist for long times, possibly indefinitely. An example is Fig. 1.16(a), where a region of nearly time-independent stripes coexists with chaotic disordered regions. Similarly, Fig. 1.17 shows several quilt-like patterns, e.g. a square lattice adjoining a hexagonal lattice, each one of which could fill the entire domain on its own. The localized region of a system separating one pattern from another is called a front. Theoretical progress can often be made by analyzing fronts as separate and simpler dynamical systems.

1.5 A strategy for studying pattern-forming nonequilibrium systems

As you now appreciate from the above examples and discussion, the scope of nonequilibrium physics is actually enormous since any system that is not in thermodynamic equilibrium is by definition a nonequilibrium system. Necessarily, the phenomena we discuss in a book must be limited and in this section we describe the kinds of nonequilibrium systems that we do and do not consider and also a strategy for investigating the systems of interest.

We will be concerned primarily with nonequilibrium systems that are maintained in a state away from thermodynamic equilibrium by the steady injection and transport of energy. Most interesting to us are systems displaying regular or nearly regular spatial structures, some examples of which we discussed in Section 1.3. We will discuss stationary spatial structures, their breakdown to persistent dynamical states that are also disordered in space, and also systems supporting propagating spatial structures. A major focus of this book is also on systems that may be investigated by precise, well-controlled laboratory experiment and for which there is a well-understood theoretical formulation. The idea is to learn about the complex phenomena of nonequilibrium systems through the study of these systems ("prepared patterns"), exploiting the close connection between theory and experiment. The ultimate goal is then to apply this knowledge to a wider range of problems ("natural patterns"), perhaps where experimental intervention such as changing parameters is not possible (e.g. the climate and many biological systems).

Figure 1.20 provides a way to understand how the systems described in this book fit into the broader scheme of sustained nonequilibrium systems. A particular nonequilibrium system can be thought of as occupying a point in a three-dimensional parameter space with axes labeled by three dimensionless parameters R, Γ, and N that we discuss in turn.

The parameter R is some dimensionless parameter like the Rayleigh number, Eq. (1.1), that measures the strength of driving compared to dissipation. For many systems, driving a system further from equilibrium by increasing R to larger values leads to chaos and then to ever-more complicated spatiotemporal states for which there is ever finer spatial structure and ever faster temporal dynamics. A canonical example of a large-R system is highly turbulent fluid flow, e.g. the flow generated behind a propeller rotating at high speeds. When discussing Fig. 1.6, we saw that the Sun's turbulent outer layer corresponds to a Rayleigh number of order 10^{12} so there are some systems for which the R-axis can span at least 12 orders of magnitude.

Some nonequilibrium systems disintegrate or change their properties when driven too strongly and so cannot be driven to arbitrarily large values along the R axis. Examples might be a laser that burns through confining mirrors if pumped too strongly, a biological system that becomes poisoned and dies if given too much

1.5 A strategy for studying pattern formation

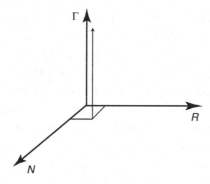

Fig. 1.20 A parameter space for categorizing sustained spatially extended nonequilibrium systems. The R-axis is the "driving" or fluid-dynamics axis that measures how far a system has been driven from equilibrium. The Γ axis is the "size" or pattern-forming axis, indicating the size of a system relative to some basic length scale such as the depth of a fluid. The N axis is the number of distinct components that interact and can be thought of as the "biological" axis since living systems have large numbers of different interacting components. Most of what is currently known about nonequilibrium systems involves regions for which at least two of the three variables R, Γ, and N are small. Most of this book will concern the regime of small N, small R, and large Γ as indicated by the thin vertical arrow.

of some nutrient like salt, or the medium in a crispation experiment that might be thrown clear from its container if shaken up and down too strongly. Also some systems have stress parameters that simply cannot be raised above some finite value. An example would be the concentration of some reagent that drives a solution of reacting chemicals out of equilibrium. The concentration cannot be increased indefinitely since, at some point, the solution becomes saturated and the reagent starts to precipitate. Since fluids are the most widely studied systems that can be driven strongly out of equilibrium, we can think of the R-axis in Fig. 1.20 as also being a "fluid dynamics" axis.

The vertical axis labeled by Γ (upper-case Greek gamma) is the "aspect ratio" axis, and indicates how large a system is compared to some characteristic length scale such as the size of some cellular structure or the depth of the medium. We can also consider this axis to be the "pattern-forming" axis since for larger Γ (bigger systems), the influence of lateral confining boundaries is reduced and the phenomenon of pattern formation becomes more clear. Nonequilibrium desktop experiments using liquid crystals as a medium have attained values of Γ as large as 1000 while numerical simulations in one-space dimension have reached $\Gamma \approx 10\,000$. These might seem like impressively large values but you should keep in mind that crystals can be considered to have a much bigger aspect ratio of order $1\,\text{cm}/10^{-7}\,\text{cm} \approx 10^7$ (ratio of macroscopic crystal size to its lattice spacing), so nonequilibrium experiments are not yet "macroscopic" compared to their characteristic length scale.

Earth's ocean and troposphere (layer of the atmosphere closest to Earth where the weather evolves) both have a depth of about 10 km and a lateral expanse of order the radius of the Earth (6400 km) and so are big nonequilibrium systems with $\Gamma \approx 600$.

The Γ-axis is important because experiments have shown that simply increasing the system size with all other parameters held fixed can induce interesting dynamics such as a transition from a stationary to chaotic behavior. This was first shown in a seminal experiment of Guenter Ahlers and Robert Behringer in 1978, when they studied the dynamics of a convecting fluid (liquid helium at the cryogenic temperature of 4 K) just above the onset of convection for several cylindrical containers whose aspect ratios Γ (ratio of radius to depth) varied from $\Gamma = 2.1$ to $\Gamma = 57$. Ahlers and Behringer then discovered that simply making a convection system larger and larger for a fixed Rayleigh number was sufficient to cause the dynamics to eventually become chaotic. This discovery has since been verified more carefully and in other nonequilibrium systems and is now considered a general, although poorly understood, feature of sustained nonequilibrium systems. We note that although many systems cannot be driven strongly from equilibrium, at least in principle all nonequilibrium systems can be made arbitrarily large. Exploring the large-Γ limit is therefore experimentally and theoretically interesting for many nonequilibrium systems.

Finally, the third axis labeled N indicates the number of distinct components that interact at each point in a given system. This number can usually be determined by inspection of the mathematical equations (if known) by simply counting the number of distinct fields. For example, a complete mathematical description of a Rayleigh–Bénard experiment involves five coupled fields – the fluid pressure $p(\mathbf{x}, t)$, the fluid temperature $T(\mathbf{x}, t)$, and the three components of the fluid's velocity field $v_i(\mathbf{x}, t)$ for $i = x$, y, and z – and so $N = 5$ for a convecting fluid. Biological, ecological, economic, and chemical systems are often characterized by large values of N so one can think of the N-axis as the "biological" axis for the space of nonequilibrium systems. Nonequilibrium systems with large values of N are perhaps the least well understood and are associated with some of the most interesting current scientific questions. Did life arise on Earth by a spontaneous self-organization in some primordial soup consisting of many chemicals? What determines the number of species in a large ecosystem? How do the many genes coordinate their dynamics and so guide the development of an organism over its lifetime? How does a human brain of 10^{10} neurons assemble itself and how do these neurons with their network of 10^{13} connections act dynamically to produce our cognitive abilities of pattern recognition, associative memory, language, and creative thinking?

Figure 1.20 suggests a simple strategy for investigating nonequilibrium systems, which is to allow only one of the three variables N, R, and Γ to become large at a time. (In contrast, the physical systems studied in an introductory nonlinear dynamics course correspond to having all three variables small or moderate at the same time, and many of the natural systems discussed in Section 1.3.1 have all three variables large at the same time.) This book is largely concerned with phenomena for large values of Γ, and small to moderate values of N and R, as indicated by the thin vertical arrow near the Γ-axis in Fig. 1.20. Although this regime might seem excessively restricted, experiments like those discussed in Section 1.3.2 show that there is an enormous richness of dynamics in this regime and so there is, if anything, an excess of phenomena to understand. Much future work will be needed, however, to explore the regimes corresponding to two or three of these variables having large values simultaneously.

In this book, we will therefore study mainly systems with large Γ and moderate values of R and N. However, there are some further assumptions to make if we are to obtain experimental systems that are as simple as possible and for which the associated theory is manageable. We will emphasize experimental systems that:

(i) **are large** in one or more spatial directions so that the influence of lateral boundary conditions on pattern formation can be reduced, simplifying subsequent theoretical analysis.
(ii) **are homogeneous** so that the pattern formation is intrinsic rather than driven by inhomogeneities. Coating the floor and ceiling of a room with flat uniform layers of copper (Fig. 1.1) was an example of how an experimental system could be made spatially homogeneous. Studying convection over a bumpy floor would be less instructive than convection over a homogeneous floor since the bumps influence the pattern formation and their influence would have to be studied as a separate problem.
(iii) **involve few fields** (small N) which reduces the mathematical and computational effort. For nearly all examples discussed in this book, N will be 6 or less.
(iv) **have local space-time interactions**, a technical mathematical assumption that the dynamical equations involve only fields and finitely many spatial and temporal derivatives of the fields. This assumption is mainly a convenience for theorists and computational scientists since it reduces the mathematical effort needed to analyze the system. Most systems discussed in Section 1.3 have such local interactions so this assumption is not a severe restriction. An example of a nonequilibrium system with nonlocal interactions would be neural tissue since a given neuron can connect to remote neurons as well as to neighboring neurons. Further, there are various time delays associated with the finite propagation speed of signals between neurons. The dynamics then depends nonlocally on information over some time into the past and the mathematical description involves delay-differential equations that can be hard to analyze.

All the systems we will discuss in detail satisfy these basic criteria. In addition the following conditions are also desirable. The systems should

(i) **be described by known equations** so that quantitative comparisons between theory, numerics, and experiment are feasible;
(ii) **be well characterized**, for example the parameters changing the behavior should be easily determined, and the geometry and boundary conditions should be accurately prescribed; and
(iii) **permit easy diagnosis**, allowing accurate quantitative measurements of one or more of the fields relevant to the pattern formation.

Historically, fluid systems have been found to approach many of these ideals. First, many fluid systems have few interacting components (velocity and pressure and sometimes temperature or concentration fields). Second, the fluid dynamics is described by mathematical equations such as the Navier–Stokes equations that experiments have confirmed to be quantitatively accurate over a large range of parameters. Third, the fluid equations involve just a few parameters such as the kinematic viscosity ν or thermal diffusivity κ and these parameters can be measured to high accuracy by separate experiments in which issues of pattern formation do not arise. Fourth, fluids are often transparent and so visualization of their spatial structure is possible at any given time. Finally, experiments have shown that many phenomena observed in non-fluid nonequilibrium experiments often have some analog in a fluid experiment so one can study general features of nonequilibrium phenomena using some fluid experiment.

Of all the possible fluid experiments, Rayleigh–Bénard convection has been especially favored in basic research because the fluid is in contact with time-independent and spatially homogeneous boundaries that are especially easy to characterize and to maintain. In other fluid experiments (e.g. Fig. 1.11 or Fig. 1.16), the fluid is set in motion by some pump or motor that oscillates and these oscillations can be an additional source of driving that complicates the identification of intrinsic pattern formation. We do note that steady technological improvements allow increasingly well-controlled experiments on more exotic systems such as chemical reactions in a gel layer (fed by opposing reservoirs) or on a carefully prepared rectangular slab of heart muscle. The possibilities for careful comparisons between theory and experiment are rapidly improving.

1.6 Nonequilibrium systems not discussed in this book

For lack of space, time, and expertise, we cannot reasonably address all the nonequilibrium systems that have been studied or even all the systems that have received the deep scrutiny of the research community. Some of the topics that we will leave out include the following:

Quenched states: If the driving of a nonequilibrium system is turned off sufficiently quickly or if a parameter describing some equilibrium system is abruptly changed to some new value (e.g. if the temperature of a liquid is quickly decreased below its freezing point), the system can "freeze" into a so-called quenched state that is often disordered and that can take a long time to return to thermodynamic equilibrium. Quenched states are technically nonequilibrium but differ from the systems discussed in this book in that they are not sustained systems. There are many interesting questions concerning how a quenched state approaches its usually ordered equilibrium limit, e.g. by the formation of small ordered domains that grow in size at certain rates. Two examples of quenched states are glass (like that found in a window) and a soap-bubble foam that coarsens over time.

Pattern formation by breakdown: Some patterns in nature are formed by some stress slowly increasing to the point that some threshold is crossed, at which point the medium relaxes quickly by creating a pattern. Examples include cracks propagating through a brittle material, electrical breakdown of an insulator from a high-voltage spark, and the occurrence of an earthquake in response to the buildup of stress in a tectonic plate.

Fully developed fluid turbulence: In our discussion of Fig. 1.20 above, we observed that fluids are one of the few continuous media that can be driven strongly out of equilibrium. There is in fact extensive theory, experiment, and applications in the fully developed fluid turbulence regime of large R, moderate N, and moderate Γ. However a reasonable discussion would be lengthy and technical to the point of almost requiring a book of its own. Also, the subject of high-Reynolds-number turbulence is sufficiently special to fluids that it falls outside our intent to discuss mainly ideas and mechanisms that apply to several nonequilibrium systems.

Adaptive systems: Economic, ecological, and social systems differ from many of the systems discussed in the book in that the rules under which the components interact change over time, the systems can "adapt" to changes in their environment. One example is the evolutionary development of language and increased intelligence in *homo sapiens* which greatly changed the rules of how humans interact with each other and with the world.

1.7 Conclusion

This has been a long but important chapter. We have introduced and discussed representative examples of pattern formation and dynamics in sustained nonequilibrium systems, and have identified questions to pursue in later chapters. The experimental results discussed in Section 1.3 are an especially valuable source of insight and direction since mathematical theory and computer simulations still lag behind experiment in being able to discover the properties of pattern-forming systems. In the following chapters, we develop the conceptual, analytical, and numerical frameworks to understand sustained nonequilibrium spatially extended systems.

1.8 Further reading

(i) A comprehensive and broad survey of pattern-formation research, although at a more advanced level than this book, is "Pattern formation outside of equilibrium" by Cross and Hohenberg [25]. This article is a good place to find discussions of pattern-forming systems that are not mentioned in this book, to see deeper discussions of pattern-forming systems, and to see many applications and discussions of theory to experimental systems.

(ii) Many examples of patterns are discussed at a non-technical level in the book by Ball *The Self-Made Tapestry: Pattern Formation in Nature* [9].

(iii) An introductory article on the large-scale structure of the Universe is "Mapping the universe" by Landy [61].

(iv) Many beautiful photographs showing the enormous diversity of snowflakes can be found in the book *The Art of the Snowflake: A Photographic Album* by Libbrecht [64].

(v) A classic paper in pattern formation in chemical reactions and the possible relevance to morphogenesis is Turing's "The chemical basis of morphogenesis" [106].

(vi) A history of pattern formation in chemical systems is given in the first chapter of Epstein and Pojman's book [34]. If you have access to a chemistry laboratory, you can explore the Belousov–Zhabotinsky reaction by following the recipe given in the appendix of the book by Ball [9].

(vii) For an introduction to low-dimensional dynamical systems see *Nonlinear Dynamics and Chaos* by Strogatz [99].

Some articles on specific topics discussed in this chapter (in addition to those referenced in the figure captions) are listed below.

(i) The thermosyphon: "Nonlinear dynamics of a convection loop: a quantitative comparison of experiment with theory" by Gorman *et al.* [41].

(ii) Cracks and fracture patterns: "How things break" by Marder and Fineberg [68].

(iii) The importance of aspect ratio in the onset of chaos: "The Rayleigh–Bénard instability and the evolution of turbulence" by Ahlers and Behringer [1].

(iv) Spiral defect chaos: "Spiral defect chaos in large-aspect-ratio Rayleigh–Bénard convection" by Morris *et al.* [77].

Exercises

1.1 **End of the Universe:** In Section 1.1, the interesting structure of the Universe was traced to the fact that the Universe was still young and evolving. But will the Universe ever stop expanding and reach an equilibrium state? If so, what kind of structure will exist in such a Universe? What might happen to the Universe in the long term has been discussed in a fascinating article "Time without end: Physics and Biology in an open universe" by the physicist

Freeman Dyson in the article *Reviews of Modern Physics* **51**, 447 (1979). Read this article and then answer the following questions:

(a) What is the difference between an "open" and "closed" universe and what are the implications for whether the universe will ever reach thermodynamic equilibrium?

(b) What experimental quantities need to be measured to determine whether our Universe is open or closed? According to current experimental evidence, is our Universe open?

(c) According to Dyson, how long will it take our Universe to reach thermodynamic equilibrium, assuming that it is open?

(d) Can life, as we know it on Earth, persist arbitrarily into the future if our Universe is open and approaching an equilibrium state?

1.2 **Just six numbers:** In the book *Just Six Numbers: The Deep Forces That Shape the Universe* (Basic Books, New York, 1999), cosmologist Martin Rees argues that our Universe can form interesting patterns – and life in particular – only because certain key parameters that describe the Universe fall within extremely narrow ranges of values. For example, one parameter is the fractional energy $\varepsilon = 0.004$ released when four hydrogen nuclei fuse to form a helium nucleus in the core of a star. If this value were just a tiny bit smaller, condensing clouds of gas would never ignite to become a star. If just a tiny bit bigger, stars would burn up so quickly that life would not have enough time to evolve.

Skim through this book and explain briefly what are the six parameters that Rees has identified as being critical to the existence of pattern formation in the Universe. Discuss qualitatively how the Universe would be different from its present form if these parameters had significantly different values.

1.3 **A sprinkling of points: pattern or not?** Like the stars in Earth's sky, some patterns are less a geometric lattice (or distortions of such a lattice) than a statistical deviation from randomness. To explore this point, assume that you are given a data file that contains the coordinates (x_i, y_i) of 4000 points in a unit square (see Fig. 1.21). Discuss how to determine whether these dots constitute a "pattern" or are "random," in which case we would not expect any meaningful structure. What are some hypotheses that a random distribution would satisfy? How would you test the consistency of the data with your hypotheses? Two possibilities to explore are a chi-squared test [89] and a wave-number spectrum.

1.4 **Properties of the Rayleigh number:** Answer the following questions by thinking about the criterion Eq. (1.1) for the onset of convection.

(a) Two identical convection systems of depth d are filled with air and mercury respectively at room temperature. Using the values in Table 1.1, determine

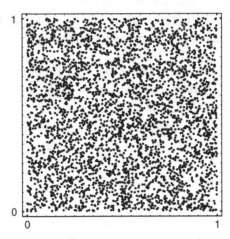

Fig. 1.21 Exercise 1.3: Do these 4000 dots in a unit square constitute a "pattern" that deviates statistically from points thrown down randomly and uniformly?

which fluid will start to convect first as the temperature difference ΔT is increased in small constant steps through onset.

(b) The bottom plate of a certain wide square convection cell is machined to have a square bump that is 5% of the fluid depth in height and four times the depth of the fluid in width as shown schematically in this figure:

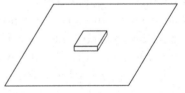

The bump is far away from the lateral walls. As the temperature difference is increased in small constant steps starting with the stable motionless fluid, where will convection first start in this system? Guess and then sketch what kind of pattern will be observed when the convection rolls first appear.

1.5 Suitability of a room for a convection experiment:

(a) Assume that the room in Fig. 1.1 has a height of $d = 3$ m, that the floor and ceiling are isothermal surfaces of temperature T_1 and T_2 respectively (both close to room temperature $T = 300$ K), and that high-precision laboratory experiments can control a temperature difference to at best about one part in 10^4. Determine whether or not this room is suitable for convection experiments.

(b) An unstated assumption for almost any laboratory experiment is that the experiment can be finished within a practical amount of time, say a week or less. Theory discussed later in the book shows that one of the slowest time scales associated with convection is the so-called vertical thermal diffusion time $t_v = d^2/\kappa$, where d is the depth of the fluid and κ is the fluid's thermal diffusivity. This is the time for a localized temperature perturbation near, say, the bottom plate to be detected by a probe near the top plate if the perturbation spreads out purely by diffusion. There is an even longer time scale called the horizontal thermal diffusion time $t_h = (L/d)^2 t_v = \Gamma^2 t_v$, which is the time taken for a localized temperature perturbation on one side of the system to be detected on the far side of the system a lateral distance L away if the perturbation again spreads out purely by diffusion. Any given convection experiment, or simulation of such an experiment, needs to span many multiples of these time scales in order for enough time to pass that transients die out and a statistically stationary state is attained.

1. What are the times t_v and t_h in units of days for air in a square room of height $d = 3$ m and width $L = 8$ m? Are these reasonable time scales for a convection experiment?
2. Answer these same questions for air in a laboratory convection apparatus with $d = 1$ mm and $L = 5$ cm.

1.6 Temperature profile and heat transport of a conducting fluid:

(a) For Rayleigh numbers in the range $0 < R < R_c$, plot the vertical temperature profile $T(z)$ of the air in Fig. 1.1. Assume $z = 0$ is the floor and $z = d$ is the ceiling.

(b) For this same regime of Rayleigh number, plot the heat flux $H = H(R)$ (heat energy per unit area per unit time) through the ceiling.

(c) To get a feeling for the order of magnitude of the heat transport, estimate the total heat transported by the air through the ceiling for a square room of height $d = 3$ m and width $L = 5$ m, when the temperature difference is the critical value ΔT_c. For comparison, a typical room heating device has a power consumption of a few kilowatts (thousands of joules per second).

(d) When $R > R_c$ so that the air in Fig. 1.1 starts to convect, discuss and sketch qualitatively how the temperature profile and total heat transport will change.

(e) Invent and explain a method to measure the instantaneous heat flux $H(t)$ experimentally for a fixed Rayleigh number R.

Hint: For a continuous medium with thermal conductivity K the heat flux is $H = -K \nabla T$ (units of energy per unit time per unit area). Assuming most of the air in the room is close to room temperature $T = 293$ K, you can use the value $K = 2.5 \times 10^{-6}$ J m^{-1} s^{-1} K^{-1} everywhere inside the room.

1.7 **Scaling of time, length, and magnitude scales for the Swift–Hohenberg equation:** To gain experience simplifying a dynamical equation and identifying dimensionless parameters, consider the following partial differential equation for a real-valued field $u(x, t)$ that depends on time t and one spatial coordinate x:

$$\tau_0 \, \partial_t u(x, t) = r u - \xi_0^4 \left(q_0^4 + 2 q_0^2 \, \partial_x^2 + \partial_x^4 \right) u - g_0 u^3. \tag{E1.1}$$

This is the so-called Swift–Hohenberg equation which we will discuss in Chapters 2 and 5 as one of the more important models of pattern formation. Eq. (E1.1) seems to have five distinct parameters, namely the time scale τ_0, the coherence length ξ_0, the critical wave number q_0, the nonlinear strength g_0, and the control parameter r.

By a clever choice of time, length, and magnitude scales t_0, x_0, and u_0, i.e. by changing variables from t, x, and u to the scaled variables τ, y, and v by the equations

$$t = t_0 \tau, \quad x = x_0 y, \quad u = u_0 v, \tag{E1.2}$$

and by redefining the parameter r to a new value $\hat{r}$, show that Eq. (E1.1) can be written in a dimensionless form with only *one* parameter:

$$\partial_\tau v = \hat{r} v - \left(1 + 2 \partial_y^2 + \partial_y^4 \right) v - v^3. \tag{E1.3}$$

This is a substantial simplification since the mathematical and numerical properties of this equation can be explored as a function of a *single* parameter $\hat{r}$.

1.8 **Applications of the Reynolds number:** For problems in which an isothermal fluid flows through a pipe or past an object like a cylinder, an analysis of the Navier–Stokes equations reveals a dimensionless stress parameter called the Reynolds number $\mathcal{R}$,

$$\mathcal{R} = \frac{vL}{\nu}, \tag{E1.4}$$

where v is a characteristic magnitude of the fluid's velocity field (say the maximum speed of the fluid before it encounters some obstacle), L is the size of the object with which the fluid interacts (e.g. the diameter of the pipe or of the cylinder), and ν is the kinematic viscosity of the fluid, the same parameter that appears in the Rayleigh number Eq. (1.1).

For small flow speeds corresponding to $\mathcal{R} < 1$, the fluid is usually laminar, i.e. time independent and without an interesting spatial structure (the stream lines are approximately parallel). For Reynolds numbers larger than about 1, laminar flows usually become unstable and some new kind of pattern or dynamics occur. When $\mathcal{R}$ becomes larger than about 1000, the fluid often becomes chaotic in time and irregular in space.

The following questions give you a chance to appreciate the many useful predictions that can be made by studying a parameter like Eq. (E1.4).

(a) Show that the Reynolds number is a dimensionless quantity and so has the same value in any system of units.
(b) For an airplane traveling at $v = 500\,\text{km/hour}$, will the air flow over the wing be laminar or turbulent?
(c) As you walk around a room, show that the air in the vicinity of your foot will be turbulent. This implies that a cockroach will need some way to locate your foot in the midst of a turbulent flow to avoid being stepped on.
(d) By flipping a coin, estimate the speed with which it falls and the speed with which it rotates and then determine whether fluid turbulence plays a role in the supposedly "random" behavior of flipping a coin to call heads or tails.
(e) From a human physiology book or from the web, find the typical speed of blood flowing through your arteries and through your heart. Does the blood flow in any of your arteries become turbulent? What about through a heart valve?
(f) When the wind blows transversely past a telephone wire, you sometimes hear an eerie whistling sound called an aeolian tune. Using the kinematic viscosity of air at room temperature from Table 1.1 and a wire diameter of $L = 2\,\text{mm}$, what is the smallest wind velocity for which you would expect to hear an aeolian tune?

1.9 **Simple experiment to demonstrate the no-slip fluid boundary condition:** To convince yourself of the fact that a fluid's velocity goes to zero at a material wall in the frame of reference of that wall, try the following simple experiment. Get a desktop fan and sprinkle some talcum powder (or any fine powder) over the blades of the fan. Then turn the fan on so that the blades rotate at high speed for several seconds and switch off the fan. Has the talcum powder been blown completely off the blades?

2
Linear instability: basics

Perhaps the most magical moment in a pattern-forming system is when a pattern first appears out of nothing, the genesis of structure. The "nothing" that one starts with is not empty space but some spatially uniform system. From our study of equilibrium systems, we are used to expecting that if the external conditions are constant in space and time, then, perhaps after some transient dynamics, the system will relax to a state that is time independent and spatially uniform on macroscopic scales. However, as the system is driven further and further out of equilibrium by turning some experimental knob in small successive steps, it is often the case that a point is eventually reached such that a spatial structure spontaneously appears. This is the beginning of pattern formation. In many examples of interest, the novel state with spatial structure develops because the spatially uniform state becomes unstable toward the growth of small perturbations.[1] Analyzing this linear instability provides a first approach to understand the beginnings of pattern formation.

In this chapter, we discuss linear instability: how to predict when a spatially uniform time-independent state becomes unstable to tiny perturbations. We also discuss some details of the growing spatial structure such as its characteristic length and time scales, and how the structure depends on symmetries of the system. As you will see, the linear stability analysis of a spatially uniform stationary state is valuable because it provides the easiest opportunity to make predictions that can be tested by experiment. It is also a useful first step toward understanding how the properties of a nonequilibrium system depend on parameters. Finally, the linear stability analysis suggests a way to classify nonequilibrium systems in regard to their pattern-forming tendencies.

---

[1] For the macroscopic finite-temperature systems discussed in this book, small perturbations are always present. Examples are spatial imperfections of the surfaces containing the nonequilibrium medium, mechanical vibrations of the experimental apparatus through its supports, tiny fluctuations in pressure, density, chemical concentration, or velocity arising from random molecular collisions, and small variations in electrical control signals arising from collisions of electrons in wires and in integrated circuits.

In the first section of the chapter, we set up a simple conceptual framework for the linear stability analysis. This framework will help us to identify some essential ways that the linear stability analysis of a pattern-forming system differs from the linear stability of fixed points of odes and maps that you might have seen in an introductory nonlinear dynamics course. These differences include the role of uniform stationary states, the simplification obtained when lateral boundaries are replaced with infinite or spatially periodic ones, and the identification of the growth rate $\sigma_\mathbf{q}$ as a function of the wave vector $\mathbf{q}$ as the central quantity of interest.

The technical aspects of the linear instability analysis for a pattern-forming system are introduced in Section 2.2 using a simple model equation known as the Swift–Hohenberg equation. Although this equation is not a precise description of any particular experimental system, it serves as a simple model that captures many of the basic features of pattern formation. From this detailed, specific calculation, we extract in Section 2.3 the key steps in a general linear stability analysis. To connect the abstract analysis back to the real world, we discuss the experimental investigation of the linear instability in pattern-forming systems, first generally, and then by a review of an historic application of linear stability theory to a laboratory experiment, namely the Taylor–Couette experiment (the system shown in Fig. 1.11). Chapter 3 will discuss another application, to chemical reaction–diffusion systems.

In the final two sections of the chapter, we introduce concepts that lay the foundation for the subsequent chapters, namely the classification of different types of linear instability in physical pattern forming systems, and the role of symmetry in the linear instability.

2.1 Conceptual framework for a linear stability analysis

Before formulating a mathematical theory of linear instability and discussing applications, we would like to set up a framework of an idealized pattern-forming system that will help us to identify the essential concepts. The idealization is attained by three simplifications: choosing a uniform geometry for the walls that confine the pattern-forming medium; replacing the lateral boundaries with ones consistent with translational symmetry; and choosing to study the linear instability toward time-independent patterns that are consistent with translational symmetry.

The phenomenon of pattern formation is most easily defined when a pattern develops from a uniform state, which is a state that has no spatial structure. Mathematically, a state is uniform if the medium is translationally invariant:[2] if the medium is translated physically by an arbitrary amount in an arbitrary direction

---

[2] A translationally invariant medium is sometimes also called homogeneous but we prefer to avoid this term since it has other meanings in the context of applied mathematics.

(mathematically the substitution $\mathbf{x} \to \mathbf{x}+\mathbf{x}_0$ is made everywhere in the description of the medium, where $\mathbf{x}_0$ is some constant vector) there is no way to distinguish the translated medium from the original medium. A medium that is translationally invariant is a mathematical idealization since only infinitely big objects can have this symmetry. Any finite object will occupy space in a different way after a translation and so cannot be translationally invariant.

An experimental nonequilibrium system such as the fluid in a convection experiment cannot be completely translationally invariant. Most importantly, the gradients of energy, momentum, or matter that drive the medium out of equilibrium "break" the translational symmetry since the properties of the medium vary along the direction of a gradient. Also, any real experiment is necessarily finite in extent which also breaks the translational symmetry.

These considerations suggest that the simplest and purest example of pattern formation by an instability would be an experiment like Fig. 2.1. A continuous medium, whose equilibrium state is uniform, is contained between two parallel uniform plates A and B whose width L is large compared to their separation d. The medium is driven out of equilibrium, usually by gradients of temperature, velocity, or chemical concentration maintained by the plates A and B. These gradients drive currents of energy, momentum, or matter from one plate to the other. The presence

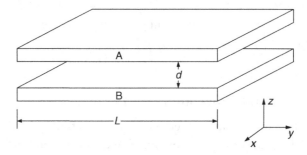

Fig. 2.1 An idealized pattern-forming system consisting of a continuous medium between two flat spatially-uniform parallel plates A and B that are separated by a distance d. The medium is driven out of equilibrium by gradients of temperature, velocity, or chemical concentrations that cause currents of energy, momentum, or mass to flow from one plate to the other. Because the plates are uniform, the gradients are normal to the plates. Coordinates parallel to the gradients (here the vertical direction z) are denoted by the vector $\mathbf{x}_\|$ and are called confined. Coordinates perpendicular to the gradients (here the horizontal directions x and y) are denoted by the vector $\mathbf{x}_\perp$ and are called extended. Pattern formation is defined to occur when there is spatial structure in the extended directions $\mathbf{x}_\perp$. It is most clearly understood when the description of the system is translationally invariant in the extended directions. Translational invariance can be achieved by assuming that the system is infinitely wide ($L \to \infty$) or is spatially periodic in the extended coordinates $\mathbf{x}_\perp$.

of such gradients and currents means the system is out of equilibrium, since in thermodynamic equilibrium they must be zero. Because the plates are uniform, the gradients are normal to the plates and independent of their position along the plates, except possibly near the edges as discussed below. In experiment, often the system is driven further and further out of equilibrium by varying one experimental knob. The corresponding property of the system that is changed we call the control parameter.

Many laboratory experiments can accurately approximate the geometry of Fig. 2.1. For example, in Rayleigh–Bénard convection (Fig. 1.1), A and B would be good thermal conductors maintained at different temperatures and the resulting temperature gradient creates a flow of energy across the fluid. In a Taylor–Couette experiment (Fig. 1.11), the plates would be rigid circular walls that rotate at different constant speeds. The shearing creates a velocity gradient and a flow of momentum from one wall to the other. In a sustained chemical reaction–diffusion experiment (Figs. 1.18(b) and (c), also Fig. 3.3), A and B would be porous walls through which chemicals can flow from reservoirs. The resulting concentration gradients lead to a flux of mass from one plate to the other. In all three cases, experimentalists can make the plates wide compared to their separation so that the assumption of uniformity within any plane parallel to the plates is good.

The natural patterns discussed in Section 1.3.1 are rarely described by the idealized geometry of Fig. 2.1. However, if the spatial extent of a pattern-forming region is large compared with the intrinsic scale of the pattern (a fairly common occurrence), the pattern in such a region can often be understood as involving small corrections to the pattern observed in an idealized geometry. So even for natural patterns, it can be useful to understand first the basic mechanism of pattern formation in an idealized geometry.

The idealized system Fig. 2.1 suggests some useful notation and vocabulary. We define the directions in the plane of the plates to be the extended directions since the size of the system in these direction will usually be large compared to the length scale of the patterns and to the separation of the plates. The corresponding coordinates will be represented by the vector $\mathbf{x}_\perp$ (pronounced "x-perp") since they are perpendicular to the direction of the gradients. In Fig. 2.1, $\mathbf{x}_\perp$ would be the two-dimensional vector (x, y), but in other situations may be one dimensional or even three dimensional. The direction normal to the plates we will call the confined direction since the distance between the plates is often small compared to their width. In Fig. 2.1, there is only one such direction (the z coordinate) but for generality we will use a vector notation here as well and denote the coordinates parallel to the gradients by the vector $\mathbf{x}_\parallel$ (pronounced "x-parallel"). With this notation, a field $u(x, y, z, t)$ can be written as $u(\mathbf{x}_\perp, \mathbf{x}_\parallel, t)$ when we want to emphasize the coordinates normal and parallel to the gradients. Similarly, we will write $\nabla_\parallel$ and $\nabla_\perp$

(pronounced as "grad-parallel" and "grad-perp") to indicate gradients with respect to the variables $\mathbf{x}_\perp$ and $\mathbf{x}_\parallel$ respectively. Finally, the boundaries that constrain the medium on the sides are called the lateral boundaries. As you might suspect, these boundaries play a secondary role in pattern formation provided that the plates are sufficiently wide compared to the scale of the pattern.

For small deviations from thermodynamic equilibrium (corresponding to gradients of small magnitude), the medium will be uniform in any plane parallel to the plates. We will call this a uniform nonequilibrium state since it is spatially constant in the extended directions, and we will see that these are the states of a pattern-forming system whose linear stability can be analyzed and understood most completely. An example of a uniform nonequilibrium state is the conducting state in a convection experiment for which the velocity field is zero everywhere and the temperature field varies linearly in space from its constant value T_1 at $z = 0$ on the warm bottom plate to its constant value T_2 at $z = d$ on the cold top plate. Figure 3.4 in the next chapter shows a numerically calculated uniform state for a chemical reaction-diffusion experiment similar to Figs. 1.18(b) and (c). The dependence on $\mathbf{x}_\parallel$ is complicated and yet the concentration fields are uniform in planes normal to the confined direction.

Once we have identified uniform nonequilibrium states as the appropriate starting point for a stability analysis, *we define pattern formation to be the formation of structure in the extended directions* $\mathbf{x}_\perp$. Thus the formation of convection rolls corresponds to the temperature, velocity, and pressure fields of the fluid becoming nonuniform within each plane parallel to the plates.

The last idealization that we make to prepare for a linear stability analysis is to eliminate the physical lateral boundaries altogether and make the system translationally invariant in the extended directions. This can be arranged theoretically in two ways. First, we can suppose the system to be infinitely wide in the lateral directions in which case there are no lateral boundaries to talk about. This is the most natural way of eliminating the boundaries, but the infinite system size can introduce mathematical difficulties, e.g. infinitely many modes can go unstable at the same time during a bifurcation, a situation that theory cannot at present handle.

A second way to make the system translationally invariant is to assume that the system "wraps around" in the extended directions so that a particle exiting one lateral boundary immediately reenters from the opposing lateral boundary. Eliminating physical boundaries in this second way is described as imposing periodic boundary conditions on the fields characterizing the medium. For a one-dimensional system, periodic boundaries would correspond to defining the system on a circle of circumference L rather than on a line; the azimuthal direction in the Taylor–Couette apparatus of Fig. 1.11 would be an experimental example. For a three-dimensional medium like that of Fig. 2.1, imposing periodic boundaries on pairs of opposing sides turns the medium topologically into the interior of a torus. Although impossible to achieve experimentally, it is a valuable simplification conceptually and easily studied by numerical simulation.[3]

Provided that the experimental plates are large in the extended directions compared to the basic length scale of the pattern, we expect (and comparisons of theory with experiment largely confirm) that only small differences arise between experiment and calculations based on an idealized geometry consistent with translational invariance.[4] This two-step approach to pattern formation – first consider the problem in the idealized geometries with translational invariance and then investigate the small corrections of actual lateral boundaries – has proven to be most instructive. We emphasize that the boundaries A and B are crucial in establishing the nonequilibrium state and cannot be idealized away via some assumption like translational invariance. As a result, the structure of the state in the confined direction must be understood prior to the onset of pattern formation via a proper formulation of the effect of these boundaries.

Three conceptual simplifications have now been introduced: the idealized geometry of Fig. 2.1, the elimination of lateral boundaries by assuming translational invariance in the extended directions, and the identification of stationary uniform nonequilibrium states as the starting point for pattern formation. The reward for introducing these simplifications is that the linear stability analysis is enormously simplified, conceptually and mathematically. The linearized problem (equations and boundary conditions) that governs the evolution of tiny perturbations of the uniform state can be solved with a single simple Fourier mode of the form

$$e^{\sigma_\mathbf{q} t} e^{i\mathbf{q} \cdot \mathbf{x}_\perp}. \tag{2.1}$$

This mode varies periodically in the extended direction $\mathbf{x}_\perp$ with wave vector $\mathbf{q}$, and grows exponentially in time with a $\mathbf{q}$-dependent complex-valued growth rate $\sigma_\mathbf{q}$. For systems that are not translationally invariant or that are linearized about a non-uniform state, only superpositions of such expressions for different wave vectors can satisfy the mathematical problem and there is no simple understanding of the growth rate in terms of a single parameter like the wave vector $\mathbf{q}$.

---

[3] A finite periodic domain, which is often used to simulate pattern-forming systems when comparing dynamics with a linear stability analysis, is not physically equivalent to an infinite domain. To see this, consider a pattern consisting of parallel straight stripes. In an infinite domain, there is no problem to rotate the stripes by a small angle about some axis perpendicular to the plane of the stripes. But for parallel stripes in a finite periodic domain, a small rotation, say about the center of the domain, will cause the ends of the stripes to move in opposite directions so that, as one stripe leaves the periodic domain, it can't continue smoothly onto a stripe from the other side. This lack of symmetry under infinitesimal rotations can be a strong constraint on the dynamics, especially in periodic domains whose size is not much bigger than the stripe wavelength.

[4] The boundaries may actually eliminate many of the steady states found in the unbounded system, as in the example of wave number selection by boundaries discussed in Section 6.4.1. The effect of the boundaries on the surviving states is small, and the steady states of the unbounded system are long-lived transients in a large but finite system.

The growth rate $\sigma_\mathbf{q}$ as a function of the wave vector $\mathbf{q}$ contains much of the useful information in a linear stability analysis of a stationary uniform state. It is the key quantity to calculate analytically or numerically as a function of the system parameters for each uniform state of interest. A given uniform state is linearly stable – all small perturbations asymptotically decay to zero – if the real part of its growth rate Re $\sigma_\mathbf{q}$ is negative for all wave vectors $\mathbf{q}$. As some parameter is varied, the uniform state becomes linearly unstable when Re $\sigma_\mathbf{q}$ first becomes positive at a critical parameter value. The instability signals a change in character of the solution and is called a bifurcation. The wave vector $\mathbf{q}_c$ for which the maximum of the real part of the growth rate, $\max_\mathbf{q}$ Re $\sigma_\mathbf{q}$, first becomes positive is called the critical wave vector of the linear instability and its magnitude q_c the critical wave number. This value q_c is one of the most important quantities obtained from the linear stability analysis since it predicts the length scale $2\pi/q_c$ of the growing perturbation. We might also expect that q_c sets the length scale of the nonlinear states just beyond the onset of the instability. This is true for the case of a supercritical or forward bifurcation of a uniform state, see Section 4.1.2.

It is useful to introduce some additional language to describe the onset of instability. The imaginary part of the growth rate evaluated at the critical wave number determines a critical frequency ω_c given by

$$\omega_c = -\operatorname{Im} \sigma_{q_c}, \tag{2.2}$$

that defines a characteristic oscillatory time scale for the growing perturbation.[5] If ω_c is nonzero, this is an oscillatory or Hopf[6] bifurcation. In many cases, $\omega_c = 0$ and the instability is toward a stationary state. This type of instability is called stationary.

The first bifurcation from the uniform state is sometimes called the primary bifurcation of the pattern-forming system. As will be discussed in Section 2.5, there turn out to be only a few physically distinct common ways that the function Re $\sigma_\mathbf{q}$ develops positive values as a parameter is varied. These different ways lead to a classification of pattern-forming systems into six classes. The amplitude equation theory discussed in Chapters 6, 7, and 10 for weakly nonlinear pattern formation just above the onset of the instability of the uniform state depends especially on which class a pattern-forming system belongs to. Successive bifurcations leading to new states are sometimes called secondary and tertiary bifurcations. You should keep in mind that not all patterns arise as a sequence of bifurcations from a uniform state, a point that we will return to in Chapter 9.

---

[5] We insert the minus sign in Eq. (2.2) so that the space-time dependence takes the conventional form $e^{i(q_c x - \omega_c t)}$ for a wave traveling in the $+x$ direction for positive ω_c.

[6] There are some additional technical restrictions for an oscillatory bifurcation to be properly called a Hopf bifurcation, but these need not concern us here.

2.2 Linear stability analysis of a pattern-forming system

In this section, we carry out our first linear stability analysis of a uniform nonequilibrium state. To avoid unnecessary mathematical details, we analyze a simple yet representative evolution equation for a single field $u(x,t)$ in a one-dimensional domain described by the coordinate x. In Chapter 3, we will study more realistic and correspondingly more complicated examples involving multiple fields that depend on two or more spatial variables.

We will study the linear stability of a uniform state belonging to an evolution equation known as the one-dimensional Swift–Hohenberg equation. As we discuss later in Chapter 5, this is a model equation that was written down, based on the insights of the authors, to capture some of the essential features of pattern-forming systems. The initial context was Rayleigh–Bénard convection. We use this model since it provides one of the simplest ways to illustrate the key steps and insights of a linear stability analysis.[7]

2.2.1 One-dimensional Swift–Hohenberg equation

The one-dimensional Swift–Hohenberg equation is the evolution equation for a single field $u(x,t)$ in a one-dimensional domain $0 \leq x \leq L$

$$\partial_t u(x,t) = (r-1)u - 2\partial_x^2 u - \partial_x^4 u - u^3. \tag{2.3}$$

All these derivatives may look intimidating but this equation is easier to understand than might appear at first glance. A more compact way to write the equation (that becomes more intuitive after gaining some experience with this type of equation) is

$$\partial_t u(x,t) = ru - (\partial_x^2 + 1)^2 u - u^3. \tag{2.4}$$

The parameter r in Eq. (2.3) or (2.4) plays a role of the control parameter. The coordinate x is an extended coordinate $x_\perp$ of the sort that we discussed in the previous section. There is no confined coordinate $\mathbf{x}_\parallel$ in this simple model. In the context of Rayleigh–Bénard convection, which motivated the invention of this model, the parameter r is related to the Rayleigh number R and you can think of the field $u(x,t)$ as the z-component of the fluid's velocity field $v_z(x, y_0, z_0, t)$ at mid-height ($z_0 = d/2$) and midway across ($y_0 = L_y/2$) a long narrow convection experiment of dimensions $L_x \times L_y \times d$ with $L_x \gg L_y > d$. In the derivation of Eq. (2.3) from the convection system, the confined direction of the convection cell

---

[7] In using the Swift–Hohenberg model to introduce the methods of linear instability, we are reversing the historical order since the form of the equation was motivated largely to reproduce the structure found in the linear instability analysis of a variety of pattern-forming system in the simplest possible equation.

is eliminated by a mathematical technique to be discussed in Chapter 6. For now, these various details are not important.

As you can easily verify, Eq. (2.3) has a simple uniform solution $u = 0$. (In a convection experiment, this solution corresponds to the uniform zero-velocity conduction state as the fluid is driven out of thermodynamic equilibrium.) As the parameter r is varied, we would like to determine the critical value r_c when the state $u = 0$ becomes linearly unstable, i.e. when the magnitude of an arbitrary infinitesimal perturbation about $u = 0$ begins to grow exponentially in time. We would also like to identify the critical wave number q_c and critical frequency ω_c associated with the onset of instability.

2.2.2 Linear stability analysis

To carry out the linear stability analysis, we denote the solution $u = 0$ as a base state u_b and ask whether the difference or perturbation field

$$u_p(x, t) = u(x, t) - u_b, \tag{2.5}$$

between an arbitrary nearby solution $u(x, t)$ and the base state will grow in magnitude over time. The perturbation u_p evolves according to the evolution equation

$$\partial_t u_p = \hat{N}[u_b + u_p] - \hat{N}[u_b], \tag{2.6}$$

where the nonlinear operator $\hat{N}$ is defined by considering the right-hand side of Eq. (2.3) to be a function of the field u:

$$\hat{N}[u] = (r - 1)u - 2\partial_x^2 u - \partial_x^4 u - u^3. \tag{2.7}$$

If the field u_p is sufficiently small, we can approximate $\hat{N}[u_b + u_p]$ in Eq. (2.6) by linearizing about u_b, that is by keeping only terms on the right-hand side of Eq. (2.6) that involve a single factor of u_p or of its spatial derivatives. We then find that an infinitesimal perturbation $u_p(x, t)$ satisfies the evolution equation

$$\partial_t u_p = \left(r - 1 - 2\partial_x^2 - \partial_x^4 - 3u_b^2 \right) u_p. \tag{2.8}$$

Notice that this equation is linear in the field u_p (for example, multiplying u_p by a constant factor leaves the equation unchanged), although the base solution u_b occurs nonlinearly. Specializing immediately to the particular base state $u_b = 0$, Eq. (2.8) becomes

$$\partial_t u_p = \left(r - 1 - 2\partial_x^2 - \partial_x^4 \right) u_p. \tag{2.9}$$

This is a linear differential equation with constant coefficients. (The coefficients are constant precisely because the base state is stationary and uniform.) We can then

2.2 Linear stability analysis

generalize a well-known result for solving linear constant-coefficient odes (see for example Section 8.11 of Ref. [55]) and argue that there is a particular solution to Eq. (2.9) that will depend exponentially on time and exponentially on space

$$u_p(x,t) = A e^{\sigma t} e^{\alpha x}. \qquad (2.10)$$

Here the constant σ is a growth rate, and σ and the constant α are possibly complex. Since partial derivatives of any order acting on an expression of the form Eq. (2.10) turn into multiplication of that expression by some algebraic factor (e.g. ∂_x^2 becomes multiplication by α^2), substituting Eq. (2.10) into the linear constant-coefficient partial differential equation Eq. (2.9) yields a simple algebraic relation that can be satisfied by choosing a particular value of σ for each value of α

$$\sigma = r - (\alpha^2 + 1)^2. \qquad (2.11)$$

The meaning of the constant α can be deduced by considering the boundary conditions that apply to the field u and to the perturbation u_p. The simplest possible cases are the idealized geometries for which the lateral boundaries are eliminated by using infinite or periodic boundaries. We discuss these two cases in turn.

(i) *Infinite boundary conditions:* If the system is infinitely large in the x coordinate ($L = \infty$), then there are no boundaries to speak of. This is consistent with a uniform state u_b that is constant everywhere in space but is not consistent with the exponential dependence $e^{\alpha x}$ in Eq. (2.10) unless the constant α is purely imaginary $\alpha = iq$ with q real. Otherwise, $e^{\alpha x}$ will diverge in one of the limits $x \to \infty$ or $x \to -\infty$, violating the assumption that the perturbation u_p is everywhere small.

(ii) *Periodic boundaries:* Alternatively, we can assume that the system is finite but periodic with length L, i.e. the system is topologically equivalent to a ring. A constant solution u_b is automatically periodic over any length but a perturbation $u_p(x,t)$ is periodic with period L only if it satisfies the periodic boundary condition

$$u_p(x,t) = u_p(x+L,t), \qquad (2.12)$$

for all times t and all positions x. The particular solution Eq. (2.10) will be periodic with period L if and only if

$$e^{\alpha x} = e^{\alpha(x+L)} \quad \text{for all } x. \qquad (2.13)$$

This implies that $e^{\alpha L} = 1$ or $\alpha L = (2\pi i)m$ for some integer m so that the quantity $\alpha = iq$ is purely imaginary, just as was the case for infinite boundaries. But now the real number q is restricted to one of the infinitely many quantized values

$$q = m \left(\frac{2\pi}{L}\right), \quad m = 0, \pm 1, \pm 2, \ldots \qquad (2.14)$$

We conclude that infinite and periodic boundary conditions are consistent with a uniform base state u_b and with a single exponential mode Eq. (2.10) provided that the mode has the form

$$u_p(x, t) = A e^{\sigma t} e^{iqx}, \qquad (2.15)$$

with q a real number. The spatial dependence of u_p is then periodic with wave number q (alternatively, with wavelength $2\pi/q$). For infinite domains, q can be any real number while for finite periodic domains, q can be any one of the infinitely many discrete values Eq. (2.14). The fact that the linearized evolution equation can be solved by a single exponential mode is the key conceptual simplification of using boundaries (infinite or spatially periodic) that are consistent with translational invariance.

Substituting $\alpha = iq$ into Eq. (2.11) gives the wave-number-dependent growth rate

$$\sigma_q = r - (q^2 - 1)^2. \qquad (2.16)$$

Eq. (2.16) says that a small-amplitude spatially periodic perturbation with wave number q (about the base solution $u_b = 0$) will grow or decay exponentially in time with a growth rate σ_q that depends on q. Because the evolution equation Eq. (2.9) is linear, a general solution can be obtained as a superposition of the particular solutions Eq. (2.10)[8]

$$u_p(x, t) = \sum_q c_q e^{\sigma_q t} e^{iqx}, \qquad (2.17)$$

where the coefficients c_q are complex numbers and where the sum goes over the discrete set Eq. (2.14) for periodic boundary conditions. (For infinite boundaries, the sum would be replaced by an integral $\int_{-\infty}^{\infty} dq \ldots$ over all wave numbers q, and Eq. (2.17) becomes a Fourier integral.) The base state $u_b = 0$ is therefore linearly stable if each exponential in Eq. (2.17) decays in the limit $t \to \infty$. This will be true if the maximum real part of all the growth rates is negative:

$$\max_q \operatorname{Re} \sigma_q < 0, \qquad (2.18)$$

which is true for negative r.

---

[8] Eq. (2.17) is none other than a Fourier analysis of the perturbation $\delta u(t, x)$. At each moment in time t, we represent the spatial dependence of δu as a Fourier sum or Fourier integral depending on whether we have periodic or infinite boundary conditions. For a constant-coefficient pde evolution equation, the time-dependent Fourier coefficient $c_q(t)$ for mode e^{iqx} has the particular form of $c_q e^{\sigma t}$, a constant c_q times an exponential in time. Since δu is real-valued, the Fourier coefficients $c_q(t)$ of positive and negative wave number $q > 0$ are complex conjugates of each other, $c_{-q}(t) = c_q^*(t)$.

2.2.3 Growth rates and instability diagram

Let us examine in detail how the growth rate σ_q determines the linear stability of the uniform base state $u_b = 0$. We saw above in Eq. (2.16) that the growth rate σ_q is given by

$$\sigma_q = r - \left(q^2 - 1\right)^2. \tag{2.19}$$

Since this expression is real-valued for all wave numbers q, the critical frequency ω_c of Eq. (2.2) is zero. Instability will occur without temporal oscillations.

We next want to determine when the maximum of the curve $\text{Re}\,\sigma_q$ versus q changes from a negative to positive value as the parameter r is varied, indicating the onset of linear instability. Since the quantity $(q^2 - 1)^2$ in Eq. (2.19) is non-negative and vanishes for $q = 1$, the maximum value of σ_q occurs for the value $q_{\max} = 1$ independently of r. The value of σ_q at its maximum is therefore

$$\max_q \text{Re}\,\sigma_q = r. \tag{2.20}$$

We have made two discoveries. First, the uniform state $u = 0$ is linearly stable when the Swift–Hohenberg parameter $r < 0$ and is linearly unstable when $r > 0$ so the critical parameter value for linear instability is $r_c = 0$.[9] Second, the critical wave number q_c at which the curve $\text{Re}\,\sigma_q$ first attains a positive value is $q_c = 1$ since this is the location of the maximum, independent of the value of r.

Additional insight can be obtained by plotting the curve $\text{Re}\,\sigma_q$ versus q for several values of r. Figure 2.2 confirms visually our analytic conclusion that the growth rate is everywhere negative if $r < 0$ (the bottom light-gray curve) but provides further insight. First, the Fourier modes e^{iqx} with wave numbers q near the maximum at $q = 1$ are the least slowly decaying in time and the modes with $q \gg 1$ are the most rapidly decaying. For $r = 0.2$ just a bit larger than $r_c = 0$, there is a narrow band of wave numbers $0.75 < k < 1.2$ centered around $q = 1$ whose corresponding Fourier modes will grow. This means that if the initial state $u_p(x, t = 0)$ is a small-amplitude noise such that all the Fourier coefficients in Eq. (2.17) are nonzero but with tiny amplitude, then a cellular pattern will start to grow out of this noise since, in Eq. (2.17), only Fourier coefficients with wave numbers close to $q_c = 1$ will grow in magnitude. Thus Fig. 2.2 suggests the beginning of pattern formation and predicts some kind of cellular structure with a characteristic wavelength $2\pi/q_c$.

An alternative and commonly used way to summarize the information in Eq. (2.16) and in Fig. 2.2 is to plot the neutral stability curve $r = r_c(q)$ for which

---

[9] The case $r = r_c$ of marginal stability requires a separate stability analysis involving higher-order terms in the expansion Eq. (2.8) to see whether small perturbations will grow or decay. However, such marginal cases are rarely interesting physically since, in practice, no experimental system can be tuned precisely to a critical value.

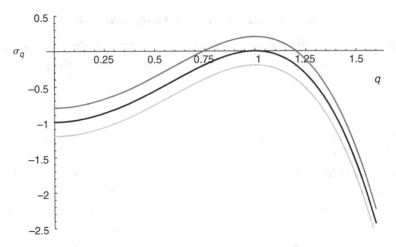

Fig. 2.2 Plot of the growth rate σ_q, Eq. (2.16), versus the wave number q for the uniform base state $u_b = 0$ of the one-dimensional Swift–Hohenberg equation Eq. (2.3) in an infinite domain. Three curves are shown for parameter values of respectively $r = -0.2$ (light gray), $r = 0$ (black), and $r = 0.2$ (dark gray). These correspond to a stable, marginally unstable, and unstable base state. The critical parameter value $r_c = 0$ and the critical wave number $q_c = 1$ are identified from the $r = 0$ curve as where $\mathrm{Re}\,\sigma_q$ first becomes zero as r is varied. For $r = 0.2$ just positive, only a narrow band of Fourier modes centered on the critical wave number can grow, predicting the appearance of a pattern with a characteristic length scale $2\pi q_c^{-1}$.

the real part of the growth rate vanishes, $\mathrm{Re}\,\sigma_q = 0$. (We can also write this as $q = q_N(r)$.) For the Swift–Hohenberg equation, Eq. (2.19) implies that the neutral stability curve for the uniform state $u_b = 0$ is given by $\sigma_q = 0$ or by the curve

$$r = \left(q^2 - 1\right)^2, \qquad (2.21)$$

which we plot in Fig. 2.3. For all points (r, q) below this curve, a perturbation of wave number q about the uniform state $u = 0$ will decay, whereas above this curve a perturbation at wave vector q will grow. The critical parameters r_c and q_c are now determined graphically by the global minimum of the neutral stability curve. Figure 2.3 again suggests that cellular patterns with wave number $q \approx q_c$ might be expected for r just larger than r_c since there is a narrow band of wave numbers for which the uniform state is unstable. Unlike Fig. 2.2, the neutral stability curve gives no information about the growth rate of the instability. However the diagram gives a single picture showing the range of modes that would grow from the uniform state, and that, loosely, might be available for pattern formation.

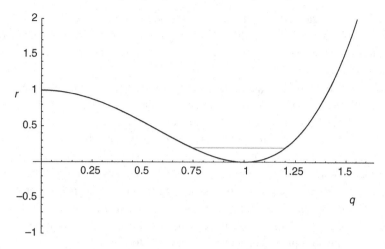

Fig. 2.3 Plot of the neutral stability curve $r = r_c(q)$, Eq. (2.21), for the uniform state $u_b = 0$ of the one-dimensional Swift–Hohenberg equation Eq. (2.3). The uniform state is stable below this curve and pattern formation can be expected above this curve. The critical parameter $r_c = 0$ and critical wave number $q_c = 1$ can be identified visually from the global minimum of this curve. The horizontal line segment indicates the band of unstable wave numbers for $r = 0.2$ and should be compared with the $r = 0.2$ curve of Fig. 2.2.

2.3 Key steps of a linear stability analysis

From the linear stability analysis of the Swift–Hohenberg equation, we can identify the following steps for carrying out the linear stability analysis of a stationary uniform state associated with some pattern-forming system.

(i) Obtain explicitly the evolution equations for the system.
(ii) Rewrite the evolution equations in dimensionless form to reduce the number of parameters **p** and to obtain the parameters in dimensionless form.
(iii) Replace the boundary conditions in the extended directions with infinite or periodic boundary conditions.
(iv) For a given vector of system parameters **p**, find explicitly at least one time-independent state $\mathbf{u} = \mathbf{u}_b(\mathbf{x}_\parallel)$ that is uniform with respect to the extended coordinates $\mathbf{x}_\perp$.
(v) Linearize the evolution equations about the uniform base state $\mathbf{u}_b$ to obtain the linear evolution equations for an infinitesimal perturbation $\mathbf{u}_p$. The coefficients of this linear differential equation will not depend on the extended coordinates $\mathbf{x}_\perp$ nor on time.
(vi) Use a particular solution of the form

$$\mathbf{u}_p = \mathbf{u}_q(\mathbf{x}_\parallel) e^{\sigma_q t} e^{i\mathbf{q} \cdot \mathbf{x}_\perp}. \tag{2.22}$$

to solve the linearized evolution equations and to obtain the wave vector dependent growth rate σ_q. If the uniform state $\mathbf{u}_b(\mathbf{x}_\parallel)$ has a nontrivial dependence on the constrained coordinates $\mathbf{x}_\parallel$, a numerical calculation may be needed to determine the functions $\mathbf{u}_q(\mathbf{x}_\parallel)$.

(vii) Analyze the function Re σ_q versus wave vector $\mathbf{q}$ visually and mathematically. The interesting features to look for are local maxima (especially the global maximum) and the wave vectors $\mathbf{q}$ corresponding to these maxima. The uniform state is linearly stable for the chosen parameters if $\max_q$ Re $\sigma_q < 0$, unstable otherwise.

(viii) Map out the linear stability of the uniform states as a function of the system parameters by repeating steps (iv) through (vii) for different values of the parameter vector $\mathbf{p}$ and possibly for different uniform states. In particular, for any parameter p of special interest, identify its critical value p_c (if one exists) defined by $\max_q$ Re $\sigma_q = 0$. Also identify the corresponding critical frequency ω_c and critical wave number q_c (or, in the case of anisotropic systems, critical wave vectors, $\mathbf{q}_c$).

Step (i) can be difficult since a detailed knowledge of the physics, chemistry, or biology of the system is needed to derive possible evolution equations and much experimental work is needed to validate proposed equations. Thus it is still not known whether the flow of a granular medium like sand can be described by continuum evolution equations, and physiologists still struggle to identify quantitatively accurate evolution equations for the electrical waves in cardiac muscle. For known evolution equations, finding a uniform state (step iv) can be hard since there is no mathematical algorithm that is guaranteed to find even a single solution of the nonlinear partial differential equations satisfied by a uniform state. In the laboratory, a uniform state can sometimes be found by starting from an equilibrium state and driving this state slowly out of equilibrium but this won't work for many natural systems such as the Sun or the weather whose parameters cannot be varied. The remaining steps are easier since they involve simple mathematical manipulations or finding the eigenvalues of a matrix. The latter is straightforward using modern numerical software.

2.4 Experimental investigations of linear stability

2.4.1 General remarks

In this section, we change our point of view and consider how to study experimentally the linear stability of a stationary uniform state.[10] Experimentalists are, in fact, often in the position of not having any theory for guidance and yet they routinely

---

[10] Similar questions and solutions arise for a computational scientist who can only run some complicated evolution code. For some codes, the nonlinearities may be defined through functions that use iterative algorithms or look-up tables so that it is not possible to linearize the dynamical equations explicitly. However, the computational scientist often has two big advantages over the experimentalist: the ability to specify periodic boundary conditions in the extended directions and the ability to set the initial conditions to any desired value.

2.4 Experimental investigations of linear stability

measure quantities such as the critical bifurcation parameter p_c, the critical wave number q_c, and the critical frequency ω_c.[11] The most progress can be made if the bifurcation of the uniform state is known experimentally to be supercritical so that nonlinear states grow continuously out of the uniform state. We will assume this to be the case. For the opposite case of subcritical bifurcations, it is straightforward to determine experimentally when the uniform state becomes unstable, but difficult to determine the critical wave number and frequency since the finite-amplitude nonlinear state may have no relation to the uniform state.

For a pattern-forming system of interest, let us assume that the experimentalist has taken the difficult step of making the medium as large as possible in the extended directions so that the lateral boundaries are likely to be a weak perturbation. There is then a possibility to compare the experimental results with a linear stability calculation based on infinite or periodic boundary conditions. We will also assume that the experimentalist has discovered by trial and error that varying some parameter p causes a stationary uniform state to undergo a supercritical bifurcation to some nonuniform patterned state. We would then like to determine the critical values p_c, q_c, and ω_c associated with the instability.

To make progress, the experimentalist must identify some appropriate order parameter $\Psi(p)$ whose measured values are able to distinguish the uniform state from the nonuniform state. It is conventional to choose a quantity that is zero in the uniform state, and nonzero in the pattern state. For Rayleigh–Bénard convection, a suitable order parameter would be the magnitude $|v_z(x_0, y_0, z_0, t)|$ of the z-component of the velocity field at some fixed point (x_0, y_0, z_0) in the fluid.[12] For media that are not optically accessible, or for experimental convenience, other order parameters can often be found that are related to the transport of heat or electrical current through the medium, since transport is often enhanced by the pattern. For example, for Rayleigh–Bénard convection the total heat transport may be convenient, often quoted in dimensionless form as the Nusselt number N (the ratio of the total heat transported to the heat transported by conduction alone). The quantity $N-1$ is then zero in the conduction state, and serves as an order parameter.

The linear instability theory discusses the evolution of a small amplitude perturbation. This is hard to measure experimentally because the signal is weak and

---

[11] Theory provides much more information in the form of the growth rate σ_q as a function of parameters. But it is difficult in most experiments to impose a periodic small-amplitude disturbance of known wave vector q and to measure how its amplitude grows over time since the experimental lateral boundaries are not generally consistent with the periodicity of the perturbation.

[12] Provided that the velocity field v is not changing too rapidly, the technique of laser Doppler velocimetry can be used to measure one or several components of the velocity at a point in the medium without perturbing the fluid. The fluid is doped with a tiny concentration of neutrally buoyant light-scattering particles such as white latex spheres about one micron in diameter. The frequency of laser light bouncing off these particles is Doppler shifted by the average velocity of the fluid in the vicinity of the spheres. The frequency shift is easily measured and converted into the velocity component along some direction.

because it is difficult to measure time-varying transients. It turns out that the quantities that characterize the linear instability point are best estimated by fitting curves to data obtained just above the onset of the instability, in the weakly nonlinear regime for which the exponentially growing perturbations have saturated to some constant small magnitude. We will discuss this nonlinear state in more detail in Chapter 4. All we need here is the simple result that, for supercritical bifurcations, the size of $\Psi(p)$ grows continuously from zero as p increases above the critical value p_c.[13]

The experimentalist therefore measures $\Psi(p)$ after transients have died out at some set of values p_i of p. Then by fitting the data points to a suitably chosen smooth curve, the critical value p_c is identified by extrapolating the fitted curve for $\Psi(p)$ to $\Psi = 0$. Often, the theory of the nonlinear state can be a guide to the fit. For example, the amplitude equation theory of Chapter 6 suggests that strengths of field variables such as a fluid velocity should grow as $(p - p_c)^{1/2}$ for p close enough above p_c, and intensity variables such as velocity squared, or convected heat

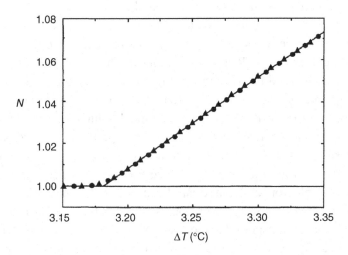

Fig. 2.4 Plot of the Nusselt number N (the total heat transport divided by the amount of heat transported by conduction) against the temperature difference across the fluid (proportional to the Rayleigh number R) in a Rayleigh–Bénard convection experiment. A linear fit to a portion of the data is used to estimate the critical Rayleigh number. Note that there is some deviation of the data from the fit for small values of the convected heat $N - 1$, often called "rounding of the transition." This rounding suggests that the bifurcation is imperfect (see Appendix 1), which could be a result of tiny experimental imperfections. (From Hu et al. [48].)

---

[13] If $\Psi(p)$ does not grow continuously, the case of a subcritical bifurcation, this method cannot be used and finding the linear instability point is more difficult. In some cases, the growth of transients is indeed studied. In other investigations, the experimentalists have cleverly stabilized the unstable small amplitude state that develops near a subcritical bifurcation.

2.4 Experimental investigations of linear stability

transport should grow linearly in $p - p_c$ for p near p_c. Figure 2.4 shows an example taken from a Rayleigh–Bénard experiment, where the heat transport, plotted as the dimensionless Nusselt number N is measured as a function of the temperature difference across the fluid, plotted as the dimensionless Rayleigh number.

The critical frequency ω_c and critical wave number q_c can be estimated by a similar data-fitting procedure. For example, the critical frequency could be estimated by measuring several different time series $u_i(t)$ of some observable (e.g. the velocity component at a fixed point in space or the total heat transport) for different parameter values p_i in the weakly nonlinear regime. For each time series i, a corresponding power spectrum $P_i(\omega)$ could be calculated and the position ω_i of the peak of largest magnitude estimated.[14] Similarly a mean wave number $\langle q \rangle_i$ could be identified from the spatial wave number spectrum $P_i(q)$, where q is the magnitude of a two-dimensional wave vector $\mathbf{q}$. Finally, the values of ω_c and q_c would be obtained by fitting the data $\{p_i, \omega_i\}$ and $\{p_i, \langle q \rangle_i\}$ to appropriate curves and then by evaluating these curves at the critical parameter value p_c, which was obtained separately and previously.

There are other subtleties that require consideration when estimating critical values. Every time the parameter p_i is changed to a new value p_{i+1} to record more data, the experimental system is perturbed and the experimentalist must wait for a transient to die out before continuing to record data. Since the time for a transient to decay is usually not known in advance, this has to be studied as a separate problem. A related difficulty is that the transient times get longer and longer the closer the parameter p gets to its unknown critical value p_c, a phenomenon known as critical slowing down. This occurs for any supercritical transition including bifurcations of odes and maps and for phase transitions in equilibrium systems. Another issue is that since the goal is eventually to compare experimental estimates with a linear stability theory based on infinite or periodic boundaries, it is sometimes necessary to obtain estimates of critical parameters for different system sizes and then extrapolate to the infinite size limit. With care and effort, these difficulties can be conquered and impressive agreement found between theory and experiment, in some cases exceeding three significant digits.

After all this work, the experimentalist has estimates of the three numbers p_c, ω_c, and q_c. But p is usually not the only parameter controlling the experiment. With further work, the critical values $p_c(s)$, $q_c(s)$, and $\omega_c(s)$ can often be traced out as functions of some secondary parameter s (or even as functions of several secondary parameters). The comparison of these functions with corresponding results

---

[14] The experimental paper by Gollub and Benson [38] discusses some of the details and subtleties of identifying the frequencies of peaks in a power spectrum. The amplitude equation theory of Chapter 6 shows that the higher harmonics in the power spectrum become smaller in magnitude the closer one approaches the value p_c so identifying the largest peak actually becomes easiest in this limit.

calculated from the linear stability analysis provide some of the strongest tests of whether theory and experiment agree with one another.

2.4.2 Taylor–Couette instability

One of the most famous comparisons of experiment and theory for the linear instability of a uniform nonequilibrium state is also the first. In a pioneering paper published in 1923, G. I. Taylor studied experimentally and theoretically the linear instability of the laminar uniform fluid state between two rotating concentric cylinders (see Fig. 1.11). He studied the onset of instability and the critical wave number q_c as a function of two parameters, the angular frequency Ω_1/ν of the rotating inner cylinder, and the angular frequency Ω_2/ν of the rotating outer cylinder.[15] Experimentally, Taylor made the axial direction as long as he possibly could to avoid end effects. He was able to attain aspect ratios Γ of 80 to 370 depending on the choice of inner radius r_i. Theoretically, Taylor assumed that the axial (extended) coordinate was periodic so he could take advantage of translational invariance. For technical reasons (he didn't have a computer in 1923!), he restricted the possible infinitesimal perturbations of the uniform fluid state to be axisymmetric, which his own experiments confirmed to be a reasonable assumption.[16]

In a typical experiment, Taylor used gears to fix the ratio $\mu = \Omega_2/\Omega_1$, injected a trace amount of ink near the inner cylinder, and then slowly increased Ω_1 in small successive steps until the uniform state became unstable. The instability was visually obvious by the spontaneous self organization of the ink into a stack of horizontal toroidal vortices that were easily seen through a glass portion of the outer cylinder. By carefully sweeping the ratio μ back and forth and by analyzing pictures of the resulting vortices, Taylor was able to estimate the critical ratio μ_c and corresponding critical wave number q_c as plotted in Figs. 2.5 and 2.6 respectively. For $\mu > 0$ (corotating cylinders), the widths of the vortices are about the same as the width $r_o - r_i$ of the fluid layer, but for $\mu < 0$ (counter-rotating cylinders), the vortices are smaller than the fluid width. The agreement between theory and experiment over a range of μ values is impressive. The situation before Taylor's paper is well summarized in his own words [102]:

> A great many attempts have been made to discover some mathematical representation of fluid instability, but so far they have been unsuccessful in every case.

---

[15] Taylor normalized both frequencies Ω_i to the fluid's kinematic viscosity ν but did not scale out the length as we did in Eq. (1.4).

[16] Theory developed years later confirmed that axisymmetric perturbations are the most "dangerous" in that they grow faster than nonaxisymmetric perturbations. An exception is counter-rotating cylinders ($\Omega_2/\Omega_1 < 0$) for which the outer cylinder is rotating much faster than the inner cylinder ($|\Omega_2/\Omega_1| \gg 1$). In this regime, there is a small systematic correction to the results calculated by Taylor based on axisymmetric perturbations.

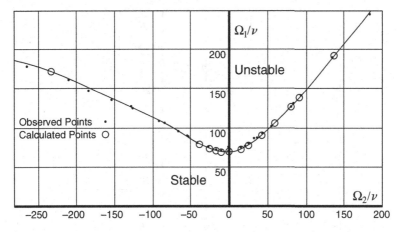

Fig. 2.5 G. I. Taylor's classic comparison of experiment with theory for the linear instability of the uniform (laminar) fluid state of a Taylor–Couette cell (see Fig. 1.11). The experimental apparatus consisted of an outer cylinder with inner radius $r_o = 4.04$ cm rotating with angular frequency Ω_2, and of an inner cylinder with outer radius $r_i = 3.55$ cm, rotating with frequency Ω_1. The cylinders were 90 cm tall so the aspect ratio was $\Gamma = 90/(4.04 - 3.55) \approx 184$. The fluid was water at close to room temperature with kinematic viscosity ν. Instability was observed visually through a glass window in the outer cylinder, when small amounts of injected ink suddenly self organized into horizontal vortices (Taylor cells). The small black points are experimental data while the larger open points are theoretical points calculated from a linear stability analysis of the Navier–Stokes equations. The smooth curve through the points is a guide to the eye, not a theoretical fit. The uniform state is stable below the curve, unstable above. The agreement is impressive, within two percent over a large range of inner and outer frequencies. (Redrawn from a figure in the paper "Stability of a viscous liquid contained between two rotating cylinders" by Taylor [102].)

Taylor's results in Figs. 2.5 and 2.6 leave no doubt that a linear stability analysis based on the Navier–Stokes equations of fluid dynamics gives a quantitative explanation of when and how the Taylor–Couette uniform state becomes unstable.

2.5 Classification for linear instabilities of a uniform state

The linear stability analysis of the Swift–Hohenberg equation in Section 2.2.2 demonstrates the importance of how the growth rate $\mathrm{Re}\,\sigma_q$ passes through zero for q near q_c as the control parameter p is increased through p_c. The critical wave number q_c tells us about the basic length scale of the pattern formation, and the band of unstable wave numbers for $p > p_c$ gives an estimate of the range of modes that are accessible for pattern formation. Analysis of many physical systems and other models suggests a useful classification scheme for various types of linear instability

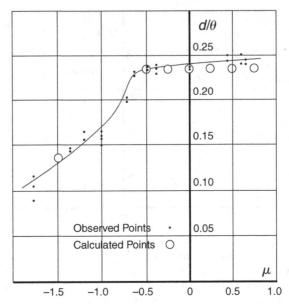

Fig. 2.6 G. I. Taylor's comparison of the predicted critical half-wavelength $\lambda_c/2 = \pi/q_c$ (large open dots) with experimental values (small black dots) as a function of the ratio $\mu = \Omega_2/\Omega_1$ of outer to inner rotation frequencies. The wavelengths were estimated by measuring the average separation between vortices in photographs that were taken of flows as close as possible to onset. The height, outer radius, and inner radius were respectively $L = 90$ cm, $r_o = 4.04$ cm, and $r_i = 3.80$ cm. Taylor's vertical label d/θ is the average width of a vortex in centimeters. Experiment and theory agree to about five percent, not quite as good as for the onset of instability in Fig. 2.5. (Source as in Fig. 2.5.)

important in pattern-forming systems.[17] This classification will also turn out to be useful for analyzing the weakly nonlinear pattern-forming tendencies of a sustained nonequilibrium system near the onset of a supercritical bifurcation of the uniform state. In the present section, we consider a system with one extended direction as in Section 2.2. In the next section, we discuss a larger number of extended directions and the role symmetry plays in these cases.

For some idealized pattern-forming system with infinite or periodic lateral boundaries, let us consider some parameter p such that a stationary uniform state of the system becomes linearly unstable when p exceeds a critical value p_c. Since we are interested in the first instability of the uniform state, we focus our attention around the maximum of the growth rate curve Re σ_q, for example around $q = 1$ in Fig. 2.2. The classification scheme is based on whether this maximum growth rate passes through zero at a zero or a nonzero value of the wave number q as the control

---

[17] This classification scheme was introduced by Cross and Hohenberg [25].

parameter p is varied, and on whether the instability is stationary or oscillatory at this point. Then the linear instabilities can be classified as belonging to one of three classes that we designate by Roman numerals as type I, type II, or type III as shown in Figs. 2.7, 2.8, and 2.9.[18] Each of these three types can be further divided into two categories depending on whether the imaginary part of the growth rate at onset given by ω_c is zero or nonzero. If zero, we call the instability stationary and add a label "s" to the type, e.g. the symbol I-s denotes a stationary type-I instability. If nonzero, we call the instability oscillatory and add a label "o," e.g. the symbol III-o denotes a type-III oscillatory instability at onset. We will concentrate on stationary instabilities in the next seven chapters since they are more relevant for the formation of time-independent structures. Chapter 10 will discuss pattern-forming oscillatory instabilities.

We now describe the classes in turn. Here and later, it is useful to introduce a dimensionless reduced bifurcation parameter

$$\varepsilon = \frac{p - p_c}{p_c}, \qquad (2.23)$$

so that $\varepsilon = 0$ defines the onset of instability, and consider the functional form of $\operatorname{Re} \sigma_q$ for ε close to zero and q close to q_c.

2.5.1 Type-I instability

For a type-I instability, the instability occurs first at a nonzero wave number, i.e. the quantity $\operatorname{Re} \sigma_q$ first becomes positive at a critical wave number $q_c > 0$. This is the case we encountered for the Swift–Hohenberg analysis. The length $2\pi/q_c$ sets the characteristic scale of the patterns. For $p > p_c$, the uniform state is unstable to perturbations over a band of wave numbers between the neutral stability values

$$q_N^-(p) < q < q_N^+(p). \qquad (2.24)$$

Assuming that $\operatorname{Re} \sigma_q$ varies smoothly as a quadratic near its greatest value and assuming a smooth variation with respect to the control parameter p, the behavior of the linear instability is pictured in Fig. 2.7.

The key aspects of the behavior of the growth rate near threshold are captured by a simple algebraic expression that turns out to be useful in defining a time-scale

---

[18] We restrict the likely types of behavior by assuming that σ_q is everywhere finite, that $\operatorname{Re} \sigma_q$ varies smoothly near its greatest value (i.e. the curve shows a quadratic maximum, not a cusp), and that $\operatorname{Re} \sigma_q < 0$ for large q. These assumptions are physically reasonable, although simple mathematical models may violate one or more of them. If these restrictions are lifted, a much wider range of behavior is possible, some of which are described in the book by Murray [79]. However, the present classification into the three main classes described here is convenient and sufficiently general.

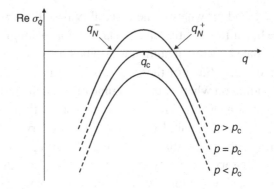

Fig. 2.7 Growth rate for a type-I instability. Re σ_q is shown as a function of q for three values of the control parameter p. Only the region of the curve near the maximum growth rate is plotted. The value of p when this maximum value passes through zero identifies the critical parameter p_c and the wave number where this happens is the critical wave number q_c.

parameter τ_0 and a length-scale parameter ξ_0 that are also relevant in the weakly nonlinear analysis.

For $q = q_c$, the growth rate passes through zero at $p = p_c$. Expanding about this value in a Taylor expansion for small ε gives

$$\sigma_{q_c} \simeq \frac{1}{\tau_0}\varepsilon, \qquad (2.25)$$

which defines a parameter τ_0 that is the characteristic time scale of the instability. A few points are worth explaining. The lowest-order constant term in the Taylor expansion is absent by virtue of our choice to expand about p_c where the growth rate is zero. On the other hand, since σ_{q_c} is "passing through zero," we make the reasonable assumption that the coefficient of the first term in the expansion is nonzero. (This is known as the transverse assumption in bifurcation theory.) The coefficient of the first-order term has the dimensions of an inverse time and so we write the coefficient as τ_0^{-1}. Finally, we ignore higher-order terms for small ε.

For a growth rate curve with a smooth maximum, we expand about the maximum value. For $p = p_c$, the maximum occurs at $q = q_c$ and so we have

$$\sigma_q(p = p_c) \approx -\frac{\xi_0^2}{\tau_0}(q - q_c)^2. \qquad (2.26)$$

The coefficient of the quadratic term is written as $-\xi_0^2/\tau_0$, which introduces the new constant ξ_0 which has dimensions of length and is called the coherence length. As we will discuss in Chapter 6, ξ_0 determines the spatial range over which some local disruption (e.g. a lateral boundary or topological defect) perturbs the surrounding pattern.

Finally, we can combine the two expansions to give

$$\sigma_q \approx \frac{1}{\tau_0}\left[\varepsilon - \xi_0^2 (q - q_c)^2\right], \quad (2.27)$$

where there will be higher-order terms of order $\varepsilon^2, \varepsilon(q-q_c), (q-q_c)^3$, etc. Note the implication of this expression that the width of the band of unstable modes $q_N^+ - q_N^-$ shrinks to zero as $\varepsilon^{1/2}$ in the limit $\varepsilon \to 0^+$.

The instability may be stationary of type I-s or oscillatory of type I-o. In the latter case, the space and time dependence is that of a standing or traveling wave.

2.5.2 Type-II instability

For a type-II instability, the growth rate at $q = 0$ is always zero. This is often the case when one of the evolving fields obeys a conservation law so that the integrated value of the field over all space is constant over time. Typical curves of $\mathrm{Re}\,\sigma_q$ for p near p_c are shown in Fig. 2.8, for the case that the growth rate is symmetric under the substitution $q \to -q$. The curves can be parameterized for small q and $p \simeq p_c$ by the form

$$\mathrm{Re}\,\sigma_q \approx D\left(\varepsilon q^2 - \frac{1}{2}\xi_0^2 q^4\right). \quad (2.28)$$

Here the constant D actually has the dimensions of a diffusion constant, and ξ_0 is again the coherence length. Note that the maximum growth rate occurs at a small

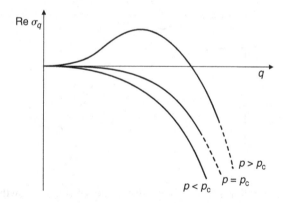

Fig. 2.8 Growth rate for a type-II instability. The real part of the growth rate $\mathrm{Re}\,\sigma_q$ is shown as a function of wave number q for three values of the control parameter p, as in Fig. 2.7. For $p < p_c$, the growth rate decreases quadratically from $q = 0$, whereas for $p > p_c$, the growth rate increases quadratically from $q = 0$, eventually decreasing at larger q. For $p = p_c$, the growth rate decreases more slowly. Note that the instability occurs first at $p = p_c$ at $q = 0$ but the wave number for the maximum growth rate occurs at a value of q that increases for $p > p_c$.

wave number, $q = q_m = \xi_0^{-1}\sqrt{\varepsilon}$. This implies a tendency toward pattern formation with a long wavelength $\xi_0 \varepsilon^{-1/2}$. Unlike a type-I instability, the characteristic length of pattern formation diverges to infinity as $\varepsilon \to 0^+$. Like a type-I instability, the width of the band of unstable wave numbers shrinks in the same limit to zero as $\varepsilon^{1/2}$, but now the band stretches to zero wave number.[19]

The instability can again be of two types, stationary of type II-s and oscillatory of type II-o.

2.5.3 Type-III instability

For a type-III instability, the growth rate is maximized at zero wave number, $q = 0$, above and below threshold as shown in Fig. 2.9. The dependence near threshold can be characterized by the expression

$$\operatorname{Re}\sigma_q \approx \frac{1}{\tau_0}\left[\varepsilon - \xi_0^2 q^2\right] \quad (2.29)$$

near onset (see Exercise 2.3 for an example). Since the maximum growth rate occurs for $q = 0$, the linear analysis predicts that there is the possibility of spatial

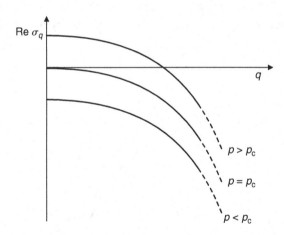

Fig. 2.9 Growth rate for a type-III instability. The maximum growth rate always occurs at $q = 0$.

---

[19] Sometimes the expansion of $\operatorname{Re}\sigma_q$ for $q \simeq 0$ is not smooth and the growth rate takes the form

$$\operatorname{Re}\sigma_q \propto \varepsilon |q| - \xi_0 q^2.$$

The maximum growth rate now occurs at $q_m \propto \varepsilon$ for $\varepsilon > 0$, and the band of unstable wave numbers similarly scales proportional to ε.

structure on a large length scale just above threshold, with a band of unstable wave numbers whose width is proportional to $\varepsilon^{1/2}$. The type-III instability is often encountered for oscillatory instabilities (i.e. type III-o). For example, an oscillatory type-III-o instability can occur for the two-chemical reaction–diffusion system we will discuss in Chapter 3. A system with a type-III-o instability typically supports nonlinear waves over a band of wave numbers of width proportional to $\varepsilon^{1/2}$ near threshold. A weakly nonlinear state near this type of instability provides a simple model to begin to understand nonlinear wave states in pattern forming system, as discussed in Chapter 10.

The significance of these three classes applies not just to the linear instability but also to the pattern state that ensues, at least for control parameter values not too far from the critical one. As will become clear in Chapters 6 and 10, just above the instability of a uniform state, the pattern formation has a universal behavior[20] that depends only on the class of linear instability, not on particular details of the evolution equations themselves. Thus systems that seem unrelated such as Rayleigh–Bénard convection, reaction–diffusion systems, and plasmas may all have similar qualitative behavior near the onset of the type-I-s instability of their uniform states. An experimental or theoretical study of one system can then give insight about other systems.

2.6 Role of symmetry in a linear stability analysis

So far, our discussion of linear instability has concerned mainly systems with one extended spatial dimension, in which case the growing disturbance is characterized by a single number, the wave number q. If there are two or more extended spatial dimensions, the dependence of the growth rate on the direction of the wave vector $\mathbf{q}$ of the unstable mode must also be considered. This is especially important if the physical medium is governed by rotational symmetries, since many distinct modes can then become unstable at the same time and with the same growth rate. In these cases, superpositions of the growing modes can occur that lead to a more complicated spatial dependence. In this section, we discuss the relation between symmetries and growth rates and the implication of this relation for pattern formation. We first discuss some consequences of symmetry using qualitative arguments, and then give brief pointers to the more formal discussion that can be found in advanced texts and in the published literature.

---

[20] The word "universal" has a special meaning in physics that arose in the study of second-order equilibrium phase transitions. "Universal" means that certain details of a certain phenomenon do not depend on microscopic details but rather on more abstract details such as the symmetry or dimensionality of the system.

2.6.1 Rotationally invariant systems

Many pattern-forming systems are isotropic in the extended directions so that the medium is unchanged by a rotation through an arbitrary angle about any axis perpendicular to these directions.[21] For example the partial differential equations that describe convection of a free fluid or that describe reacting and diffusing chemicals in the geometry of Fig. 2.1 are unchanged by a rotation about the vertical axis. (The lateral boundaries may break this symmetry, but, as we have discussed before, it is valuable to study first the idealized situation for which the effects of the lateral boundaries are ignored.)

A simple model for a pattern-forming system that is rotationally invariant in two dimensions is the two-dimensional Swift–Hohenberg equation for the evolution of a scalar field u that is a function of two space coordinates:

$$\partial_t u(x,y,t) = ru - (\partial_x^2 + \partial_y^2 + 1)^2 u - u^3. \tag{2.30}$$

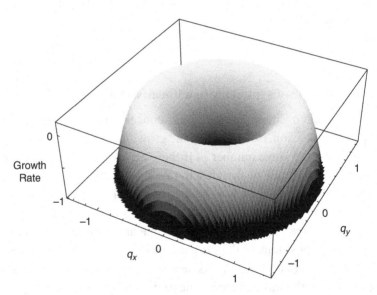

Fig. 2.10 Plot of the growth rate $\operatorname{Re}\sigma_\mathbf{q}$ for a type-I pattern forming system that is rotationally invariant in the xy plane. The surface plot is shaded according to the value of $\operatorname{Re}\sigma_\mathbf{q}$ on a grey scale with light corresponding to larger values. The actual result plotted is for Eq. (2.30) with $r = 0.1$.

---

[21] Mathematically, this corresponds to rotating the coordinate system about the origin, i.e. making the substitution $\mathbf{x} \to \mathbf{O}\mathbf{x}$ everywhere in the mathematical description of the medium, where $\mathbf{O}$ is an orthogonal (length-preserving) matrix.

2.6 Role of symmetry in a linear stability analysis

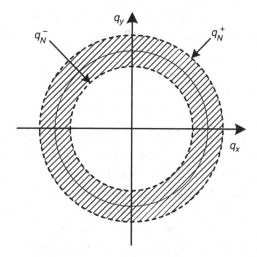

Fig. 2.11 Contour plot of Re $\sigma_\mathbf{q}$ for a type-I instability in a system with rotational symmetry in the two extended directions $\mathbf{x}_\perp$. For simplicity, just the Re $\sigma_\mathbf{q} = 0$ curves are plotted, i.e. the neutral stability wave numbers $q_N^\pm(p)$ for some choice of $p > p_c$. The hatched region shows where Re $\sigma_\mathbf{q} > 0$. Sufficiently close to onset, nonlinear states can be constructed as a superposition of modes Eq. (2.22) whose wave vectors lie in this shaded annular region.

The growth rate Re $\sigma_\mathbf{q}$ as a function of the wave vector $\mathbf{q} = (q_x, q_y)$ for this equation with a value $r = 0.1$ slightly above threshold is shown in Fig. 2.10. This is the two-dimensional generalization of the upper curve in Fig. 2.7. (It is too cumbersome to show all three curves on the same plot for the two-dimensional case.) The implications of the symmetry can be clarified by a contour plot of Re $\sigma_\mathbf{q}$ as a function of $\mathbf{q} = (q_x, q_y)$. For a rotationally symmetric system, these contours will be circles in the (q_x, q_y) plane. In Fig. 2.11, we just show the neutral stability contours given by Re $\sigma_\mathbf{q} = 0$ for a particular value of $p > p_c$. (These correspond to the points $q_N^\pm$ for the one-dimensional case.) The annular region between the neutral contours corresponds to the range of growing modes, Re $\sigma_\mathbf{q} > 0$, in the linear analysis. This annular strip generalized the range of wave numbers $0.75 < k < 1.2$ for the $r = 0.2$ curve in the plot of the growth rate for the one-dimensional Swift–Hohenberg equation in Fig. 2.2.

The rotational symmetries lead to a degeneracy of the unstable modes above onset: for any wave vector $\mathbf{q}$ that is unstable, modes with wave vectors related to this one by a rotational symmetry are also unstable with the same growth rate. (The mode with wave vector $-\mathbf{q}$ does not give a new solution but instead combines with the mode at $\mathbf{q}$ to give a real solution, but other symmetry-related wave vectors do.) In the linear analysis, solutions formed out of superpositions of the symmetry-related modes also grow at the same rate, so that an enormous range of patterns appears possible. As Chapter 4 discusses, this degeneracy is eliminated or strongly reduced by nonlinearities in the system. The nonlinearities thus act as a pattern selection mechanism.

2.6.2 Uniaxial systems

An example of a translationally invariant but not isotropic medium is a nematic liquid crystal. This is a liquid of identical rod-like molecules. In the nematic phase, the molecules are preferentially aligned parallel to one another so that the average alignment defines a particular orientation in the liquid called the director $\hat{\mathbf{n}}$.[22] The liquid is said to be uniaxial, since there is a single preferred orientation. Translating the liquid crystal in any direction preserves the properties of the medium since the liquid still looks locally like many rods preferentially oriented the same way. But at any given point, the medium is anisotropic, with properties (e.g. the dielectric constant) that depend on the direction of measurement relative to the director. Rotating the medium around the director leaves the medium unchanged but not so for axes that are not parallel to the director since this changes the mean orientation of the molecules.

Convection experiments with liquid crystals and with the geometry of Fig. 2.1 have been carried out such that the molecules preferentially line up along a particular direction in the xy plane. (This can be achieved by treating the surfaces of the horizontal plates in a special way that forces the nematic axis to be parallel to a particular direction in the plate.) The pre-convection state remains spatially uniform, but is no longer isotropic in the plane. However, the system still has inversion symmetry ($\mathbf{x}$ and $-\mathbf{x}$ are equivalent).[23]

The possible forms of $\mathrm{Re}\,\sigma_\mathbf{q}$ slightly above threshold for such a uniaxial system with director along the x-axis are shown in Fig. 2.12.[24] The corresponding contour plots for $\mathrm{Re}\,\sigma_\mathbf{q} = 0$ are shown in Fig. 2.13. For some parameter values,[25] convection may occur first for a wave vector aligned along the preferred direction: $\mathbf{q}_c = \pm q_c \hat{\mathbf{x}}$, as in Figs. 2.12(a) and 2.13(a). (It is also possible for the maximum growth rate to be along the y axis.) In these cases, for values of the control parameter slightly above threshold, the range of unstable wave vectors form small

---

[22] Since the molecular alignment of the rods does not have any sense (backwards and forwards are equivalent), the director $\hat{\mathbf{n}}$ is not a vector but a unit line segment that indicates just an orientation.
[23] In a system without this symmetry, disturbances with wave vectors $\mathbf{q}$ and $-\mathbf{q}$ are not equivalent. Once the "forward" and "backward" directions are not equivalent, we would typically expect the disturbance to propagate in one direction or the other, leading to a type-"o" oscillatory instability of the uniform state.
[24] The actual equation used to construct the plots is discussed in Exercise 2.7.
[25] Convection in this system is actually driven by an AC applied voltage, rather than by a time-independent thermal gradient. The frequency of the applied voltage provides a second control parameter that can be tuned independently of the voltage (which plays the role of the temperature difference in Rayleigh–Bénard convection). We discuss electroconvection further in Chapter 6, see Fig. 6.2.

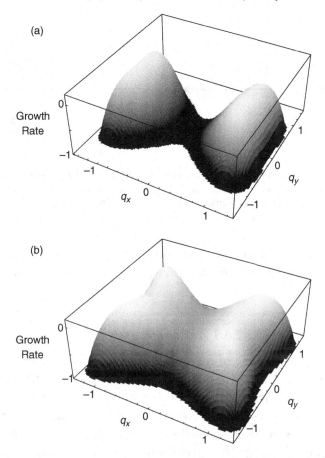

Fig. 2.12 Growth rate Re $\sigma_{\mathbf{q}}$ as a function of the wave vector $\mathbf{q} = (q_x, q_y)$ for a type-I instability in a uniaxial (non-isotropic) system. The preferred direction is taken to be the x-axis and the control parameter has a value slightly above critical. There are two possibilities: (a) the growth rate is maximum at two points, either on the q_x or q_y axes ($q_y = 0, q_x = \pm q_c$ as shown, or $q_x = 0, q_y = \pm q_c$); or (b) the growth rate is a maximum at four symmetry-related points $\mathbf{q}_c = (\pm \cos\theta, \pm \sin\theta)$.

elliptical regions around two points, as in Fig. 2.13(a). Another possibility is that the maximum growth rate and critical wave vectors lie at an angle θ with respect to the x-axis that is neither 0 nor $\pi/2$, as shown in Figs. 2.12(b) and 2.13(b). There will then be four symmetry-related values of $\mathbf{q}_c$ given by the vectors $(\pm q_c \cos\theta, \pm q_c \sin\theta)$.

Just above onset, there are elliptical regions of growing modes around these critical wave vectors (although the axes of the ellipses are not necessarily aligned with the directions of $\mathbf{q}_c$). Above threshold, we expect patterns formed out of modes near these elliptical regions, either from a plus/minus pair or from all four regions.

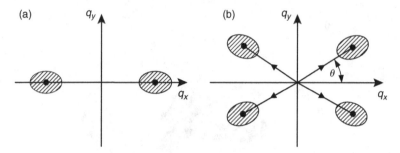

Fig. 2.13 Two possible forms of the neutral stability contours Re $\sigma_\mathbf{q}$ for a type-I instability in a uniaxial system as in Fig. 2.12. The neutral stability contours for p slightly greater than p_c are ellipses centered on $\mathbf{q}_c$. In (b), the axes of the ellipses will in general not be parallel nor perpendicular to the critical wave vector $\mathbf{q}_c$.

These possibilities will lead to patterns with different geometries than ones formed from the annulus of unstable modes as in Fig. 2.11 for the rotationally invariant system.

2.6.3 Anisotropic systems

Finally we consider the case of no rotational symmetry in the plane of extended directions. We do not know of any experimental examples of instabilities in fluid or reaction diffusion systems that show this absence of symmetry but we include it for completeness. To show a stationary type-I instability, the system must retain invariance under $\mathbf{x} \to -\mathbf{x}$ in the extended directions (parity symmetry), as discussed in footnote 23. Thus the instability will occur first at a pair of critical wave vectors $\mathbf{q}_c$ at some particular orientation given by $\pm(q_c \cos\theta, q_c \sin\theta)$, as shown in Fig. 2.14. The contours of maximum growth rate slightly above threshold will be ellipses with major and minor axes at no special orientation relative to $\mathbf{q}_c$. This case is identical to Fig. 2.13(b) if only modes near one plus/minus pair of ellipses are excited.

2.6.4 Formal discussion

Symmetry is key to understanding much of nature, and mathematical tools to describe symmetry and understand its consequences are well developed and have had many successes. An example that might be familiar to some readers is quantum mechanics, where symmetry plays a vital role in classifying quantum states, simplifying calculations of states and energies, and understanding energy level degeneracy. The mathematical formulation of symmetry, known as group theory, was largely developed in this context. The linear stability analysis in pattern-forming systems deals with the properties of linear operators, such as the quantity acting on u_p on the right-hand side of Eq. (2.8), just as quantum mechanics does.

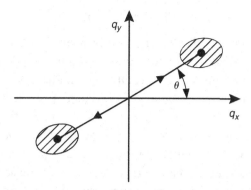

Fig. 2.14 Neutral stability contours Re σ_q for a type-I instability in a system without rotational symmetry. The neutral stability contours for p slightly greater than p_c will be ellipses about $\mathbf{q}_c$. The axes of the ellipses will in general not be parallel nor perpendicular to the critical wave vector $\mathbf{q}_c$.

As you might expect, the group-theory formalism is a valuable and useful tool in pattern formation.

We introduced perhaps the simplest result of group theory using algebraic arguments, namely the result that, for a translationally invariant system, the exponentially growing modes may be taken as single Fourier modes in space. In the language of group theory, we might have phrased the argument by saying that translations in the plane are symmetries of the linear operator of the instability analysis so that the growing solutions form a representation of the translation group. For this elementary result, the mathematical formalism is not really necessary. However, as the symmetries become richer, for example including rotational symmetries or temporal symmetries that are involved once oscillatory instabilities are studied, the formalism becomes more essential. In the study of pattern formation, the linear analysis is just the first step, and it is also crucial to understand the nonlinear saturation of the exponential growth. This is the topic of later chapters in this book. The application of the tools of group theory to this topic is known as equivariant bifurcation theory. We will proceed using intuitive arguments, rather than these formal tools.

Another interesting situation where degeneracy can arise is by varying two system parameters simultaneously. (For example, we could vary the inner and outer Reynolds numbers for the Taylor–Couette system in Fig. 1.11.) It may then be possible to arrange for two distinct maxima of the curve Re σ_q to become zero at the same time. For such a codimension-two bifurcation, the possibilities for pattern formation are particularly rich since, for example, there will be two different annuli in the isotropic case of Fig. 2.11. Symmetry arguments, and the tools of equivariant bifurcation theory, then play a particularly important role in identifying combinations of unstable modes that are relevant in the nonlinear state.

2.7 Conclusions

This chapter has discussed the linear stability analysis of a stationary uniform nonequilibrium state for systems with idealized infinite or periodic lateral boundaries. The analysis is usually a tractable, although not necessarily analytic or easy, scheme for investigating even complicated systems provided that the evolution equations and appropriate boundary conditions on the confined coordinates are known. The analysis can predict for what parameter values the uniform state will become unstable to small-amplitude perturbations and predict some details of the growing instability such as its characteristic length and time scales and the spatial structure in the confined direction. For a supercritical bifurcation, these length and time scales as well as the symmetry of the neutral stability surface play an important role in the weakly nonlinear pattern formation just above the onset of instability. The classification into six different types of instabilities (Section 2.5), together with the discussion of symmetries (Section 2.6), unifies the pattern formation of diverse systems, at least for large aspect ratio systems undergoing a supercritical bifurcation of a uniform state. In this chapter, we have used a simple, perhaps artificial example to illustrate the issues and techniques. In the next chapter, we apply what we have learned to the example of reacting and diffusing chemicals.

2.8 Further reading

(i) The paper by Taylor [102] discussed in Section 2.4.2 is a readable example of theoretical linear stability analysis and its application to an experimental system.
(ii) A second classic paper on linear instability in fluid dynamics is Rayleigh's analysis of the convection instability [91].
(iii) The books *Hydrodynamic and Hydromagnetic Stability* by Chandrasekhar [18] and *Hydrodynamic Stability* by Drazin and Reid [32] provide advanced discussions of many examples of linear instability in fluid and plasma systems.
(iv) For an account of the role of symmetry in instability theory see *Pattern Formation: An Introduction to Methods* by Hoyle [47].
(v) A recent book on the unifying role that symmetry plays in pattern formation and dynamics in general is *The Symmetry Perspective: from Equilibrium to Chaos in Phase Space and Physical Space* by Golubitsky and Stewart [39].

Exercises

2.1 **Confined and extended directions:** Describe a nonequilibrium experiment such that the confined coordinates $\mathbf{x}_\parallel$ are two-dimensional and the extended coordinate $x_\perp$ is one-dimensional. Can you describe an experiment such that the extended coordinates $\mathbf{x}_\perp$ are three-dimensional?

Exercises

2.2 Linear stability of the uniform state for a different Swift–Hohenberg model: For what range of r is the uniform base state $u = 0$ linearly stable if the nonlinear term $-u^3$ in Eq. (2.3) is replaced with the cubic term $u(\partial_x u)^2$?

2.3 Linear stability analysis of a one-variable reaction–diffusion system: Perform a linear stability analysis for the one-variable one-dimensional reaction–diffusion system

$$\partial_t u(x,t) = f(u) + D\, \partial_x^2 u, \qquad (\text{E2.1})$$

about a constant base solution $u = u_b$ (which therefore satisfies $f(u_b) = 0$).

(a) Show that the growth rate σ_q is largest for the wave number $q = 0$.
(b) What are the conditions on $f(u)$ for instability of the fixed point?
(c) Explain why parts (a) and (b) still hold if the concentration field $u(\mathbf{x}, t)$ depends on two spatial variables $\mathbf{x} = (x, y)$ so that the evolution equation has the form

$$\partial_t u(\mathbf{x}, t) = f(u) + D \nabla^2 u, \qquad (\text{E2.2})$$

where $\nabla^2 = \partial_x^2 + \partial_y^2$.

2.4 Linear stability analysis of an ode system: This exercise illustrates a linear stability analysis in the simpler context of ordinary differential equations. Consider the system of ordinary differential equations known as the Lorenz model[26]

$$d_t X = -\sigma(X - Y), \qquad (\text{E2.3a})$$
$$d_t Y = rX - Y - XZ, \qquad (\text{E2.3b})$$
$$d_t Z = -bZ + XY, \qquad (\text{E2.3c})$$

where σ and b are positive constants and r is the control parameter. These equations, used by Lorenz in his pioneering investigation of chaos, may be derived as a simple truncated-mode description of convection. We will use Eqs. (E2.3) just as a mathematical model.

(a) Show that $X = Y = Z = 0$ is a fixed point (time-independent solution) of Eqs. (E2.3) for all values of σ, r, and b.
(b) Now consider the equations linearized about this fixed point, and look for a solution in the form $\delta X, \delta Y, \delta Z \propto e^{\lambda t}$ to show that the fixed point is stable for $r < 1$ and unstable for $r > 1$. Is the instability stationary or oscillatory?

---

[26] For conciseness, we use here and later in the book the notation d_t to denote the time derivative d/dt. This is by analogy to the notation ∂_t used to represent the partial derivative $\partial/\partial t$.

(c) For $r > 1$, show that there are additional fixed point solutions $X = Y = \sqrt{b(r-1)}$, $Z = r - 1$.

(d) By an appropriate linearization, derive a cubic equation that gives the growth rate for perturbations about these new fixed points. For $\sigma = 10$ and $b = 8/3$, show that the nonzero fixed points become unstable for $r > 24.74$. What is the nature of the instability when it occurs: stationary, oscillatory?

2.5 **Linear stability analysis of the Kuramoto–Sivashinsky equation:** In this problem you investigate the linear instability of the uniform state in an equation known as the Kuramoto–Sivashinsky equation. This is a one-dimensional evolution equation for the real field $u(x, t)$

$$\partial_t u = -r\partial_x^2 u - \partial_x^4 u - u\, \partial_x u. \tag{E2.4}$$

Here r is the control parameter.

(a) Show that $u = 0$ is a solution for all values of r. This is the uniform solution.

(b) Show that as r passes from negative to positive values, the uniform solution becomes unstable toward solutions of the form $u \propto e^{\sigma_q t} e^{iqx}$ with positive growth rate σ_q. Calculate how σ_q depends on q and r.

(c) What type of instability is this in the classification scheme of Section 2.5?

(d) Show that by a suitable rescaling of the variables x, t, and u, Eq. (E2.4) can be written in a form that does not depend on the magnitude of r. Write down the equation for both positive and negative values of r.

(e) The scaled version of the equation has no explicit control parameters. However, for a finite domain $0 \le x \le l$ and $r > 0$, with periodic boundary conditions or boundary conditions of $u = \partial_x u = 0$ at each boundary, the length of the domain in the scaled units acts as a control parameter. Show that this is proportional to $r^{1/2} l$, so that this combination of parameters of the original system acts as the control parameter.

The Kuramoto–Sivashinsky equation, which we discuss further in Section 5.5, has received much attention since the solutions that evolve from unstable uniform states for sufficiently large domains (large l) are not steady solutions with spatial periodicity but are chaotic in time and disordered in space. Equation (E2.4) has therefore served as a model equation for the phenomenon of spatiotemporal chaos.

2.6 **Pattern formation in a liquid of square molecules:** Suppose we could find a system with molecules in the shape of squares that all lie in a horizontal plane of two extended coordinates, and whose sides are parallel to one another. What would the figure corresponding to Fig. 2.13 look like for a pattern forming instability in this case? How many qualitatively different possibilities are there?

2.7 **Uniaxial Swift–Hohenberg equation:** The particular equation used to construct the plots in Fig. 2.12 was a modified Swift–Hohenberg equation

$$\partial_t u = \left[r - (\partial_x^2 + \cos^2\theta)^2 - (\partial_y^2 + \sin^2\theta)^2 \right.$$
$$\left. - b(\partial_x^2 + \cos^2\theta)(\partial_y^2 + \sin^2\theta) \right] u - u^3, \quad \text{(E2.5)}$$

where θ is a fixed angle and where b is an anisotropy parameter with $|b| < 2$. The values used in the figure were $r = 0.1$, $\theta = 0$, and $b = -0.5$ and $r = 0.1$, $\theta = 3\pi/16$, and $b = -1$.

(a) Show for $b < 2$ that the linear growth of a disturbance at wave vector $\mathbf{q}$ is maximum at $\mathbf{q} = (\pm \cos\theta, \pm \sin\theta)$.
(b) Make plots with Mathematica or some other graphing program for Re $\sigma_{\mathbf{q}}$ to elucidate the role of the parameters b. Use, for example, $r = 0.1$, $\theta = 3\pi/16$, and try various values of $b < 2$.
(c) [Optional] Simulate the full dynamical equation in a square domain with periodic boundary conditions for these values of r, b, and θ. Then explore whether the patterns that you get are what you might expect based on your linear stability analysis. Chapter 12.1 explains some of the details needed to simulate an evolution equation like Eq. (E2.5).

2.8 **Linear stability analysis of a discrete coupled-map system:** A coupled map lattice (abbreviated CML) is a discrete-time discrete-space model of a spatially extended dynamical system. In this problem, you carry out a linear stability analysis of the uniform states of a one-dimensional CML on an infinite integer lattice. In Exercise 9.9 of Chapter 9, you will have a chance to explore its dynamics, especially spatiotemporal chaotic states.

Consider the CML defined by the expression

$$u_i^{t+1} = f(u_i^t) + D\big(f(u_{i+1}^t) - 2f(u_i^t) + f(u_{i-1}^t)\big), \quad \text{(E2.6)}$$

where the function $f(u)$ is the quadratic

$$f(u) = au(1-u), \quad \text{(E2.7)}$$

associated with the logistic map. At each successive integer time $t = 0, 1, \ldots$, the state of this dynamical system is an infinity of real numbers u_i^t corresponding to the values of a u variable that is associated with each integer lattice point $i = \ldots, -1, 0, 1, \ldots$ Eq. (E2.6) is then an evolution equation that evolves a known state u_i^t at time t to a future state u_i^{t+1} at time $t + 1$. The spatial coupling parameter D determines how strongly the neighboring lattice values $u_{i\pm 1}^t$ are coupled to the value u_i^t at a given site. For zero coupling $D = 0$, the CML turns into an infinity of independent one-dimensional maps $u_i^{t+1} = f(u_i^t)$, one at each lattice site.

(a) Show that the uniform solution $u_i^t = u^*$ is a fixed point of the CML if and only if u^* is a fixed point of the map f so that

$$u^* = f(u^*). \tag{E2.8}$$

(b) Derive and state an analytic criterion in terms of the function $f(u)$ and of the coupling constant D for when the uniform fixed point $u_i^t = u^*$ becomes linearly unstable. Discuss whether a nonzero spatial coupling constant D (which can be positive or negative for this problem) enhances or suppresses instability of the fixed point.

(c) What is the spatial structure of the most unstable mode for D negative?

(d) Assume that the function $f(u)$ maps the unit interval into itself so that $0 \le f(u) \le 1$ for $0 \le u \le 1$. Assume also that, at time $t = 0$, all the lattice values are initialized with values inside the unit interval so that $0 \le u_i^0 \le 1$. For what values D of the coupling constant will all lattice values u_i^1 at time $t = 1$ remain in the unit interval?

Restricting the spatial coupling D to these values is a simple way to ensure that the solution is bounded for all time, although other values of D might be consistent with bounded solutions for particular choices of $f(u)$.

(e) Explain how to generalize the classification of type-I, type-II, and type-III instabilities (Section 2.5), both stationary and oscillatory, to CMLs.

(f) There are many other interesting questions related to CMLs that you may enjoy exploring. For example, if the isolated map Eq. (E2.7) undergoes a periodic-doubling sequence to chaos as some parameter a is varied, how does the spatial coupling D and the system size N affect this sequence? In particular, does the infinitely fine detail of the bifurcation diagram of the logistic map survive spatial coupling in the thermodynamic limit $N \to \infty$? Some details are given in Waller and Kapral [110].

2.9 **Linear stability analysis in a finite geometry:** In Section 2.2.2, we studied the linear instability of the uniform state for the one-dimensional Swift–Hohenberg equation in an infinite domain, or for a finite domain with periodic boundary conditions. This problem guides you through an analysis that illustrates how the onset of instability can be suppressed by boundary conditions on a finite domain that correspond to physical side walls.[27] The suppression or enhancement of a bifurcation is an important effect that shows up in many laboratory experiments and that can complicate the comparison of experiments with a linear stability analysis based on infinite or periodic boundaries.

Consider the one-dimensional Swift–Hohenberg equation for a field $u(x, t)$,

$$\partial_t u = \left(r - \left(\partial_x^2 + 1 \right)^2 \right) u - u^3, \tag{E2.9}$$

on the finite interval $[-l/2, l/2]$ with the boundary conditions

$$u = \partial_x u = 0, \quad \text{at } x = -l/2, x = l/2. \tag{E2.10}$$

(a) Derive the evolution equation and boundary conditions for an infinitesimal perturbation $u_p(x, t)$ of the zero homogeneous state $u = 0$.

(b) Since the equation for u_p is linear with time-independent coefficients, we can seek solutions that are exponential in time with growth rate σ:

$$u_p = e^{\sigma t} f(x). \tag{E2.11}$$

The state $u = 0$ is marginally unstable when $\sigma = 0$. Show that the function f of the marginally stable mode must satisfy the time-independent equation

$$rf = \left(\partial_x^2 + 1 \right)^2 f, \tag{E2.12}$$

with boundary conditions $f = 0$ and $\partial_x f = 0$. (See Eq. (E2.10).) Equation (E2.12) is an eigenvalue problem where r is the eigenvalue and f is the eigenfunction; the linear operator $(\partial_x^2 + 1)^2$ acting on f is analogous to multiplication of a vector by a matrix and the right side can be expected to be parallel to f only for special choices of the parameter r. The resulting eigenvalues r determine the onset of instability in the finite geometry.

(c) Since Eq. (E2.12) is an equation with constant coefficients, the solutions are sinusoidal functions of x. Because of the symmetric formulation of the problem about $x = 0$, the solutions will either be even (pure cosine) or odd (pure sine). Substituting $\cos(qx)$ or $\sin(qx)$ into Eq. (E2.12) shows that q must take on one of two values:

$$q_\pm = \sqrt{1 \pm \sqrt{r}}. \tag{E2.13}$$

---

[27] More specifically, the boundary conditions are related to a so-called no-slip boundary condition in which the fluid velocity is zero at a side wall in the rest frame of the wall.

(d) The previous result and linearity suggests that the general symmetric solution of Eq. (E2.12) will be of the form

$$f(x) = \begin{cases} A_+ \cos(q_+ x) + A_- \cos(q_- x), & \text{even}, \\ A_+ \sin(q_+ x) + A_- \sin(q_- x), & \text{odd}, \end{cases} \quad (E2.14)$$

where A_+ and A_- are constants. Use Eq. (E2.14) and the boundary conditions Eq. (E2.10) to derive the transcendental equation

$$q_+ \tan\left[(l/2)q_+\right] = q_- \tan\left[(l/2)q_-\right], \quad (E2.15)$$

that relates the parameters r and l.

(e) For a given system size l, the onset of pattern formation will correspond to the first positive root of the transcendental equation Eq. (E2.15). Show that, for large system sizes l, the first positive root is given approximately by:

$$r_c \approx \left(\frac{2\pi}{l}\right)^2. \quad (E2.16)$$

Graphical plots of your transcendental equation may help you to see what kind of approximations are needed to get this result.

Equation (E2.16) answers the original question of how finite boundaries modify the onset of a type-I-s instability for boundary conditions corresponding to those of a fluid. It is indeed harder to initiate the instability and the critical parameter value increases roughly as $1/l^2$ as the system size l is decreased. This is a specific prediction that can be tested experimentally.

(f) For a length $l = 10\pi$, calculate numerically the critical value r_c to three digits, compare this answer with Eq. (E2.16), and plot the corresponding eigenfunction Eq. (E2.14). Mathematica or a similar program will facilitate these calculations.

(g) Harder: Using Mathematica or a similar program, calculate and plot the solutions r versus l of your transcendental equation and so confirm the overall accuracy of the approximation Eq. (E2.16). Is it possible for the uniform state $u = 0$ to become unstable to a pattern for arbitrarily small system sizes l?

2.10 **Non-normal stability of a fixed point:** All infinitesimal perturbations of a stable fixed point decay asymptotically to zero. However, it is sometimes possible for a tiny perturbation of a stable fixed point to increase substantially in magnitude before ultimately decaying exponentially away, in some cases becoming so large that instability might occur through some nonlinear mechanism. This situation can arise when the $N \times N$ Jacobian matrix $\mathbf{J}(\mathbf{u}_0)$ that determines the linear stability of some fixed point $\mathbf{u}_0$ has eigenvectors that are not orthogonal. One can show that a matrix **J** has non-orthogonal eigenvectors when the matrix is non-normal, which means it satisfies the condition

$$\mathbf{JJ}^\dagger \neq \mathbf{J}^\dagger \mathbf{J}, \tag{E2.17}$$

where $\mathbf{J}^\dagger$ denotes the complex conjugate of the matrix transpose of **J**. (Diagonal, symmetric, Hermitian, and unitary matrices are classes of normal matrices but not all normal matrices are of these types.) Non-normal matrices have been found to occur in various stability problems associated with pattern-forming systems, especially fluid dynamic problems for which there is net transport of fluid through the system such as flow of water through a pipe. The growth of small perturbations by a non-normal mechanism to large magnitudes has been suggested by the theoretical meteorologist Brian Farrell to play a role in the formation of cyclones.

As a simple example of non-normal stability of a fixed point, consider the following dynamical system $d\mathbf{u}/dt = \mathbf{f}(\mathbf{u})$ with quadratic nonlinearities:

$$d_t u_1 = -2\varepsilon u_1 + \left(u_1^2 + u_2^2\right)^{1/2} u_2, \tag{E2.18}$$

$$d_t u_2 = -\varepsilon u_2 + u_1 - \left(u_1^2 + u_2^2\right)^{1/2} u_1. \tag{E2.19}$$

The positive parameter ε is considered to be small and corresponds to the inverse Reynolds number $1/\mathcal{R}$ of a fluid problem in the limit of large $\mathcal{R}$.

(a) Show that the zero fixed point $\mathbf{u}_0 = (u_1, u_2) = (0, 0)$ is always linearly stable.
(b) Show that the 2×2 Jacobian matrix **J** of the evolution equations $d_t(\delta \mathbf{u}) = \mathbf{J}\, \delta \mathbf{u}$ linearized around $\mathbf{u}_0$ is non-normal.
(c) Construct an explicit analytical solution of the linearized evolution equations (about $\mathbf{u}_0$) that passes through the initial data $\delta u_1(0) = \varepsilon u_{10}$ and $\delta u_2(0) = 0$ for some nonzero constant u_{10}. Show that, no matter how small $\varepsilon > 0$, the perturbation $\delta u_2(t)$ grows in magnitude to order u_{10} over a time scale ε^{-1} before decaying to zero.
(d) Using Mathematica or a similar program, integrate the above nonlinear evolution equations numerically for $\varepsilon = 0.01$ and for different initial conditions of the form $\mathbf{u}_0 = (u_{10}, 0)$, where the constant u_{10} varies over successive powers of ten ranging from 10^{-6} to 10^{-1}. Summarize the effect of nonlinearities on the non-normal linear growth.

You can learn more about non-normality and its role in fluid instability in the article "Hydrodynamic stability without eigenvalues" by Trefethen et al. [105].

3
Linear instability: application to reacting and diffusing chemicals

In Section 2.2 of the previous chapter we used a simplified mathematical model Eq. (2.3) to identify the assumptions, mathematical issues, and insights associated with the linear stability analysis of a stationary uniform nonequilibrium state. In this chapter, we would like to discuss a realistic set of evolution equations and the linear stability analysis of their uniform states, using the example of chemicals that react and diffuse in solutions (see Fig. 1.18). Historically, such a linear stability analysis of a uniform state was first carried out in 1952 by Alan Turing [106]. He suggested the radical and highly stimulating idea that reaction and diffusion of chemicals in an initially uniform state could explain morphogenesis, how biological patterns arise during growth. Although reaction–diffusion systems are perhaps the easiest to study mathematically of the many experimental systems considered in this book, they have the drawback that quantitative comparisons with experiment remain difficult. The reason is that many chemical reactions involve short-lived intermediates in small concentrations that go undetected, so that the corresponding evolution equations are incomplete. Still, reaction–diffusion systems are such a broad and important class of nonequilibrium systems, prevalent in biology, chemistry, ecology, and engineering, that a detailed discussion is worthwhile.

The chapter is divided into two halves. In the first part, we introduce the simple model put forward by Turing, and give a careful analysis of the instability of the uniform states. In the second part, we apply these ideas to realistic models of experimental systems.

3.1 Turing instability

Realistic equations that describe chemical reactions in experimental geometries are complicated to formulate and difficult to investigate. We will postpone a discussion of realistic experimental chemical pattern forming systems to Section 3.2. The same was true in the 1940s when Alan Turing was thinking about morphogenesis. These

3.1 Turing instability

difficulties did not stop Turing who, in the tradition of great theoretical science, set as his goal not the quantitative explanation of morphogenesis but the discovery of a clear plausible mechanism that could guide researchers in how to think about such a complex phenomenon. Indeed, the opening paragraph of his 1952 paper begins with these classic words[1]

> In this section a mathematical model of the growing embryo will be described. This model will be a simplification and an idealization, and consequently a falsification. It is to be hoped that the features retained for discussion are those of greatest importance in the present state of knowledge.

We will follow Turing in his 1952 paper and examine analytically the linear stability analysis of the simplest possible reaction–diffusion system that forms a pattern from a uniform state. The analysis will lead to several insights, some unexpected. One insight is that at least two interacting chemicals are needed for pattern formation to occur. Second is Turing's most surprising insight, that diffusion in a reacting chemical system can actually be a destabilizing influence that causes instability. This is contrary to intuition since diffusion by itself smooths out spatial variations of a concentration field and so should prevent pattern formation. A third insight is that the instability caused by diffusion can cause the growth of structure at a particular wavelength. This provides a possible mechanism for producing patterns like the segmentation patterns in the developing fly embryo, the periodic arrangement of tentacles around the mouth of the Hydra organism, or zebra stripes. A fourth insight (which was not clearly stated until after Turing's paper) is that pattern formation in a chemical system will not occur unless the diffusion coefficients of at least two reagents differ substantially. The difficulty of satisfying this condition for chemicals in solution partially explains why nearly 40 years passed after Turing's paper before experiments were able to demonstrate the correctness of his ideas.

3.1.1 Reaction–diffusion equations

Since you can show in Exercise 2.3 that an evolution equation for a single chemical species does not show interesting pattern formation, we will study the Turing model[2]

---

[1] The technical level of Turing's paper is about the level of this chapter and we encourage you to track down and read this visionary paper. The paper has many bold and interesting ideas that draw upon Turing's interdisciplinary thinking about biology, chemistry, and mathematics. His paper is also interesting from a historical point of view, to see what facts Turing used to develop his hypotheses. For example, Turing could only speculate about how an organism knew how to grow since the role of DNA would only be announced a year later in 1953. The last section of the paper mentions one of the first simulations on a digital computer and Turing states his belief that these new computers (which he helped to invent) will be important for future scientific research.

[2] In his paper, Turing examined two kinds of models, spatially coupled odes that modeled discrete biological cells and coupled pdes of the form that we analyze here, that treated the tissue as a continuous medium.

for two reacting and diffusion chemicals of the form

$$\partial_t u_1 = f_1(u_1, u_2) + D_1 \partial_x^2 u_1, \tag{3.1a}$$

$$\partial_t u_2 = f_2(u_1, u_2) + D_2 \partial_x^2 u_2, \tag{3.1b}$$

or in vector form

$$\partial_t \mathbf{u} = \mathbf{f}(\mathbf{u}) + \mathbf{D} \partial_x^2 \mathbf{u}, \tag{3.2}$$

where we have introduced a diagonal 2×2 diffusion matrix D defined by

$$\mathbf{D} = \begin{pmatrix} D_1 & 0 \\ 0 & D_2 \end{pmatrix}. \tag{3.3}$$

Equations (3.1) describe the evolution of two concentration fields $u_i(x, t)$ that are functions of one space coordinate and time. The nonlinear functions $f_i(u_1, u_2)$ are the reaction rates of the two chemicals while the D_i are the corresponding diffusion coefficients. The simplest possible model is obtained by assuming that there is no prior spatiotemporal structure in the system so that the functions f_i and the diffusion coefficients D_i do not depend explicitly on time t or on position x. For simplicity, we further assume that the diffusion coefficients are constants and so do not depend on the field values u_i. These assumptions are all quite reasonable for many experimental situations. For ease of presentation, we also make the simplification to the one-dimensional case. For higher dimensions, if we assume that the evolution equations have rotational symmetry, the one-dimensional Laplacian ∂_x^2 becomes a higher-dimensional Laplacian ∇^2. The analysis in this case is not significantly harder since the combination qx in the spatial dependence assumed in Eq. (3.8) below becomes everywhere a dot product $\mathbf{q} \cdot \mathbf{x}$ with a wave vector $\mathbf{q}$, and the wave-number squared q^2 becomes the quantity $\mathbf{q} \cdot \mathbf{q}$.

As we will see more explicitly in Section 3.2, Eqs. (3.1) cannot accurately describe a sustained nonequilibrium chemical system since they lack a transverse confined coordinate along which reactants can be fed into the system and products removed. The neglect of a confined coordinate is a major simplification. Indeed, an accurate treatment of the feed direction introduces complicated spatial structure of the sort shown in Fig. 3.4, which makes the linear stability analysis considerably harder. Conceptually, you can think of the Turing model Eq. (3.2) as attempting to describe pattern formation along a single line parallel to the plates A and B in Fig. 2.1. Actually, many early experiments on pattern formation in chemical reactions could be rather well approximated by ignoring the confined coordinate. Typically these experiments were done using a thin layer of chemicals in a Petri dish, or chemicals soaked in filter paper. The variation of chemical concentration across the layer or thickness of filter paper is perhaps small in these experiments (typically the conditions are not well controlled, so this is just an assumption). A reduction to equations describing just the spatial variation in the plane (Eqs. (3.1) but with $\partial_x^2 \to \nabla^2 = \partial_x^2 + \partial_y^2$) would be a reasonable approximation. However, since in these experiments there is no feed of refreshed chemicals to sustain the reactions, any pattern formation or dynamics is a transient and eventually the system would approach a uniform chemical equilibrium. This difficulty may be hidden in the simple reduced equations (3.1) by approximating some dynamical chemical concentrations as constants in the reaction term $\mathbf{f}(\mathbf{u})$.

3.1.2 Linear stability analysis

We now perform the linear stability analysis of uniform solutions of the two-chemical reaction–diffusion model Eqs. (3.1). Turing's surprising and important discovery was that there are conditions under which the spatially uniform state is stable in the absence of diffusion[3] but can become unstable to nonuniform perturbations precisely because of diffusion. Further, for many conditions the instability first occurs at a finite wavelength and so a cellular pattern starts to appear.

The following discussion is not mathematically difficult but has many details. You will likely best appreciate the discussion if you take your time and derive the results for yourself in parallel with the text. The goal of the discussion is to derive, and then to understand physically, conditions that are sufficient for the real parts of all growth rates to be negative. When these conditions are first violated and instability occurs, it is then important to think about the values of the wave numbers corresponding to the fastest growing modes.

We begin by assuming that we have somehow found a stationary uniform base solution $\mathbf{u}_b = (u_{1b}, u_{2b})$. This satisfies the Turing model with all partial derivatives set to zero, leading to $\mathbf{f}(\mathbf{u}_b) = \mathbf{0}$ or

$$f_1(u_{1b}, u_{2b}) = 0, \tag{3.4a}$$

$$f_2(u_{1b}, u_{2b}) = 0. \tag{3.4b}$$

These are two nonlinear equations in two unknowns. Finding a uniform solution can be hard since there is no systematic way to find even a single solution of a set of nonlinear equations. Numerical methods such as the Newton method (see Section 12.4.2 and Exercise 12.16) can find accurate approximations to solutions of nonlinear equations but only if a good guess for a solution is already known. For two nonlinear equations like Eqs. (3.4), a graphical way to find solutions that is sometimes useful is to plot the nullclines of each equation. An equation of the

---

[3] Diffusion can be suppressed in several ways. Mathematically, we simply set the diffusion coefficients to zero. Experimentally, we can stir the chemicals to eliminate spatial nonuniformity.

form $f_1(u_1, u_2) = 0$ defines an implicit relation $u_2 = g_1(u_1)$ between the two variables u_1 and u_2 called the nullcline of that equation. If the functions $f_i(u_1, u_2)$ are sufficiently simple, their nullclines $g_i(u_1)$ can sometimes be found explicitly and then plotted on a single plot with axes labeled by u_1 and u_2. Any intersection of the two nullclines is then a solution of the nonlinear equations. We will make substantial use of nullclines in Chapter 11 when we discuss two-variable models of excitable media. Exercise 12.16 also explains how to estimate nullclines numerically.

By linearizing about the base state $\mathbf{u}_b$, you can show that an arbitrary infinitesimal perturbation $\mathbf{u}_p(x,t) = (u_{p1}(x,t), u_{p2}(x,t))$ of the base state will evolve in time according to the following linear constant-coefficient evolution equations

$$\partial_t u_{p1} = a_{11} u_{p1} + a_{12} u_{p2} + D_1 \partial_x^2 u_{p1}, \tag{3.5a}$$

$$\partial_t u_{p2} = a_{21} u_{p1} + a_{22} u_{p2} + D_2 \partial_x^2 u_{p2}. \tag{3.5b}$$

The constant coefficients a_{ij} come from the 2×2 Jacobian matrix $\mathbf{A} = \partial \mathbf{f}/\partial \mathbf{u}$ evaluated at the constant base solution $\mathbf{u}_b$,

$$a_{ij} = \left.\frac{\partial f_i}{\partial u_j}\right|_{\mathbf{u}_b}. \tag{3.6}$$

The mathematical structure of Eqs. (3.5) can be clarified by writing them in vector form

$$\partial_t \mathbf{u}_p = \mathbf{A} \mathbf{u}_p + \mathbf{D} \partial_x^2 \mathbf{u}_p, \tag{3.7}$$

where $\mathbf{D}$ is the 2×2 diffusion matrix previously introduced in Eq. (3.3). Because Eq. (3.7) is linear with constant coefficients and because the boundaries are periodic or at infinity, we can use translational symmetry to seek a particular solution $\mathbf{u}_p(x,t)$ that is a constant vector $\mathbf{u}_q$ times an exponential in time times an exponential in space

$$\mathbf{u}_p = \mathbf{u}_q e^{\sigma_q t} e^{iqx} = \begin{pmatrix} u_{1q} \\ u_{2q} \end{pmatrix} e^{\sigma_q t} e^{iqx}, \tag{3.8}$$

with growth rate σ_q and wave number q. Note that both components of the perturbation vector $\mathbf{u}_p$ have the same dependence on time and space since only with this assumption can the spatial and temporal dependencies be eliminated completely from the linearized evolution equations and a simple solution found.

If we substitute Eq. (3.8) into Eq. (3.7), divide out the exponentials, and collect some terms, we obtain the following eigenvalue problem

$$\mathbf{A}_q \mathbf{u}_q = \sigma_q \mathbf{u}_q, \tag{3.9}$$

where the 2 × 2 real matrix $\mathbf{A}_q$ is defined by

$$\mathbf{A}_q = \mathbf{A} - \mathbf{D}q^2 = \begin{pmatrix} a_{11} - D_1 q^2 & a_{12} \\ a_{21} & a_{22} - D_2 q^2 \end{pmatrix}. \tag{3.10}$$

Equation (3.9) tells us that the growth rate σ_q and constant vector $\mathbf{u}_q$ form an eigenvalue–eigenvector pair of the matrix $\mathbf{A}_q$, and that there is one such 2 × 2 eigenvalue problem for each wave number q. The eigenvalue problem for a given q has generally two linearly independent eigenvectors that we will denote by $\mathbf{u}_{iq}$ for $i = 1, 2$. If the corresponding eigenvalues are σ_{iq}, the particular solution with wave number q will have the form

$$\left(c_{1q} \mathbf{u}_{1q} e^{\sigma_{1q} t} + c_{2q} \mathbf{u}_{2q} e^{\sigma_{2q} t} \right) e^{iqx}, \tag{3.11}$$

where the coefficients c_{iq} are complex constants that depend on the initial perturbation at $t = 0$.[4] This solution decays if $\operatorname{Re} \sigma_{iq} < 0$ for $i = 1, 2$. An arbitrary perturbation $\mathbf{u}_p(x, t)$ is a superposition of expressions like Eq. (3.11) over all wave numbers q. The uniform solution $\mathbf{u}_b$ is stable if both eigenvalues σ_{iq} have negative real parts for all wave numbers q, i.e. if $\max_i \max_q \operatorname{Re} \sigma_{iq} < 0$.

The characteristic polynomial for the eigenvalue problem Eqs. (3.9) and (3.10) can be written

$$0 = \det(\mathbf{A}_q - \sigma_q \mathbf{I}) = \sigma_q^2 - (\operatorname{tr} \mathbf{A}_q)\sigma_q + \det \mathbf{A}_q, \tag{3.12}$$

where $\operatorname{tr} \mathbf{A}_q$ denotes the trace (sum of diagonal elements) of the matrix $\mathbf{A}_q$ and $\det \mathbf{A}_q$ the determinant. The eigenvalues are then σ_{1q} and σ_{2q} given by

$$\sigma_q = \frac{1}{2} \operatorname{tr} \mathbf{A}_q \pm \frac{1}{2} \sqrt{(\operatorname{tr} \mathbf{A}_q)^2 - 4 \det \mathbf{A}_q}. \tag{3.13}$$

The regions of stability (both $\operatorname{Re} \sigma_q$ negative) and instability (at least one $\operatorname{Re} \sigma_q$ positive) in the $\operatorname{tr} \mathbf{A}_q$–$\det \mathbf{A}_q$ plane are shown in Fig. 3.1. From this figure or from the expression Eq. (3.13) a simple criterion can be derived that determines when the real parts of both eigenvalues are negative: *the trace of the matrix must be negative and the determinant of the matrix must be positive*. For the matrix $\mathbf{A}_q$ in Eq. (3.10), these criteria for stability take the explicit form

$$\operatorname{tr} \mathbf{A}_q = a_{11} + a_{22} - (D_1 + D_2)q^2 < 0, \tag{3.14a}$$

$$\det \mathbf{A}_q = (a_{11} - D_1 q^2)(a_{22} - D_2 q^2) - a_{12} a_{21} > 0. \tag{3.14b}$$

If both conditions hold for all wave numbers q, the stationary uniform base state $\mathbf{u}_b$ is linearly stable.

---

[4] A real solution can be obtained as usual by adding the complex conjugate solution.

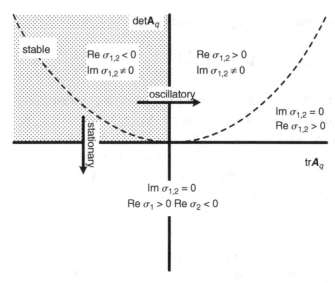

Fig. 3.1 Stability regions of the Turing system in the $\operatorname{tr} \mathbf{A}_q$–$\det \mathbf{A}_q$ plane. The plot shows regions of different characteristics of the two eigenvalues σ_1 and σ_2 calculated from Eq. (3.12). The parabola $\det \mathbf{A}_q = \frac{1}{2} \operatorname{tr} \mathbf{A}_q$ divides the plane into two halves such that above this curve the two eigenvalues are complex (and are complex conjugates of one another), and below this curve both eigenvalues are real. The shaded region is the stable region, $\operatorname{Re} \sigma_{1,2} < 0$. Over the unshaded portion there is at least one eigenvalue with positive real part. A stationary instability occurs passing through the negative $\operatorname{tr} \mathbf{A}_q$ axis to negative values of $\det \mathbf{A}_q$, whereas an oscillatory instability occurs passing through the positive $\det \mathbf{A}_q$ axis to positive values of $\operatorname{tr} \mathbf{A}_q$.

For N interacting chemicals whose dynamics satisfy equations analogous to Eq. (3.2), we would need to solve a $N \times N$ eigenvalue problem Eq. (3.9) for each wave number q. For $N \geq 3$, analytical criteria that all the eigenvalues of a $N \times N$ matrix $\mathbf{A}_q$ have negative real parts become cumbersome to work with[5] and so the case $N = 2$ that Turing discussed in his paper hits the mathematical sweet spot of being manageable and leading to interesting results. For experiments with $N > 3$ reacting chemicals, it is usually easiest to study the corresponding model by numerical methods. With modern computers and modern numerical algorithms, it is straightforward to find all the eigenvalues of a $N \times N$ matrix quickly for N as large as 10 000. As you can imagine, it would be exceedingly difficult to map out the reaction rates for so many interacting chemicals. Progress in studying the

---

[5] In Appendix 2 of his book [79], Murray discusses some necessary and sufficient analytical criteria that all the eigenvalues of a real $N \times N$ matrix have negative real parts. The Routh–Hurwitz criterion states that a certain sequence of determinants from size 1 to N all have to be positive. Determinants are difficult to work with symbolically since they involve a sum of $N!$ products of matrix elements.

linear stability of chemical systems with large N is therefore limited by scientific knowledge, not by the ability to calculate eigenvalues.

We now discuss the physical meaning and implications of the mathematical criterion Eqs. (3.14) in the context of pattern formation.[6] There are many abstract symbols and equations here and some careful thinking is needed to see how to extract some physical insights. A first step is to identify the precise scientific question of interest, not just the technical mathematical question of when some base state $\mathbf{u}_b$ becomes linearly unstable. Turing's insight was that diffusion of chemicals may somehow cause a pattern-forming instability. If so, then the starting point is to imagine that somehow the diffusion has been turned off (mathematically by setting the diffusion coefficients or the wave number q to zero, experimentally by stirring the solutions at high speed) and then we slowly turn on the diffusion to see if instability ensues. If we adopt this as our strategy, then we need to assume that the reacting chemicals form a stable stationary state in the absence of diffusion. Setting the diffusion constants D_i to zero in Eqs. (3.14), we obtain the following criteria for linear stability of the uniform state in the absence of diffusion

$$a_{11} + a_{22} < 0, \tag{3.15a}$$

$$a_{11}a_{22} - a_{12}a_{21} > 0. \tag{3.15b}$$

If these criteria are satisfied, a well-mixed two-chemical solution will remain stable and uniform. Comparing Eq. (3.15a) with Eq. (3.14a) and remembering that diffusion constants D_i and the quantity q^2 are non-negative, we conclude that

$$\operatorname{tr} \mathbf{A}_q = a_{11} + a_{22} - (D_1 + D_2)q^2 < a_{11} + a_{22} < 0, \tag{3.16}$$

so the trace of the matrix $\mathbf{A}_q$ is always negative. We conclude that the only way for diffusion to destabilize the uniform state is for the second criterion Eq. (3.14b) to become reversed so that the determinant of $\mathbf{A}_q$ becomes negative.

The next step is therefore to figure out when the determinant $\det \mathbf{A}_q$ changes sign from positive to negative. Equation (3.14b) tells us that the determinant $\det \mathbf{A}_q$ is a parabola in the quantity q^2 that opens upwards, being positive for $q^2 = 0$ by Eq. (3.15b) and positive for large q^2. A condition for linear instability in the presence of diffusion is then obtained by asking when the minimum value of this parabola first becomes negative. Setting the derivative of $\det \mathbf{A}_q$ with respect to q^2 to zero, we learn that the minimum occurs at the wave number q_m given by (see also Exercise 3.1)

$$q_m^2 = \frac{D_1 a_{22} + D_2 a_{11}}{2 D_1 D_2}. \tag{3.17}$$

---

[6] Our discussion here follows that of a paper by Segel and Jackson [94] that clarified Turing's original analysis.

The corresponding value of det $\mathbf{A}_q$ at this minimum is

$$\det \mathbf{A}_{q_m} = a_{11}a_{22} - a_{12}a_{21} - \frac{(D_1 a_{22} + D_2 a_{11})^2}{4 D_1 D_2}. \qquad (3.18)$$

This expression is negative when the inequality

$$D_1 a_{22} + D_2 a_{11} > 2\sqrt{D_1 D_2 (a_{11}a_{22} - a_{12}a_{21})} \qquad (3.19)$$

is satisfied. The term inside the square root is positive because of Eq. (3.15b). As a corollary, Eq. (3.19) implies that

$$D_1 a_{22} + D_2 a_{11} > 0, \qquad (3.20)$$

which can also be deduced directly from Eq. (3.17) since q_m^2 is a non-negative real number. From Eqs. (3.15a) and (3.20), we see that one of the quantities a_{11} and a_{22} must be positive and the other negative. For concreteness, let us choose $a_{11} > 0$ and $a_{22} < 0$ in the subsequent discussion. Then Eq. (3.15b) further implies that the quantities a_{12} and a_{21} must also have opposite signs.

Equation (3.19) is a necessary and sufficient condition for linear instability of a uniform state that is stable in the absence of diffusion Eqs. (3.15). As some experimental knob is turned, the matrix elements a_{ij} will change their values smoothly through their dependence on the experimental parameter. (Again, diffusion constants D_i can be considered constant for many experiments and so usually do not play the role of an easily varied bifurcation parameter.) At some parameter value, the inequality Eq. (3.19) may become true and the uniform state will become unstable to perturbations growing with a wave number close to the value q_m in Eq. (3.17).

The condition Eq. (3.19) can be expressed alternatively in terms of two diffusion lengths

$$l_1 = \sqrt{\frac{D_1}{a_{11}}} \quad \text{and} \quad l_2 = \sqrt{\frac{D_2}{-a_{22}}}, \qquad (3.21)$$

in the form

$$q_m^2 = \frac{1}{2}\left(\frac{1}{l_1^2} - \frac{1}{l_2^2}\right) > \sqrt{\frac{a_{11}a_{22} - a_{12}a_{21}}{D_1 D_2}}. \qquad (3.22)$$

This implies that the length l_2 must be sufficiently larger than the length l_1. Now our assumption that $a_{11} > 0$ implies that chemical 1 enhances its own instability and so could be called an activator. Similarly, since $a_{22} < 0$, chemical 2 inhibits its own growth and could be called an inhibitor. The necessary condition $l_2 > l_1$ for a Turing instability is then sometimes referred to as "local activation with long-range inhibition."

The condition $l_2 > l_1$, when expressed in the equivalent form $D_2/D_1 > (-a_{22}/a_{11})$ partly explains why experimentalists had such a hard time finding a laboratory example of a Turing instability. The diffusion coefficient D_2 of the inhibitor has to exceed the diffusion coefficient D_1 of the activator by a factor $(-a_{22})/a_{11}$ which can exceed 10 for some realistic models of reaction–diffusion experiments. Since the diffusion coefficients of most small ions in water have the same value of about 10^{-9} m²/s, some ingenuity is required to create a Turing instability. Experimentalists found that one way to achieve a large disparity in diffusion coefficients was to introduce a third molecule (starch in the CDIMA case discussed in Section 3.2) that was fixed to an immobile matrix in the solution (the walls of the porous gel). The effective diffusion coefficient for a chemical that reversibly binds to this immobile molecule is substantially smaller than that for chemicals that do not bind.

The criteria Eqs. (3.15) and Eq. (3.19) for the instability of a uniform state are rather abstract and so we now apply these criteria to a simple two-variable mathematical model known as the Brusselator[7] to illustrate the ideas.

Etude 3.1 The Brusselator reaction–diffusion model
The Brusselator is a reaction–diffusion model that describes the evolution of two chemical concentrations $u_1(x, t)$ and $u_2(x, t)$

$$\partial_t u_1 = a - (b+1)u_1 + u_1^2 u_2 + D_1 \partial_x^2 u_1, \tag{3.23a}$$

$$\partial_t u_2 = b u_1 - u_1^2 u_2 + D_2 \partial_x^2 u_2. \tag{3.23b}$$

The parameters a, b, D_1, and D_2 are positive constants. Although invented with the goal of understanding the Belousov–Zhabotinsky reaction of Fig. 1.18(a) and although successful in producing uniform oscillations and traveling waves, this model was intended not to describe a specific chemical experiment but to show how an invented plausible sequence of chemical reactions could reproduce qualitative but difficult to understand features of actual experiments. The context in which this model was invented suggests assigning the following parameter values

$$a = 1.5, \quad D_1 = 2.8, \quad D_2 = 22.4, \tag{3.24}$$

and varying the parameter b as the bifurcation parameter. We now show how to predict analytically for what value of b a uniform base state becomes linearly unstable.

A stationary uniform base state $\mathbf{u}_b = (u_{1b}, u_{2b})$ can be found by looking for solutions of Eqs. (3.23) with all partial derivatives set to zero. It is straightforward

---

[7] Two of the more widely studied models of reaction–diffusion dynamics are named after the geographical location where the model was invented. Thus the Brusselator is named after Brussels, in Belgium, and the Oregonator is named after the state of Oregon in the USA.

to show that there is only one uniform state given by

$$u_{1b} = a, \quad u_{2b} = \frac{b}{a}. \tag{3.25}$$

Please keep in mind that it is rarely this easy to find the stationary uniform state of some set of nonlinear evolution equations!

By linearizing the Brusselator model around this base state, you can show that the Jacobian matrix $\mathbf{A} = \partial \mathbf{f}/\partial \mathbf{u}$ is given by

$$\mathbf{A} = \begin{pmatrix} a_{11} & a_{12} \\ a_{21} & a_{22} \end{pmatrix} = \begin{pmatrix} b-1 & a^2 \\ -b & -a^2 \end{pmatrix}. \tag{3.26}$$

The off-diagonal elements have opposite signs as required for a Turing instability, but the diagonal elements have opposite signs only if

$$b > 1. \tag{3.27}$$

When this inequality holds, we conclude that chemical 1 is an activator ($a_{11} > 0$) and that chemical 2 is an inhibitor ($a_{22} < 0$).

The uniform state is stable in the absence of diffusion when Eqs. (3.15) hold

$$\operatorname{tr} \mathbf{A} < 0 \implies b < 1 + a^2 = 3.25, \tag{3.28a}$$

$$\det \mathbf{A} > 0 \implies a^2 > 0. \tag{3.28b}$$

Only the first condition leads to a constraint, that the parameter b must be smaller than the value 3.25. Using the matrix elements Eq. (3.26) and the fact that $a_{11}a_{22} - a_{12}a_{21} = a^2$, the criterion for linear instability Eq. (3.19) can be manipulated into the form

$$b \geq \left(1 + a\sqrt{\frac{D_1}{D_2}}\right)^2. \tag{3.29}$$

The critical value b_c is determined by equality and the parameter values Eq. (3.24) imply

$$b_c \approx 2.34. \tag{3.30}$$

The corresponding wave number q_c at instability is given by Eq. (3.17)

$$q_c = \sqrt{\frac{D_1 a_{22} + D_2 a_{11}}{2 D_1 D_2}} \approx 0.435, \tag{3.31}$$

which corresponds to a wavelength of $2\pi/q_c \approx 14.5$.

3.1 Turing instability

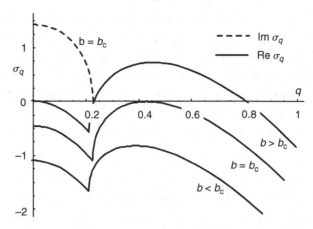

Fig. 3.2 Solid curves: plots of the maximum real part of the growth rate, $\max_i \mathrm{Re}\,\sigma_{iq}$ versus wave number q, for infinitesimal perturbations of the uniform state Eq. (3.25) of the Brusselator model Eqs. (3.23), for the choice of parameters Eq. (3.24). Three curves are plotted corresponding to a b-parameter below instability ($b = 0.6b_c \approx 1.40$), to the critical value ($b = b_c \approx 2.34$), and above instability ($b = 1.4b_c \approx 3.28$). The critical wave number $q_c \approx 0.435$ is identified as the wave number for which the maximum real part of the growth rate first becomes zero. The wave number corresponding to the fastest growing mode increases slowly with increasing b. The dotted line is the imaginary part of the eigenvalue that has the maximum real part for $b = b_c$. The imaginary parts for the other b values are not shown since they are nearly identical.

The onset of instability can be understood visually by plotting the maximum growth rate curve $\max_i \mathrm{Re}\,\sigma_{iq}$ as a function of the wave number q for values of the parameter b below, equal to, and above the critical value b_c (see Fig. 3.2). The maximum growth rate can be calculated explicitly as the maximum of the real part of the two eigenvalues σ_{iq} associated with each wave number q (recall the discussion associated with Eq. (3.11)). We summarize the results in Fig. 3.2. The growth rate σ_q is actually complex for small q (the dotted line in the figure indicates the imaginary part of the eigenvalue with the largest real part) but becomes real for larger q. In particular, the imaginary part is zero near the peak corresponding to the fastest growing mode. For a given parameter b, note how the curve $\max_i \mathrm{Re}\,\sigma_{iq}$ has a kink – the slope changes discontinuously – because the eigenvalues switch from complex to real at this point.

Since the Brusselator does not describe an actual experiment, you may wonder whether it is possible to test independently the above predictions of the critical parameter b_c and critical wave number q_c. This can be done by numerical methods. It is straightforward to write a computer code that integrates the evolution equations Eqs. (3.23) in a large periodic interval. The parameter values could then be set to those of Eq. (3.24) and the initial conditions of the fields u_1 and u_2 set to be the uniform values Eq. (3.25) plus some random noise of tiny amplitude. For $b < b_c$, the small-amplitude noise should decay exponentially and the fields will converge toward their uniform values. For b just larger than b_c, the uniform state should be unstable and a cellular structure with wave number close to Eq. (3.31) should appear. What happens in the long term as the exponential growth starts to saturate is not predicted by the linear stability analysis but would be revealed by the numerical integration. The unstable uniform state could evolve onto a stationary, periodic, quasiperiodic, or chaotic attractor.

3.1.3 Oscillatory instability

The analysis of the previous section focused on the Turing instability, an instability that first occurs with a wave number $q \neq 0$. However, the analysis also shows that the two reaction–diffusion equations may also undergo an oscillatory type-III-o instability at $q = 0$. This occurs when tr $A_{q=0}$ passes to a positive value while det A_q is positive for all q. As mentioned in Section 2.5.3, since on the unstable side of a type-III-o transition there is a band of growing oscillatory modes with wave numbers centered around zero, we might expect this system to support spatially uniform nonlinear oscillations and long wavelength nonlinear wave states. We discuss the weakly nonlinear oscillation or wave states that occur near a supercritical type-III-o instability in Chapter 10.

In fact, at the same time Turing was doing his theoretical work on reaction–diffusion systems in Britain, the chemist B. P. Belousov in the Soviet Union was observing oscillating chemical reactions in his laboratory. For many years, this work was not believed and was rejected for publication since chemical oscillations were thought to be inconsistent with the idea that a closed system of mixed chemicals must relax monotonically to equilibrium. Later, A. M. Zhabotinsky continued the investigation and published work on both spatially uniform oscillations and wave states from the late 1960s onwards. It is now a common demonstration experiment to mix chemicals in a shaken test tube or stirred beaker, and watch the color periodically change (from blue to red and back for a modern version of the reaction used by Belousov). The shaking or stirring effectively mixes the chemicals, eliminating spatial inhomogeneities so that only a spatially uniform ($q = 0$) oscillation is seen. In an unstirred Petri dish on the other hand, beautiful patterns of propagating waves are seen, as in Fig. 1.18.

In most cases, experimental chemical systems showing oscillations or waves are not near a linear instability and the oscillations or waves are highly nonlinear so that a description based on the linear modes is not quantitatively useful. We return to this type of system in Chapter 11 where other theoretical tools more suited to the highly nonlinear regime are developed to study the phenomena.

3.2 Realistic chemical systems

We now turn to realistic systems of reacting and diffusing chemicals that experimentalists use to study pattern formation in chemical systems. As is often the case, the apparatus is quite complex to approach the ideal conditions for which the phenomena are most cleanly seen, and which allow a quantitative comparison with theory. An additional difficulty in observing the Turing instability is that the diffusion constants of some of the chemical participants must usually differ by a large ratio, which is hard to arrange for chemical reactions between small molecules in solution. This criterion does not apply for the study of oscillations and waves. The dynamical equations describing the evolution of the chemicals are also much more complicated than the simple two-variable Turing system, and we spend some time explaining how the equations are deduced.

3.2.1 Experimental apparatus

Figure 3.3 shows schematically the design of an experiment that studied pattern formation of reacting and diffusing chemical solutions. (Two patterns from this experiment were previously discussed in Section 1.3.2, see Figs. 1.18(b) and (c).) The geometry is similar to that of Fig. 2.1 but with several refinements. The flat parallel circular plates are made of a transparent porous glass through which chemicals can diffuse and that allows visual observation of the pattern between the plates. The thin cylindrical volume between the plates is filled with a transparent uniform porous gel whose pores are so small (about 80 Å) that they suppress fluid motion. This simplifies the experiment conceptually since the pattern formation is due only to chemicals reacting and diffusing. The gel also renders the pattern visible by changing color according to the concentration of one of the reaction products. A system of reservoirs, pumps, and continuously fed stirred tank reactors (CSTRs) provides a constant flow of fresh reagents across the outer surfaces of plates A and B. As a result, these outer boundaries are surfaces of constant chemical concentrations for each of the reagents. The chemicals diffuse through the glass into the gel where they react, and reaction products diffuse back out into the flowing solutions where they are swept away and permanently removed. Thus the outer boundaries are also surfaces of zero concentration for the reaction products. The diameters of the plates (about 25 mm) are over 100 times larger than the typical length scale of the cellular patterns (about 0.2 mm) so that the system is approximately translationally invariant in the extended directions. In the actual experiment, no influence of the lateral boundaries was observed for the instability and resulting patterns, although a systematic study was not carried out by varying the diameter of the gel.

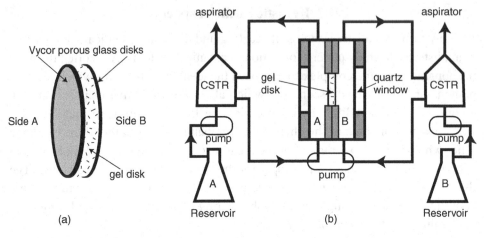

Fig. 3.3 Schematic design of an experiment to study the pattern formation of reacting and diffusing chemicals in solution. (a) The chemical reactions take place near the middle of a uniform transparent porous gel 2 mm thick, that is confined between two wide (25 mm), thin (0.4 mm), uniform, porous, and transparent glass plates A and B. (b) Details of how the chemicals from two reservoirs are fed to the gel. Reservoirs A and B contain mixtures of chemicals that are inert by themselves but react when combined. The contents of these reservoirs are pumped into a "continuously fed stirred tank reactor" or CSTR where the solutions are mixed thoroughly so that the concentrations are spatially uniform. The contents of each CSTR are then pumped to provide a steady flow at known concentrations past the outer sides of the porous plates of part (a). The chemicals diffuse through the porous plate into the gel, react, and reaction products diffuse out and are swept away. Pattern formation within the gel is visualized through a transparent quartz window. The gel is kept at a constant temperature throughout any given experiment. The chemical concentrations in the reservoirs or the temperature can be used as bifurcation parameters. (From Ouyang and Swinney [86].)

3.2.2 Evolution equations

With this experiment in mind, we now discuss how to derive the evolution equations that mathematically describe the experiment. The small pores of the gel in Fig. 3.3 suppress any fluid motion so an evolution equation is not needed for the velocity, which is zero everywhere. It also turns out that the diffusion of heat is so fast compared to the diffusion of chemicals that the temperature field can be assumed to be constant and so does not evolve. The state of the system at any given time t is therefore given by the values at each point in space of continuously varying concentration fields $u_i(\mathbf{x}, t)$, which have the meaning of the local concentration of the ith chemical at point $\mathbf{x}$ at time t. Note that the concentrations can be treated as continuous variables because the pattern formation occurs on a length scale of millimeters that is huge compared to the mean free path of collisions between molecules, of order nanometers.

The evolution equations for the system, together with mathematical descriptions of the boundaries and initial values for the concentration fields, determine how the concentration fields change from one moment in time to the next. The concentration fields change their values by two mechanisms. Chemical reactions change concentrations of reagents and of products according to the concentration values at each point in space. Diffusion by molecular collisions decreases the values of concentration fields where they are locally larger than surrounding values. We discuss these in turn and then combine their contributions to get the final evolution equations.

Let us first consider just the effects of chemical reactions by assuming that the chemical concentrations are spatially uniform so that diffusion can be ignored. Then the rate of reaction $v(t)$ for some chemical reaction is defined in terms of the time derivatives of concentrations and of stoichiometric coefficients. For example, let us consider a binary chemical reaction in which a moles of molecules labeled A and b moles of molecules labeled B react to produce c moles of molecules labeled C and d moles of molecules labeled D. In standard notation, this reaction would be written in the form

$$a\,\mathrm{A} + b\,\mathrm{B} \rightarrow c\,\mathrm{C} + d\,\mathrm{D}. \tag{3.32}$$

The coefficients a, b, c, and d are the stoichiometric coefficients for molecules A, B, C, and D respectively. By definition, the reaction rate $v(t)$ for the entire reaction is the non-negative quantity given by

$$v(t) = -\frac{1}{a}\frac{d[\mathrm{A}]}{dt} = -\frac{1}{b}\frac{d[\mathrm{B}]}{dt} = \frac{1}{c}\frac{d[\mathrm{C}]}{dt} = \frac{1}{d}\frac{d[\mathrm{D}]}{dt}. \tag{3.33}$$

The notation [X] denotes the concentration of molecule X.

For simple chemical reactions in gases or solutions, the reaction rate $v(t)$ is given by the law of mass action, which states that the reaction rate is proportional to the product of powers of reactant concentrations giving a rate law

$$v(t) = k[\mathrm{A}]^{m_\mathrm{A}}[\mathrm{B}]^{m_\mathrm{B}}, \tag{3.34}$$

with the powers $m_\mathrm{A} = a$ and $m_\mathrm{B} = b$ given by the stoichiometry factors. (Chemists call these powers the orders of the reaction with respect to the concentrations.) The positive proportionality constant k is called the rate constant. Reactions obeying this law are called elementary reactions. For reactions that are not elementary, the functional form Eq. (3.34) still sometimes applies but with exponents m_A and m_B whose values may not be simply related to the stoichiometry and so need to be deduced from experiments. In the most general case, the reaction rate may be some arbitrary nonlinear function of the concentrations. Note that if the concentration of B is so large that it can be treated as constant, Eq. (3.34) takes the form

$$v(t) = -\frac{1}{a}\frac{d[\mathrm{A}]}{dt} = k_1[\mathrm{A}]^{m_\mathrm{A}}. \tag{3.35}$$

The effective rate constant k_1 now depends on the concentration of B and so can be varied as a control parameter.

Let us next consider the effects of diffusion without chemical reaction. In this case, the concentration of each chemical species i can only change if there is flux of that species. This is described mathematically by the conservation equation

$$\partial_t u_i = -\nabla \cdot \mathbf{j}_i, \qquad (3.36)$$

which states that the rate of change of the concentration u_i at a point $\mathbf{x}$ is given by the negative of the total flux of u_i into an infinitesimal region surrounding that point, given by the divergence of a current $\mathbf{j}_i$. We must now determine the current. Associated with the ith chemical species there is a chemical potential $\mu_i(x, t)$ that is the thermodynamic variable conjugate to the concentration field $u_i(x, t)$. Gradients in the chemical potentials drive currents of the chemicals, and, in turn, these currents can be related to gradients in the concentrations. To a good approximation, a gradient in the concentration of the ith species drives a current only of the ith concentration.[8] Thus we have the current $\mathbf{j}_i$ of u_i at the point $\mathbf{x}$ is proportional to the gradient in the concentration of u_i:[9]

$$\mathbf{j}_i = -D_i \nabla u_i. \qquad (3.37)$$

The positive number D_i is the diffusion coefficient for u_i and has SI units of m^2/s. These two equations can be combined to yield a diffusion equation

$$\partial_t u_i = \nabla \cdot (D_i \nabla u_i) = D_i \nabla^2 u_i, \qquad (3.38)$$

where the last expression $D_i \nabla^2 u_i$ holds if the diffusion coefficient is constant, a good approximation for many experiments.

By combining the effects of reaction and diffusion, Eqs. (3.33), (3.34), and (3.38), we conclude that the evolution equations for the concentration fields $u_i(x, t)$ take the general reaction–diffusion form

$$\partial_t u_i = f_i(\{u_j\}) + D_i \nabla^2 u_i, \qquad (3.39)$$

with one such equation for each chemical concentration. Here the ith chemical diffuses with a constant diffusion coefficient D_i and the reaction rates f_i are nonlinear functions of the chemical concentrations. For simple rate laws of the form Eq. (3.34), the f_i are multinomials in the concentrations u_i but more complicated nonlinear functions are common.

---

[8] This statement is not as obvious as it might seem at first. For example, in convection of a binary fluid mixture, a gradient in the temperature can drive a concentration current in addition to an energy current, a phenomenon known as the Soret effect.

[9] The direction of the concentration current $\mathbf{j}$ is the negative of the gradient since a chemical flows in the direction from larger to smaller concentration values.

3.2 Realistic chemical systems

To obtain a unique solution of the evolution equations Eq. (3.39), further information is needed in the form of initial values of the concentration fields at some starting time t_0 and of mathematical conditions that describe how the boundaries constrain the fields. Since the reactor geometry of Fig. 3.3 has been constructed in such a way that the contents of the reservoirs flow quickly past the outer surfaces of plates A and B, to a good approximation each concentration field $u_i(x, t)$ corresponding to a reagent has a constant positive value on this outer surface equal to the concentration in the corresponding reservoir. The concentration of a reagent is zero on the opposing plate since the flowing solutions sweep away any of the chemical that reaches that side. For the same reason, the concentration fields corresponding to products are zero on the outer surfaces of both plates. Finally, the chemicals are sealed in by the lateral boundary of the gel and so all the concentration fields satisfy a zero-flux condition $\hat{\mathbf{n}} \cdot \mathbf{j}_i = -D_i \hat{\mathbf{n}} \cdot \nabla u_i = 0$ at each point on the lateral boundary, where $\hat{n}$ is the unit vector normal to the lateral boundary at a given point. As we discussed in Section 2.1, these no-flux lateral conditions would typically be replaced by infinite or periodic boundary conditions when carrying out a linear stability analysis.

The derivation of equations such as Eq. (3.39) involves various approximations that are less well justified than those used to derive the evolution equations for fluids. With a few exceptions, a simple rate law of the form Eq. (3.34) holds only for elementary reactions in dilute solutions or for ideal gases. Whether some particular reaction is elementary can be difficult to establish experimentally. Further, the identification of the reaction mechanism – the sequence of elementary steps that lead from reagents through intermediates to final products – often requires separating the important reactions (those that are slower and so rate limiting) from a much larger list of possible reactions that produce various short-lived and often unknown intermediate molecules. This separation is a rather ad hoc procedure since there is no small parameter that can be exploited in a perturbation theory to improve the validity systematically.

In contrast, the fundamental approximation leading to the evolution equations for a fluid (the Navier–Stokes equation) is that the flow varies spatially over much larger distances than the microscopic scale set by the mean free path for molecular collisions. This is an excellent approximation for typical laboratory fluid experiments whose spatial variations are millimeters or larger, and can be improved, if necessary, by increasing the size of the experiment.

We illustrate the above points with the following Etude, which discusses the reaction diffusion equations used to describe a particular experimental reaction system known as the CDIMA reaction. (For another example, see Exercise 3.9.) The Etude also discusses a numerical calculation using the CDIMA reaction of the state that is uniform with respect to the extended coordinates for a realistic gel and for experimentally reasonable chemical gradients. A surprise is that the structure along the transverse confined coordinate is surprisingly complicated (see Fig. 3.4), too much so to handle by analytical means.

Etude 3.2 Evolution equations for the chlorine dioxide–iodine–malonic acid CDIMA reaction

Following the work by the chemists I. Lengyel, G. Rábai, and I. Epstein [62, 63], let us write down the evolution equations for the pattern-forming chlorine dioxide–iodine–malonic acid reaction (abbreviated CDIMA) that has been studied experimentally. In Fig. 3.3, reservoir A would contain chlorine dioxide ClO_2 and iodine I_2 which do not react together while reservoir B would contain malonic acid $CH_2(COOH)_2$ (abbreviated as MA). The reaction between the ClO_2, I_2, and MA molecules produces further reactants such as the iodide ion I^- and the chlorite ion ClO_2^-, and also produces products that take no further part in the reaction. The iodide concentration $[I^-]$ is visualized with an immobile starch indicator S that is embedded in the gel and that turns blue reversibly upon binding to iodide.

By comparing theory and experiment for stirred CDIMA reactions such that diffusion did not play a role, the chemists proposed a simplified reaction mechanism consisting of the following four reactions:

$$MA + I_2 \rightarrow IMA + I^- + H^+, \tag{3.40a}$$

$$ClO_2 + I^- \rightarrow ClO_2^- + \tfrac{1}{2}I_2, \tag{3.40b}$$

$$ClO_2^- + 4I^- + 4H^+ \rightarrow Cl^- + 2I_2 + 2H_2O, \tag{3.40c}$$

$$S + I_2 + I^- \rightleftharpoons SI_3^-. \tag{3.40d}$$

Since the reactions are sustained out of nonequilibrium, with new reagents ClO_2, I_2, and MA constantly being supplied and reaction products steadily being removed, we can assume that the reverse reactions for the first three equations proceed at a negligible rate. The reversible formation of the starch complex SI_3^- in Eq. (3.40d) plays a doubly important role in the pattern formation. First, this is the colored indicator that actually allows the pattern to be seen. Second, because this complex is fixed to the gel, the effective diffusion constants of the iodine and iodide are reduced since these molecules become immobile for the fraction of the time that they are bound to the starch. As we suggested in Section 3.1, significantly different diffusion coefficients of at least two reactants are a necessary condition for the linear instability of a uniform state. This condition would be hard to attain without the immobile starch since the diffusion coefficients of small ions in solution are all comparable.

3.2 Realistic chemical systems

A comparison of theory with experiment for the stirred CDIMA reaction suggests the following respective reaction rates $r_j(t)$:

$$r_1 = \frac{k_{1a}\,[\text{MA}]\,[\text{I}_2]}{k_{1b} + [\text{I}_2]}, \tag{3.41a}$$

$$r_2 = k_2\,[\text{ClO}_2]\,[\text{I}^-], \tag{3.41b}$$

$$r_3 = k_{3a}\,[\text{ClO}_2^-]\,[\text{I}^-]\,[\text{H}^+] + \frac{k_{3b}\,[\text{ClO}_2^-]\,[\text{I}_2]\,[\text{I}^-]}{h + [\text{I}^-]^2}, \tag{3.41c}$$

$$r_4 = k_{4a}\,[\text{S}]\,[\text{I}^-]\,[\text{I}_2] - k_{4b}\,[\text{SI}_3^-]. \tag{3.41d}$$

The various parameters are determined by fits to experimental data. The reaction rates r_2 in Eq. (3.41b) and r_4 in Eq. (3.41d) have the simple form expected of an elementary reaction but the other two have more complicated nonlinear dependencies on the concentrations. This complexity can be partly understood as arising from the elimination of short-lived intermediate products from the rate equations, along the lines discussed in Exercise 3.9. You should keep in mind that these reaction rates are plausible deductions from empirical data rather than obtained from first principles by a theoretical argument. It is even possible that the functional form of these expressions could change in the future since research still continues on how best to quantify the CDIMA system.

Using the definition of reaction rate Eq. (3.33) and allowing the chemicals to diffuse, we obtain the following six coupled evolution equations for the CDIMA pattern-forming system:

$$\partial_t\,[\text{ClO}_2] = -r_2 + D_{\text{ClO}_2}\,\nabla^2\,[\text{ClO}_2], \tag{3.42a}$$

$$\partial_t\,[\text{ClO}_2^-] = r_2 - r_3 + D_{\text{ClO}_2^-}\,\nabla^2\,[\text{ClO}_2^-], \tag{3.42b}$$

$$\partial_t\,[\text{MA}] = -r_1 + D_{\text{MA}}\,\nabla^2\,[\text{MA}], \tag{3.42c}$$

$$\partial_t\,[\text{I}_2] = -r_1 + \frac{1}{2}r_2 + 2r_3 - r_4 + D_{\text{I}_2}\,\nabla^2\,[\text{I}_2], \tag{3.42d}$$

$$\partial_t\,[\text{I}^-] = r_1 - r_2 - 4r_3 - r_4 + D_{\text{I}^-}\,\nabla^2\,[\text{I}^-], \tag{3.42e}$$

$$\partial_t\,[\text{SI}_3^-] = r_4. \tag{3.42f}$$

The linear combination of reaction rates in each equation follows from the corresponding stoichiometry in the reaction mechanism Eq. (3.40). There is no diffusion term in Eq. (3.42f) for the starch-triiodide complex since the starch is immobile. The values of the five diffusion coefficients have to be determined by experiment. Together with the parameters in the reaction rates Eq. (3.41), this system is described by a total of 13 parameters (which can be reduced to five dimensionless parameters by changing to dimensionless units of space, time, and concentration). In contrast, two dimensionless parameters – the Rayleigh number R and Prandtl number σ – are needed to characterize a Rayleigh–Bénard convection experiment. Each of the equations in Eqs. (3.42) also requires boundary conditions and initial data to complete the mathematical description. As noted before, the boundary conditions have a simple form, being constant on the plate surfaces or having a zero flux on the lateral boundaries of the gel.

If no reactions were to occur within the gel, we would expect the concentrations [ClO_2], [MA], and [I_2] to interpolate linearly between their boundary values on either side of the plates. Such linear profiles would be analogous to the linear temperature profile of the conducting uniform state in Rayleigh–Bénard convection. In fact, the reactions in the interior of the gel produce a much more complicated set of profiles for the chemicals, with a z dependence that cannot be calculated analytically. As an example, we show in Fig. 3.4 the concentration profiles deduced from a numerical calculation based on the above evolution equations with experimentally estimated parameters and with experimentally plausible concentrations in reservoirs A and B. The stationary concentrations are plotted as a function of a dimensionless confined variable $x_\parallel = z$, whose value is $z = 0$ at the surface of plate B and $z = 1$ at the surface of plate A. The six concentration fields are constant and uniform in each plane transverse to the z direction. This complicated structure – a direct consequence of the tight coupling to the strong chemical gradients imposed by the reservoirs – constitutes the "spatially uniform solution" that would be the starting point of a full linear stability analysis. Pattern formation would then be the occurrence of spatial structure in the concentration fields within each plane transverse to z.

Besides the explicit example Eqs. (3.42) of realistic evolution equations, perhaps the most important conclusion of this Etude is that the stationary uniform solution of a sustained nonequilibrium system can have a surprisingly rich structure in the confined directions even before pattern formation occurs in the extended directions. This structure can be quite hard to calculate and by no means can we always do this analytically.

3.2.3 Experimental results

Examples of the stationary Turing patterns observed in the experimental system described in Section 3.2.1 have already been shown in Figs. 1.18(b) and (c). The variety of patterns – stripes, hexagons, spots, and spatiotemporal chaos – is not something that can be discussed using linear theory, and we will discuss these nonlinear states in later chapters such as Chapters 4, 7, and 9. A quantitative analysis of the onset of a reaction–diffusion chemical pattern is shown in Fig. 3.5. The

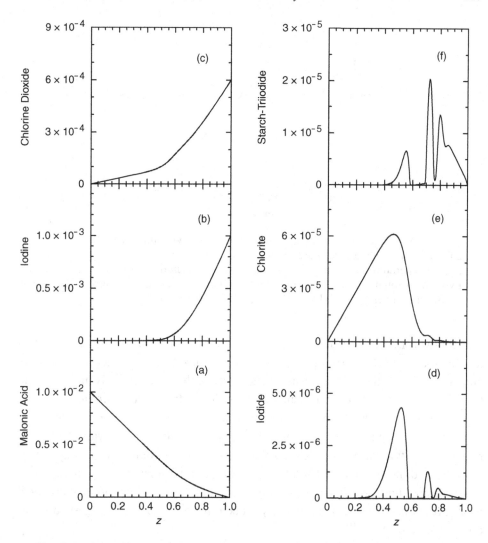

Fig. 3.4 The uniform solution may have complicated structure in the confined directions: numerically calculated profiles (chemical concentrations as a function of the confined coordinate $x_\parallel = z$) for the stationary uniform state of the CDIMA reaction, Eqs. (3.40), in the reactor geometry of Fig. 3.3. The gel was assumed to have a thickness of $d = 0.3$ cm and the z coordinate measures the fractional distance across the gel, with $z = 0$ corresponding to plate B and $z = 1$ corresponding to plate A. The boundary conditions are $[MA]_L = 1 \times 10^{-2}$ M at the left boundary $z = 0$, and $[I_2]_R = 1 \times 10^{-3}$ M and $[ClO_2]_R = 6 \times 10^{-4}$ M at the right boundary $z = 1$. All other boundary conditions are zero concentration. Especially for the intermediates like iodide and the starch-triiodide complex, the profiles have a surprisingly complicated rapidly varying spatial structure. The concentrations are spatially uniform in each plane of constant z. (From Setayeshgar and Cross [95].)

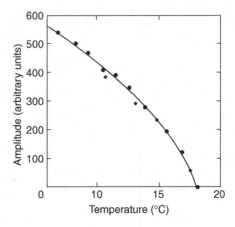

Fig. 3.5 Amplitude of a chemical pattern at onset, showing the linear instability of a uniform state at a temperature of about 18 °C. In this case, the instability occurs as the temperature is lowered. The amplitude was measured as the magnitude of the two-dimensional Fourier transform of a visualization of the pattern in the vicinity of the pattern periodicity. (From Ouyang and Swinney [84].)

plot follows the scheme discussed in Section 2.4. The order parameter used is the magnitude of the two-dimensional Fourier transform integrated over a wave number band near the peak intensity. This is zero in the ideal spatially uniform state (noise in the experiment or measurement would contribute a small value), and is a good measure of the strength of the pattern. The control parameter used in the experiment was the temperature, which is easy to control and to vary. However, rate constants typically change their values by different amounts as the temperature is varied and these amounts need to be measured experimentally if a quantitative link is desired between this control parameter and the parameters of the theoretical model. Such measurements have not yet been made. Figure 3.5 suggests a linear onset at around 18 °C.[10]

For other chemical combinations, oscillatory wave states can be observed as discussed in Section 3.1.3 and shown in Fig. 1.18(a). If the states are sustained in a continuously fed reactor similar to Fig. 3.3, the controlled conditions allow quantitative measurements to be made. Figure 3.6 shows results from a reaction–diffusion chemical experiment in which the order parameter is "intensity contrast," an optical measure of the difference in indicator concentration between wave crests and troughs. The measurements suggest a linear onset at a sodium bromate concentration [$NaBrO_3$] of about 0.018 M, although there is a small amount of hysteresis at

---

[10] When we investigate the nonlinear theory (see Section 7.3.3 and Fig. 7.2), we will discover that a hexagonal pattern is not expected to grow continuously from the uniform state. Instead, either an intervening stripe state at small amplitudes or a discontinuous jump to nonzero amplitude is expected. Perhaps one or the other occurs on a finer scale than is resolved by the data.

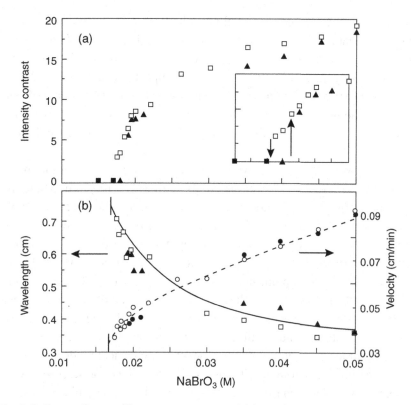

Fig. 3.6 Onset of an oscillatory wave pattern: (a) intensity contrast across a spiral wave front as a function of $NaBrO_3$ concentration. The onset is almost continuous at a concentration of about 0.018M, but, as shown in the inset, there are small jumps and a small amount of hysteresis; (b) the wave propagation velocity and wavelength do not go to zero at onset (note offset zeros in this plot). (Redrawn from Tam et al. [101].)

the onset (so that the bifurcation is slightly subcritical, rather than supercritical). The fact that the wavelength and velocity of the waves (and therefore also the frequency) tend to nonzero values at the onset represents an experimental demonstration that this transition is a type-I-o instability.

3.3 Conclusions

With the material of this and the previous chapter, you now have the tools to look at a key aspect of pattern formation in diverse systems, namely the linear instability from the uniform state toward a state with spatial, and perhaps temporal, structure. As we have seen, the linear stability analysis tells us about the characteristic length and time scales that might be expected in the pattern. Conversely, a prediction from a theoretical model of a precise spatial periodicity that is observed in experiment can help to validate the basic equations of the model. Such validation is often a serious issue in many real-world examples of pattern formation. In Chapter 2, we presented the basic formalism in the context of a simple toy example, Eq. (2.4). In the present chapter, we described the more complex and realistic example of reacting and diffusing chemicals in a three-dimensional medium with realistic boundary conditions. The exercises that follow will give you more practice in applying these ideas.

After all this work, we are left with a physically unrealistic solution, namely a perturbation $\mathbf{u}_p(\mathbf{x}, t)$ that grows exponentially in time to a magnitude that violates the assumption of small amplitude that was used to derive the linearized evolution equations (by neglecting higher-order terms). Furthermore, as we saw in the preceding chapter, we are left with too many growing modes. For the case of an isotropic base state, the perturbation can be an arbitrary superposition of modes Eq. (2.22) whose wave vectors $\mathbf{q}$ can point in any direction (Fig. 2.11). The next step in our discussion of pattern formation is therefore to include the effects of nonlinearity. A naive expectation would be that nonlinearity saturates the exponential growth of the modes at some finite value without changing the spatial structure too much, so that the state is at least qualitatively reminiscent of the critical mode at onset. In the presence of symmetries, nonlinearity also serves to reduce the degeneracy giving pattern selection. In later chapters, we will discuss first the qualitative aspects of nonlinear effects (Chapter 4), and then analytical techniques to make these ideas more precise and quantitative (Chapters 6 and 7).

3.4 Further reading

(i) Alan Turing's paper "The chemical basis of morphogenesis" [106] presents many of the basic ideas of pattern formation.
(ii) Chapters 14–17 of the book *Mathematical Biology, Third Edition* by Murray [79], describe the Turing instability in some detail, as well as implications of linear stability analysis to various mechanisms of pattern formation in biological systems.
(iii) *An Introduction to Nonlinear Chemical Dynamics: Oscillations, Waves, Patterns, and Chaos* by Epstein and Pojman [34] gives an extended discussion of pattern formation and dynamics in chemical systems.

Exercises

3.1 **Wave number of fastest growth rate in the Turing model:** The wave number q_m given by Eq. (3.17) was obtained from the condition that the criterion for linear stability Eq. (3.14b) first became reversed.

(a) Show that q_m is also the wave number that maximizes the growth rate σ_q at the onset of instability.

(b) Show that q_m is no longer the wave number of the fastest growing mode once the system is a small but finite distance beyond the onset of instability.

Hint: you might find it easiest to start from Eq. (3.12).

3.2 **A two-chemical Turing instability cannot be an oscillatory bifurcation:** A Turing instability is a linear instability at a finite wave number of a stationary uniform state that is stable in the absence of diffusion. Show that a reaction–diffusion system of two chemicals of the form Eqs. (3.1) cannot be oscillatory at onset, i.e. it cannot be of type I-o with Im $\sigma_{q_c} \neq 0, q_c > 0$.

For a greater challenge, show that a Turing instability with three chemicals *can* be oscillatory.

3.3 **Type of instability for two coupled reaction–diffusion equations:** Consider the following reaction–diffusion evolution equations

$$\partial_t u = au + \partial_x^2 u - b\, \partial_x^2 v - \left(u^2 + v^2\right)(u + cv), \quad \text{(E3.1a)}$$

$$\partial_t v = av + b\, \partial_x^2 u + \partial_x^2 v - \left(u^2 + v^2\right)(v - cu), \quad \text{(E3.1b)}$$

for the fields $u(x, t)$ and $v(x, t)$ on the real line, where the parameters a, b, and c are arbitrary real constants. Derive a condition for the linear instability of the zero base solution $\mathbf{u}_b = (u_b, v_b) = (0, 0)$, and determine whether the instability is of type I-s, I-o, or III-o.

Note: This is not a Turing problem, you do not have to assume that the base solution is stable in the absence of diffusion.

3.4 **Stability of the stirred Brusselator:** For the stirred Brusselator given by Eqs. (3.23) without the diffusion terms, plot in the a–b parameter plane the region of stability of the uniform solution, the region (if any) of its instability via an oscillatory instability, and the region (if any) of its instability via a stationary instability.

3.5 **Nullclines of the stirred Brusselator:** For the stirred Brusselator of the previous problem:

(a) Plot the nullclines $u_2 = g_1(u_1)$ and $u_2 = g_2(u_1)$ of Eqs. (3.23a) and (3.23b) for $a = 1$ and $b = 2.5$. Use this plot to confirm graphically that the nullclines intersect at a single point so that there is only one uniform solution $\mathbf{u}_b$. Draw horizontal and vertical arrows in each sector defined by the intersecting nullclines to indicate the direction of the evolution $d_t u_1$ and $d_t u_2$.

(b) Numerically solve for the time evolution for the parameters $a = 1$ and $b = 2.5$. After transients, you should find a periodic orbit (limit cycle). Superimpose this orbit on the nullcline plot. Is the resulting diagram consistent with what the nullclines and evolution arrows tell you?

3.6 **Linearized Brusselator equations:** For the Brusselator Eqs. (3.23):

 (a) Derive the linearization of the reaction rates Eq. (3.26) for small perturbations about the stationary uniform solution.

 (b) Determine and plot the neutral stability curve $b = b(q)$ of the uniform state Eq. (3.25) for the parameter values Eq. (3.24).

 (c) Using b as the control parameter and defining $\varepsilon = (b - b_c)/b_c$ with b_c the value at onset, determine the numerical value of the coherence length ξ_0 (see Eq. (2.27)) for the parameter values Eq. (3.24) and compare that value with the critical wavelength $2\pi/q_c \approx 14.4$ (see Eq. (3.31)).

3.7 **Type-III-s instability:** Convince yourself that a type III-s instability (a stationary instability first occurring at $q = 0$) is not possible for Eqs. (3.1) One possibility is to show that when a stationary instability occurs at $q = 0$, the system is already unstable toward modes with $q \neq 0$.

3.8 **Type-III-o instability in the Brusselator:** Can the Brusselator model Eqs. (3.23) show a type III-o instability? If so, find an example of parameter values for which this happens.

3.9 **The Oregonator:** A more realistic, but still simplified, model of the Belousov–Zhabotinsky reaction is given by the following five reactions:

$$BrO_3^- + Br^- + 2H^+ \to HBrO_2 + HOBr, \qquad (E3.2a)$$

$$HBrO_2 + Br^- + H^+ \to 2HOBr, \qquad (E3.2b)$$

$$BrO_3^- + HBrO_2 + 3H^+ + 2Ce^{3+} \to 2HBrO_2 + 2Ce^{4+} + H_2O, \quad (E3.2c)$$

$$2HBrO_2 \to BrO_3^- + HOBr + H^+, \qquad (E3.2d)$$

$$Z + Ce^{4+} + ?H_2O \to hBr^- + Ce^{3+} + ?HCOOH$$
$$+ ?CO_2 + ?H^+, \qquad (E3.2e)$$

where h is some number (the stoichiometry factor) and the last reaction is a reaction of Ce^{4+} with a mixture of $CH_2(COOH)_2$ (malonic acid) and $CHBr(COOH)_2$ (bromomalonic acid) written as Z. The question marks "?" in Eq. (E3.2e) denote numerical factors depending on h that are not necessary for this exercise.

 (a) Use the law of mass action (that the reaction rate satisfies Eq. (3.34) with exponents equal to the stoichiometric coefficients of the reactants on the left-hand side) to derive dynamical evolution equations for the three concentrations

$$u = [\text{HBrO}_2], \tag{E3.3a}$$

$$v = \left[\text{Ce}^{4+}\right], \tag{E3.3b}$$

$$w = \left[\text{Br}^-\right]. \tag{E3.3c}$$

Assume that the concentrations $[\text{H}^+]$, $[\text{BrO}_3^-]$, $[\text{CH}_2(\text{COOH})_2]$, $[\text{CHBr}(\text{COOH})_2]$, $[\text{Ce}^{3+}]$, and $[\text{H}_2\text{O}]$ are constant. The resulting three odes constitute the Oregonator model of the Belousov–Zhabotinsky reaction. The name comes from the University of Oregon, where a group of research chemists first proposed and analyzed this model.

(b) By an appropriate rescaling of time t and of the various concentrations, show that the rescaled equations can be written in the form

$$\eta \, \partial_t u = qw - uw + u - u^2, \tag{E3.4a}$$

$$\partial_t v = u - v, \tag{E3.4b}$$

$$\eta' \, \partial_t w = -qw - uw + bv. \tag{E3.4c}$$

The constants η, η', q, and b are related to the reaction rates and to the concentrations that are held constant.

(c) For some parameter values, the inequalities $\eta' \ll \eta \ll 1$ hold. Assuming in this case that the left side of Eq. (E3.4c) is approximately zero, show that the variable w can be eliminated so that Eqs. (E3.4) reduce to a two-variable model

$$\eta \, \partial_t u = u - u^2 - bv \frac{u - q}{u + q}, \tag{E3.5a}$$

$$\partial_t v = u - v. \tag{E3.5b}$$

This illustrates how a non-polynomial reaction rate can arise from the elimination of a slowly changing variable.

3.10 **Time scales in realistic reaction diffusion experiments:** Small molecules in water all have about the same diffusion constant of $D \approx 2 \times 10^{-9} \, \text{m}^2/\text{s}$. By extending the discussion of Exercise 1.5(b) on lateral diffusion times, estimate the minimum observation time in days for an experiment as in Fig. 3.3 that involves a gel that is 25 mm in diameter. Compare your answer with the longest observation time of about a week in the Ouyang and Swinney experiment.

3.11 **Turing instability in a finite one-dimensional domain with no-flux boundaries:** Consider a one-dimensional Turing system Eq. (3.1) with parameters that are just at the critical values that give the onset of a stationary Turing pattern in an infinite domain. Now confine the system to a domain of length L with no-flux boundary conditions $\partial_x \mathbf{u} = \mathbf{0}$ at $x = 0$ and at $x = L$. Show that as L is varied, instability occurs in the finite domain whenever L takes the value $m\pi/q_c$ with q_c the critical wave number in the infinite domain (given by q_m, Eq. (3.17) with the right-hand side evaluated at their onset values).

This result is atypical of pattern-forming systems. In general, as we will see in Chapter 6, there is a systematic suppression of the onset in finite domains with realistic (not periodic) boundary conditions.

3.12 **Onset of instability for reaction–diffusion equations with no-flux boundaries in a two-dimensional system with arbitrary shape:** Consider the general reaction diffusion system in two spatial dimensions

$$\partial_t \mathbf{u} = \mathbf{f}(\mathbf{u}) + \mathbf{D}\nabla^2 \mathbf{u}, \qquad (E3.6)$$

with $\mathbf{D}$ a diagonal diffusion matrix and $\nabla^2 = \partial_x^2 + \partial_y^2$ with parameters chosen to be at the Turing instability point as in Exercise 3.11. In the following you find a condition for instability in a finite two-dimensional domain with no-flux boundaries of arbitrary shape as the domain size L increases (at fixed shape).

Let us assume that we can first solve a scaled "vibrating drumhead" eigenvalue problem

$$\nabla_X^2 \psi_m + \lambda_m \psi_m = 0 \qquad (E3.7)$$

with Neumann boundary conditions

$$\hat{\mathbf{n}} \cdot \nabla_X \psi = 0, \qquad (E3.8)$$

for a domain of the given shape but of some reference size (e.g. maximum chord of size unity). The λ_m are eigenvalues characteristic of the shape of the boundary, and the $\psi_m(\mathbf{X})$ are the corresponding eigenfunctions. There will be some smallest positive eigenvalue λ_0, and an infinite sequence of larger ones extending to infinite value. Next, consider solutions to the linear instability problem of the form

$$u_i^{(m)}(\mathbf{x}) = \bar{u}_i^{(m)} \psi_m(\mathbf{x}/L) e^{\sigma_m t}. \qquad (E3.9)$$

Show that the growth rate σ_m is the same as that of a mode of the infinite system with wave number $\sqrt{\lambda_m}/L$. Also show that $u_i^{(m)}(\mathbf{x})$ satisfies Neumann boundary conditions in a system which is the reference system scaled up by the factor L. Hence show that instability in the finite domain occurs at the same parameter values as in the infinite domain whenever $L = \sqrt{\lambda_m}/q_c$ with q_c the critical wave number of the infinite domain.

3.13 **Turing instability in a cellular system:** Turing also studied the instability in a discrete model appropriate to a cellular system. For a ring of cells he used a model with two concentrations U_n, V_n defined as a function of an index n around the ring of l cells

$$d_t U_n = f(U_n, V_n) + D_1 (U_{n+1} - 2U_n + U_{n-1}), \quad \text{(E3.10a)}$$
$$d_t V_n = g(U_n, V_n) + D_2 (V_{n+1} - 2V_n + V_{n-1}), \quad \text{(E3.10b)}$$

where the last terms give the transfer of chemicals to adjacent cells at a rate proportional to the differences in concentrations. Suppose that the linearizations of f and g around the steady-state solutions are

$$f(U_n, V_n) \simeq a_{11} \delta U_n + a_{12} \delta V_n, \quad \text{(E3.11)}$$
$$g(U_n, V_n) \simeq a_{21} \delta U_n + a_{22} \delta V_n. \quad \text{(E3.12)}$$

(a) Show that the ansatz $\delta U_n, \delta V_n \propto e^{iqn} e^{\sigma t}$ can be used to reduce the equations for the linear stability analysis to a simple matrix problem as in the continuous case.

(b) What are the possible values of q and what restricted range of q is sufficient to give all possible solutions?

(c) Find the modified criteria for the Turing instability to take place in this system.

(d) Describe the nature of the disturbances in U_n and V_n at the onset of the $q = \pi$ mode.

4
Nonlinear states

The linear stability analysis of Chapter 2 predicts that a small perturbation about a uniform state will grow exponentially in magnitude when the uniform state becomes unstable. Over time, the magnitude of a perturbation will grow so large that the nonlinear terms that were neglected when deriving the linearized evolution equation can no longer be ignored.[1] These nonlinear terms play a fundamental role in the resulting pattern formation: they saturate the exponential growth, and they select among different spatial states. *It is the essential role of nonlinearity in a spatially extended system that makes the study of pattern formation novel and hard.*

We can gain a great deal of insight about the nonlinear regime of pattern formation by considering spatially periodic patterns. This is natural when considering the fate of a single exponentially growing Fourier mode of the linearized evolution equations associated with a linear stability analysis. Nonlinearities in the evolution equations for the system generate spatial harmonics (Fourier modes with wave vectors $n\mathbf{q}$ with n an integer) of this growing mode so that the finite-amplitude solution maintains the periodicity over the length $2\pi/q$. A key role of the nonlinearity is to quench the exponential growth of the solution, leading to steady spatially periodic solutions for a stationary instability, and nonlinear oscillations or waves for an oscillatory instability. If this steady or periodic solution is to be physically relevant, we must also require that it be stable with respect to small perturbations. Thus we will study the existence and stability of steady or oscillatory spatially periodic (for $q_c \neq 0$) solutions.

The growth and saturation of a single mode can be expressed conveniently in terms of a complex-valued multiplicative factor that is called the amplitude of the mode (see Section 4.1.1 below). Close to the threshold of the instability, expansion techniques based on the assumption of a small amplitude lead to a quantitative

---

[1] For example, when going from the evolution equation Eq. (2.3) to the linearized evolution equation Eq. (2.9) for a small perturbation u_p, we discarded the terms $-3u_b u_p^2 - u_p^3$ that were nonlinear in u_p and so negligible compared with u_p for small u_p.

theory of the existence of a nonlinear spatially periodic solution, and of its stability with respect to perturbations of the same spatial periodicity. These are in fact the tools of bifurcation theory that you have probably seen in a class or text on dynamical systems. Appendix 1 gives a brief overview of the elementary ideas of bifurcation theory relevant to the following discussion. More generally, the amplitude equation formalism provides a quantitative description of pattern formation near the onset of instability, even for patterns that are more complex than periodic stripes, that evolve in time, and for general boundary conditions. We will discuss this quantitative theory with some applications in Chapters 6, 7, and 8. The phase of the complex amplitude is itself an interesting field since it is closely connected with translational symmetries of the system. When the spatial structure of a pattern changes slowly over a distance that is large compared to the cellular structure of the pattern, an evolution equation for the phase called a phase equation can be derived that is valid even when the system is not close to the onset of instability and when nonlinear effects are no longer small. Chapter 9 discusses these phase equations and some of the insights that can be derived from them.

In this chapter, we first use general ideas to study the existence of stripe patterns, i.e. patterns with spatial variation of the form $e^{\pm i\mathbf{q}\cdot\mathbf{x}}$ plus harmonics. Stripe patterns might arise in the context of a strictly one-dimensional system, such as in the demonstration calculation of Section 2.2.1, or for a physical system with just one extended and two or three confined dimensions, or for a system with two or more extended directions where we only seek solutions that are spatially periodic in one of the extended directions.

We next consider the stability of stripe patterns with respect to general classes of perturbations. Physical realizability requires stability for all small perturbations. This criterion restricts realizable spatially periodic solutions to a narrower range of possible wave numbers than the criterion for existence. As various control parameters are changed, this band of stable wave numbers traces out what is called the stability balloon of the stripe solutions.[2] For example, the Prandtl number σ and Rayleigh number R can be varied independently in a Rayleigh–Bénard experiment, and so one gets a three-dimensional balloon-like region of linearly stable stripe wave numbers in the space with axes σ, R, and wave number k.

The stability balloon provides a basic conceptual starting point for the study of more complex patterns including the parameter dependence of patterns and of the range of observed wavelengths. For example, there may be a choice of parameters for which all the nonlinear stripe states are unstable. In this case, a time-dependent

---

[2] The stability balloon for stripe patterns of a Rayleigh–Bénard convection system is often called the Busse balloon in honor of the fluid dynamicist Fritz Busse. Over many years, Busse and his collaborators investigated in great detail the stability balloons for stripes as a function of Rayleigh number and Prandtl number. His calculations yielded many valuable insights about pattern formation for convection and for other type-I-s systems.

state or a stationary state with a new geometry (say curved stripes or a hexagonal lattice) is likely to be observed experimentally. Provided that the evolution equations are known to a reasonable accuracy and that the associated numerical work is not too hard, calculating stability balloons for various system parameters and comparing their properties with experiments is the next most useful step toward understanding a pattern-forming system, after the linear stability analysis of the uniform state that we discussed in the previous two chapters.

We next discuss more general nonlinear stationary solutions that are periodic for two extended directions and discuss how they are represented mathematically and how they differ by symmetry. Here the interesting question is the competition between different two-dimensional lattices, say rectangles and hexagons, and stripes.[3] This question dramatically illustrates the importance of nonlinearity in selecting amongst different states since the lattice states are simply linear combinations of stripes with different orientations and all such superpositions are equally good solutions of the linearized evolution equations. The question of whether say stripes or hexagons are expected in an experimental or naturally occurring physical system is one that has been asked in many different areas of science and engineering. We will see in Chapter 7 that, at least in the weakly nonlinear regime near threshold, general arguments may be brought to bear on this question. Further from onset, one can again calculate stability balloons of linearly stable wave vectors for lattices of a given symmetry, using symmetries suggested by mathematical arguments or by experiment.

Although different ideal symmetric states can be readily constructed theoretically or studied numerically by imposing periodic boundary conditions on the extended coordinates, the observation of ideal states in the laboratory relies on careful choices of the cell geometry and of the experimental protocol. Typically, patterns in laboratory experiments and in nature will be disordered. These patterns may locally take the form of stripe or lattice states, but the orientation of these local structures may vary over the system, as can be seen in Fig. 4.8 below and in other figures from Chapter 1. The lack of order might be a residual effect of the growth of perturbations from uncontrolled random initial conditions or from localized irregularities at different points in the system, perhaps on the walls (e.g. a small bump or thermal inhomogeneity). Alternatively, the disorder may reflect the long-range influence of the system's lateral boundaries, which are rarely compatible with the stripe or lattice structure of the idealized infinite system. If a region of disorder occupies

---

[3] We will use the term "lattice" to refer generally to two-dimensional lattices such as squares and hexagons, in distinction to stripe states. Strictly, stripes also are a lattice – a one-dimensional lattice. Three-dimensional lattices do not play a major role in our discussion, since in most pattern-forming systems, at least one spatial dimension is confined.

only a small fraction of the pattern's area and is one of several often observed characteristic types, it is called a defect. Defects with a stability implied by topological arguments are of particular interest since their structure and dynamics can often be understood by general arguments. On the other hand, there might be large disordered regions over which no regular structure exists. As you might guess, these are hard to classify and to understand.

In the present chapter, we introduce these ideas largely through qualitative arguments. In subsequent chapters, we will revisit many of them using more formal tools, including amplitude equations in Chapters 6–8, and phase equations in Chapter 9. Our discussion in all these chapters will mainly concern pattern formation via a supercritical type-I-s instability, since this is the type most thoroughly studied by experiment, and is also the case for which the mathematical details are more easily worked out. Later in Chapter 10, we will develop a qualitative and quantitative understanding of type-I-o and type-III-o oscillatory instabilities.

4.1 Nonlinear saturation

The linear stability analysis of Chapter 2 leaves us with exponentially growing solutions, which are unphysical at long times. (Long means compared with $1/\sigma_q$, the reciprocal of the exponential growth rate.) The cause of the problem is clear: we linearized the equations by assuming a tiny perturbation. This assumption is good at short times and for an initial condition that has a small magnitude, but at long times the nonlinear terms left out in the linear approximation become important. One effect of nonlinearity is to quench the exponential growth. A system just beyond the onset of a stationary instability might then eventually approach a time-independent state solution such that the nonlinear terms have the same magnitude as the linear terms and so balance them.[4] We call the resulting solution a saturated nonlinear steady state. If the control parameter is only slightly above threshold, a small contribution of the nonlinear terms is sufficient to balance the small linear growth rate, and we expect the stationary solution to have a magnitude that grows continuously from zero as the control parameter is increased from the onset value. The linear instability of the base state is then called a supercritical bifurcation. The reasons for this nomenclature will become clearer shortly (also see Appendix 1). Such a bifurcation is also known as a forward bifurcation, or a continuous or second-order transition. For supercritical bifurcations, the weakly nonlinear state for a control parameter just above the bifurcation value is close in structure to the growing linear mode, and is particularly accessible to theoretical approximation.

---

[4] A nonlinearity might also cause saturation but result in a time-dependent state, say oscillatory or chaotic. In this case, the linear growth terms balance the nonlinear terms only on average. We do not discuss this more complicated situation in this chapter.

On the other hand, the nonlinear terms may initially serve to enhance the growth rate further. In this case, even if the control parameter is only slightly larger than the threshold value, the magnitude of the disturbance will typically grow to a large value before saturation occurs. (Physically we do not expect growth to an infinitely big value.) The resulting nonlinear state may not have any simple relationship to the linearly growing mode and theoretical treatment is much harder. For example, the saturated state may be chaotic even though the linear instability was a stationary instability. The linear instability of the base state is then called a subcritical bifurcation, also called a backward bifurcation, or a discontinuous or first-order transition. Our focus in this chapter is mainly on the more easily treated supercritical case.

For oscillatory instabilities, the same ideas apply except that the saturated nonlinear state will have a periodic time dependence and be a state of nonlinear oscillations or waves.

4.1.1 Complex amplitude

We can gain useful initial insights into the effects of the nonlinearity by retaining the idea that the size of the perturbation to the uniform solution is small, but continuing the perturbative expansion to higher order than the first term giving the linear approximation. Near enough to the threshold of the linear instability, the first few higher-order terms can provide a quantitatively accurate description of the pattern formation.

Let us first ask what happens to a single exponentially growing Fourier mode for a system with a stationary instability, for example Eq. (2.15) or (2.22). First, note that the exponentially growing solution $\mathbf{u}(\mathbf{x}, t)$ with wave vector $\mathbf{q}$ can be written as[5]

$$\mathbf{u}(\mathbf{x}, t) = A(t) e^{i\mathbf{q} \cdot \mathbf{x}_\perp} \mathbf{u_q}(\mathbf{x}_\parallel) + \text{c.c.} \tag{4.1}$$

The time-dependent amplitude of the perturbation $A(t)$ satisfies the equation

$$d_t A = \sigma A. \tag{4.2}$$

Equation (4.2) results from the linearization. We can think of this approximation as the first term in an expansion of the dynamics in the size of the perturbation, now represented by A. Thus, we can learn useful things by continuing the expansion in Eq. (4.2) to higher order in A.

In Eq. (4.1), *it is useful to take A to be a complex number* $|A|e^{i\Phi}$. The magnitude $|A|$ of A gives us the size of the perturbation. What does the phase Φ of A tell

---

[5] The notation "c.c." appearing after an addition sign means "complex conjugate of the preceding expression." For Eq. (4.1), "c.c." therefore stands for $A^* e^{-i\mathbf{q} \cdot \mathbf{x}_\perp} \mathbf{u_q^*}(\mathbf{x}_\parallel)$. The symbols A^* and $\mathbf{u_q^*}$ denote the complex conjugates of A and $\mathbf{u_q}$ respectively.

4.1 Nonlinear saturation

us? By rewriting the first two factors in Eq. (4.1) as

$$Ae^{i\mathbf{q}\cdot\mathbf{x}_\perp} = |A|e^{i\mathbf{q}\cdot(\mathbf{x}_\perp + \hat{\mathbf{q}}\Phi/q)}, \qquad (4.3)$$

(dropping the time dependence for now and with $\hat{\mathbf{q}}$ a unit vector in the direction of the wave vector), we see that a change $\Phi \to \Phi + \delta\Phi$ in the phase by an amount $\delta\Phi$ is the same as a displacement of the solution through a distance $-\delta\Phi/q$ along the direction of the wave vector $\hat{\mathbf{q}}$. The phase tells us the position of the pattern! Note that a displacement $\delta\mathbf{x}$ perpendicular to the wave vector $\mathbf{q}$ (along the stripes) does not lead to any change in the structure.

What about the dynamics of $|A|$ and Φ? Remember that for a stationary instability the growth rate σ is real. Equation (4.2) can therefore be rewritten as the two equations

$$d_t|A| = \sigma|A| \quad \text{and} \quad d_t\Phi = 0. \qquad (4.4)$$

(To get these equations, substitute $A = |A|e^{i\Phi}$ into Eq. (4.2), multiply both sides by $e^{-i\Phi}$ and take real parts of both sides to get the first equation, imaginary parts of both sides to get the second equation.) The dynamics of the magnitude of A capture the exponential growth or decay given by σ. The phase does not change with time which is not surprising since we do not expect any time-dependent translation for a stationary instability.

We can now proceed by expanding the evolution equation for A to higher order in the magnitude of A. There are two ways of doing this. One is to substitute Eq. (4.1) into the evolution equation for $\mathbf{u}(\mathbf{x}, t)$ and use a perturbation technique to derive the evolution equation for A to some order in the small quantity $|A|$. This approach depends on taking a specific physical system or model, and Exercise 4.5 gives you the chance to work through the details for the relatively simple case of the Swift–Hohenberg equation.

A second approach is to use casual arguments based on smoothness assumptions and symmetry. This is the approach we take in this chapter since it can be completed (up to some unknown coefficients) without knowing the details of a specific system and the resulting evolution equation often provides a fruitful starting point. The symmetry of the system leads to constraints on the form of the complex amplitude's evolution equation. In particular, the translational invariance of the system leads to the conclusion that the evolution equation for A must be invariant under a constant change in the phase of A. This means that if we define $\bar{A} = Ae^{i\Delta}$ with Δ any real-valued constant, then the evolution equation for $\bar{A}$ must be the same as that for A. This is true since the phase change just shifts the pattern and, by the assumed translational invariance of the system, we know that a shifted pattern must follow the same dynamics as the original. That *the phase of the complex amplitude captures symmetry aspects of the pattern* is an extremely important insight for analyzing the dynamics of patterns. We will take advantage of this insight numerous times later in the book.

Expanding the equation of motion for A to higher order now gives

$$d_t A = \sigma A - \gamma A^2 A^* + \cdots, \qquad (4.5)$$

where γ is a constant and the dots ... denote higher-order terms that are negligible provided that the magnitude $|A|$ is small enough. Recognizing that $AA^* = |A|^2$, we obtain a more commonly written form

$$d_t A = \sigma A - \gamma |A|^2 A + \cdots. \qquad (4.6)$$

In developing the expansion Eq. (4.5), we use the rule that each term can contain non-negative integer powers of A and A^*.[6] After writing down the possible terms, we can choose to simplify the expression via the replacement $AA^* \to |A|^2$ but we must first develop the expansion using this rule.

Note that there are no quadratic terms in Eq. (4.6) because none of the terms AA, AA^*, or A^*A^* satisfies the requirement of translational invariance, that the equation remain unchanged by the substitution $A \to Ae^{i\Delta}$. The term $|A|A$ is not allowed since it is not a product of integer powers of A and A^*. The cubic term in Eq. (4.6) is thus the unique leading order correction to the linear approximation.

We will usually be interested in this equation near threshold where the growth rate approaches zero, $\sigma \to 0$. It is then useful to expand σ in the control parameter p about the critical value $p = p_c$. We introduce the dimensionless reduced control parameter

$$\varepsilon = \frac{p - p_c}{p_c}, \qquad (4.7)$$

which is zero ($\varepsilon = 0$) at the onset of linear instability $p = p_c$, and is small near onset when p is close to the critical value p_c. We next expand σ about $\varepsilon = 0$ where by definition $\sigma = 0$ for the onset of a stationary instability. Since σ has the dimensions of inverse time, we write this expansion as

$$\sigma = \tau_0^{-1} \varepsilon + O(\varepsilon^2), \qquad (4.8)$$

with τ_0 a constant with the dimensions of time. (Note that ε and τ_0 are the same parameters that were introduced in Section 2.5.) Equation (4.8) implies that the time constant τ_0 can be explicitly calculated from the dependence of the growth rate $\sigma(p)$ on the parameter p via the expression

$$\tau_0^{-1} = \left.\frac{d\sigma}{d\varepsilon}\right|_{\varepsilon=0} = p_c \left.\frac{d\sigma}{dp}\right|_{p=p_c}. \qquad (4.9)$$

---

[6] This is because the nonlinear terms are generated from powers of **u** and its derivatives, which from Eq. (4.1) will lead to this type of expression.

4.1 Nonlinear saturation

Assuming that the instability occurs for increasing p, the constant τ_0 is positive. Equation (4.6) now becomes

$$\tau_0 \, d_t A = \varepsilon A - g|A|^2 A + \cdots, \tag{4.10}$$

with $g = \gamma \tau_0$.

If the growth rate σ is real-valued near onset – the case for a stationary instability – we can argue from symmetry that, in almost all cases, the coefficient g of the cubic nonlinearity will be real-valued. (The minus sign in Eq. (4.10) is included by convention, and g may be positive or negative depending on the details of the system.) This is because we expect a stationary instability only if the physical system is unchanged under the inversion $\mathbf{x}_\perp \to -\mathbf{x}_\perp$. Without this symmetry, a spatially periodic state is likely to propagate, either to the right or to the left. We might still be able to set the propagation speed at threshold to zero by carefully tuning a second control parameter (we varied one parameter to tune the system to the threshold of instability), but this is a special case that would almost never be found experimentally in the most common situation that an experimentalist varies just one parameter at a time. (This special case is a codimension-two bifurcation in the language of Section 2.6.) The reflection symmetry of the physical system translates into the requirement that the evolution equation for A must be unchanged by the substitution $A \to A^*$, and this leads to the constraint that g must be real.

It is worth understanding the details of this argument to gain experience with manipulating amplitudes under symmetry transformations. The amplitude is introduced as

$$\mathbf{u}(\mathbf{x}, t) = \left(A(t) e^{i\mathbf{q} \cdot \mathbf{x}_\perp} + A^*(t) e^{-i\mathbf{q} \cdot \mathbf{x}_\perp} \right) \mathbf{u}_\mathbf{q}(\mathbf{x}_\|), \tag{4.11}$$

where we have explicitly included the complex conjugate to make $\mathbf{u}$ a real field, and we have assumed that $\mathbf{u}_\mathbf{q}(\mathbf{x}_\|)$ can be chosen to be real. Now suppose we look at the same physical system, but with a new horizontal coordinate defined by $\bar{\mathbf{x}}_\perp = -\mathbf{x}_\perp$. The dynamical equations with all variables expressed as functions of $\bar{\mathbf{x}}$ will be unchanged, as follows from the reflection symmetry. We suppose that $\mathbf{u}$ is comprised of variables that do not change under this operation, such as a position rather than an x-velocity. (The argument can be generalized to other cases if needed.) Then we have

$$\mathbf{u}(\bar{\mathbf{x}}, t) = (A(t) e^{-i\mathbf{q} \cdot \bar{\mathbf{x}}_\perp} + A^*(t) e^{i\mathbf{q} \cdot \bar{\mathbf{x}}_\perp}) \mathbf{u}_\mathbf{q}(\mathbf{x}_\|), \tag{4.12}$$

$$= (\bar{A}(t) e^{i\mathbf{q} \cdot \bar{\mathbf{x}}_\perp} + \bar{A}^*(t) e^{-i\mathbf{q} \cdot \bar{\mathbf{x}}_\perp}) \mathbf{u}_\mathbf{q}(\mathbf{x}_\|). \tag{4.13}$$

A new amplitude $\bar{A} = A^*$ has been introduced to get the same equation for the introduction of the amplitude in the new coordinate system. The physics must "look the same" with respect to the new coordinates and amplitude. Thus the equation of motion for $\bar{A}$ must be the same as for A or, more loosely, the equation of motion for A must not change under the substitution $A \to A^*$.

Our informal discussion of the saturation of the growing mode near a type I-s bifurcation has introduced the concept of the complex amplitude A. We will see in Chapter 6 that this quantity can be generalized to include a slowly varying spatial dependence which then provides a powerful tool to study many aspects of pattern formation near onset. The same type of arguments we have used for the stationary instability can be made for oscillatory instabilities. In fact, the same equation for the complex amplitude, Eq. (4.5), is obtained but now g (and of course σ) are complex numbers. This is the topic of Exercises 4.6 and 4.7.

4.1.2 Bifurcation theory

We can now use Eq. (4.10) to begin the investigation of the effects of nonlinearity. To study these issues, we ignore the higher-order terms in Eq. (4.10) and use the resulting equation to investigate the existence and stability of the solutions saturated by the nonlinearity. We argue that it makes sense to retain the cubic term, but not higher-order terms, if the control parameters of the system are near the ones giving linear instability so that the growth rate σ is small (or has a small real part in the oscillatory case). Then the leading linear term on the right-hand side of Eq. (4.10) with a small coefficient might be comparable with the next order cubic term, whereas further terms in the expansion remain negligible. Thus an expansion in the complex amplitude is useful near the threshold of pattern formation.

Since the parameters τ_0, ε, and g in Eq. (4.10) are real, the evolution equation for the phase of A is still $d_t \Phi = 0$ (see Eq. (4.4) above) so the phase is constant. For a translationally invariant system, the value of the constant phase cannot matter and so we can set $\Phi = 0$, in which case the amplitude A becomes a real-valued variable. Equation (4.10) then becomes the slightly simpler evolution equation for a real amplitude $A(t)$

$$\tau_0 \, d_t A = \varepsilon A - g A^3 + \cdots. \tag{4.14}$$

We emphasize, though, that generally one cannot assume the amplitude to be real and that the phase will be time dependent.

We can now ask about the existence and stability of solutions for Eq. (4.14). We seek solutions where the growth has ceased[7] $d_t A = 0$. For any given value of the control parameter ε, there can be two symmetrically related nonzero solutions

---

[7] It is not too hard to show that the only bounded nontransient solutions of any one-variable ode like Eq. (4.14) are fixed points so there is no need to consider oscillatory or chaotic solutions.

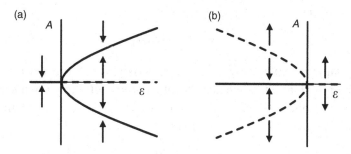

Fig. 4.1 Existence and stability of stationary nonlinear saturated amplitude solutions $A(t)$ near onset according to the amplitude equation Eq. (4.10). Panel (a) plots the possible solutions A_0 given by Eq. (4.15) as a function of the reduced parameter $\varepsilon = (p - p_c)/p_c$ for the forward-bifurcation case of $g > 0$. Panel (b) shows the case for a backward bifurcation, with $g < 0$. The solid and dashed lines represent linearly stable and unstable solutions respectively. The arrows show the evolution of A according to Eq. (4.14) and give useful insight into the dynamics.

$A = \pm A_0$ that satisfy

$$A_0^2 = \varepsilon/g. \tag{4.15}$$

Remember that $\varepsilon > 0$ signifies the instability of the uniform state. Since A_0^2 is necessarily positive, Eq. (4.15) shows us that the nonzero saturated solution exists on the unstable side of the bifurcation for $g > 0$, but on the stable side for $g < 0$. We can now understand the motivation for the terms "forward bifurcation" and "backward bifurcation" for these two cases, as shown in Fig. 4.1. For g positive, the new solution grows in the forward direction of $p > p_c$ (taking the convention that the instability occurs as the control parameter is increased), whereas for g negative, the new solution develops in the backward direction. The magnitude of the nonlinear solution increases as

$$|A_0| = |g|^{-1/2} \sqrt{|\varepsilon|}, \tag{4.16}$$

with increasing $|\varepsilon|$ on the positive-ε side for the forward bifurcation, and on the negative-ε side for the backward bifurcation. This square-root dependence on the control parameter near threshold will appear many times throughout this book, and is a behavior that has been verified in many experiments and simulations near onset. The shape of the solution curves in Fig. 4.1 leads to the nomenclature pitchfork bifurcation for this type of instability.

The discussion so far has concerned the existence of solutions. But we can go further and deduce a useful fact: the new branch of solutions is necessarily unstable for the backward bifurcation but may be stable for the forward bifurcation. This is easy to see by studying a particular perturbation

$$A(t) = A_0(1 + \delta(t)), \qquad (4.17)$$

of the steady saturated solution A_0 and then linearizing in the tiny perturbation $\delta(t)$. (When working with amplitude equations, it is often useful to write a perturbation in terms of a multiplicative factor of the base state, here $A_0\delta(t)$.) Substituting into the evolution equation Eq. (4.14), we find that

$$\tau_0 d_t \delta = \left(\varepsilon - 3gA_0^2\right)\delta = -2gA_0^2\delta, \qquad (4.18)$$

where we used Eq. (4.15) to obtain the second equality. The perturbation δ therefore decays exponentially for $g > 0$, but grows exponentially for $g < 0$. The nonlinear saturated solutions for $g < 0$ are therefore themselves linearly unstable (see Fig. 4.1). This implies that the corresponding solutions Eq. (4.11) of the original evolution equation are linearly unstable and so will not be seen in most experiments.

The case of $g > 0$ is more complicated. Our analysis has only shown stability with respect to a particular class of perturbations Eq. (4.17). There may be instability toward other perturbations such as the growth of other modes at different wave vectors. We will return to this question in Section 4.2.1 but note for now that the new nonlinear solution may be stable for a forward bifurcation. Of course, a single demonstration of instability is sufficient to demonstrate the instability for the backward bifurcation.

Our analysis as summarized in Fig. 4.1 is a simple example of bifurcation theory. The theory sets up a formalism for predicting the properties of the nonlinear state that develops from linear instabilities, and is also a useful preliminary step in calculating the quantitative properties of the nonlinear states. The connection between symmetries and the algebraic form of simple evolution equations near the instability or bifurcation point ("near" in the control parameter and in the small amplitude of the perturbation), and the connection between the nature of the bifurcation (forward or backward) and the stability of the new solutions, illustrate the type of results that are derived in bifurcation theory.

There is an important difference between the bifurcation of a spatially uniform state and the bifurcation of a fixed point of some arbitrary set of odes, as discussed in most introductory texts on nonlinear dynamics: the bifurcation to stripes in a system with translational symmetry is always of the pitchfork type as shown in Fig. 4.1. The analog to the transcritical bifurcation in simple bifurcation theory (see Appendix 1) does not exist in this case. Such a case occurs when there is a quadratic nonlinearity in the equation for the growth of the mode as in Eq. (A1.2) of Appendix 1, which, as we have seen, is forbidden for a stationary instability at nonzero wave number in a translationally invariant system. We will see in Section 7.4 that the transition to a hexagonal state may be transcritical.

4.1.3 Nonlinear stripe state of the Swift–Hohenberg equation

The discussion in the previous section of the saturation of exponential growth by nonlinearity was rather general and so abstract. Here, we make the discussion more concrete by constructing explicitly a saturated nonlinear periodic solution to lowest order near onset for the rather simple Swift–Hohenberg evolution equation. We introduced this equation in Section 2.2 as a simple model of a pattern-forming system whose uniform state undergoes a type-I-s instability.

In one spatial dimension, the Swift–Hohenberg equation has the form

$$\partial_t u = ru - \left(\partial_x^2 + 1\right)^2 u - u^3. \tag{4.19}$$

We found in Chapter 2 that, as the control parameter r is increased from negative values, the uniform state $u = 0$ becomes linearly unstable at $r = 0$ to an exponentially growing mode that we can write as[8]

$$u \propto e^{rt} \cos x. \tag{4.20}$$

with critical wave number $q_c = 1$. We wish to understand the nonlinear saturation of this state for small r. Note that we have chosen a mode whose wave number is the critical wave number. In Exercise 4.11, you may work out the corresponding analysis for a general wave number q, with the $\cos x$ term in Eq. (4.20) replaced with $\cos(qx)$.

A natural guess for the steady saturated solution $u(x)$ is to look for a solution with the same periodicity as the perturbation,

$$u = a_1 \cos x, \tag{4.21}$$

with the coefficient a_1 to be determined. To see whether this ansatz is a solution, we substitute it into Eq. (4.19) with $\partial_t u = 0$ to get

$$0 = \left(ra_1 - \frac{3}{4}a_1^3\right)\cos x - \left(\frac{1}{4}a_1^3\right)\cos(3x). \tag{4.22}$$

Here we have rewritten $\cos^3 x$ (generated by the u^3 term) as a sum of $\cos x$ and $\cos(3x)$ terms. Since the Fourier modes $\cos(nx)$ are linearly independent, the coefficient of each mode on the right-hand side must be zero. This gives two equations. From the coefficient of $\cos x$, we find the first equation

$$ra_1 - \frac{3}{4}a_1^3 = 0. \tag{4.23}$$

---

[8] We could use the complex notation $u \propto e^{rt}(e^{ix} + \text{c.c.})$ of Eq. (4.14) but for calculating numbers, rather than displaying symmetry aspects, the real notation is simpler.

This equation has three solutions, the solution $a_1 = 0$, (which is just the unstable uniform state, and the two solutions

$$a_1 = \pm\sqrt{4r/3}, \qquad (4.24)$$

which seem to fix the amplitude of the nonlinear state. However, the second equation obtained by setting the coefficient of $\cos 3x$ to zero gives a contradiction that a_1 must be zero:

$$-\frac{1}{4}a_1^3 = 0 \implies a_1 = 0. \qquad (4.25)$$

This inconsistency shows that the original guess Eq. (4.21) is incomplete: the $\cos x$ generates a harmonic $\cos(3x)$ via the cubic nonlinearity and there is nothing in Eq. (4.21) to balance that term. The inconsistency can be removed by adding a small amount of $\cos 3x$ to the original ansatz so that now

$$u \simeq a_1 \cos x + a_3 \cos(3x). \qquad (4.26)$$

Substituting this new ansatz into Eq. (4.19) with $\partial_t u = 0$ and setting the coefficient of $\cos 3x$ to zero now gives instead of Eq. (4.25) the equation

$$(r - 64)a_3 - \frac{1}{4}a_1^3 = 0. \qquad (4.27)$$

Since Eq. (4.24) implies that $a_1 = O(r^{1/2})$ as $r \to 0$, we have $a_3 = O(r^{3/2})$, a factor of r smaller than a_1 close to onset. Furthermore, adding this small term $a_3 \cos(3x)$ to the ansatz for u causes only a small $O(r)$ correction to the coefficient a_1 that we calculated with the assumption that $a_3 = 0$. To see this, you can verify that, after substituting Eq. (4.26) into Eq. (4.19) with $\partial_t u = 0$ and reducing products of trigonometric terms to sums of harmonics, the first equation Eq. (4.23) becomes

$$ra_1 - \frac{3}{4}a_1^3 = \frac{3}{4}a_1^2 a_3 + \frac{3}{2}a_1 a_3^2. \qquad (4.28)$$

The left side is as before and its terms are of order $r^{3/2}$ near onset. Close enough to onset, the two terms on the right-hand side are at least a factor of r smaller than the left-hand side and so can be ignored, again leading to Eq. (4.23). For example, for $a_3 = O(r^{3/2})$ the term $a_1^2 a_3 = O(r^{5/2})$ is a factor of r smaller than the left-hand-side terms, and the term $a_1 a_3^2 = O(r^{7/2})$ is smaller still. Including the right side in Eq. (4.28) causes the value of a_1 to change by an amount of order r compared to the value when the right-hand side is set to zero.

We have resolved the inconsistency arising from considering the $\cos(3x)$ coefficient. But now when we substitute the new ansatz Eq. (4.26) into the stationary Swift–Hohenberg equation, the nonlinearity produces the additional harmonics $\cos(5x)$, $\cos(7x)$, and $\cos(9x)$ and we will need to add small portions of these terms to Eq. (4.26) for consistency. It should now be apparent that, in fact, all odd modes must be included in the ansatz

$$u = \sum_{n \text{ odd}} a_n \cos(nx), \qquad (4.29)$$

and that the nth coefficient a_n will go to zero as $r^{n/2}$ as r approaches zero. For sufficiently small r, Eq. (4.24) is the correct expression for a_1 and the coefficients a_n can be calculated iteratively.

The above discussion has effectively proven the existence of stationary nonlinear stripe solutions of the Swift–Hohenberg equation close enough to onset, at least for stripes with the critical wave number $q_c = 1$. The solution has the form

$$u = \pm\sqrt{4/3}\, r^{1/2} \cos x + O\!\left(r^{3/2}\right) \cos(3x), \qquad (4.30)$$

and consists of a dominant term that has the same shape as the linear mode, that can take either sign, and that grows in magnitude from threshold as $r^{1/2}$. These results are consistent with a pitchfork bifurcation. In addition, there are spatial harmonic corrections, with amplitudes that grow with r progressively more slowly with increasing order of the harmonic.

Further from onset, the explicit form of the spatially periodic nonlinear steady states must usually be calculated numerically. A conceptually simple numerical strategy is the Galerkin method, which represents each unknown stationary field as a finite linear combination of basis functions with unknown coefficients. (Equation (4.26) is an example where only two basis functions of Fourier modes are used.) Substituting the linear combinations into the stationary evolution equations and then expanding the resulting expressions themselves into a finite linear sum of basis functions leads to a set of nonlinear equations for the coefficients. A numerical method such as Newton's method (see Section 12.4) can then be used to find approximate values for the coefficients from the set of nonlinear equations. Fourier modes with the spatial periodicity of the unstable modes are appropriate basis functions for the transverse spatial dependence $\mathbf{x}_\perp$. For the spatial dependence in the confined direction(s) $\mathbf{x}_\parallel$, any convenient complete set of functions may be used.

4.2 Stability balloons

4.2.1 General discussion

To determine the physical relevance of the steady, nonlinear, spatially periodic states formed above the instability of the spatially uniform state, we must in turn test their stability against all small perturbations, not just those of the same spatial periodicity as we did in Section 4.1.2. Unlike that analysis in Chapter 2, the base state $\mathbf{u}_b(\mathbf{x}_\perp, \mathbf{x}_\|)$ about which we linearize to study stability is no longer constant in the extended coordinates $\mathbf{x}_\perp$ but spatially periodic in these coordinates. (As before, the dependence of the base state on the confined variables $\mathbf{x}_\|$ usually does not have any simple form.) If you review the discussion leading up to Eq. (2.8) in Chapter 2, you will see that, in general, the linear evolution equation for an infinitesimal perturbation about this base state will no longer have constant coefficients, but coefficients that are functions of the base state $\mathbf{u}_b$ and so are themselves spatially periodic in the vector $\mathbf{x}_\perp$.

How to solve linear evolution equations with spatially periodic coefficients was first worked out by Felix Bloch in the early days of quantum mechanics. (In 1928, Bloch solved the Schrödinger equation, which is also a linear evolution equation, to understand the behavior of electrons in a spatially periodic crystalline lattice of atoms.) Bloch's analysis tells us that solutions to the linear evolution equation for perturbations must have the form

$$\delta\mathbf{u} = e^{\sigma(\mathbf{Q},\mathbf{q})t} e^{i\mathbf{Q}\cdot\mathbf{x}_\perp} \mathbf{u}_\mathbf{Q}(\mathbf{x}_\perp, \mathbf{x}_\|), \tag{4.31}$$

which generalizes the simple sinusoidal solution for the uniform base state Eq. (2.22) and is called a Bloch state.[9] The Bloch states are parameterized by a wave vector $\mathbf{Q}$, which is analogous to the wave vector $\mathbf{q}$ of Eq. (2.22). The function $\mathbf{u}_\mathbf{Q}(\mathbf{x}_\perp, \mathbf{x}_\|)$ depends on $\mathbf{Q}$ and has the same periodicity with respect to the extended coordinates $\mathbf{x}_\perp$ as the base state $\mathbf{u}_b$. For example, if the base state is a periodic stripe state with wave vector $\mathbf{q}$, then the function $\mathbf{u}_\mathbf{Q}$ will be periodic with wave vector $\mathbf{q}$. The growth rate will depend on the two vectors $\mathbf{Q}$ and $\mathbf{q}$, so we can write $\sigma = \sigma(\mathbf{Q}, \mathbf{q})$.

Because the linear stability of stripe states have been more thoroughly studied theoretically and experimentally than the linear stability of other spatially periodic patterns, we will restrict our discussion in the rest of this section to the linear stability of stripe patterns. It is straightforward but more tedious to work out similar details for nonlinear states that are periodic in two or more extended variables such as the two-dimensional lattice states described in the next section. In the following, we will also assume for simplicity that the two extended coordinates are the x and y coordinates and that the single confined coordinate is z so $\mathbf{x}_\perp = (x, y)$ and $\mathbf{x}_\| = z$. For a stripe base state with wave vector $\mathbf{q}$, we will align our xy coordinates so that x points in the direction of $\mathbf{q}$, which means that y points along the length of the stripes. With these assumptions, the vector $\mathbf{Q} = (Q_x, Q_y)$ is a two-dimensional wave vector, and the function $\mathbf{u}_\mathbf{Q}(x, y, z)$ is periodic in the variable x with period $2\pi/q$.

For a given stripe base state with wave vector $\mathbf{q} = (q, 0)$ and for given system parameters, we want to determine if the real part of the growth rate $\text{Re}[\sigma(\mathbf{Q}, \mathbf{q})]$

---

[9] A similar expression arises in the Floquet theory for the stability of time-periodic states of some system of odes.

ever becomes positive (implying linear instability) as the wave vector $\mathbf{Q}$ is varied over all possible values. For a stripe base state, it turns out conveniently that we have to investigate only a restricted range

$$-\frac{q}{2} < Q_x \leq \frac{q}{2}, \tag{4.32}$$

for the x-component of $\mathbf{Q}$. This follows from the observation that, since $u_\mathbf{Q}(x, y, z)$ is periodic in x with wave number q, the function

$$e^{imqx} u_\mathbf{Q}, \tag{4.33}$$

is also periodic in x with wave number q for any integer m. So a Bloch state Eq. (4.31) can be written as another Bloch state

$$e^{\sigma t} e^{i((Q_x - mq)x + Q_y y)} \left[e^{imqx} u_\mathbf{Q} \right], \tag{4.34}$$

with the x component of $\mathbf{Q}$ shifted by an arbitrary integer multiple of q. For example, a Bloch-state perturbation with wave vector $\mathbf{Q} = (2q/3, Q_y)$ outside the range Eq. (4.32) can be expressed as some other Bloch state with wave vector $\mathbf{Q} = (2q/3 - q, Q_y) = (-q/3, Q_y)$ inside this range, provided we redefine the function $u_\mathbf{Q}$ to be the function $e^{iqx} u_\mathbf{Q}$.[10]

Equation (4.31) allows us to identify the following classes of instabilities of the stripe state according to the values of the wave vector $\mathbf{Q}$:

(i) $\mathbf{Q}$ may be zero, which means the perturbation does not change the basic spatial periodicity of the pattern;
(ii) Q_x may equal $q/2$ ($\mathbf{Q}$ is "on the zone boundary" in the language of solid state physics) so that the spatial period in the x-direction is doubled at the instability. The y component Q_y may be zero (no structure in the y-direction) or have a general value;
(iii) The instability may occur at long wavelength, with $\mathbf{Q} \to 0$;
(iv) $\mathbf{Q}$ may take on a general value, with Q_x incommensurate with the base wave number.

In addition, there may be different instability types characterized by different behavior under discrete symmetries. For example, if the dynamical equations and base state are unchanged by the parity transformation $x \to -x$, then perturbations with $Q_x = 0$ or $Q_x = q/2$ can be classified according to the parity of $u_\mathbf{Q}$ as an even or odd function under this transformation. (A general Q_x eliminates the $x \to -x$ symmetry, and $u_\mathbf{Q}$ does not reflect the parity symmetry for these values.) Finally the instability may be stationary or oscillatory.

Clearly the stability analysis about a spatially periodic base state allows for a much larger range of possibilities than the analysis about the uniform state. The

---

[10] In signal processing, a similar restriction of the range of frequencies in the power spectrum analysis is known as the Nyquist range, and in the study of waves in periodic structures in solid state physics the range is known as the first Brillouin zone [7].

situation for the stability of lattice states such as hexagonal or square patterns is even more complicated so that a careful analysis of the different possibilities using the symmetry of the system and the tools of group theory is a useful first step.

Conceptually, the steps for mapping out the set of stable wave numbers are as follows. After having identified the possible types of instabilities, for each control parameter p of interest, and for each wave number q for which a nonlinear saturated stripe state is known to exist for the given p value (these q lie inside the neutral stability curve Eq. (2.24)), we find the largest growth rate $\text{Re}\,[\sigma(\mathbf{Q};q)]$ over all possible Bloch wave vectors $\mathbf{Q}$ and over the different types of instabilities at each $\mathbf{Q}$. We can then determine the range of q

$$q_S^-(p) < q < q_S^+(p), \qquad (4.35)$$

for which the base state is linearly stable, $\max_\mathbf{Q} \text{Re}[\sigma(\mathbf{Q};q)] < 0$. (The subscript S indicates an instability that bounds stable wave numbers.) This range will be smaller than the range for which stationary periodic solutions exist, Eq. (2.24).[11] The region bounded by $q_S^\pm(p)$ as p varies is known as the stability balloon of the stripe states and is indicated by the shaded regions in Fig. 4.2.

The spatially periodic state becomes unstable to a particular perturbation at $q_S^\pm(p)$. As p varies, instabilities of different symmetries and geometrical form may bound the stability region. Some of the instabilities are long wavelength instabilities that appear in the limit of vanishing Bloch wave number $Q \to 0$. These derive from the nature of the broken symmetry and so are universal for a given type of pattern, as we will see in Section 9.1.1. For a stripe state, such instabilities are the Eckhaus instability, which corresponds to a long-wavelength longitudinal perturbation (Q_x small, $Q_y = 0$), and the zigzag instability, which is a long-wavelength transverse distortion ($Q_x = 0$, Q_y small). Whether these two instabilities are the ones that form the boundary of the stability balloon depends on details of the system and parameter values.

Although some details of the stability balloon can be determined analytically near onset, most often the stability balloon is determined numerically by extending the Galerkin approach. Since the function $\mathbf{u}_\mathbf{Q}$ in Eq. (4.31) has the same period as the spatially periodic state $\mathbf{u}_q$, it can be expanded in the same set of functions. The linearized evolution equations about $\mathbf{u}_q$ can then be written as linear coupled odes for the expansion amplitudes of $\mathbf{u}_\mathbf{Q}$. These can be readily solved for the growth rates $\sigma(\mathbf{Q})$ using standard numerical linear algebra packages. This rather complex procedure is illustrated in the following Etude, which calculates the transverse instability of the Swift–Hohenberg equation.

---

[11] In most cases studied, the stable wave numbers form a single connected domain, but in a few cases two or more disconnected ranges are found for some values of p.

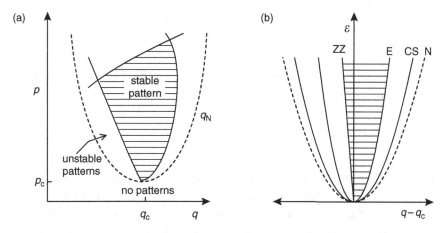

Fig. 4.2 (a) A representative stability balloon for a stripe state with wave number q. The vertical axis is a system parameter p such that, as p is increased, the uniform state undergoes a type-I-s instability. The shaded region denotes stable nonlinear stationary stripe states bounded by instabilities labeled by the curves $q_S^\pm(p)$. As the parameter p is increased for fixed q, or as the wave number q is increased or decreased for fixed p, the stripe states become unstable to different instabilities. (b) Near threshold (small reduced parameter ε), the instability boundaries have a universal form, with the Eckhaus (E) and cross-stripe (CS) instabilities having a parabolic dependence on q, and the zigzag (ZZ) instability having a linear dependence on $q - q_c$. Depending on the system, either the Eckhaus or the cross-stripe instability may bound the stable region for $q > q_c$.

Etude 4.1 Zigzag instability in the Swift–Hohenberg model

For the Swift–Hohenberg equation in two spatial dimensions (which we assume have infinite extent)

$$\partial_t u = ru - \left(\partial_x^2 + \partial_y^2 + 1\right)^2 u - u^3, \quad (4.36)$$

we determine the linear stability of stationary nonlinear stripe solutions $u_q(x)$ to transverse perturbations. We obtain analytical results by going so close to onset that the stripe state and other fields can be approximated by a single Fourier mode.

In Exercise 4.11, you show that the coefficient a_1 in Eq. (4.24) for a saturated nonlinear stripe state $u_q \simeq a_q \cos(qx)$ with general wave vector $q\hat{\mathbf{x}}$ satisfies

$$a_q^2 = \frac{4}{3}\left(r - (q^2 - 1)^2\right). \quad (4.37)$$

Within this single-mode approximation, stripes exist whenever the right-hand side is non-negative, which gives the neutral stability band

$$\sqrt{1 - \sqrt{r}} \le q \le \sqrt{1 + \sqrt{r}}. \quad (4.38)$$

For r sufficiently small, this can be approximated as

$$1 - \sqrt{r}/2 \leq q \leq 1 + \sqrt{r}/2. \tag{4.39}$$

We now expand the Bloch form of the perturbation Eq. (4.31) using a Fourier expansion of the periodic function $u_\mathbf{Q}$.

$$\delta u \simeq e^{\sigma t} e^{i\mathbf{Q}\cdot\mathbf{x}} \sum_{n=1}^{\infty} (c_+ e^{inqx} + c_- e^{-inqx}). \tag{4.40}$$

A transverse perturbation of the base state varies only in the y direction (perpendicular to the stripes) so that the perturbation vector has the form $\mathbf{Q} = Q\hat{\mathbf{y}}$. In this case, it is simplest to use the parity symmetry of the base state $a_q \cos(qx)$ to argue that we may use two separate Bloch forms, one for "even" transverse perturbations for which the periodic function $u_\mathbf{Q}(x)$ is an even function of x

$$\delta u \simeq e^{\sigma t} e^{iQy} \sum_{n=1}^{\infty} c_n \cos(nqx), \tag{4.41}$$

and a similar form for odd perturbations

$$\delta u \simeq e^{\sigma t} e^{iQy} \sum_{n=1}^{\infty} s_n \sin(nqx). \tag{4.42}$$

In these equations, $\sigma(Q, q)$ gives the growth rate. We have used the reflection (even) symmetry of the base state $a_q \cos(qx)$ to argue that the perturbation must be even or odd in x. To simplify the calculation further, we will truncate the Galerkin expansion of $u_\mathbf{Q}$ to a single mode ($c_n = s_n = 0$ for $n > 1$), which is consistent with the calculation of the base solution. The Bloch forms now become

$$\delta u = e^{\sigma t} e^{iQy} \cos(qx) \quad \text{and} \quad \delta u = e^{\sigma t} e^{iQy} \sin(qx), \tag{4.43}$$

respectively for the even and odd cases.

The perturbations δu satisfy the Swift–Hohenberg equation linearized about the base solution u_q

$$\partial_t \delta u = r \, \delta u - (\partial_x^2 + \partial_y^2 + 1)^2 \, \delta u - 3 u_q^2 \, \delta u. \tag{4.44}$$

Substituting Eq. (4.43) and $u_q = a_q \cos(qx)$ with Eq. (4.37) yields for the even perturbation

$$\sigma = r - (q^2 + Q^2 - 1)^2 - 3\left[r - (q^2 - 1)^2\right], \tag{4.45}$$

and for the odd perturbation

$$\sigma = r - (q^2 + Q^2 - 1)^2 - \left[r - (q^2 - 1)^2\right]. \tag{4.46}$$

Here we reduced the product of Fourier modes to a sum of Fourier modes via identities such as

$$\cos^2(qx)\cos(qx) = \tfrac{3}{4}\cos(qx) + \tfrac{1}{4}\cos(3qx), \qquad (4.47)$$

and dropped the higher harmonic terms $\cos(3qx)$ *and* $\sin(3qx)$ *terms to be consistent with the single-mode truncation scheme.*

The most interesting result is for the odd perturbation, since the r terms cancel to give

$$\sigma = 2(1-q^2)Q^2 - Q^4. \qquad (4.48)$$

If we interpret this expression as having a fixed stripe wave number q while varying the perturbation wave vector $\mathbf{Q}$ *to look for instability, we can think of Eq. (4.48) as a parabola* $\sigma = Q^2(c-Q^2)$ *in the quantity* Q^2 *with constant coefficient* $c = 2(1-q^2)$. *This parabola always has positive values (* $\sigma > 0$ *so instability) if* $c > 0$ *(which implies* $q < 1$*) and has no positive values (* $\sigma \leq 0$ *so stability) if* $c \leq 0$ *(which implies* $q \geq 1$*). We conclude that stripes with wave numbers* $q < 1$ *are always unstable to this odd transverse perturbation, while stripes with wave numbers* $q \geq 1$ *are stable, at least with respect to this specific perturbation. Equation (4.48) further predicts that the instability first occurs at long wavelengths (long compared to the stripe wavelength), corresponding to a small value of Q. Indeed, for q slightly less than unity,* $\sigma(Q)$ *is positive only for small Q. So as q decreases below 1 and passes into the unstable range, the instability develops as a small-Q long-wavelength instability. This is the zigzag instability. Figure 4.3(a) shows an experimental example of how the zigzag instability grows from a zigzag-unstable stripe state, although most of the evolution shown is already in a nonlinear regime.*

Close to threshold, the instability boundaries may be calculated using the amplitude equation formalism to be introduced in Chapter 6, and they take on a universal form as discussed there and in Chapter 7. As well as the Eckhaus and zigzag instabilities, a cross-stripe instability,[12] in which the stripes become unstable toward the growth of a set of stripes at a different orientation and wave number, may be important. The boundaries of the Eckhaus and cross-stripe instabilities vary as

$$q_S^{\pm}(p) - q_c \propto \varepsilon^{1/2}. \qquad (4.49)$$

The proportionality constant for the Eckhaus instability is such that the ratio of the width of the Eckhaus stable band to the existence band (given by q_N) is always $1/\sqrt{3}$ near threshold, independent of the details of the system. The proportionality constant for the cross-stripe instability depends on the specific system, and either the

---

[12] The instability was discussed first in the context of Rayleigh–Bénard, where it was called the cross-roll instability.

Eckhaus or the cross-stripe instability may form the boundary to the stable region for $q > q_c$. For a rotationally invariant system, the zigzag boundary varies linearly with ε for small ε, and always forms the boundary of the stable region on the small-q side near threshold.

Experiments and simulations have shown that stability balloons are much more widely useful than suggested by the derivation in terms of the stability region of spatially periodic states in a laterally infinite system. Often applying the stability criteria to a locally defined wave vector gives a good guide to the behavior, even in small geometries or for disordered states. The question of what stationary patterns might form can be crudely formulated in terms of what states may be generated from the range of wave vectors that fall within the stability balloon. The onset of time dependence can similarly often be associated with the local wave vector in some portion of the system passing outside of the stable band. The local wave vector distribution in the pattern may sometimes be predicted from other considerations, therefore providing a route to predicting the onset of time dependence in the system. For example, in geometries in which the number of stripes is constrained by the boundaries, the wave number will be approximately constant as the control parameter is varied, and a boundary of the stability balloon may be encountered along a vertical path in Fig. 4.2. On the other hand, if the system size is slowly varied under the same circumstances, the wave number changes at a fixed control parameter as the fixed number of stripes are compressed or stretched. In this case, the stability balloon is traversed horizontally. The instabilities encountered on a slowly growing domain may be important in biological systems such as young fish with stripes that grow bigger as they age. Thus the stability balloon is often a useful first guide to the range of behavior expected as system parameters are varied.

The stability balloon does have limitations that experimentalists and computational scientists should consider. One limitation is that the stability boundaries are calculated for an idealized infinite domain. A finite domain or a domain with certain boundary conditions might shift or suppress some of the linear instabilities; the onset and manifestation of long-wavelength instabilities can be especially sensitive to the size and shape of an experimental domain. Another limitation is that the balloon only takes into account the effects of tiny perturbations: instability to other states might occur for finite perturbations. An example is the spiral defect chaos state in Rayleigh–Bénard convection that we show in Fig. 1.15. This state is observed experimentally for parameter values that correspond to stable stationary stripe states according to the stability balloon. Indeed, numerical calculations of the convection evolution equations confirm the stability of the ideal stripe state in geometries consistent with this state (e.g. periodic boundary conditions), but show that, even in these idealized geometries, most initial conditions evolve into the spiral chaos state rather than a stripe state. Thus multistability is known to occur with periodic boundary conditions. With other boundary conditions, the multistability may disappear, and only the chaotic state may survive, or there may be multistability between the chaotic state and a different time-independent state that is consistent with the boundary conditions and may be spatially disordered.

4.2.2 Busse balloon for Rayleigh–Bénard convection

The stability balloon for stripe states in Rayleigh–Bénard convection – called ideal roll states by fluid dynamicists – was worked out by Fritz Busse and coworkers via numerical Galerkin calculations over a number of years starting around 1965 and represents the most detailed theoretical study of the existence and stability of stripe states of any physical system. These researchers also performed extensive experiments to verify their many predictions. This work pioneered what we might label the "second phase" of pattern-formation research, in which the novel implications of nonlinearity in the phenomenon were pursued. (We would characterize the linear stability studies of Taylor, Rayleigh, Turing, Chandrasekhar, and others as the first stage.) In honor of Busse and his many contributions, the stability balloon in the context of convection is called the "Busse balloon."

Many different types of stripe instabilities are encountered in convection and these are classified in several ways. One is through the properties of the Bloch wave vector **Q** in Eq. (4.31). For example, if instability first occurs when **Q** has a small magnitude, one has a so-called long-wavelength instability such that perturbations first grow on length scales large compared to the roll wavelength. Instability may also occur first for finite-size **Q** – with a wave number Q comparable to the roll wave number q – and these are called finite-wavelength instabilities (also short-wavelength instabilities). As we saw in the previous section, the linear instabilities can also be classified by the orientation of **Q** with respect to **q**: longitudinal if the vectors are parallel, transverse if the vector are orthogonal, or *skew* if the vectors are neither parallel nor perpendicular. One can distinguish the instabilities further by whether the growth rate σ at onset is real or complex, and in addition through discrete parity and inversion properties of the function $\mathbf{u_Q}$ in Eq. (4.31).

Because of these different length scales and symmetries, the different instabilities have distinct geometric shapes and their visual appearance has led to a rich nomenclature. Figure 4.3 shows two experimental examples in which an initial stripe state was created[13] in a large-aspect-ratio convection experiment such that the initial

---

[13] We recommend that the reader look through the original paper by Busse and Whitehead for the experimental details. Briefly, a physical template with the desired stripe wavelength was created by attaching parallel strips of opaque black tape to a large sheet of glass. While the Rayleigh number was below critical, a strong light source was shone through the template which caused a small local heating of the fluid in the form of thermal stripes. The Rayleigh number was then increased above the critical value, at which point the thermal imprinting strongly favored convection rolls to form at the desired wavelength.

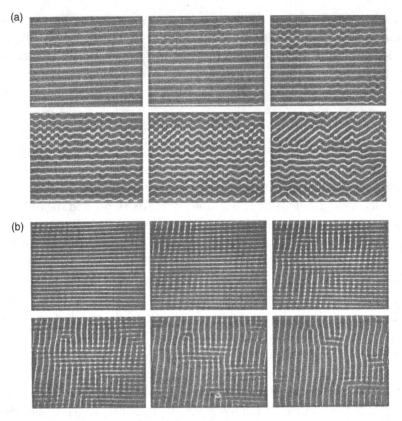

Fig. 4.3 Shadowgraph images of evolving convection stripe states whose initial wave numbers lie outside the stability balloon for a silicon oil of Prandtl number $\sigma \approx 1.0 \times 10^2$ in a cell of aspect ratio $\Gamma \approx 160$. (a) For Rayleigh number $R = 3600 \approx 2R_c$ and initial wave number $q \approx 2.8$, the stripes are unstable to the long-wavelength transverse zigzag instability. At successive times (left to right, first row then second row), small inhomogeneities cause the instability to grow at different points in the fluid. The instability rapidly reaches a nonlinear regime for which the zigzags have a rather short wavelength and large amplitude. Time intervals between panels are 9, 10, 10, 26, and 72 minutes respectively. (b) For $R = 3000 \approx 1.7R_c$, and for an initial wave number $q = 1.64$, the stripes are unstable to a cross-roll instability. Again inhomogeneities in the cell cause the instability to have a nonuniform appearance. The final pattern (lower right panel) has a new average wave number that is distinctly smaller than the original wave number. This illustrates how a new wavelength can be introduced by a linear instability. [Adapted from Busse and Whitehead, *J. Fluid Mech.* **47**, 305 (1971).]

wave number lay outside the stability balloon and so the stripes become unstable to distinct instabilities. Figure 4.3(a) follows the evolution of the transverse zigzag instability to a nonlinear disordered state consisting of roll patches that are oriented at about 45° with respect to the original roll state. The analytical calculation (see

Etude 4.1 above for the Swift–Hohenberg model as a simple example and the more general discussion in Section 7.1.4) shows that the zigzag should first grow at long wavelengths but the instability rapidly enters a nonlinear regime that causes the zigzags to appear at a finite wavelength, and to increase in magnitude to the point that zigzags from neighboring rolls can connect, leading to the structure in the last lower right panel. Similarly, in Fig. 4.3(b), the stripes are unstable to the cross-roll instability and the final result are new stripes that are oriented 90° with respect to the original stripes *and* that have a substantially smaller wave number.

Figure 6.6 in Chapter 6 shows schematically how the stationary long-wavelength longitudinal Eckhaus instability affects the underlying stripe patten, by compressing and expanding the separation between successive rolls. Figure 9.5(d) shows schematically the effect of a stationary, long-wavelength, skew instability, which causes the spacing of adjacent rolls to contract or expand like the Eckhaus but along a direction that makes a nonzero angle with the roll axis. Finally Fig. 4.5(b) shows a snapshot of an experimental roll state that has become unstable to an oscillatory, finite-wavelength, transverse instability called the oscillatory instability. This turns out to be a type-I-o instability and the transverse structure on each roll is observed to propagate along the length of the roll.

For all of these instabilities, it is often difficult to compare the theoretical geometrical shape of the linear instability with experimental data since many Bloch states with adjacent wave vectors **Q** typically become unstable at once (so one observes a complicated superposition growing out of noise rather than a single mode), and the perturbations are often not visible experimentally until they have become so large that nonlinear effects distort the Bloch states via saturation and selection.

Two examples of Busse balloons are shown in Fig. 4.4, for a fluid like air at room temperature with a Prandtl number of 0.71, and for water at elevated temperatures, which has a Prandtl number of 7.0. There are several features to learn from such stability balloon diagrams. For each value of the control parameter (the Rayleigh number R in this case), there is a range of wave numbers q for stable stationary rolls but only up to some maximum value of R. Above this maximum value (which depends on system parameters, note how stable rolls persist for higher Rayleigh numbers for the large Prandtl number case panel (b)), all roll states are unstable and either some other stationary state will be observed or a time-dependent state will occur. Many fluid pattern-forming systems have stability balloons restricted to small values of the control parameter, but other systems such as chemical reaction–diffusion systems often have stable stationary stripe patterns for parameter values far from the linear instability of the uniform state.

Another feature of the Busse balloon is that the shaded region of stable rolls is bounded by particular instability curves. One can therefore predict from these instabilities the geometric shape and some dynamical properties of roll states that

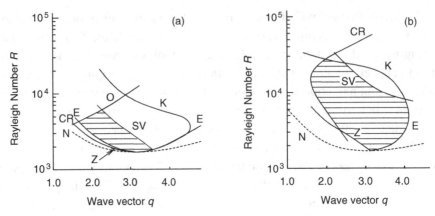

Fig. 4.4 Stability balloons for stripe states (called ideal roll states by fluid dynamicists) of wave number q in Rayleigh–Bénard convection for different values of the Rayleigh number R, see Eq. (1.2). Panel (a) is for a Prandtl number of 0.71 (say air at room temperature), while (b) is for a Prandtl number of 7.0 (say water near 40 °C). The hatched regions indicate where convection rolls in an infinite domain are linearly stable. The stability region is bounded by various instabilities that are named according to the geometry of the perturbation that begins to grow: E Eckhaus (stationary, long-wavelength, and longitudinal); Z zigzag (stationary, long-wavelength, and transverse); SV skew-varicose (stationary, long-wavelength, skew); K knot (stationary, finite-wavelength, and transverse); O oscillatory (oscillatory, finite-wavelength, and transverse); and CR cross-roll (stationary, finite-wavelength, and transverse). The dashed line is the neutral stability curve, above which the uniform (thermal conduction) state becomes unstable to perturbations of wave number q. (From Cross and Hohenberg [25].)

have become unstable. For example, in Fig. 4.4(b), Eckhaus and zigzag instabilities (labeled E and Z respectively in the figure) limit the stable range for small R, while the skew-varicose (SV) and cross-roll (CR) instabilities cause instability of rolls for larger R. These are all stationary instabilities such that Im $\sigma = 0$ at onset. In contrast, for lower Prandtl number fluids such as Fig. 4.4(a), the so-called oscillatory instability (labeled O) becomes important at larger values of R (around 5×10^3). As the name implies, this instability is of oscillatory type with Im $\sigma \neq 0$ at onset. When this instability occurs, transverse ripples are observed that propagate along rolls, and these ripples have a frequency that is predicted by the linear stability analysis (although these values are not traditionally indicated on the Busse balloon). Figure 4.5(b) shows a snap-shot of straight rolls that were unstable to the oscillatory instability.

For finite-wavelength instabilities like the cross-roll, knot, and oscillatory instabilities, the linear stability analysis can also predict that a new length scale might appear, namely the finite wavelength of the most unstable (fastest growing) Bloch state. For short-wavelength instabilities, there is no reason why the wavelength

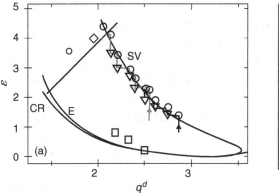 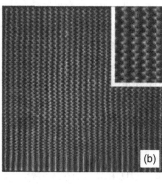

Fig. 4.5 (a) Comparison of the Busse balloon with experiment for a Rayleigh–Bénard convection experiment in a large square domain (aspect ratio $\Gamma = 50$) and for a fluid with Prandtl number $\sigma = 1.03$. The horizontal axis is the dimensionless variable qd (roll wave number q times the depth of the fluid) while the vertical axis is the reduced control parameter $\varepsilon = (R - R_c/R_c)$. The continuous lines are the theoretical instability boundaries while the discrete symbols are experimental data. The instability labels CR, E, and SV are the same as in Fig. 4.4. The arrows indicate the experimental parameter path taken while increasing ε. Symbols have the following meaning: open circles, before skew-varicose instability; upside-down triangles, after skew-varicose instability; diamonds, onset of oscillatory instability; squares, cross-roll instability for decreasing ε. (b) Picture of convection rolls whose wave number q is unstable to the oscillatory instability. [Adapted from the Ph.D. thesis of B. B. Plapp, Cornell University, Ithaca, New York, 1997.]

of the fastest growing mode must itself lie within the Busse balloon and so an interesting dynamics can arise if an initial set of stripes is unstable to a new set of stripes whose wave number is itself unstable. Something like this happens in the domain chaos state of Fig. 1.15(b), in which a small region of rolls in a rotating convection experiment becomes unstable to new rolls of the same wave number that are oriented 60° away, and these new rolls are themselves unstable to the same mechanism and so there is a never ending dynamics.

Figure 4.5 shows results from experiments that were designed to test the Busse balloon. In these experiments, a state of nearly uniform straight parallel rolls was formed by first tilting the whole convection apparatus away from the horizontal. This sets up a large scale circulating flow in the tilt direction that tends to align the rolls preferentially, and that also eliminates roll curvature and defects induced by the lateral walls. The apparatus is then returned to the horizontal position to study the Rayleigh–Bénard configuration. The skew-varicose instability line, for example, is traced out by increasing ε from an initially stable straight roll state. When the instability boundary is crossed, the skew-varicose instability develops, and eventually leads to the elimination of a pair of rolls through the production of a pair of dislocation defects, and so a decrease in the wave number q. Since this instability boundary increases in ε as q decreases, the control parameter can be further raised until the boundary is reached again. Other protocols lead to information on the other instabilities, although with less detail. Figure 4.5(b) shows the effects of the oscillatory instability. This is a type-I-o instability to waves of undulation of the rolls that propagate along the rolls. The waves are traveling in the upwards direction in the figure, but on other runs they might propagate in the opposite direction. The observation that the amplitude of the oscillations is largest toward the upper end of the figure can be understood from the discussion of type-I-o instabilities in Section 10.4.2.

Other experiments on less well controlled initial conditions have shown that the stability balloon gives a useful phenomenological understanding of approximately straight roll regions in the disordered patterns encountered in most experimental convection geometries. In particular, the balloon gives an approximate account of the observed range of wave numbers. In addition, the range in control parameter for which steady states exist, and the types of dynamics found when the pattern is stressed beyond these limits, are predicted well. Thus the stability balloon provides a useful organizational principle for the nonlinear states in this experimental system.

4.3 Two-dimensional lattice states

So far, we have assumed that the observed patterns are based on the stripe state that develops from a single Fourier mode that grows from the unstable uniform state. If the physical system has rotational symmetry, for example full rotational symmetry in the horizontal plane as in Fig. 2.11 or the discrete symmetry as in Fig. 2.13(b), stripes at different orientations may begin to grow.[14] We will focus on the case of full rotational symmetry in the plane, Fig. 2.11. The linear stability analysis then gives us a critical circle of unstable modes defined by $|\mathbf{q}| = q_c$. The growing solution near threshold may be a superposition of a few or of an infinite number of modes with wave vectors on or near this circle. Once the amplitude of the modes becomes large enough, their interaction through the nonlinearity of the evolution equations will lead to a competition between them. The ultimate result of this competition will be a pattern that may be ordered or disordered, static or dynamic.

Since it is not possible to catalog all possible solutions to nonlinear pdes mathematically nor explore all possibilities experimentally, we need some strategy to

---

[14] As usual, we imagine the third spatial dimension to be involved in maintaining the system out of equilibrium, and so we do not consider the possibility of symmetries spanning three dimensions. The extension to three dimensions would not be difficult, and indeed we may well have more intuition of this case based on the extensive prior study of equilibrium crystals. (See for example Chapter 4 in the book by Ashcroft and Mermin [7].)

guess the possible patterns. It is again useful to consider first a subset of possible nonlinear states, in particular those states that retain some subset of continuous or discrete translational and rotational symmetries out of the original symmetry. Patterns that retain symmetry under a reduced set of translations are the stripe state (discrete translational symmetry in one direction, full translational symmetry in the other, see Fig. 4.6(a)) and the two-dimensional lattices for which only discrete translational symmetries remain. The lattices are further distinguished by the allowed discrete rotational symmetries about each lattice point. A classical result is that only n-fold rotational symmetries with $n = 2, 3, 4$, and 6 are consistent with a periodic planar structure.[15] Patterns with the symmetry of a two-dimensional periodic lattice are often observed in experiment, either as ordered patterns throughout the whole domain (see Fig. 1.14, panel (b)), or as local structures in disordered patterns (see Fig. 1.17, panels (b) and (e)). There are also exotic quasicrystalline patterns, analogous to those observed in certain crystals, that retain rotational symmetries but are not periodic in space (see Fig. 1.16, panel (c)).

Just as was the case for our discussion of stripe states earlier in this chapter (see especially Section 4.1.3), for each lattice of a given symmetry, we want first to establish if stationary nonlinear saturated states exist. For those states that exist, we then want to determine if they are physically realizable by being linearly stable to arbitrary infinitesimal perturbations. Close to onset, existence and stability can often be calculated analytically using the amplitude equation formalism of the next few chapters. Further from onset, existence and stability can be determined by numerical methods, leading to a generalization of the stability balloons discussed in Section 4.2.1.

A new aspect of the stability test for lattices is to determine whether one kind of lattice state (stripes would be included) is unstable specifically toward the formation of a different kind of lattice state. For example, a calculation might show that a stripe state is unstable to the growth of an orthogonal set of stripes in which case a square lattice state might ultimately form. Alternatively, a square lattice state may be found to be unstable toward the decay of stripes in one direction and the further growth of stripes in the orthogonal direction, a situation that might produce a stripe state. In such cases, we would say that "the stripes are unstable toward squares," or that "the squares are unstable toward stripes," although the linear analysis cannot predict the structure of the final nonlinear saturated state, which must be determined by experiment or simulation.

It is insightful to write down explicit mathematical expressions for representative stationary nonlinear saturated lattice states and plot their geometric structure (Fig. 4.6). Before doing so, we observe that nonlinear states that form just above

---

[15] See Ashcroft and Mermin [7] for a proof of this.

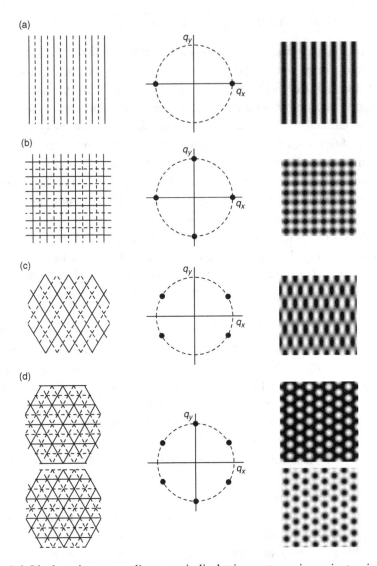

Fig. 4.6 Ideal stationary nonlinear periodic lattice patterns in an isotropic system: (a) stripes; (b) squares formed from two perpendicular sets of stripes; (c) orthorhombic state formed from two sets of stripes at a general angle; (d) hexagons formed from three sets of stripes at 120°. The left column shows the component stripes for each pattern. The middle column shows the corresponding wave vectors of modes on the critical circle $q = q_c$. The right column is a gray-scale density plot of the pattern $u(\mathbf{x}_\perp, \mathbf{x}_{\|,0})$ for some fixed value $\mathbf{x}_{\|,0}$ of the confined coordinate and suggests what might be observed in an optical experiment. Close enough to onset, a spatial power spectrum of a panel in the right column would produce the dots in the middle column since harmonics and sum and difference of peaks would be fainter in this regime. There are two inequivalent hexagonal states as shown in (d). If the harmonic and sum and difference terms are ignored in Eq. (4.54), the two states are related by the symmetry $\mathbf{u} \to -\mathbf{u}$ (black interchanged with white in the gray-scale plot). However, this is not true in general.

a supercritical type-I-s instability of a uniform state are composed of modes with wave vectors $\mathbf{q}$ on the critical circle $q = q_c$. The simplest periodic lattices that can be formed from wave vectors of equal magnitude are the hexagonal state (6-fold symmetry), the square state (4-fold symmetry), and the orthorhombic state (only 2-fold symmetry) based respectively on 6, 4, and 2 wave vectors on the critical circle $q = q_c$. (The middle column of Fig. 4.6 shows the critical circle with locations of the corresponding wave vectors.) In the following, we assume that the state of the system is described by a vector field $\mathbf{u}(\mathbf{x}_\perp, \mathbf{x}_\parallel)$ with two extended coordinates $\mathbf{x}_\perp = (x, y)$ and zero or one confined coordinate $\mathbf{x}_\parallel$.

Expressions for representative nonlinear periodic states sufficiently close to threshold are then the following. We use the subscripts S, Q, O, and H respectively to denote the stripe, square, orthorhombic, and hexagonal states. In the mathematical expressions, the dots ... represent sums and differences of modes (which includes spatial harmonics) that are generated by the nonlinearity but that are negligibly small close enough to onset. (The example in Section 4.1.3 explains why the terms generated by nonlinearities are negligible near onset.)

(i) Stripe state, Fig. 4.6(a):

$$\mathbf{u} = A_S \mathbf{u}_q(\mathbf{x}_\parallel) \left[e^{i\mathbf{q}_1 \cdot \mathbf{x}_\perp} + \text{c.c.} \right] + \cdots. \tag{4.50}$$

The wave vector $\mathbf{q}_1 = q_c \hat{\mathbf{q}}_1$ is the wave vector of the stripes normal to the direction $\hat{\mathbf{q}}_1$. The real parameter A_S sets the magnitude of the field in the saturated state[16] and is calculated from the dynamical equations.

(ii) Square state, Fig. 4.6(b):

$$\mathbf{u} = A_Q \mathbf{u}_q(\mathbf{x}_\parallel) \left[e^{i\mathbf{q}_1 \cdot \mathbf{x}_\perp} + e^{i\mathbf{q}_2 \cdot \mathbf{x}_\perp} + \text{c.c.} \right] + \cdots. \tag{4.51}$$

Here $\mathbf{q}_1 = q_c \hat{\mathbf{q}}_1$, $\mathbf{q}_2 = q_c \hat{\mathbf{q}}_2$ and the two wave vectors are perpendicular so $\hat{\mathbf{q}}_1 \cdot \hat{\mathbf{q}}_2 = 0$. The real-valued parameter A_Q again sets the magnitude of the saturated state.

(iii) Orthorhombic state, Fig. 4.6(c):

$$\mathbf{u} = A_O \mathbf{u}_q(\mathbf{x}_\parallel) \left[e^{i\mathbf{q}_1 \cdot \mathbf{x}_\perp} + e^{i\mathbf{q}_2 \cdot \mathbf{x}_\perp} + \text{c.c.} \right] + \cdots. \tag{4.52}$$

Here $\mathbf{q}_1 = q_c \hat{\mathbf{q}}_1$, $\mathbf{q}_2 = q_c \hat{\mathbf{q}}_2$ and the directions $\hat{\mathbf{q}}_1$ and $\hat{\mathbf{q}}_2$ are neither parallel or perpendicular. The nonlinear magnitude of the orthorhombic state is determined by the real parameter A_O. Note that we may write the bracketed term as

$$e^{i\mathbf{q}_1 \cdot \mathbf{x}_\perp} + e^{i\mathbf{q}_2 \cdot \mathbf{x}_\perp} + \text{c.c.} = 4 \cos(\mathbf{Q}_1 \cdot \mathbf{x}_\perp) \cos(\mathbf{Q}_2 \cdot \mathbf{x}_\perp), \tag{4.53}$$

with $\mathbf{Q}_1 = \frac{1}{2}(\mathbf{q}_1 + \mathbf{q}_2)$ and $\mathbf{Q}_2 = \frac{1}{2}(\mathbf{q}_1 - \mathbf{q}_2)$ so that $\mathbf{Q}_1 \cdot \mathbf{Q}_2 = 0$. Thus the nodes of the pattern form a rectangular mesh, and the pattern will have the visual appearance of rectangles, as is apparent in the grey-scale plot in Fig. 4.6(c). The wave vectors made from linear combinations of $\mathbf{q}_1$ and $\mathbf{q}_2$ form a centered rectangular lattice.

---

[16] More strictly: the magnitude of the deviation of the field from the spatially uniform solution.

(iv) Hexagonal state, Fig. 4.6(d):

$$\mathbf{u} = A_H u_q(\mathbf{x}_\|) \left[e^{i\mathbf{q}_1 \cdot \mathbf{x}_\perp} + e^{i\mathbf{q}_2 \cdot \mathbf{x}_\perp} + e^{i\Phi_3} e^{i\mathbf{q}_3 \cdot \mathbf{x}_\perp} + \text{c.c.} \right] + \cdots. \qquad (4.54)$$

Here $q_1 = q_2 = q_3 = q_c$ and the angle between any two of the unit vectors $\hat{\mathbf{q}}_1$, $\hat{\mathbf{q}}_2$, and $\hat{\mathbf{q}}_3$ is 120°. The magnitude of the hexagonal nonlinear state is given by the real number A_H. Note that in Eqs. (4.50)–(4.52), we could have multiplied each exponential term by an arbitrary phase factor $e^{i\Phi_i}$. However, these phases can be absorbed into a redefinition of the origin of coordinates, and do not change the pattern except for this translational shift. For Eq. (4.54), the two degrees of freedom given by choosing the position of the origin may be used to absorb two phase factors, e.g. for the first two terms, but then the third term will have a phase factor $e^{i\Phi_3}$. This phase is not related to any symmetries, and its value is prescribed by the dynamical equations. In fact, we will see in Etude 7.4 in Section 7.3 that there are two possible choices of this phase giving two inequivalent hexagonal states, as shown in Fig. 4.6(d).

More exotic states with these translational and rotational symmetries can also be formed. For example, consider the state based on equal amplitudes of the eight modes with wave vectors distributed on the $q = q_c$ circle as shown in Fig. 4.7(a). The state produced also has square symmetry, but has a spatial periodicity that is larger than $2\pi/q_c$. This can be seen by observing that the eight wave vectors, and all their possible sums and differences, can be written as linear combinations with integer coefficients of the vectors[17] $\mathbf{q}_1 = q_1 \hat{\mathbf{x}}$ and $\mathbf{q}_2 = q_1 \hat{\mathbf{y}}$ with $q_1 = q_c/\sqrt{5}$. Correspondingly, the spatial period is $2\pi \sqrt{5} q_c^{-1}$ even though the dominant peaks in the spatial power spectrum will be the spots on the critical circle with wave number q_c. The spatial structure takes the appearance of a superlattice, as shown in the gray-scale plot of Fig. 4.7(a). Similarly, hexagonal superlattice states may be constructed, as in Fig. 4.7(b).

Only n-fold symmetry axes with $n = 2, 3, 4$, and 6 are consistent with a periodic lattice in two dimensions. However, in the context of crystalline materials, researchers discovered nonperiodic crystals whose X-ray diffraction pattern (a measure of its spatial power spectrum) showed sharp discrete peaks (indicating a regular order) and rotational symmetries for other values of n such as $n = 5$. Such structures are called quasicrystals. Analogous structures may readily be constructed at pattern-forming instabilities and indeed had been hypothesized in the context of fluid instabilities (where they were called regular structures [92]) before their discovery by the materials scientists. This type of quasiperiodic structure may be constructed from the superposition of N Fourier modes with wave vectors that are equally distributed over the upper half of the critical circle $q = q_c$ circle (the

---

[17] The vectors $\mathbf{q}_1$ and $\mathbf{q}_2$ are then the basis vectors of the lattice in $\mathbf{q}$-space, called the reciprocal lattice, that contains all the wave vectors represented in the pattern. The reciprocal lattice is shown as the empty circles in Figs. 4.7(a) and (b).

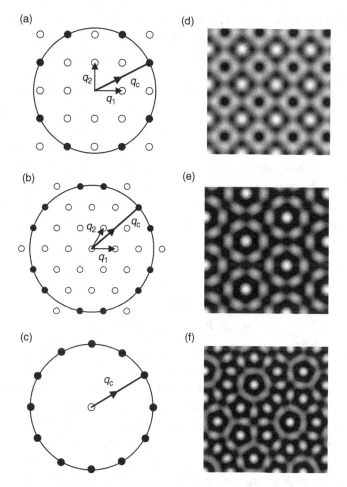

Fig. 4.7 More complicated patterns made from superpositions of modes on the critical circle: (a) a superlattice with 4-fold symmetry; (b) a superlattice with 6-fold symmetry; (c) a quasicrystal with 12-fold symmetry. In each case, the first panel shows the wave vectors of the modes used, and the second panel the resulting patterns. In (a), there are eight wave vectors on the $q = q_c$ circle that go unstable together, but they produce a 4-fold symmetry. The primitive basis vectors of the reciprocal lattice may be chosen to be $\mathbf{q}_1$ and $\mathbf{q}_2$, which have magnitude $q_c/\sqrt{5}$. A similar construction with 12 vectors leads to the hexagonal superlattice in (b). On the other hand, 12 vectors equally spaced around the circle as in (c) give a quasiperiodic pattern.

complex conjugate modes fill in the lower half circle) like this:

$$\mathbf{u} = A_C \mathbf{u}_q(\mathbf{x}_\parallel) \sum_{i=1}^{N} \left[e^{i\Phi_i} e^{i\mathbf{q}_i \cdot \mathbf{x}_\perp} + \text{c.c.} \right] + \cdots. \quad (4.55)$$

Here $q_i = q_c$, $\hat{\mathbf{q}}_i \cdot \hat{\mathbf{q}}_1 = (i-1)\pi/N$, and the numbers Φ_i are real phases. Figure 4.7(c) shows an example with $N = 6$ that gives a quasicrystalline pattern with 12-fold rotational symmetry, while Fig. 1.16(b) is an experimental example of this pattern.

For readers who are not familiar with equilibrium phase transitions (as often discussed in undergraduate physics and engineering courses on thermodynamics), it is worth pointing out that, in our discussion of possible states, there is little that is specific to nonequilibrium systems and an analysis of the possible equilibrium phase transitions of a two-dimensional periodic structure would follow similar arguments, e.g. in studying solidification or the formation of liquid crystal states. In that context, the approach is known as Landau theory. Indeed, many of the ideas and methods developed in Landau theory may be transferred to the present discussion, and, in turn, some insights gained in pattern formation may be useful in the context of equilibrium phases.

4.4 Non-ideal states

4.4.1 Realistic patterns

The ideal ordered state of straight parallel stripes or of a regular lattice over the whole system is rarely seen in experiment or nature. More typical is the type of disordered pattern shown in Fig. 4.8 which is taken from a convection experiment. Over much of the system, the local structure consists of stripes as expected from the analysis of the ideal infinite system. However, the stripes are curved rather than straight and the orientation of the stripes is different in different parts of the system. Experiments like Fig. 4.8 and related fluid simulations strongly suggest that the lateral boundaries play an important role in inducing the disordered pattern since the stripes tend to approach the boundaries orthogonally.[18]

Since disordered patterns are common, we would like some quantitative tools to describe them. At some point $\mathbf{x}$ in a region where stripes are apparent, their orientation and periodicity can be defined by a local wave vector $\mathbf{q}(\mathbf{x})$. This quantity is not precisely defined mathematically in a disordered pattern, but in practice is a useful construction. We could find $\mathbf{q}(\mathbf{x})$ over much of Fig. 4.8 for example, by measuring the spacing of adjacent stripes with a ruler, giving a local wavelength or inverse wave number. The direction of $\mathbf{q}$ is then determined as the local normal to the stripes. Many other algorithms have been proposed as practical schemes of

---

[18] Remarkably, although the perpendicular orientation is often observed in a wide variety of experimental and numerical systems, there is no clear understanding of why this is so. Indeed, the result does not seem to be a strict boundary condition, but rather a common occurrence. In addition, the effect is weaker near threshold, and close enough to threshold straight stripes with an orientation unaffected by the boundaries are sometimes seen (see Fig. 1.15(a)), which is consistent with a prediction of the amplitude equation formalism discussed in the next chapter.

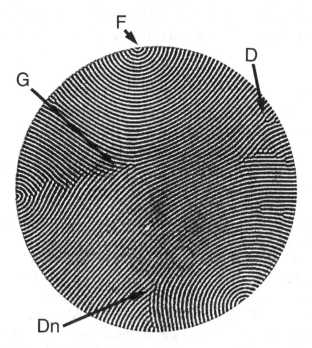

Fig. 4.8 Shadowgraph image of a time-dependent state of disordered convection rolls in gaseous carbon dioxide (Prandtl number $\sigma \approx 0.96$) near onset (Rayleigh number $R = 1.16 R_c$) in a cylindrical geometry of aspect ratio $\Gamma = 78$. The lateral walls tend to orient the rolls so that they are nearly perpendicular to the lateral wall which causes defects to form in the bulk of the pattern. Examples of common defects are marked: D for a dislocation, Dn for a disclination, F for a focus singularity, and G for a grain boundary. (From Morris et al. [77].)

estimating the local wave vector, for example ones based on wavelet analysis, and the degree to which different schemes agree can be used to estimate the accuracy of the constructions.

Just as in the completely ordered state, the wave vector **q** is not a complete description of the pattern, since we still need to fix the position of the stripes. To do this, we generalize the idea of the phase variable introduced in Section 4.1.1. The appearance of stripes corresponds to a structure that locally has a spatial periodicity. We can therefore write an approximate expression for the variables defining the patterns as[19]

$$\mathbf{u}(x,y) \simeq \mathbf{u}_q(\phi(x,y)) \quad \text{with} \quad \nabla \phi = \mathbf{q}(x,y). \tag{4.56}$$

[19] The phase ϕ is related to the phase Φ of the complex amplitude by $\phi = \Phi + q_c x$, assuming that the reference stripes used to set up the amplitude equation are normal to the x-direction.

Here $\mathbf{u}_q$ is the same function that would define an ideal stripe state at wave number q. For example, for ideal stripes normal to the x-axis we would have

$$\mathbf{u}(x, y) = \mathbf{u}_q(qx). \tag{4.57}$$

The expression for $\mathbf{u}(x, y)$ in Eq. (4.56) is not exact if $\mathbf{q}$ is varying in space, but will be a good approximation for slowly varying $\mathbf{q}$, which is the case if locally the stripes are well defined. This introduction of a phase variable whose gradient gives the local wave vector is an important concept. In Chapter 9, we will generalize Eq. (4.56) to time-dependent situations to study the dynamics of patterns in terms of the phase dynamics.

Although the pattern over much of the system in Fig. 4.8 locally appears as stripes, there are regions of more complex structure where a stripe pattern cannot be clearly identified – overlapping stripes, stripes that terminate, sharply curved regions, and others. There are some regions of more complex structure, either near points or lines, that reappear in different locations, and in different experiments and systems. Indeed many have geometrical structures that are familiar from other branches of physics, such as the study of solids and liquid crystals. Such motifs are called defects. These defects are also seen in lattice states, although their structure is more complicated there. Since the defects can be identified as the regions where the structure is not locally periodic in space, Eq. (4.56) is not a good approximation here, and the phase ϕ and wave vector $\mathbf{q}$ are not well defined at these regions. Defects are therefore often described as singularities where the wave vector or phase cannot be defined. (The singular behavior refers only to the simplified description in terms of $\mathbf{q}$ and ϕ. The underlying physical fields such as fluid velocities, temperature, or concentration always vary smoothly everywhere.)

Some defects have a topological character, which makes the task of studying them easier. A topological defect is a defect that can be identified from the behavior of the pattern away from the defect, where the stripes are well defined and so the phase description applies. Common examples in the stripe state are a dislocation, where a stripe-pair ends at some point in the system; a disclination, which is a point where stripes of different orientations come together; and a grain boundary, which is a line separating two regions or domains of different stripe orientation. Since topological defects are a prominent feature of patterns, we now discuss some of the ideas in more detail.

4.4.2 Topological defects

The nature of the topological arguments used to discuss defects can be illustrated by the target or focus defect shown in Fig. 4.9. A simple target is an axisymmetric pattern of stripes. The wave vector is normal to the stripes, and points out from the

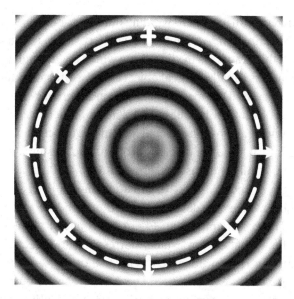

Fig. 4.9 A topological defect called a target. The arrows denote the local wave vector **q**. The dashed line shows a possible choice of contour circling the defect for defining the winding number. The presence of the target can be deduced by the fact that the local wave vector rotates by an amount 2π while tracing once a circuit around the contour.

center. At the center, the direction of the stripes is not defined, and so the center of the target is singular at the level of the wave vector description. The presence of the center can be predicted purely from the behavior far away where the wave vector is well defined and varies smoothly. Consider for example, the behavior of the wave vector around the dashed contour in Fig. 4.9. If we follow the contour around, say in an anticlockwise direction, the orientation of the wave vector also rotates in the anticlockwise direction, by an amount 2π in a complete circumnavigation. The number of 2π rotations in a single circumnavigation is called the winding number. In Fig. 4.9, the winding number of the wave vector is $+1$, with a plus sign because the winding has the same sense as the direction of going around the contour.

This winding number is a topological characteristic of the defect. Its value does not depend on the geometry of the contour since any closed contour that circles the center of the target just once, say a contour of different radius or a non-circular loop, will also produce a winding number of $+1$. The nonzero winding number necessarily implies the singular behavior somewhere inside the contour, since we can imagine shrinking the contour while restricting it to regions where **q** is well defined. The winding number is preserved as the contour shrinks, and eventually we end up with a tiny contour around which the wave vector rotates by the winding number, which is a singular behavior.

162   *Nonlinear states*

A topological defect with a nonzero winding number for the direction of the wave vector is called a disclination. An interesting subtlety arises because the sign of $\hat{\mathbf{q}}$ has no meaning, all that matters is the line along which the normal to the stripes lies. So it is better to treat $\hat{\mathbf{q}}$ as a mathematical object called a director, which is a headless vector such that $-\hat{\mathbf{q}} \equiv \hat{\mathbf{q}}$. This means that the winding number for $\hat{\mathbf{q}}$ can take on half-integral as well as integral values. The defects associated with $\hat{\mathbf{q}}$ are in fact similar to those found in a two-dimensional nematic liquid crystal, whose states are specified by a director $\hat{\mathbf{n}}$. (See the discussion of uniaxial systems in Section 2.6.2.) Various disclination configurations are shown in Fig. 4.10. In addition to the $+1$ disclination in Fig. 4.9, there is a $+\frac{1}{2}$ disclination shown in Fig. 4.10(a) in which the direction of the wave vector (denoted by the double headed arrow) rotates by π around the core. The configurations in Fig. 4.10(b) and (c) also yield a nonzero winding number for $\hat{\mathbf{q}}$, but do not usually correspond to defects in a stripe state since the wave number (proportional to the inverse of the separation between the lines in the figure) varies over a large range and so can lie outside the stability balloon. For such a large distortion, the stripe state typically evolves into other defects. For example, the configuration in Fig. 4.10(c) will usually evolve into a defect consisting of one or more grain boundaries (see Fig. 4.12) along the directions where the stripe orientation changes most rapidly.

Another type of topological defect characteristic of locally striped patterns is the dislocation, which can be identified as a point in the stripe pattern where a stripe-pair ends. Unlike the disclination, there is no large variation of the direction of the wave vector at large distances from the dislocation, so the winding number of the

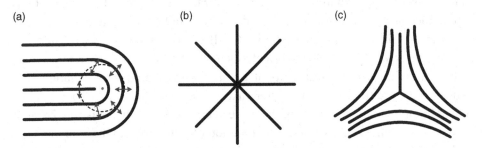

Fig. 4.10 Examples of topological defects known as disclinations that have a nonzero winding number in the wave vector direction. The thick lines show the direction along the stripes and the double headed arrows indicated the direction of a wave vector. (a) A disclination with winding number $+\frac{1}{2}$. As the dotted contour is circumnavigated, the normal to the stripes (denoted by the double headed arrow since the direction but not the sign is significant) rotates in the same direction but at half the rate giving a total rotation of π. (b) A putative configuration for a $+1$ disclination; (c) A putative configuration for a $-\frac{1}{2}$ disclination. In (b) and (c), the lines are along the stripes. These configurations are not consistent with a stripe wavelength that is roughly fixed and are not commonly observed.

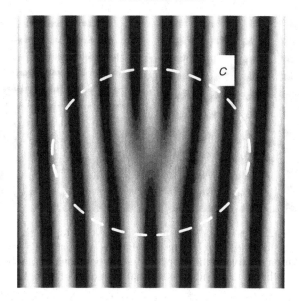

Fig. 4.11 A topological defect known as a dislocation, where a pair of stripes ends. Notice that there is one less spatial period in the lower half of the figure (below the dislocation) than there is in the upper half. The dashed line shows a possible choice of the integration contour C for defining the winding number via Eq. (4.58).

wave vector is zero. However, a different winding number W can be defined for the dislocation in terms of a phase variable rather than in terms of the angle of the wave vector. The dislocation winding number is defined by

$$W = \frac{1}{2\pi} \oint_C \nabla\phi \cdot d\mathbf{l} = \frac{1}{2\pi} \oint_C \mathbf{q} \cdot d\mathbf{l}, \qquad (4.58)$$

where the line integral is taken around any closed contour surrounding the point (see Fig. 4.11). Since the quantity $e^{i\phi}$ must be single valued away from the defect, the number W must be an integer. A nonzero value of W implies the existence of one or more dislocations within the contour, by the same sort of argument that we used above.

Disclinations and dislocations are point defects since the singularity in the coarse-grained description occurs at a point. There are also line defects, for which the singularity extends along some finite region of the pattern. An example of such a line defect is a grain boundary where two domains of stripes with different orientations come together along a line. (This line is also called a domain wall.) Figure 4.12 shows two examples of grain boundaries. Panel (a) shows the case where the boundary is at some general angle relative to the stripes. Panel (b) is a perpendicular grain boundary, for which case the boundary is normal to one set of stripes.

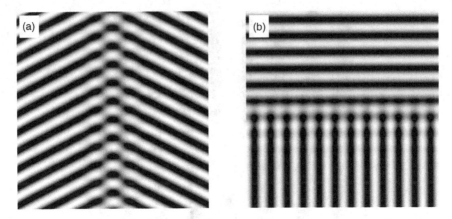

Fig. 4.12 Two examples of grain boundaries, which are topological line defects. (a) Two sets of stripes with different orientations meet along a line that has a general orientation relative to the stripes. (b) A perpendicular grain boundary is a special case for which the boundary is normal to one set of stripes.

4.4.3 Dynamics of defects

The motion of defects is important in pattern selection, the transient relaxation from an initial condition to a final steady state, and in situations where there is persistent dynamics. This is because defects can be dynamic in situations where the rest of the pattern is held stationary, by boundaries or by constraints from other regions of the pattern. If the defects are well separated or few in number, it might be possible to develop a theoretical understanding of the pattern dynamics directly in terms of the defect dynamics. Phenomena such as the nucleation of defects (or of defect pairs if there are topological constraints), their interaction and motion, and annihilation on close approach, are often apparent in experiment and simulation. On the other hand, conditions for a stationary defect configuration may determine properties of the pattern in the ordered region far away from the defect, and so issues of pattern selection may sometimes be addressed by finding these conditions.

As an example, consider the dislocation in Fig. 4.11. It may move in the vertical direction which is called dislocation climb. If the dislocation moves upwards in the figure, the number of periods in the upper half plane decreases and similarly a downwards motion increases the number of periods in the lower half plane. Thus dislocation climb will increase or decrease the number of spatial periods in the pattern and so will change the mean wave vector. The climb motion of the dislocation is in turn found to depend on the mean wave number q_b of the surrounding pattern (the "background" wave number). The dislocation will move up if $q_b > q_d$ or down if $q_b < q_d$, where q_d is a particular wave number that we might call the dislocation selected wave number. The value of q_d depends on the system and the control parameter. Dislocation climb tends to shift the wave number toward q_d so that the dislocation is stationary only if the wave number is equal to q_d. Dislocation climb is studied further in Section 8.1.2.

On the other hand, horizontal motion of the dislocation in Fig. 4.11 through a stationary background roll configuration requires the successive breaking off and reconnection of the stripes. This so-called glide motion will be jerky and so more difficult to characterize mathematically than the climb motion.

The climb dynamics of dislocations tends to move the wave number to q_d, and is a mechanism of wave number selection in the nonlinear state. There are other instances of wave number selection through defects. For example, the focus or target disclination shown in Fig. 4.10(a) is only stationary if the wave number far away tends to a particular value q_f, whose value is again dependent on the system and parameters. For other wave numbers, the target will either emit or suck in stripes. This is discussed further in Section 9.1.2, and the general question of wave number selection is discussed in Section 9.1.4.

4.5 Conclusions

This chapter has presented a mainly qualitative introduction to the vital role that nonlinearity plays in pattern formation. The linear stability analysis is useful in suggesting the beginnings of pattern formation, the physics involved in the structure formation, and the length and time scales of the patterns. But the linear stability cannot provide information on the actual patterns that will be observed. Nonlinearity is essential in saturating the exponential growth of the linear analysis, and selecting between many states that are typically degenerate in the linear analysis. Important steps in the analysis of the effects of nonlinearity are studying the existence of solutions, and then the stability of these solutions. Since it is not possible to list all possible solutions that might exist, we are led to look at certain simple classes of solutions, such as ideal stripe and lattice states, and versions of these periodic with disorder of a special kind, namely defects (particularly topological defects).

The ideas presented in this chapter build on ones you may have encountered before. In particular, bifurcation theory, developed in the context of nonlinear odes or low-dimensional dynamical systems, is useful in understanding the nonlinear states that might occur near threshold. The ideas developed in the study of crystals in solid state physics provide suggestions for type of patterns that might arise and how to treat their symmetries. The important role of defects, particularly topological defects, also arises in the solid state and other ordered equilibrium systems, and sophisticated tools have been developed to classify them mathematically (see further reading below).

Many of the questions of pattern formation, such as wave number and pattern selection, have been introduced in this chapter. In the succeeding chapters, we develop and apply the amplitude equation formalism and the nonlinear phase diffusion equation that can be used to address these questions quantitatively.

4.6 Further reading

(i) *Nonlinear Dynamics and Chaos: With Applications to Physics, Biology, Chemistry and Engineering* by Strogatz [99] is an excellent introduction to nonlinear dynamics in systems described by odes and bifurcation theory.
(ii) For further discussion of the Busse balloon in convection discussed in Section 4.2.2 see the paper by Busse and Whitehead [16] and the review "Nonlinear properties of convection" by Busse [15].
(iii) There are many standard textbooks in solid state physics that provide an introduction to lattices, for example *Solid State Physics* by Ashcroft and Mermin [7].
(iv) *Principles of Condensed Matter Physics* by Chaikin and Lubensky [17] is a good source for more information on Landau theory (see in particular Chapter 4) and topological defects (Chapter 9).
(v) The article "Natural patterns and wavelets" by Bowman and Newell [14] describes the application of wavelet theory to characterizing patterns. You can also find references to general discussions of wavelets there.
(vi) Topological defects play an important role in ordered equilibrium phases. The review article "The topological theory of defects in ordered media" by Mermin [74] describes the mathematical homotopy theory behind the "winding number" classification of defects.
(vii) We only discussed dislocations in stripe states. For a theoretical and experimental discussion of dislocations in hexagonal patterns see "Defects in roll-hexagon competition" by Ciliberto *et al.* [22].

Exercises

4.1 **Definition of the amplitude for stripes:** We could perversely have chosen to define the amplitude A in terms of the expression

$$\mathbf{u}(\mathbf{x},t) = \left[iA(t)e^{i\mathbf{q}\cdot\mathbf{x}_\perp} - iA^*(t)e^{-i\mathbf{q}\cdot\mathbf{x}_\perp}\right]\mathbf{u}_\mathbf{q}(\mathbf{x}_\parallel), \quad (E4.1)$$

rather than by Eq. (4.1).

(a) In this case, how does A change under the space reflection $\mathbf{x}_\perp \to -\mathbf{x}_\perp$?
(b) Show that the coefficient g of the lowest-order nonlinear term is still real in the evolution equation for this new A.

Exercises

4.2 **Real formulation of the amplitude:** Suppose we had used a real formulation of the amplitude, writing

$$\mathbf{u} = a(t)\cos(\mathbf{q} \cdot \mathbf{x}_\perp)\mathbf{u_q}(\mathbf{x}_\parallel), \qquad (E4.2)$$

rather than Eq. (4.1).

(a) Explain how to generalize the ansatz Eq. (E4.2) to include displacements of the stripes.

(b) Explain why no quadratic terms can appear in the amplitude equation for this a.

4.3 **Higher-order terms in the amplitude equation:** The discussion just after Eq. (4.5) explained that the symmetry of translational invariance prevents terms quadratic in the amplitude A or its complex conjugate A^* from appearing in the amplitude's evolution equation. Extend this discussion a bit further:

(a) Discuss whether any even power (term proportional to $A^m(A^*)^n$ with $m+n$ an even integer) can occur in Eq. (4.5).

(b) Determine all possible fifth-order terms that can appear in Eq. (4.5).

4.4 **Harmonics:** Why is there no $\cos(2x)$ term in Eq. (4.26)? What about a $\sin(3x)$ term?

4.5 **Values of σ, τ_0, and g for the amplitude equation of the Swift–Hohenberg equation:** For the exponentially growing solution at the critical wave number

$$u(x, t) = A(t)e^{ix} + \text{c.c.}, \qquad (E4.3)$$

with amplitude A, derive the amplitude equation Eq. (4.10) with explicit values for τ_0, ε, and g if u satisfies the Swift–Hohenberg equation Eq. (4.19).

4.6 **Complex amplitude for an oscillatory instability at zero wave number:** In anticipation of the study of oscillatory instabilities in Chapter 10, work through the discussion of Section 4.1.1 for the case of a type-III-o instability with complex-valued growth rate at onset (Im $\sigma_{q_c} \neq 0$). You must now use the invariance of the system under a time translation to argue for the invariance of the equation for A to a constant phase change.

(a) Derive an equation analogous to Eq. (4.5) for the complex amplitude of the mode going unstable at a $q = 0$ (type-III-o) instability. You should find the same equation but with g (and σ) complex. Note: For a type-III-o instability, the amplitude $A(t)$ is now defined by the expression

$$\mathbf{u}(\mathbf{x}, t) = A(t)e^{-i\omega_c t}\mathbf{u_q}(\mathbf{x}_\parallel) + \text{c.c.},$$

where ω_c is given by Eq. (2.2).

(b) Now derive equations for $d_t|A|$ and $d_t\Phi$. Show that the equation for $|A|$ is the same as Eq. (4.14) for the real amplitude, except for the replacements $\sigma \to \text{Re}\,\sigma$ and $g \to \text{Re}\,g$. Interpret the phase equation in the linear (small $|A|$) approximation, and then when the leading-order nonlinearity is included. Show that the same pictures for the bifurcation hold, Fig. 4.1.

4.7 **Complex amplitude for type-I-o oscillatory instability at finite wave number:** Repeat the first part of Exercise 4.6 for the case of the growth of a single Fourier mode $e^{i\mathbf{q}\cdot\mathbf{x}_\perp}$ at an oscillatory instability with a finite wave number $q_c > 0$.

4.8 **Traveling or standing waves:** For an oscillatory instability at nonzero wave number q in a one-dimensional system with inversion symmetry the modes at q and $-q$ go unstable together and with the same frequency ω_q. A small amplitude solution constructed from these modes is

$$u(x,t) \simeq A_+ e^{i(qx-\omega_q t)} + A_- e^{i(-qx-\omega_q t)} + \text{c.c.} \tag{E4.4}$$

(a) Argue that the complex amplitudes will satisfy equations of the form

$$d_t A_+ = \sigma A_+ - g_1 |A_+|^2 A_+ - g_2 |A_-|^2 A_+, \tag{E4.5a}$$

$$d_t A_- = \sigma A_- - g_1 |A_-|^2 A_- - g_2 |A_+|^2 A_-, \tag{E4.5b}$$

where σ, g_1, and g_2 are complex in general.

(b) Now consider two classes of solutions: traveling waves $A_+ = A_T \neq 0$, $A_- = 0$ (or the opposite); and standing waves $A_+ = A_- = A_S$. (In the later case, any phase difference between A_+ and A_- can be eliminated by a shift of coordinates.)

1. While the saturated solutions are defined by no exponential growth of $A_\pm$, there may be a phase evolution $A_\pm = |A_\pm|e^{-i\Omega t}$ corresponding to a shift in the frequency from the onset frequency $\omega = \omega_q + \Omega$. Find the magnitudes and frequency shifts $|A_T|, \Omega_T$ and $|A_S|, \Omega_S$ in the saturated nonlinear states.
2. Find the condition for a forward or backward bifurcation of the traveling and standing wave states.
3. Show that for the backward bifurcation the nonlinear state is unstable in either case.
4. For the case where the bifurcations to traveling and standing waves are both forward, show that within the description of Eqs. (E4.5) either the standing wave is stable and the traveling wave is unstable, or vice versa, depending on properties of g_1 and g_2.

Exercises

4.9 Nonlinear saturation in the Swift–Hohenberg equation: In the calculation of the nonlinear saturation for the Swift–Hohenberg equation we started with a cosine linear mode Eq. (4.20). What would we have found if we instead started with the linear solutions

(a) $u \sim e^{\varepsilon t} \sin x$,
(b) $u \sim e^{\varepsilon t} e^{ix}$ (take care!)?

4.10 Coefficient of next higher harmonic in stationary nonlinear state: Calculate the value of the coefficient a_3 in Eq. (4.26) to order $r^{3/2}$.

4.11 Nonlinear saturation at general wave numbers:

(a) Repeat the derivation of Eq. (4.30) for a general wave number q by calculating the coefficient a_q in the expression

$$u = a_q \cos(qx) + \cdots, \qquad (E4.6)$$

such that this expression satisfies the time-independent one-dimensional Swift–Hohenberg equation sufficiently close to onset. Equation (E4.6) represents the truncation of a Galerkin approximation of u to a single mode.

(b) Determine the order of the next harmonic $\cos(3qx)$ close to onset: does it scale as $r^{3/2}$, which is the case for $q = 1$ in Eq. (4.30)?

4.12 Lack of field inversion symmetry does not give a transcritical bifurcation for a stripe state: In simple bifurcation theory we associate a pitchfork bifurcation with a system that is symmetric under the change in sign of the dynamical variable, e.g. $u \to -u$. Consider, however, the generalized Swift–Hohenberg equation for $u(x,t)$ in one spatial dimension

$$\partial_t u = ru - \left(\partial_x^2 + 1\right)^2 u - g_2 u^2 - u^3, \qquad (E4.7)$$

which for nonzero g_2 does not have this symmetry.

(a) By considering the ansatz

$$u = \sum_{n=0}^{\infty} a_n \cos(nx), \qquad (E4.8)$$

show that the bifurcation remains of pitchfork character and calculate a_1 to lowest order in an expansion in small r.

(b) What is the condition on g_2 for the bifurcation to remain supercritical?
(c) Calculate a_0 and a_2 to lowest nontrivial order in r.
(d) Show that the two nonlinear solutions are not in general related simply by a change of sign.

Note that in two dimensions, a transcritical transition to a hexagonal state is indeed obtained for nonzero g_2.

4.13 **Growth and saturation in a modified Swift–Hohenberg equation:** Consider the equation (which actually describes Rayleigh–Bénard convection when the top and bottom plates are poor thermal conductors)

$$\partial_t u = ru - (\nabla^2 + 1)^2 u + \nabla \cdot \left[(\nabla u)^2 \nabla u \right] \tag{E4.9}$$

for $u(x, y, t)$ a real field in a two-dimensional space and r the control parameter. (These types of models are discussed in Section 5.2.)

(a) What is the growth rate $\sigma(\mathbf{q})$ of a small perturbation at wave vector $\mathbf{q}$ from the uniform state $u = 0$? What is the critical wave number q_c and the critical value of the control parameter r_c giving the first instability as r is raised?

(b) For a nonlinear saturated stripe solution

$$u = a \cos x + \cdots \tag{E4.10}$$

find out how the amplitude a varies with $r - r_c$ near threshold to lowest non-trivial order.

(c) Repeat the calculation for a square solution

$$u = a(\cos x + \cos y) + \cdots. \tag{E4.11}$$

In these expressions, the $\cdots$ denote higher-order terms.

4.14 **Phases in the lattice states:** Determine the change in the phase factors Φ_j for each component mode in the expression

$$\mathbf{u} = A\mathbf{u}_q(\mathbf{x}_\parallel) \sum_j e^{i(\mathbf{q}_j \cdot \mathbf{x}_\perp + \Phi_j)} \tag{E4.12}$$

for the stripe, square, and hexagonal states sketched in Fig. 4.6 for shifts of the coordinate origin to positions $(1, 0)$, $(0, 1)$, and $(1, 1)$. You may assume $|\mathbf{q}_j| = 1$.

4.15 **Hexagonal patterns:** Consider the superposition of modes

$$u(x, y) = e^{i(\mathbf{q}_1 \cdot \mathbf{x} + \Phi_1)} + e^{i(\mathbf{q}_2 \cdot \mathbf{x} + \Phi_2)} + e^{i(\mathbf{q}_3 \cdot \mathbf{x} + \Phi_3)} + \text{c.c.} \tag{E4.13}$$

with $\mathbf{x} = (x, y)$ and the wave vectors $\mathbf{q}_1, \mathbf{q}_2, \mathbf{q}_3$ forming an equilateral triangle

$$\mathbf{q}_1 + \mathbf{q}_2 + \mathbf{q}_3 = 0, \quad |\mathbf{q}_i| = q \tag{E4.14}$$

and where the Φ_i are the possible phases of the modes. Since q sets the scale of the pattern, let's choose $q = 1$, and orient our axes so that $\mathbf{q}_1 = (1, 0)$.

(a) Show that by redefining the origin of coordinates we may set Φ_1 and Φ_2 to zero, so that the form of the pattern (outside of translations and rotations) is determined by a single phase variable $\Phi_3 = \Phi$.

(b) Using Mathematica, Matlab, or some other convenient plotting environment make some contour or density plots of the field $u(x, y)$ for various choices of Φ.

(c) As we will discuss later in Section 7.3.3, the amplitude equation analysis near the onset of the instability shows that the sum of the phases $\Phi_1 + \Phi_2 + \Phi_3$ (and so the reduced phase Φ) will asymptotically evolve toward 0 or π. Plot the patterns for these two values (choosing $\Phi_1 = \Phi_2 = 0$). Describe the locations of the maxima and minima of the two patterns.

(d) How are the patterns for $\Phi = 0$ and $\Phi = \pi$ related in the approximation Eq. (E4.13)?

4.16 **Quasicrystalline states:** Write down the superposition of modes with wave number $q = 1$ giving quasicrystalline states with 8-fold and with 10-fold rotational symmetry. Using Mathematica, Matlab, or some other convenient plotting environment plot the resulting patterns.

4.17 **Stability of square state:** Show, using a Galerkin method truncated to the lowest modes, that the square state in Exercise 4.13 is stable for Bloch wave vector $\mathbf{Q} = 0$ perturbations.

4.18 **Eckhaus instability in the Swift–Hohenberg equation:** Calculate the stability of the solutions to the Swift–Hohenberg equation near onset to a longitudinal perturbation in the one-mode approximation, following the methods of Etude 4.1. Focus on small-Bloch-wave vectors $\mathbf{Q}$ and show by expanding up to $O(Q^2)$ that there is an eigenvalue of the form $\sigma = \alpha(r, q)Q^2 + \cdots$ Find the value of $q - 1$ when α passes through zero, signaling instability, for small r and compare with the neutral stability value $q_N - 1$. You will probably find the complex notation Eq. (4.40) reduces the complexity of the algebra.

4.19 **Galerkin calculation of the stability balloon:** Set up the calculation of the linear stability analysis for general perturbation wave vector $\mathbf{Q}$ of the nonlinear state of the Swift–Hohenberg equation calculated in Exercise 4.11 using the Galerkin expansion Eq. (4.40) truncated to the lowest mode. Derive the equation that needs to be solved to find the growth rate σ. Leaving the equation in the form of a determinant set to zero is a good way to eliminate messy algebra. Solving this equation in the general case is not informative, but you might find particular limits (e.g. small $\mathbf{Q}$) interesting.

4.20 **Stability balloon in Rayleigh–Bénard convection:** Investigate the original literature to discover the geometrical nature and symmetry characteristics of the various instabilities bounding the stability balloon for Rayleigh–Bénard convection shown in Fig. 4.4.

4.21 **Dislocations:** Use Mathematica, Matlab, or some other convenient plotting environment to plot the field

$$u(x, y) = e^{i[\theta(x,y)+\Phi]} e^{ix} + \text{c.c.}, \qquad (E4.15)$$

where $\theta(x, y)$ is the polar angle of the point (x, y) and Φ is some constant, which represents a dislocation defect in a stripe state of unit background wave vector. Verify that the winding number defined in Eq. (4.58) is indeed unity. What happens to the dislocation as you vary Φ? You might notice that the field is actually singular at the origin – the value depends on how the origin is approached. How could you change the expression (E4.15) to make the field nonsingular?

5
Models

In this chapter, we discuss model evolution equations that serve as a bridge between the previous chapter on the qualitative properties of nonlinear saturated states and later chapters on the amplitude equation and phase diffusion equation, which provide a way to understand many quantitative details of pattern formation. Model equations are a natural next step after the previous chapter because they demonstrate a nontrivial insight, that many details of experimental pattern formation can be understood without having to work with the quantitatively accurate but often difficult evolution equations that describe pattern formation in say a liquid crystal or some reaction–diffusion system. Analytical and numerical calculations show that if a simplified model contains certain symmetries (say rotational, translational, and inversion), has structure at a preferred length scale (generated say by a type-I instability), and has a nonlinearity that saturates exponentially growing modes, the model is often able to reproduce qualitative features, and in some cases quantitative details, observed in experiments.

The same insight that mathematically simplified models can have a rich pattern formation is useful for later chapters because model equations provide a more efficient way to carry out and test theoretical formalisms than would be the case for quantitatively accurate evolution equations. We have already used the Swift–Hohenberg model in this way, for example to calculate how the growth rate σ_q of an unstable mode depends on the perturbation wave number q (Eq. (2.11)), or to calculate approximate stationary nonlinear stripe solutions near onset (Eqs. (4.21) and (4.24)). Model equations can also usually be studied numerically more thoroughly than the fully quantitative equations. A generalized Swift–Hohenberg model described below, that describes the evolution of a convection stripe pattern near onset, has just two fields (versus five for experimental convection) that depend on only two spatial variables (versus three for experimental convection). As a result, this model allows aspect ratios and observation times to be investigated that are an order of magnitude larger than would be possible by integrating the Boussinesq equations in a three-dimensional domain. The dependence of the dynamics on parameters can also be examined for many more parameter values than would be the case for the Boussinesq equations.

But model equations are much more important than just being mathematically simpler and numerically more efficient, they often provide valuable scientific intuition. The reason is that many pattern-formation problems have so many details that it is rarely possible to understand which detail is the cause of some observed behavior. For example, researchers have succeeded in reproducing essentially all the features of the spiral defect chaos state of Fig. 1.15(a) by numerical integrations of the three-dimensional Boussinesq equations that quantitatively describe a convecting fluid. While replication of an experimental state by simulation is satisfying, the simulations do not answer some of the most basic questions: why do spirals, rather than stripes, occur in the first place? Why is the dynamics chaotic? Why does the spiral defect chaos change to some other spatiotemporal state as the Rayleigh and Prandtl numbers are varied? And can we expect spiral defect chaos in systems other than a convecting fluid? We will see below that a state similar to spiral defect chaos occurs in a generalized Swift–Hohenberg model that has just two extended variables and no confined variables (see Section 5.2.3). It is then unlikely that the structure of the convecting fluid in the vertical coordinate is essential for understanding spiral defect chaos.

Similarly, the Turing model discussed in Chapter 3 is a classic example of how a simplified model can provide a deep scientific insight. When Turing published his ideas in 1952 as a proposed mechanism for biological morphogenesis, no one knew what fundamental evolution equations to write down that might describe morphogenesis, and the expectation was that any mathematical description would be extremely complicated because of the known complexity of cells and of how cells interact with one another. Turing's model and his related linear stability analysis gave a brilliant insight, that the essence of morphogenesis – the emergence from a uniform state of a patterned structure with a preferred length scale – could be understood in terms of just *two* diffusing and reacting morphogens in a homogeneous medium. Even though Turing's paper provided no quantitative details, it suggested the radical idea that morphogenesis could have a simple origin and the paper stimulated much theoretical and experimental work.

Because model equations are usually based on just symmetry and instability arguments, they are often not constrained by basic conservation laws of energy, momentum, and mass so there is greater flexibility to add or remove terms that can test various hypotheses. For example, we will discuss how it is easy to modify the Swift–Hohenberg equation to be potential, in which case the only nontransient dynamics are time-independent states, or to be nonpotential, in which case periodic, quasiperiodic, or chaotic behavior is possible. Other simple modifications can eliminate an inversion symmetry so that hexagons rather than stripes occur near onset, break a chiral symmetry which would correspond to a constant rotation of a convection experiment and that leads to spatiotemporal chaos near onset, mimic the effects of horizontal plates of poor thermal conductivity which cause square lattice states to form, make the primary bifurcation subcritical which allows localized states to occur, and so on.

Of course to some degree, any theoretical description of a real system is a model that involves various approximations that might be controlled and valid in some limit, or uncontrolled or more phenomenological, and so a complete description of models will be no simpler than a complete description of pattern formation. In this chapter, we discuss representative models that have been important in the study of pattern formation, with an emphasis on models that have been proposed to understand qualitative features, rather than those that are derived by controlled approximations and so are intended to apply quantitatively to experimental phenomena. These models are simplified compared to the fully quantitative evolution equations, not in having a "few degrees of freedom" as is the case for the logistic map and Lorenz equations but by having fewer spatial dimensions and fewer fields.

A final purpose of this chapter is to give the reader a sense of how one discovers or invents model equations. Pattern-formation models have diverse origins. Some of them were guessed at using scientific intuition, some of them were accidentally discovered when using perturbation theory to derive approximate solutions from fully quantitatively evolution equations, and others were derived by putting together the simplest ingredients consistent with known symmetries and linear instabilities. We begin our discussion with an example of this last case, the Swift–Hohenberg model that was originally invented and studied in the context of convection but since has been applied to other topics such as laser patterns, material science, and orientation maps of the mammalian visual cortex.

5.1 Swift–Hohenberg model

We have already used the Swift–Hohenberg model several times in this book to illustrate some of the formal techniques for analyzing pattern formation. Studying the Swift–Hohenberg model before studying more realistic evolution equations actually reverses the historical development of pattern formation in that most analytical techniques were first developed in the context of fully quantitative evolution equations and only later were simplified models derived and investigated.[1] The great diversity of phenomena observed in convection experiments near onset together

---

[1] It is interesting to note that Swift and Hohenberg's original motivation to write down their equation in 1977 was not to study pattern formation, but to understand to what extent the nonequilibrium transition from a uniform to nonuniform convecting state was similar to an equilibrium phase transition. Only several years after this model was published did numerical calculations (some shown in Fig. 5.2) reveal its unexpectedly rich spatiotemporal dynamics, which in turn motivated others to study this equation as a model of pattern formation.

with the ease with which many of these phenomena could be investigated by simple changes to the original Swift–Hohenberg equation have made this equation one of the better examples of how model equations can be developed and studied to understand pattern formation.

In this section, we discuss the Swift–Hohenberg equation in greater detail. We explain how this equation can be derived heuristically from symmetry arguments and from the assumption of a type-I-s transition, we show that the equation is potential, we illustrate the richness of its dynamics with some numerical simulations, and we discuss how to compare experimental parameter regimes of convection with parameter regimes of this equation. In the following section, we discuss generalizations of the Swift–Hohenberg equation that take new physical phenomena into account, often by adding terms that break some symmetry such as a rotational or inversion symmetry.

5.1.1 Heuristic derivation

The starting point of a heuristic derivation is to assume that we are interested in a stationary type-I-s instability of some uniform state in a system that is itself rotationally invariant in a two-dimensional plane. The uniformity of the base state together with the rotational invariance of the system implies that the growth rate $\sigma_\mathbf{q}$ of an infinitesimal sinusoidal perturbation of the uniform state with two-dimensional wave vector $\mathbf{q}$ can only depend on the magnitude q of the wave vector but not on its orientation. (It may be helpful to review the related discussion in Section 2.1.) We can then expand this growth rate to lowest order about the onset of the primary instability and about the finite critical wave number q_c to obtain the following expression:

$$\sigma_\mathbf{q} \simeq p - c(q - q_c)^2. \tag{5.1}$$

Here p is the control parameter, and we have subtracted off constants so that the primary instability of the uniform state occurs at $p = 0$ with wave number $q = q_c$ as p increases from negative values. For a type-I-s instability, the constant c must be positive if Eq. (5.1) is to have a local maximum at $q = q_c$ for $p = 0$.

If we multiply both sides of Eq. (5.1) by the single Fourier mode

$$u(\mathbf{x}, t) = e^{\sigma_\mathbf{q} t} e^{i\mathbf{q} \cdot \mathbf{x}_\perp}, \tag{5.2}$$

and observe that $\sigma_\mathbf{q} u = \partial_t u$ while $q^2 u = (-\nabla^2) u$ (where $\nabla^2 = \partial_x^2 + \partial_y^2$ is the two-dimensional Laplacian), then the growth rate Eq. (5.1) can be interpreted as an evolution equation for a Fourier mode:

$$\partial_t u(t) = \left(p - c\left(\sqrt{-\nabla^2} - q_c\right)^2\right) u, \tag{5.3}$$

and this becomes a starting point for obtaining an evolution equation for some general field $u(\mathbf{x}, t)$ related to a type-I-s pattern-forming system.

Equation (5.3) has two serious flaws as a possible more general evolution equation. One is associated with the expression $\sqrt{-\nabla^2}$ which is too difficult to work with unless used inside the Fourier expansion of some field.[2] Fourier expansions are generally not convenient for pattern-forming problems since the Fourier representation of some nonlinear term in an evolution equation involves convolutions (sums of products of Fourier coefficients) that are difficult to work with analytically. A second difficulty is that Fourier representations are best suited for problems on periodic or infinite spatial domains, and such domains exclude most of the experimentally interesting situations.

The difficulty of obtaining an evolution equation that does not require a Fourier representation can be traced to the wave number magnitude q that appears in Eq. (5.1); the difficulty would go away if we could rewrite this equation in terms of q^2, which corresponds to the familiar Laplacian $-\nabla^2$ of a field which is easy to work with analytically and numerically. Equation (5.1) can in fact be expressed in terms of q^2 if we are willing to allow a small systematic error. As discussed in Chapter 2, sufficiently close to onset (here corresponding to the parameter p being positive and arbitrarily small), only wave numbers q close to the critical wave number q_c occur. Thus sufficiently near onset, $q + q_c \approx 2q_c$, so we can approximate the right-hand side of Eq. (5.1) as follows:

$$\sigma_\mathbf{q} \simeq p - c(q - q_c)^2 \simeq p - c\left(\frac{q + q_c}{2q_c}\right)^2 (q - q_c)^2, \qquad (5.4)$$

$$\simeq p - \frac{c}{4q_c^2}(q^2 - q_c^2)^2. \qquad (5.5)$$

Equation (5.5) reduces to Eq. (5.1) to lowest order in $q - q_c$ on expanding about q_c and suggests that the linear terms in the evolution equation for some real field $u(x, y, t)$ could be[3]

$$\partial_t u = ru - \left(\nabla^2 + 1\right)^2 u, \qquad (5.6)$$

where we have chosen spatial units such that $q_c = 1$, chosen time units such that $c/(4q_c^2) = 1$, and then defined $r \propto p$ to eliminate any constant factors in the first term on the right-hand side. (Problem 1.7 on page 53 gives you a chance to work out these scalings.)

---

[2] There are several ways to define a real-space "fractional derivative" like $\sqrt{-\nabla^2}$ and they all involve infinite sums of integer powers of the Laplacian, too awkward to work with analytically and impractical to work with numerically. Section 5.3 does discuss the possibility of using the exact growth rate $\sigma_\mathbf{q}$ in a model equation, but the calculations are practical only for Fourier representations in periodic domains.

[3] The expression $(\nabla^2 + a)^2$ for a a constant should be understood as an abbreviation for $(\nabla^2)^2 + 2a\nabla^2 + a^2$, which is called a generalized constant-coefficient biharmonic operator.

For $r > 0$, Eq. (5.6) yields exponentially growing solutions

$$ae^{i\mathbf{q}\cdot\mathbf{x}_\perp}e^{\sigma_q t} + \text{c.c.},\tag{5.7}$$

with growth rate

$$\sigma_q = r - (q^2 - 1)^2.\tag{5.8}$$

For finite solutions to exist at long times, some nonlinear term that decreases the growth rate must be added to Eq. (5.6). Often the cubic term $-u^3$ is used since this is arguably the simplest nonlinearity that is consistent with the inversion symmetry $u \to -u$ which is often present in physical systems (for example the Boussinesq equations describing convection).[4] This cubic term yields the Swift–Hohenberg equation

$$\partial_t u(x, y, t) = ru - \left(\nabla^2 + 1\right)^2 u - u^3.\tag{5.9}$$

By an appropriate choice of units for the field u, the coefficient of the cubic term can be assumed to have the value one.

To model systems on a realistic finite spatial domain, conditions for the field u must be given at each point on the boundaries of the domain. Since Eq. (5.9) involves fourth-order spatial derivatives (from the Laplacian squared), two arbitrary conditions need to be specified at each boundary point to have a well-defined initial-boundary-value problem. If $\hat{\mathbf{n}}$ denotes the unit vector normal to the boundary at a given point (and that points away from the interior for a simply connected domain), then the boundary conditions

$$u = 0 \quad \text{and} \quad (\hat{\mathbf{n}} \cdot \nabla)u = 0,\tag{5.10}$$

are commonly used. Empirically, numerical simulations with these boundary conditions reproduce the tendency often seen in large-aspect-ratio experiments that unconstrained stripes approach the boundary orthogonally. Analytically, one can show that these boundary conditions reproduce a result to lowest order of the amplitude equation formalism discussed in Section 6.2.4 of the next chapter, that the amplitude goes to zero at the boundary for stripes that are constrained to be parallel to a straight boundary. But other boundary conditions are possible and have been investigated with the goal of understanding how constraints at the lateral boundaries affect pattern formation in the domain's interior.

Equation (5.9) with boundary conditions Eq. (5.10) is an attractively simple model and numerical investigations show that interesting patterns occur that are similar to those observed near onset in large convection experiments. However, it should be clear that the above derivation involved several ad hoc uncontrolled

---

[4] A quadratic term $\pm u^2$ is not by itself saturating, since this enhances the growth of one sign of u. The behavior of the equation with the combination $\alpha u^2 - u^3$ is discussed later.

steps so quantitative agreement with experiments is not guaranteed, even close to onset. Firstly, replacing Eq. (5.1) by Eq. (5.5) introduces terms into the growth rate that are higher order in the quantity $q - q_c$ and these presumably will not match the higher order terms for a real physical system. The effect of such higher-order terms on the model's dynamics is not known in advance. However, since larger wave numbers q correspond to smaller length scales, we might expect the model to reproduce slow long-wavelength (small q) spatial modulations of a stripe pattern well, but some short length scale behavior less reliably, such as perhaps stripe pinch-off events in which a stripe is eliminated, leaving one or two dislocations in its place. Secondly, the form of the nonlinear term is motivated purely by simplicity rather than by accuracy. Although there is considerable flexibility in what nonlinear terms to use (see the examples in Section 5.2.1 below), a choice has to be made carefully since changes in the functional form or even in the numerical values of coefficients of a given functional form can lead to stripe dynamics (the case for the Swift–Hohenberg equation) or to rectangular or hexagon lattice dynamics. (As an example, see Exercise 5.6 on page 204.) Finally, although the form of the equation is motivated by an expansion of the growth rate about threshold $r \simeq 0$, the model behavior is often investigated for larger values of r.

5.1.2 Properties

An important property of the Swift–Hohenberg is that it has so-called potential dynamics (also called in various contexts relaxational dynamics or a gradient flow). This means that there is a potential (also called a Lyapunov functional) which is a functional[5] of the Swift–Hohenberg field $u(x, y, t)$ that has the property of decreasing monotonically during the dynamics. The potential is thus analogous to the total energy (potential and kinetic) of a frictionally damped ball in a potential, or to the free energy of a thermodynamic system that is nearly in equilibrium. Just as we know that the motion of a damped ball will eventually cease with the ball at a minimum of the total energy, the evolution of the Swift–Hohenberg field will eventually cease, leaving a pattern that corresponds to some local minimum of the potential. The potential greatly simplifies the analysis of the equation but actually detracts from its use as a representative model of pattern formation since sustained nonequilibrium systems are generally not expected to show potential dynamics.[6]

---

[5] A functional is some mathematical rule that associates a number with a function. Some simple examples of functionals are the definite integral of a function or its global maximum over some domain. Most of the functionals discussed in pattern formation are integrals over some domain of some nonlinear expression containing fields, and spatial derivatives of the fields, related to some pattern-forming system.

[6] One reason is that potential dynamics represents a strong mathematical constraint and there is no known reason, say related to conservation laws or some symmetry, why nonequilibrium systems should satisfy such a constraint. A second reason is that experiments and simulations of stripe-forming systems show that a sustained

We define the potential $V[u]$ by the following area integral over the domain of interest[7]

$$V = \iint dx\,dy \left\{ -\tfrac{1}{2}ru^2 + \tfrac{1}{4}u^4 + \tfrac{1}{2}\left[\left(\nabla^2 + 1\right)u\right]^2 \right\}, \qquad (5.11)$$

and claim that, for periodic boundary conditions or for the boundary conditions Eq. (5.10), V evolves according to the equation

$$\frac{dV}{dt} = -\iint dx\,dy\,(\partial_t u)^2. \qquad (5.12)$$

Equation (5.12) implies that whenever $\partial_t u \neq 0$ anywhere in the system, the potential is decreasing, which establishes that the Swift–Hohenberg equation has potential dynamics. Since you can show in Exercise 5.2 that Eq. (5.11) is bounded below by the quantity $-\tfrac{1}{4}r^2 S$ with S the area of the domain, V cannot decrease forever and the dynamics must become constant at long times.

Potential dynamics with a potential bounded from below therefore has the implication that *all* initial states eventually become time-independent. This means that the Swift–Hohenberg model with periodic boundaries or with the boundary conditions Eq. (5.10) cannot have periodic, quasiperiodic, or chaotic dynamics.[8] However, the potential dynamics does not rule out the possibility that the transient leading to some stationary state might have a complicated temporal dependence that could appear to be periodic or chaotic if observed over an insufficiently long time. Indeed, the Belousov–Zhabotinsky chemical reaction discussed in Section 3.1.3 is an example of a pattern-forming system with potential dynamics since the experiment is a closed system (no influx of reagents, no efflux of chemical products) and so the system must eventually reach thermal equilibrium. (The free energy is the potential for this system.) But the transient toward a stationary state is long lived and the short time dynamics is periodic to good accuracy. The first reports of oscillatory dynamics were initially doubted, because of the mistaken thought that a monotonically decreasing potential implies that the concentrations must themselves monotonically change over time, which is not the case.

---

time dependence can occur quite close to onset and such a time dependence is not consistent with potential dynamics.

[7] Expressions like V are not just snatched out of thin air, there is a considerable history of researchers thinking about such potentials in the context of soft condensed matter physics, such as the theory of liquid crystals. In fact, the dynamics of the Swift–Hohenberg model has several similarities to the dynamics of smectic liquid crystals.

[8] That the asymptotic dynamics of the Swift–Hohenberg model with appropriate boundary conditions must be stationary is a rigorous result. However, this conclusion does not necessarily hold for simulations of Eq. (5.9) which necessarily involves approximations of the mathematics (Chapter 12 discusses some of these issues). And indeed numerical integrations with time or spatial resolutions that are too coarse, or with a time integration algorithm that is unstable, can lead to oscillatory or even chaotic dynamics that have nothing to do with the mathematical solutions.

To establish Eq. (5.12), take the time derivative of both sides of Eq. (5.11) to get

$$\frac{dV}{dt} = \iint dx\,dy \left\{ \left(-ru + u^3\right)\partial_t u + \left[\left(\nabla^2 + 1\right)u\right]\left[\left(\nabla^2 + 1\right)\partial_t u\right]\right\}, \quad (5.13)$$

and twice integrate by parts the term $\left[(\nabla^2 + 1)u\right]\nabla^2 \partial_t u$ to transfer the operation of ∇^2 from $\partial_t u$ to the quantity $(\nabla^2 + 1)u$. The first integration by parts creates a vector term $\left[(\nabla^2 + 1)u\right]\nabla u$ that is evaluated on the domain's boundary, and the second integration by parts creates a vector term $\partial_t u\, \nabla\left[(\nabla^2 + 1)u\right]$ that is also evaluated on the domain's boundary. You may work through the details in Exercise 5.2, where you show that the boundary (surface) terms vanish for periodic boundary conditions and for the boundary conditions Eq. (5.10). What remains after the two integration by parts is the expression

$$\frac{dV}{dt} = \iint dx\,dy \left\{ \left[-ru + u^3 + \left(\nabla^2 + 1\right)^2 u\right]\partial_t u \right\}. \quad (5.14)$$

The bracketed term is just $-\partial_t u$ so that Eq. (5.14) indeed becomes Eq. (5.12). The Swift–Hohenberg equation itself can be written in terms of the functional derivative $\delta V/\delta u$ of V

$$\partial_t u = -\frac{\delta V}{\delta u}, \quad (5.15)$$

which shows directly that the dynamics of u "runs down" the potential, by analogy to Newton's second law of motion $m(d\mathbf{v}/dt) = -\nabla V$ for a particle of mass m responding to a potential force $\mathbf{F} = -\nabla V$.

The existence of a potential leads to many useful deductions. An example is the question of the competition between two patterns, e.g. between a stripe state and hexagonal lattice state, if both are present in the system with a wall or domain boundary between them. We can argue that if there is any motion of the wall, it must be in the direction that increases the fraction of the pattern with the lower value of the potential density $v = V/S$ (the potential per unit area).[9] The contribution to the potential density from the domain wall itself does not change as the wall translates. Thus the parameter value for which the two states have equal potential densities can be used to identify the point at which the preferred pattern switches from one to the other.

Two caveats should be stated. The first is that the result applies *only in the context of an experiment in which the competition between bulk saturated regions, in contact via a domain wall, occurs*. Other experimental conditions, such as the

---

[9] We can see from the definition Eq. (5.11) that, for patterns that are periodic, the potential is an approximately extensive quantity, whose value for some field is approximately proportional to the area for large areas, since Eq. (5.11) simply adds up the identical contributions of each identical unit cell and there will be many unit cells in a large domain. So the appropriate quantity to compare for two competing periodic states that may occupy different areas in some domain is the intensive potential per unit area, i.e. the potential density.

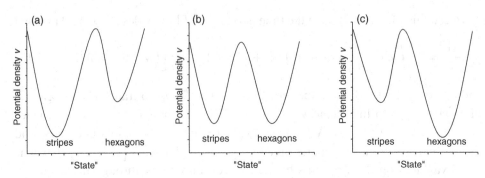

Fig. 5.1 Competition between patterns for a system with potential dynamics. The variation of the potential density v between states of ideal stripes and hexagons is shown for the case that (a) the stripe state has the lower potential density, (b) the stripe and hexagon states have the same potential density, and (c) the hexagon state has the lower potential density. In general, there will be a potential density maximum or *barrier* between the two minima so that dynamics from the higher minimum to the lower cannot be inferred. Also a dynamical path that connects one state with another (e.g. in (a), the higher hexagon minimum to the lower stripe potential minimum) can be a long winding path, corresponding to a long-lived transient with complex spatiotemporal structure.

growth from small initial conditions, may give different results. This is because in general there is a potential barrier between the two ideal states (see Fig. 5.1) and only in special physical circumstances, for example the domain wall between two coexisting regions, is there a dynamical path between the two states that flows monotonically down the potential. Secondly, the motion may be impeded, for example by pinning of the wall to the stripes themselves, in which case there may be no motion even if the potentials are different. This situation, which is discussed briefly in Section 9.2.1 in the context of "nonadiabatic effects," would then give a finite range of parameters for coexistence.

A second application of the potential is to the question of wave number selection, the precise value of the wave number in a stripe state or unit cell size in the lattice states. Again we can argue that any local dynamics that mediates between two ideal states (that occupy large portions of the system) will evolve so as to favor the state with a lower value of the potential $v(q)$, which will then dominate the integral that forms the potential. (Here $v(q)$ is the potential density evaluated for the ideal periodic state with wave number q.) In the case of potential dynamics, different dynamical mechanisms that allow the wave number to change, such as dislocation motion, boundary relaxation etc., will all tend to yield the *same* wave number, namely the one that minimizes $v(q)$. For systems without a potential, there is no such argument, and different dynamical mechanisms may lead to different wave numbers. Conversely, this leads to the following conclusion: *it is possible to test a* sustained nonequilibrium system experimentally for the existence of a potential – without knowing the evolution equations of the system – by examining whether different wave number selection mechanisms lead to the same wave number.

The motion of defects can also be considered using the potential. If we consider the climb of an isolated dislocation in a large system (see Section 4.4.2), the dislocation must move in a direction that changes the wave number to a value that decreases the potential, in accordance with Eq. (5.15). Thus a question that is generally hard – which way a dislocation moves – becomes easy to answer for an equation with a potential. Similarly a target defect also discussed in Section 4.4.2 will be stationary if the asymptotic wave number is the one that minimizes the potential.

The importance of the Swift–Hohenberg equation in the discussion of pattern formation points out an interesting fact, that some of the intriguing and difficult questions associated with pattern formation come from the complicated geometrical structure of the patterns and not particularly from the fact that the systems are far from equilibrium. Even for equilibrium systems, the understanding of stripe systems is far less advanced than that of two- and three-dimensional lattices. (Examples of equilibrium stripe systems are smectics and diblock copolymers.) In addition to the range of questions that arise in general for stripe structures, we must come to grips with the nonequilibrium aspects of the systems for stripe patterns far from equilibrium. The original Swift–Hohenberg equation is not of use here and so various modifications that eliminate the potential nature of the dynamics have been proposed.

5.1.3 Numerical simulations

The Swift–Hohenberg equation is too difficult to solve analytically except in simple cases such as time-independent stripe or lattice states in periodic domains close to onset, and so numerical simulations[10] have played an important role in exploring what kinds of dynamics are possible. Since the equation is potential, the main insights learned concern how some initial state relaxes to a steady state pattern, and the role of the lateral boundaries in influencing the pattern. We illustrate with a pair of examples.

Figure 5.2 shows some simulations in large square and rectangular geometries with the boundary conditions Eq. (5.10). Note that these simulations, which were performed over twenty-five years ago at the time of writing, were carried out to times of order 10^4 characteristic time scales in domains with aspect ratios of 16 or larger. Comparably long computations on systems such as the fluid and heat equations for

---

[10] Chapter 12, especially Section 12.3.3 and Exercise 12.10 (E12.16), indicates some of the details associated with integrating the Swift–Hohenberg and Kuramoto–Sivashinsky models that have a biharmonic operator.

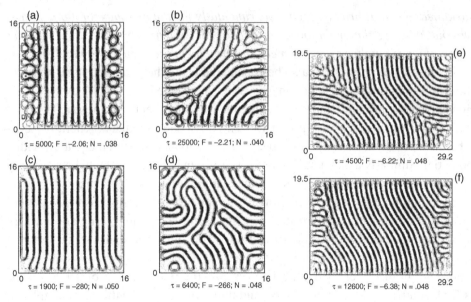

Fig. 5.2 Simulations of the Swift–Hohenberg equation: (a), (b), (e), and (f) for $r = 0.1$ and (c), (d) for $r = 0.9$. The initial conditions were parallel stripes for (a) and (c) and random for the others. The values of τ give the time of the snapshot, F is the value of the potential at that time (the V of Eq. (5.11)), and N is mean square average of u. Panel (e) is still evolving in time, eventually giving the pattern in (f). The other states have reached a steady state to within the resolution of the simulations. (From Greenside et al. [42].)

three-dimensional Rayleigh–Bénard convection are only just becoming feasible on the most powerful parallel supercomputers. Insights gained from these simulations include the tendency of the stripes to approach the boundaries at a perpendicular orientation, either through a short set of cross-stripes near the boundary (panels (a) and (f)) or by forcing curvature of the stripes over the scale of the system. The wide variety of defects induced by the boundaries and curvature is also apparent. The values of the potential (called F in the figure) can be used to compare the "relative stability" of the various patterns that may occur in the same geometry and at the same value of the control parameter. Since the potential can only decrease in the dynamics, a state of lower F cannot evolve into one with higher F. On the other hand, the system may get stuck in a non-optimal state such as panel (d) which has a larger value of F than the pattern in panel (c).

Simulations of the Swift–Hohenberg equation can also be used to study the quantitative aspects of the approach to steady state from random initial conditions, as shown in Fig. 5.3. The simulations show that the pattern evolves through states that can be characterized as domains of relatively well ordered stripes with an average domain size that increases steadily as time evolves; the pattern is said

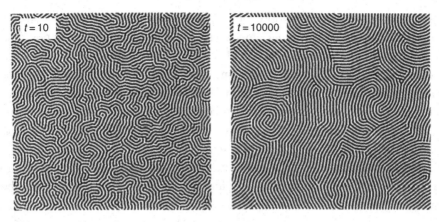

Fig. 5.3 Simulation of the Swift–Hohenberg equation in a large periodic geometry of size 256 × 256 with $r = 0.25$, starting from random initial conditions. The increase in the average domain size from time 10 to time 10 000 is called "coarsening." (From Elder *et al.* [33].)

to coarsen. In the simulations shown, the typical size ξ_d of a domain, which can be obtained by a numerical Fourier analysis of the field $u(\mathbf{x}, t)$, is found to grow as a power law $\xi \propto t^p$ with p about 0.2. Whether this is a true description of the coarsening process and whether the exponent p has different values for nonpotential models are topics that remain under active discussion at the time of writing.

5.1.4 Comparison with experimental systems

Although simple models such as the Swift–Hohenberg equation cannot be expected to give a quantitatively accurate description of an experimental system, the practical question arises of how to choose the parameters of this or other models to correspond to a given experiment. For example, in the Swift–Hohenberg equation we might ask: "What value of the parameter r, what aspect ratio of a circular cell, and what integration time should be used to model a Rayleigh–Bénard convection experiment for Rayleigh number $R = 2000$ in a cylindrical cell of aspect ratio $\Gamma = 30$ for ten horizontal diffusion times?" A first step would be to match the length and time units of the model and experimental system so that observation times and domain sizes correspond. For the Swift–Hohenberg model, Eq. (5.9), length is measured in units such that $q_c = 1$, whereas in a Rayleigh–Bénard experiment, length might be measured in units of the cell depth, in which case $q_c \simeq 3.12$. In this case, an experimental aspect ratio Γ would have to be divided by 3.12 (made smaller) to obtain the aspect ratio for a corresponding model equation simulation.

A second step would be to choose model parameters to match the growth near the linear instability for the particular physical system of interest as well as possible.[11] (The value of the field at saturation is determined by normalization conventions, and does not lead to any further matching conditions.) If this is done, the model equation should give a good description of the slow space and time modulations of a regular pattern, and a comparison of the model and convection dynamics may reveal that some details (say of a spatiotemporal chaotic state) are not sensitive to details on "fast" length and time scales such as defect nucleation and annihilation.

To perform the matching we expand the growth rate of a mode at wave number q near the critical wave number $q_c = 1$ and for small r for the Swift–Hohenberg equation to give

$$\sigma_q \simeq r - 4(q - 1)^2. \tag{5.16}$$

On the other hand, the growth rate near a general type-I-s instability is parameterized in terms of parameters τ_0 and ξ_0 in Eq. (2.27). We want to match these expressions for some particular value of the reduced control parameter ε. Since threshold for the Swift–Hohenberg equation is at $r = 0$, we write

$$r = \lambda \varepsilon, \tag{5.17}$$

where λ is the proportionality constant to fix. Dividing Eq. (5.16) through by λ leads to

$$\lambda^{-1}\sigma_q = \varepsilon - 4\lambda^{-1}(q - 1)^2. \tag{5.18}$$

This should be compared with the general expression for a type-I-s instability written in the form

$$\tau_0^{-1}\sigma_q = \varepsilon - (\xi_0 q_c)^2 (q/q_c - 1)^2. \tag{5.19}$$

Matching parameter values gives

$$(q_c \xi_0)^2 = 4\lambda^{-1} \quad \text{and} \quad \tau_0 = \lambda^{-1}. \tag{5.20}$$

The parameter τ_0 is measured in some "Swift–Hohenberg time unit," and we are still free to choose the relationship of this unit to the physical time unit so that τ_0 matches the corresponding value for the physical system. However, the product $\xi_0 q_c$ is dimensionless (its value does not change with a change in length units) so this is a quantity that should be matched between the two equations. For Rayleigh–Bénard convection, $\xi_0^2 \simeq 0.148$ and $q_c \simeq 3.12$ giving $\xi_0 q_c \simeq 1.20$, and so to match a Swift–Hohenberg equation to this system λ must have the approximate value 2.78 to make these two expressions equal. We conclude that the Swift–Hohenberg parameter

---

[11] More formally we would match the coefficients of the amplitude equation to be introduced in Chapters 6 and 7 that quantitatively describe this regime.

value should be about three times bigger than the numerical value of the reduced Rayleigh number $\varepsilon = R/R_c - 1$

$$r \simeq 2.78\varepsilon, \qquad (5.21)$$

for a model simulation to correspond to some convection experiment near onset. Similar comparisons can be made for other stripe-forming systems.

The behavior of the Swift–Hohenberg equation depends on the single parameter r and so there are no other parameters (except the system size) to tune. In more complicated examples such as the generalized models studied in the next section, there may be more parameters to tune. In this case, a third step to match parameters can be carried out, which is to compare the stability balloons of the model and of the experimental system. Here some choices must be made since the model will only have a few parameters, and so a match to the whole stability balloon will not be possible.

5.2 Generalized Swift–Hohenberg models

In this section, we introduce some of the generalizations that have been made to the Swift–Hohenberg model to incorporate additional physics. As in the motivation for the original equation, the key consideration in developing these enhancements is the symmetry of the new physical effect that it is desired to include.

5.2.1 Non-symmetric model

The original Swift–Hohenberg equation is symmetric under inversion of the field $u \to -u$. Physical systems without this symmetry[12] can be modeled by including a quadratic nonlinearity γu^2, which is one of the simplest possible terms that eliminates this symmetry

$$\partial_t u = ru - (\nabla^2 + 1)^2 u + \gamma u^2 - u^3. \qquad (5.22)$$

For reasons to be discussed in Section 7.3.3, in two dimensions the uniform solution $u = 0$ undergoes a transcritical bifurcation to a hexagonal pattern. Equation (5.22) can be used to study several questions related to the competition between stripe and hexagonal states such as wave number selection, front propagation, and defect motion.

---

[12] An example would be a convection system for which the temperature difference between the bottom and top plates is so large that some fluid parameters such as the fluid viscosity start to vary significantly with position. This is called "non-Boussinesq convection," and the experimental observation of hexagons near onset is a common indicator of such convection.

5.2.2 Nonpotential models

The potential nature of the Swift–Hohenberg equation can be eliminated by changing the form of the nonlinearity. One nonpotential version is

$$\partial_t u = ru - (\nabla^2 + 1)^2 u + \alpha \nabla \cdot \left[(\nabla u)^2 \nabla u \right] + \beta (\nabla u)^2 \nabla^2 u. \tag{5.23}$$

Unlike most Swift–Hohenberg models, Eq. (5.23) can actually be derived systematically near onset as the nonlinear equation that describes convection between thin poorly conducting plates, for example a layer of liquid mercury convecting between thin glass plates. An interesting feature of such convection is that the critical wave number q_c becomes small as the plate conductivity becomes small so that the size of a convection roll becomes much larger than the depth of the fluid. This is a convection example of what is already familiar with reaction–diffusion systems, that the basic length scale of some pattern is not always determined by the geometry of the container, for example by the depth of the fluid. For $\beta = 0$, the equation is potential, but for general values of α and β, there is no known potential. Equation (5.23) with $\alpha = 0$ can be used as a simple Swift–Hohenberg model without a potential.

An even simpler nonpotential model is

$$\partial_t u = ru - (\nabla^2 + 1)^2 u - (\nabla u)^2 u. \tag{5.24}$$

Numerical simulations of Eq. (5.24) in large square domains (starting with various initial conditions) always lead to states that become time independent. Thus nonpotential dynamics is necessary but not sufficient for oscillatory or chaotic dynamics to be observed. Usually some specific mechanism, like the mean flow discussed in the following subsection, must be included to observe sustained time-dependent states.

5.2.3 Models with mean flow

As we will discuss later in Section 9.1.3, an additional ingredient in the physics of pattern formation in fluid systems such as Rayleigh–Bénard convection is a mean flow that tends to advect the pattern, and is in turn driven by slow spatial modulations of the pattern such as curvature of stripes and gradients of the wave number. The mean flow introduces new physical behavior since it acts as a long-range coupling between different regions of the pattern. The Swift–Hohenberg equation can be modified to include this physics by adding a second field $\mathbf{V}(\mathbf{x}, t)$ representing the velocity of the mean flow, which then advects or carries along the field u representing the stripes. The advection is incorporated by modifying the dynamical equation for $u(\mathbf{x}, t)$ to include an advection term $\mathbf{V} \cdot \nabla u$ in the time

5.2 Generalized Swift–Hohenberg models

derivative

$$\partial_t u \to \partial_t u + \mathbf{V} \cdot \nabla u, \tag{5.25}$$

where ∇ is the two-dimensional gradient (∂_x, ∂_y). Including an advection term can be done for either the original or generalized Swift–Hohenberg equations, e.g. the original Swift–Hohenberg equation becomes

$$\partial_t u + \mathbf{V} \cdot \nabla u = ru - (\nabla^2 + 1)^2 u - u^3. \tag{5.26}$$

The mean flow $\mathbf{V}$ is an incompressible two-dimensional velocity field, and can therefore be written in terms of a stream function ζ

$$\mathbf{V} = (V_x, V_y) = \nabla \times (\zeta \hat{\mathbf{z}}) = (\partial_y \zeta, -\partial_x \zeta). \tag{5.27}$$

Thinking of x and y as horizontal coordinates, this stirring flow in the horizontal plane corresponds to a vertical vorticity Ω given by

$$\Omega \hat{\mathbf{z}} = \nabla \times \mathbf{V} = -\nabla^2 \zeta \, \hat{\mathbf{z}}. \tag{5.28}$$

The mean flow $\mathbf{V}$ is in turn driven by distortions of the pattern, so we must now find these driving terms. In the spirit of the type of arguments that went into the original Swift–Hohenberg equation, we seek the simplest terms that are consistent with the symmetries of the problem such as rotational invariance and inversion symmetry $u \to -u$. If we formulate the argument in terms of the vorticity, we need to construct an axial vector in the z-direction, and the only vector we can form out of the scalar field u is ∇u and its derivatives. To get an axial vector, we must form a cross product out of these vectors. Since $\nabla u \times \nabla u$ is zero, the cross product must involve different (scalar) derivatives of ∇u. The form with the smallest number of derivatives is $\nabla u \times \nabla(\nabla^2 u)$. This string of arguments leads to the driving term for the mean flow

$$\Omega = -g\hat{\mathbf{z}} \cdot \nabla u \times \nabla(\nabla^2 u), \tag{5.29}$$

with g a constant that specifies the strength of the mean flow. (Roughly, g is inversely proportional to the fluid Prandtl number, and becomes negligible in the limit of large Prandtl number.) The mean flow velocity is calculated from this using Eq. (5.28) to write an equation that can be solved for the stream function

$$\nabla^2 \zeta = g\hat{\mathbf{z}} \cdot \nabla u \times \nabla(\nabla^2 u), \tag{5.30}$$

and then using Eq. (5.27) to evaluate the mean flow velocity.[13] With some work, one can show that Eq. (5.29) reproduces Eq. (9.16) in a weakly nonlinear treatment of the equations, which further confirms that this is a sensible expression to use.

Notice that Eq. (5.30) is reminiscent of the Poisson equation for an electrostatic potential with the right-hand side acting as the source charge density. We know that electrostatics involves long-range interactions, and so this equation suggests that adding the mean flow equation will introduce long-range coupling into the pattern formation problem.

Equation (5.29) specifies the vorticity instantaneously in terms of the local pattern. On the other hand we know that a flow will have some inertia, and so will continue to respond for some time after the application of a driving force. To include this inertia, sometimes a dynamical equation is used for the vorticity

$$\tau_v \, \partial_t \Omega + \Omega = -g \hat{\mathbf{z}} \cdot \nabla u \times \nabla(\nabla^2 u), \tag{5.31}$$

which gives a vorticity field that relaxes to the value specified by the right-hand side on some time scale specified by τ_v. Usually we are interested in applying the generalized Swift–Hohenberg equation to look at the slow evolution of the pattern over time scales that are long compared with the basic time scales of the equation. In these situations, the time derivative term in Eq. (5.31) will be small compared with the second term on the left-hand side and can be neglected.

The coupled equations (5.26) and (5.29) or (5.31) are complicated, and most investigations of their properties have involved numerical simulations. The coupling to the mean flow eliminates the potential nature of the equations, and one interesting result that has been studied in some detail is the existence of spatiotemporal chaos with visual structure quite similar to the spiral defect chaos seen in Rayleigh–Bénard and other experiments.

5.2.4 Model for rotating convection

If a horizontal convecting fluid layer is physically rotated with a constant angular frequency about an axis perpendicular to the layer, experiments show that a state of spatiotemporal chaos called "domain chaos," shown in Fig. 1.15, is found arbitrarily close to threshold for sufficiently large rotation rates. (Domain chaos will be discussed further in Section 9.2.4.) This discovery is enormously interesting since chaotic states near supercritical bifurcations of uniform states are rare, and, for such

---

[13] A natural boundary condition for Eq. (5.30) would be that the mean flow field vanishes, $\mathbf{V} = 0$ but this imposes two conditions at each boundary point, $\partial_y \zeta = \partial_x \zeta = 0$, while only one condition is needed for a Poisson problem. While one could choose arbitrarily a boundary condition such as $\zeta = 0$ or $\hat{\mathbf{n}} \cdot \nabla \zeta = 0$, another possibility is to replace the Laplacian ∇^2 with a biharmonic ∇^4 which would allow two conditions at each boundary point to be imposed. Only a small physical error is incurred with such a substitution.

5.2 Generalized Swift–Hohenberg models

states, researchers can investigate spatiotemporal chaos using the weakly nonlinear amplitude equation formalism that will be discussed in the following three chapters. The dynamics leading to the chaos can be associated with a reduction in the symmetry of the system since clockwise and counter-clockwise rotations of the pattern are no longer equivalent. Since symmetries motivated the form of the Swift–Hohenberg equation, we might expect a modified form of the equation to describe the rotating system, leading perhaps to chaotic dynamics whereas the unmodified equation cannot show chaos. This is indeed the case. The resulting model is of great interest since it provides a way to study an example of spatiotemporal chaos without using the full fluid equations.

There are no new linear terms that can be constructed deriving from this reduced symmetry,[14] and this is consistent with the physical observation that stationary stripe states remain solutions of the rotating Rayleigh–Bénard system (stationary in a frame corotating with the container). At the nonlinear level, however, we can find new terms that would necessarily have zero coefficients when forward and backward rotations are equivalent. The simplest nonlinear term has a structure identical to the driving term of the vorticity, which should not be surprising since rotation and vorticity have the same structure of an axial vector in the $\hat{z}$-direction. These considerations motivate the following equation

$$\partial_t u(x, y, t) = \left(r - \left(\nabla^2 + 1\right)^2\right) u - u^3$$
$$+ g_2 \hat{z} \cdot \nabla \times \left[(\nabla u)^2 \nabla u\right] + g_3 \nabla \cdot \left[(\nabla u)^2 \nabla u\right], \quad (5.32)$$

where $\hat{z}$ is a unit vector pointing along the positive-z-axis (perpendicular to the xy plane of the fluid layer), r is the usual control parameter, and the parameters g_2 and g_3 are new constants. The term with coefficient g_2 is the new term that reflects the asymmetry between clockwise and anticlockwise rotations, and the constant g_2 should be proportional to the angular frequency of rotation, at least for not too large rotation rates. The term with coefficient g_3 maintains all the symmetries of the conventional Swift–Hohenberg equation (and was used in the modified equation (5.23)). This term is convenient to include here since it allows the tuning of the angle of the cross-roll instability, which is the instability leading to the break down of the stripe state to spatiotemporal chaos. Several exercises in Chapter 7 apply the amplitude-equation formalism to understand the properties of Eq. (5.32) and we encourage you to explore this model further after reading that chapter.

---

[14] This is not true near the lateral boundary, where the normal to the boundary provides an orientation reference. In this case, new linear terms arise, such that, for large enough rotation rates, an entirely new linear instability occurs to a state of traveling convection rolls (a type-I-o instability) radially localized near the boundary.

5.2.5 Model for quasicrystalline patterns

The Swift–Hohenberg equation is formulated to describe patterns with a single length scale, set by the critical wave number q_c of a type-I-s instability. A simple modification to incorporate two incommensurate length scales leads to the Lifshitz–Petrich equation that has quasiperiodic spatial patterns. As was discussed in Section 4.3, quasiperiodic patterns were first realized in atomic systems, where they are called quasicrystals. In nonequilibrium pattern-forming systems, they have been observed in the Faraday crispation instability where a pattern of standing waves develops on the surface of a fluid shaken vertically. Patterns that are quasiperiodic in space are found if the shaking has a time dependence comprised of *two* frequencies, as shown in Fig. 1.16. It is easy to see how this leads to two length scales, through the resonance condition of the dispersion relation of the surface waves with the driving frequencies (actually half the driving frequencies, since this is a parametric instability).

The Lifshitz–Petrich equation takes the form

$$\partial_t u = ru - (\nabla^2 + 1)^2(\nabla^2 + q^2)^2 u + \gamma u^2 - u^3. \tag{5.33}$$

(It turns out that the symmetry-breaking term γu^2 must be included, as in Section 5.2.1, since the stabilization of the quasicrystal pattern rests on three-wave interactions generated by such a term.) The parameter q sets the second length scale q^{-1}. Quasicrystalline patterns are expected when this length is chosen to be an appropriate incommensurate ratio to the other length scale (given by the unit wave number $q = 1$ in the term $(\nabla^2 + 1)^2$ on the right-hand side of Eq. (5.33)). An example of the stationary patterns generated by Eq. (5.33) is shown in Fig. 5.4. Note that, by rescaling the u variable, the control-parameter set can be reduced to the combination r/γ^2 and q.

5.3 Order-parameter equations

Our motivation for the Swift–Hohenberg equation was to find a simple partial differential equation that reproduces some of the features of the linear instability spectrum, that includes nonlinear terms to saturate the growth, and that could be solved efficiently on finite spatial domains with boundary conditions of experimental relevance. The desire to obtain a simple real-space (local) form for the linear terms when expressed in terms of differential operators led to uncontrolled approximations of how the linear growth rate σ_q depended on the wave number q. Further uncontrolled approximations were associated with the cubic nonlinear term, which was chosen purely for simplicity. If we relax some of these constraints and instead use a Fourier-space formulation that is best suited for periodic spatial domains, a

5.3 Order-parameter equations

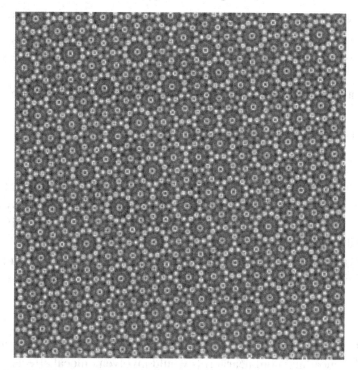

Fig. 5.4 A 12-fold symmetric quasiperiodic pattern formed by the Lifshitz–Petrich equation (5.33). Parameters used were $r/\gamma^2 = 0.015$ and $q = 2\cos(\pi/12)$. (After Lifshitz and Petrich [66].)

better quantitative account of the behavior of real physical systems near threshold can be achieved with an evolution equation that is still substantially simpler than the fully quantitative evolution equations. The resulting equation is sometimes called an order-parameter equation.

The order parameter in Fourier space $\psi_\mathbf{q}(t)$ can be introduced as the coefficient of the linear mode at wave vector $\mathbf{q}$, Eq. (2.22), at the critical parameter $p_c(\mathbf{q})$ for that wave vector (giving $\mathrm{Re}\,\sigma_\mathbf{q} = 0$), so that the deviation from the uniform solution becomes

$$\mathbf{u}_p(\mathbf{x},t) = \sum_\mathbf{q} \psi_\mathbf{q}(t) \mathbf{u}_\mathbf{q}(\mathbf{x}_\parallel) e^{i\mathbf{q}\cdot\mathbf{x}_\perp} + \text{c.c.} + \cdots. \tag{5.34}$$

This is an expansion in the linear modes, and so is expected to be useful near threshold. We have written the expansion in terms of a sum over modes (a Fourier series) as would be appropriate for a finite geometry with periodic boundary conditions. In an infinite geometry, a Fourier integral would be used. The $\cdots$ in Eq. (5.34) represent correction terms that are small near threshold. An evolution equation for $\psi_\mathbf{q}(t)$ can be derived in the spirit of a Galerkin expansion by substituting Eq. (5.34) into the physical equations and collecting equations of the mode coefficients. Since the evolution of $\psi_\mathbf{q}(t)$ is slow near threshold and for q near critical, the coefficients of the spatial harmonics are calculated quasistatically, that is as the values given by time-independent equations and the instantaneous values of $\psi_\mathbf{q}(t)$. Truncating this procedure at cubic order in the expansion leads to an equation of the form

$$\frac{d\psi_\mathbf{q}}{dt} = \sigma_\mathbf{q} \psi_\mathbf{q}(t) + \sum_{\mathbf{q}_1 \mathbf{q}_2} \bar{K}(\mathbf{q}, \mathbf{q}_1, \mathbf{q}_2) \psi_{\mathbf{q}_1}(t) \psi_{\mathbf{q}_2}(t) \psi_{\mathbf{q}-\mathbf{q}_1-\mathbf{q}_2}(t). \qquad (5.35)$$

Here $\sigma_\mathbf{q}$ is the *exact* linear instability spectrum. The interaction kernel $\bar{K}(\mathbf{q}, \mathbf{q}_1, \mathbf{q}_2)$ is some complicated function that must be calculated for each physical system. The wave vector $\mathbf{q} - \mathbf{q}_1 - \mathbf{q}_2$ of the third field in the nonlinear term is fixed by the translational invariance of the system.

We could attempt to write this equation in terms of the real-space order parameter

$$\psi(\mathbf{x}_\perp, t) = \sum_\mathbf{q} \psi_\mathbf{q}(t) e^{i\mathbf{q} \cdot \mathbf{x}_\perp}. \qquad (5.36)$$

This would give an equation reminiscent of the Swift–Hohenberg equation except that both the linear and nonlinear terms would involve nonlocal effects

$$\partial_t \psi(\mathbf{x}_\perp, t) = \int L(\mathbf{x}_\perp, \mathbf{x}'_\perp) \psi(\mathbf{x}'_\perp, t) \, d^2\mathbf{x}'_\perp$$
$$- \iiint K(\mathbf{x}_\perp, \mathbf{x}'_\perp, \mathbf{x}''_\perp, \mathbf{x}'''_\perp) \psi(\mathbf{x}'_\perp, t) \psi(\mathbf{x}''_\perp, t) \psi(\mathbf{x}'''_\perp, t) \, d^2\mathbf{x}'_\perp \, d^2\mathbf{x}''_\perp \, d^2\mathbf{x}'''_\perp, \qquad (5.37)$$

so that the resulting equation is not too informative. The Swift–Hohenberg equation is obtained if we approximate the linear terms as a simple form in terms of $\psi(\mathbf{x}'_\perp, t)$ and its derivatives, and if the interaction kernel is replaced by a completely local function $\delta(\mathbf{x}_\perp - \mathbf{x}'_\perp)\delta(\mathbf{x}_\perp - \mathbf{x}''_\perp)\delta(\mathbf{x}_\perp - \mathbf{x}'''_\perp)$. The form of these equations is sufficiently general that the mean flow effects common in fluid systems could be included as part of the interaction kernels $\bar{K}$ and K. However, it is useful to separate out these long-range effects and to include them explicitly in ways analogous to Section 5.2.3, leaving the explicit interaction kernel $K(\mathbf{x}_\perp, \mathbf{x}'_\perp, \mathbf{x}''_\perp, \mathbf{x}'''_\perp)$ short range (nonzero only for $\mathbf{x}'_\perp, \mathbf{x}''_\perp, \mathbf{x}'''_\perp$ close to $\mathbf{x}_\perp$).

Equations (5.35) and (5.37) are derived as weakly nonlinear expansions near threshold and for q near q_c (although the linear term is correct for all q). However for Rayleigh–Bénard convection, where these equations have been compared with more complete calculations, they often give quite accurate results even up to twice the critical Rayleigh number. Figure 5.5 shows the good agreement for the stability

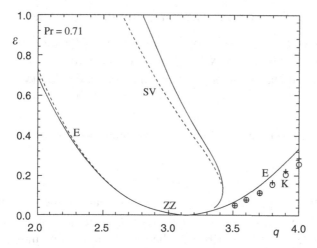

Fig. 5.5 Comparison of the stability balloon for Rayleigh–Bénard convection for a fluid with Prandtl number 0.71 calculated with the order-parameter equation (5.35) (dashed lines and open circles) and the more complete Galerkin calculation (solid lines and crosses). The instabilities are skew-varicose (SV), Eckhaus (E), zigzag (ZZ), and knot (K). (From Decker and Pesch [31].)

Fig. 5.6 Comparison of spiral defect chaos in (a) numerical simulations of the order-parameter equation (Eq. (5.35) supplemented with mean flow terms) and (b) experiment, for a fluid with Prandtl number 0.7 at a Rayleigh number $R/R_c = 2$. In (a) the temperature field at midplane, and in (b) the shadowgraph image is shown. (Source as for Fig. 5.5.)

balloon of Rayleigh–Bénard convection for a fluid of Prandtl number 0.71 (see Section 4.2.2), and Fig. 5.6 shows the remarkably good reproduction of the spiral defect chaos state. Quantitative measures such as the variation of the mean wave number with Rayleigh number are also well reproduced.

5.4 Complex Ginzburg–Landau equation

So far we have concentrated on model equations for stripe patterns arising from a type-I-s instability. The complex Ginzburg–Landau equation is a model evolution equation with complex coefficients for a complex-valued field $A(\mathbf{x}, t)$ that describes the properties of nonlinear oscillations and waves such as might develop from a type-III-o instability (oscillatory instability at zero wave number). The equation can be written in a simple form that eliminates unnecessary constants,

$$\partial_t A = A + (1 + ic_1)\nabla^2 A - (1 - ic_3)|A|^2 A, \tag{5.38}$$

where the only two parameters, c_1 and c_3, are real numbers. Commonly used boundary conditions are periodic boundaries and the condition $A = 0$.

Equation (5.38), often described as the CGLE, has been studied in one, two, and three spatial dimensions, usually with periodic boundary conditions or with the condition that $A = 0$ on one or more boundaries. Unlike the Swift–Hohenberg equation, the complex Ginzburg–Landau equation can be derived by a systematic multiple scales perturbation analysis for the weakly nonlinear behavior near a type-III-o instability as described in Section 10.2.2, and so there is an increased likelihood of obtaining quantitative agreement with some experimental details sufficiently close to onset. On the other hand, in contrast to the corresponding equation for the weakly nonlinear behavior near a type-I-s instability (the real amplitude equation Eq. (6.9) to be introduced in Chapter 6), which had already been studied in other branches of physics such as the theory of the phase transition to superfluidity and superconductivity, the complex version had not been studied before. As analytical and numerical investigations of the CGLE accumulated, researchers realized that this equation has a rich and interesting set of dynamics, and started studying this equation intensely in its own right as a model of the type of complex behavior that can occur in nonequilibrium systems.

The behavior of Eq. (5.38) is rich and diverse. However, our discussion here will be brief since this equation will be described in detail in Chapter 10. The limit $c_1 = c_3 = 0$ yields the real Ginzburg–Landau equation. In the opposite limit of $c_1^{-1}, c_3^{-1} \to 0$, when the imaginary terms dominate, the equation becomes the nonlinear Schrödinger equation, an equation that is famous in applied mathematics for being analytically solvable in one spatial dimension with soliton solutions. The nonlinear Schrödinger equation also is used in the study of Bose condensation and superfluidity at zero temperature, where the field A then has the physical significance of the wave function of the Bose condensate, and the equation is known as the Gross–Pitaevski equation. The nonlinear Schrödinger equation also shows up in the study of light pulses propagating down fiber optics and therefore is fundamental to the transmission of information through the Internet.

If $c_1 = 0$ and $c_3 \neq 0$, Eq. (5.38) reduces to a simple example of a "λ–ω" system, much studied in the applied mathematics literature. Perturbing about the real case $c_3 = 0$ then provides an analytic approach to studying the properties of the complex case. Solutions of Eq. (5.38) in which the field magnitude $|A|$ is time independent can be related through length and time rescaling to solutions for different parameter values c_1' and c_3', satisfying

$$\frac{c_1' + c_3'}{1 - c_1' c_3'} = \frac{c_1 + c_3}{1 - c_1 c_3}. \tag{5.39}$$

This allows a restricted set of solutions for the λ–ω equation to be transferred to the more general case.

5.5 Kuramoto–Sivashinsky equation

As one of the simplest continuum models that has spatiotemporal chaos, the Kuramoto–Sivashinsky equation has been widely used to develop and to test ideas about how to quantify spatiotemporal chaos. The original form is the evolution equation for a scalar field $u(x, t)$

$$\partial_t u = -\partial_x^2 u - \partial_x^4 u - u\,\partial_x u, \tag{5.40}$$

on a spatial interval $[0, L]$ of length L, which is the only parameter. The most commonly used boundary conditions are periodic boundaries, such that $u(x, t) = u(x + L, t)$ for all x, or the conditions $u = \partial_x u = 0$, which are similar to the conditions Eq. (5.10) used for the Swift–Hohenberg equation.

Kuramoto derived this equation to describe the dynamics on the unstable side of the Eckhaus instability in a stationary stripe state. In this context, $u = \partial_x \phi$ is the gradient of a phase field $\phi(x, t)$ that describes the positions of the stripes. Equation (5.40) can alternatively be written as an evolution equation for the phase

$$\partial_t \phi = -\partial_x^2 \phi - \partial_x^4 \phi - \tfrac{1}{2}(\partial_x \phi)^2. \tag{5.41}$$

This is a nonlinear phase diffusion equation with a *negative* diffusion term $-\partial_x^2 u$ that causes instability, and with a fourth-order derivative term that quenches the instability at short length scales. Independently and about the same time as Kuramoto, Sivashinsky derived Eq. (5.40) to describe the dynamics of a combustion front, with u the displacement of the front from a straight line and x the coordinate along the front. (Think of a Bunsen burner that has been beaten with a hammer to form a long one-dimensional slot through which the flame emerges.) This equation also arises when studying the flow of a thin film of liquid down an inclined plane, in which case $u(x, t)$ describes the height of the liquid film along a horizontal line that is parallel to the base of the plane.

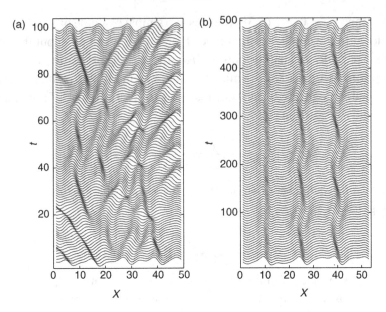

Fig. 5.7 Numerical space-time solutions $u(x, t)$ of the one-dimensional Kuramoto–Sivashinsky equation, Eq. (5.40), for the boundary conditions $u = \partial_x u = 0$. For both panels, the spatial coordinate x is the horizontal axis while time t increases upwards along the vertical axis. (a) For a domain of size $L = 50$, most initial conditions evolve into a spatiotemporal chaotic state with Lyapunov fractal dimension $D \approx 8.8$. Spatial curves $u(x, t_i)$ are plotted every $\Delta T = 1$ time units starting at time $t = 50\,000$ so this is an approximately nontransient state. (b) For a slightly larger cell with $L = 54$, most initial conditions lead to a periodic attractor of period $T = 127.6$ time units. (From Tajima and Greenside [100].)

The Kuramoto–Sivashinsky equation can also be derived as a limiting case of the complex Ginzburg–Landau equation when the parameters c_1 and c_3 are chosen to be close to an interesting stability boundary. We will see in Chapter 10 that the nonlinear oscillation and wave solutions of the CGLE become unstable for $c_1 c_3 > 1$ through an Eckhaus-like instability known as the Benjamin–Feir instability. Again a reduction to a phase description in terms of the phase ϕ of the complex amplitude is possible near this instability, and the Kuramoto–Sivashinsky equation in the form Eq. (5.41) is then the lowest-order dynamical equation for the evolution of this phase. The interesting dynamics seen in numerical simulations has led to widespread interest in Eq. (5.40), independently of these physical contexts.

Writing Eq. (5.40) in the form

$$\partial_t u = \tfrac{1}{4} - \left(\partial_x^2 + \tfrac{1}{2}\right)^2 u - u\,\partial_x u, \tag{5.42}$$

reveals some similarities to the one-dimensional version of the Swift–Hohenberg equation, Eq. (5.9). The linear term $[1/4 - (\partial_x^2 + 1/2)^2]u$ is the same as in Swift–Hohenberg for $r = 1/4$ except the critical wave number here has a value of 1/2 instead of 1. From our discussion in Section 5.1, we see that the uniform state $u = 0$ is unstable to the growth of cellular structures over a band of wave numbers from $q = 0$ to $q = 1$ with a maximum growth rate at $q = 1/2$.

In fact, stable stationary nonlinearly saturated spatially periodic structures have been shown to exist over a band of wave numbers. However numerical simulations show that most initial conditions do not lead to one of these periodic states (even over long simulation times) but rather to a persistent time-dependent state that is chaotic according to numerical estimates of the largest Lyapunov exponent λ_1 and by the observation of broad power spectra of time series. Although Eq. (5.40) is simple to write down and easy to simulate numerically, it is still extremely difficult to study analytically. In particular, no one has yet rigorously proved that any solution of the Kuramoto–Sivashinsky equation is chaotic. For most scientists, this is a moot point since the properties of the presumably chaotic dynamics are empirically statistically stationary and the chaos endures for as long as any one has had the patience to look (over 10^6 time units in some simulations). Still, it would be useful to know whether spatiotemporal chaos truly exists or whether the observed simulations correspond to long-lived transients.

Sometimes a "generalized Kuramoto–Sivashinsky" equation is used, in which a linear term $-\eta u$ with control parameter η is added:

$$\partial_t u = -\eta u - \partial_x^2 u - \partial_x^4 u - u\, \partial_x u. \tag{5.43}$$

The $u = 0$ state is now stable for $\eta > 1/4$, and, as η decreases toward zero, this state undergoes a type-I-s instability at $\eta = 1/4$ with a critical wave number of $q_c = 1/\sqrt{2}$. For large spatial domains, there is an interesting transition as η is decreased from $\eta = 1/4$, for which the dynamics is periodic, to $\eta = 0$, for which the dynamics is chaotic, via a mechanism called spatiotemporal intermittency. Here the periodic state breaks up into a mix of disordered lower-amplitude regions (called "chaotic") and ordered higher-amplitude regions (called "laminar"). The chaotic regions become a larger and larger fraction of the domain until, at $\eta = 0$, the entire domain is chaotic.

5.6 Reaction–diffusion models

The focus of Turing's work was on instabilities to stationary spatial structure (type-I-s), but we have already seen in Chapter 3 that the simple two-variable model can lead to an instability to spatially uniform oscillations (type-III-o), and this gives propagating waves above onset. Reaction–diffusion models are therefore used to understand the oscillatory and wave phenomena in chemical systems, such as shown in Fig. 1.18(a). As we discussed in Section 3.2.2, there is not a systematic method for precisely formulating the reaction terms in a chemical system, since there may be many unknown intermediate complexes. Thus much of the work in this area involves simplified models of a reduced set of chemical concentrations, such as the Brusselator model introduced in the Etude 3.1 on page 105. The reaction rates often involve phenomenological nonlinear functions, rather than the products of powers of concentrations as expected for elementary reactions from the Law of Mass Action, with the more complicated function representing the effect of short-lived intermediaries not included in the equations retained in the model.

The combination of reaction (a rate of change of a field depending in a nonlinear manner on the component fields) and diffusion (a rate of change of a field at a point depending on nearby field values) is a general concept that applies to diverse systems in physics, chemistry, biology, and other fields. Reaction–diffusion models are therefore widespread, even outside the field of reacting and diffusing chemical (or gene product) concentrations. For example, a simple model in neurobiology known as the Wilson–Cowan model simplifies the complex neuronal structure to two populations: a population $E(\mathbf{x}, t)$ of excitory neurons whose activity stimulates the activity of other neurons to which they are connected; and a population $I(\mathbf{x}, t)$ of inhibitory neurons, whose activity tends to reduce the activity of connected neurons. The rate of change of each population will depend in a nonlinear way on a combination such as $aE - bI$ with a and b positive constants that quantify the degree of connectivity of the two sets of neurons. This gives the reaction terms. The diffusion terms come naturally from the spatial distribution of the neurons and the connections between them. The complex connections can be reduced to simple Laplacian operators for phenomena in which E and I vary over distances large compared to the range of connections.

As we will study in detail in Chapter 11, a pair of nonlinear reaction–diffusion equations can lead to propagating pulses. As a consequence, simple reaction–diffusion models have been used to understand dynamic phenomena in excitable media such as the propagation of electrical pulses in nerve fibers and in heart tissue. The "diffusion" in these equations actually represents the effects of electrical resistance,[15] and one of the variables is an electrical voltage across the cell membrane. As you can readily appreciate, nerve fibers or heart tissues, are complex structures, and various levels of models are used to get approximate descriptions. The simplest models are two-variable descriptions such as the FitzHugh–Nagumo model described in Section 11.1.3, with one variable that represents the voltage across the membrane, the other variable represents an ion concentration. The

---

[15] In the presence of a nonzero resistivity, charges placed at a point will spread diffusively.

original Hodgkin–Huxley equation, which we will discuss in Section 11.1.1, is a more complicated and more quantitatively accurate reaction–diffusion equation that was motivated by careful experiments on squid axons. It has four variables, a membrane voltage and three gating variables which represent the time- and voltage-dependent probabilities for channels in the nerve membrane to open that selectively allow potassium and sodium ions to flow. As a wealth of experimental information has accumulated since the Hodgkin–Huxley model was published in the 1950s, models involving more ions and gating variables have been developed to improve agreement with the data, with some models involving more than 20 coupled nonlinear pdes. These complicated modern mathematical equations are a good example of how the distinction between a model and a theory can become blurred.

5.7 Models that are discrete in space, time, or value

We have formulated most of our discussion of pattern formation in terms of partial differential equations that describe continuously varying media. Other kinds of models have been widely studied, for example models that are discrete in space such as a network of coupled odes, that are discrete in space and time such as a network of coupled maps, or even cellular automata models that are discrete in time, space, and in their field values. The use of models with discrete features may be motivated by the attempt to describe particular physical systems (for example, pattern formation in a biological system may take place at a scale comparable with a cell size in which case a discrete model may be more appropriate than a continuum one) or simply for the numerical convenience. In the latter case, if the results are used as a basis for understanding pattern formation in continuum systems, the role of the discretization must be carefully investigated since the numerical mesh may have a geometric structure that can bias the numerical results. For example, a square grid of mesh points introduces an anisotropy such that numerical stripe solutions may incorrectly align along a grid axis, contrary to the rotational isotropy of the mathematical model and of the domain on which the problem is studied. Of course, at some level, essentially all numerical simulations involve some kind of discretization, as we discuss briefly in Chapter 12. Testing the convergence of the numerical results as some discretization is refined is an essential part of any simulation.

5.8 Conclusions

The preceding sections have shown that model equations can capture a surprisingly broad range of pattern formation, despite the fact that these equations are greatly simplified compared to a full quantitative description. The reduced dimensionality and reduced number of fields makes the model equations more analytically tractable. They are also much easier to simulate, for example in larger domains, for longer times, and for a greater choice of parameter values. Especially important is that it is easy to turn on or off specific physical effects in a model equation that would be difficult to do with a more complete description.

Model equations are, in fact, used heavily in nearly all frontiers of science to obtain insights into complex phenomenon. The role of model equations in current science represents a substantial shift in the philosophy of what it means to "understand" some experiment. Before computers became widely available, understanding meant obtaining analytical solutions of fully quantitative evolution equations whose structure could be related to fundamental conservation laws of mass, momentum, and energy. But the observations of diverse nonequilibrium phenomena such as those discussed in Chapter 1 show that the spatiotemporal structure of even the simplest nonequilibrium systems are so complex that it is pretty much useless to write down analytical solutions that describe such complexity, even if this were possible. Instead, scientists often use their mathematical skills to discover simplified mathematical models of the kind discussed in this chapter and elsewhere in the book, and then use a combination of analytical and numerical insights to understand their properties.

5.9 Further reading

(i) An extensive review of the complex Ginzburg–Landau equation is "The world of the complex Ginzburg-Landau equation" by Aranson and Kramer [6].
(ii) *Theory and Applications of Coupled Map Lattices* by Kaneko [52] describes pattern formation and chaos in coupled map lattices. A recent compilation of reviews edited by Chazottes and Fernandez on the same topic is *Dynamics of Coupled Map Lattices and of Related Spatially Extended Systems* [20].

Exercises

5.1 **Analysis of symmetries in a model equation:** Consider the following nonlinear partial differential equation

$$\partial_t u(x, y, t) = ru - \left(\nabla^2 + 1\right)^2 u + \nabla \cdot \left(u^2 \nabla u\right), \quad \text{(E5.1)}$$

for a real-valued field $u(x, y, t)$ in an infinite two-dimensional plane, where $\nabla^2 = \partial_x^2 + \partial_y^2$ is the two-dimensional Laplacian. Justifying your answers, explain whether this equation is

(a) time-translational invariant under the substitution $t \to t+c$ for a constant c

(b) translationally invariant under $\mathbf{x} \to \mathbf{x} + \mathbf{c}$ for a constant vector $\mathbf{c}$
(c) space-inversion symmetric under $\mathbf{x} \to -\mathbf{x}$
(d) time-inversion symmetric under $t \to -t$
(e) field-inversion symmetric under $u \to -u$
(f) rotationally invariant under $\mathbf{x} \to \mathbf{M}\mathbf{x}$, where $\mathbf{M}$ is an orthogonal (length-preserving) matrix such that $\mathbf{M}^T \mathbf{M} = I$.

Why is it important to know about such symmetries?

5.2 **Potential of the Swift–Hohenberg equation:**

(a) For a spatial domain with periodic boundary conditions and for a domain with the boundary conditions Eq. (5.10), show that Eq. (5.12) follows from Eq. (5.11) provided that $u(\mathbf{x}, t)$ evolves according to the Swift–Hohenberg equation, Eq. (5.9). To carry out the multivariate version of "integration by parts," you will want to use the vector identity

$$\nabla \cdot (f \nabla g) = \nabla f \cdot \nabla g + f \nabla^2 g, \tag{E5.2}$$

and the two-dimensional form of Gauss's law

$$\iint_V \nabla \cdot \mathbf{v} \, dA = \int_S \mathbf{v} \cdot \hat{\mathbf{n}} \, dl, \tag{E5.3}$$

that relates the "volume" integral of the divergence of a vector field $\mathbf{v}(x, y)$ over the two-dimensional domain to the "surface" integral of $\mathbf{v}$ over the closed boundary of the domain, where $\hat{\mathbf{n}}$ is the unit vector that is locally normal to the boundary and pointing to the exterior direction.

(b) Show that the potential $V[u(x, y, t)]$, Eq. (5.11), is bounded below by the quantity $(-1/4)r^2 S$, where S is the area of the domain. This lower bound implies that the monotonic decrease of V implied by Eq. (5.12) cannot continue forever.

5.3 **Fastest growing mode of the Swift–Hohenberg model does not minimize the potential:** Consider the one-dimensional Swift–Hohenberg equation for a field $u(x, t)$ in an infinite domain that is close to onset, with $r \ll 1$. Using a "Galerkin approximation" truncated at second order for a uniform stripe state at wave number q,

$$\psi \simeq a_1 \cos(qx) + a_3 \cos(3qx), \tag{E5.4}$$

find the value of the wave number q_m correct to $O(r^2)$ that minimizes the potential per unit length $v(q)$. Show that q_m is *not* the same as the wave number q_{max} for the maximum growth rate of a small perturbation from the uniform $u = 0$ state. (Hint: evaluate a_1, a_3, and q_m to sufficient order in r by minimizing the potential with respect to each parameter.)

Your result has the interesting implication that the wave number selected by the fastest growing mode is not the wave number preferred by the potential so that a nonlinear evolution of the wave number is expected.

5.4 **Model integration times and experimental observation times:** Based on the discussion of Section 5.1.4, if a convection experiment is observed for a time T in seconds, for what length of time T' should a numerical simulation of the Swift–Hohenberg equation be carried out to match the experimental observation time?

Note: Recall that for the dimensionless Boussinesq equations, time is measured in units of the vertical thermal diffusion time d^2/κ, where d is the thickness of the convecting fluid layer and κ is the fluid's thermal diffusivity. Assume for this problem that the convecting fluid has a depth of $d = 1$ mm with a mean temperature of $20\,°C$ so you can use the data in Table 1.1 on page 8.

5.5 **Mean flow vanishes for straight or radial rolls:** Show that the right side expression of Eq. (5.30), which drives the mean flow Eq. (5.27), vanishes for stripe solutions $u(\mathbf{k} \cdot \mathbf{x})$ (where $\mathbf{k}$ is some constant wave vector) and vanishes for radially oriented rolls of the form $u(kr)$ in a polar coordinate system (r, ϕ). Thus mean flows arise from deviations from straight or radial stripes, as well as from variations in the local roll amplitude.

5.6 **A model with rectangular cell patterns:** Consider the generalized Swift–Hohenberg equation (5.23) with $\alpha = 1$ and $\beta = 0$.

$$\partial_t u(t, x, y) = ru - \left(\nabla^2 + 1\right)^2 u + \nabla \cdot \left((\nabla u)^2 \nabla u\right). \quad (E5.5)$$

(a) Show that Eq. (E5.5) with periodic boundary conditions or with the boundary conditions Eq. (5.10) has potential dynamics so that nontransient oscillatory or chaotic behavior is not possible.

(b) Sufficiently close to onset ($0 < r \ll 1$), a Fourier mode expansion of a stationary state $u(x, y)$ can be approximated by the lowest-order Fourier modes:

$$u(x, y) = A_x \cos(q_x x) + A_y \cos(q_y y). \quad (E5.6)$$

Substitute Eq. (E5.6) into Eq. (E5.5) and obtain expressions for the coefficients A_x and A_y to lowest order, in terms of the parameter r and the wave numbers q_x and q_y. Determine numerically the shape of the region in the $q_x q_y$ plane over which solutions exist for $r = 1/4$. The special case $q_x = q_y$ gives square cells.

Is it necessary to consider rectangular patterns of other symmetry, e.g. involving just sine modes or a combination of sine and cosine modes?

(c) Show that for small r, the square solution with $q_x = q_y = 1$ and $A_x = A_y = A_Q$ (with A_Q to be calculated by minimizing the potential) has a lower potential density than a stripe solution with wave number $q_x = 1$ given by $A_x = A_S$, $A_y = 0$ (with A_S to be calculated by minimizing the potential density). Also show that as a function of A_x and A_y the stripe state is a *saddle* of the potential density, whereas the square state is a *minimum*, so that squares would be expected to be the physical state near onset.

Why might rectangles also be seen, even though they have a higher potential than squares?

(d) Advanced and open ended: Study numerically the properties of Eq. (E5.5) in a large periodic square box and in a large square box with boundary conditions Eq. (5.10). Starting from small-amplitude random initial conditions, what kinds of patterns form? Are square or rectangular cells prevalent? How do the rectangular cells change (if at all) near the lateral boundaries? Modify Eq. (E5.5) to break rotational symmetry as in Section 5.2.4 and study empirically whether sustained dynamical states now occur and what they look like. Following the discussion on page 189, study numerically the dynamics of rectangular cells in the presence of a mean flow. Is a sustained dynamics observed? If so, what are its qualitative properties?

5.7 **Long-lived complex transients in pattern-forming systems:** Read the three papers

(a) Order, disorder, and phase turbulence, B. Shraiman, *Phys. Rev. Lett.* **57**, 325 (1986)
(b) Are attractors relevant to turbulence?, J. Crutchfield and K. Kaneko, *Phys. Rev. Lett.* **60**(26), 2715 (1988)
(c) Size-dependent transition to high-dimensional chaotic dynamics in a two-dimensional excitable medium, M. Strain and H. Greenside, *Phys. Rev. Lett.* **80**, 2307 (1998)

and summarize what they say about the possible role of "supertransients" in sustained nonequilibrium systems. A supertransient is a transient complex spatiotemporal state whose average decay time to an asymptotic stationary or periodic state increases rapidly, typically exponentially, with the system size L. It is an open research question whether long-lived seemingly chaotic spatiotemporal states observed in experiments and in simulations of large domains are truly chaotic or just supertransients. The question is more than academic since strategies for controlling chaos (that use small perturbations to convert a chaotic behavior to a periodic behavior) may not work for a supertransient.

An advanced challenge for the reader would be to derive analytically (rather than deduce indirectly by simulation) the dependence of the average transient time on the system size L, say for some one-dimensional pattern-forming model.

Note that in all three papers, it was essential that the authors used model equations to investigate supertransients since simulation of quantitatively accurate evolution equations like the Boussinesq equations would have not been possible for such long times and for such big domains.

5.8 **Heuristic derivation of the one-dimensional Schrödinger equation:** The trick of interpreting a growth-rate equation Eq. (5.1) as an evolution equation Eq. (5.3) for a plane wave has a long history in mathematics and science. A particularly important scientific example was a heuristic derivation of the Schrödinger equation, the fundamental evolution equation of quantum mechanics that in principle can describe all properties of atoms, molecules, and materials. The flavor of the derivation is the following. From classical mechanics, we know that a point particle of mass m and speed v has kinetic energy

$$E = \frac{1}{2}mv^2 = \frac{p^2}{2m}, \tag{E5.7}$$

where we have rewritten the kinetic energy in terms of the momentum $p = mv$ of the particle. Through guesses and experiments which showed that particles could act like waves and vice versa, scientists came to believe that the energy E and momentum p of a quantum particle of mass m were respectively proportional to the angular frequency ω and wave number k of its wave-like properties as follows:

$$E = \hbar\omega, \quad \text{and} \quad p = \hbar k, \tag{E5.8}$$

where $\hbar$ is the key physical constant of quantum mechanics known as Planck's constant. (Although the relations in Eq. (E5.8) appear mathematically trivial, they represent profoundly deep and nonobvious physical insights since they relate particle properties like E and p to wave-like properties.) Finally, experiments like the 1927 Davison–Germer experiment (scattering of an electron beam from the surface of a crystal) suggested that free particles acted like plane waves of the form $\exp(i(kx - \omega t))$.

(a) Substitute Eq. (E5.8) into Eq. (E5.7) to obtain a "growth rate" equation (actually, a dispersion relation) $\omega = \omega(k)$.
(b) By multiplying both sides of your growth rate equation with the plane wave $u(x, t) = \exp(i(kx - \omega t))$ and by following the steps that led to Eq. (5.3), deduce a partial differential equation for the evolution of the

field u. This is the free-particle Schrödinger equation. Since experiments show that the principle of superposition holds for particle waves, there is no need to explore what nonlinear terms to add to this linear evolution equation.

(c) Explain why this derivation does not work for a real-valued plane wave of the form $\sin(kx - \omega t)$, so that you need to use the complex form of a plane wave. Your analysis helps to explain why the Schrödinger equation naturally involves complex numbers, even though it is an equation that describes nature (and very accurately at that).

6

One-dimensional amplitude equation

Chapter 4 discussed the evolution of infinitesimal perturbations of a uniform state into saturated, stationary, spatially periodic solutions. By restricting attention to such solutions, we were able to study the effects of the nonlinearities, using analytical methods near threshold and numerical methods further from threshold. However, most realistic geometries do not permit spatially periodic solutions since these solutions are usually not compatible with the boundary conditions at the lateral walls. Even if periodic solutions are consistent with some finite domain, they do not exhaust all the possible patterns. As we have seen in Section 4.4, typically patterns have the ideal form (stripes, hexagons, etc.) only over small regions and these ideal forms are distorted over long length scales or disrupted in localized regions by defects. In addition, the distortions and defects are often time-dependent.

In this chapter, we introduce the amplitude equation formalism which provides a powerful and broadly useful method to study spatial and temporal distortions of ideal patterns. The formalism represents a substantial conceptual and technical simplification in that, near onset and for slowly varying distortions of periodic patterns, the evolution of the many fields $\mathbf{u}(\mathbf{x}, t)$ that describe some physical system (e.g. temperature, velocity, and concentration fields) can be described quantitatively in terms of the evolution of a *single* scalar complex-valued field $A(\mathbf{x}, t)$ called the amplitude.[1] The evolution equation for the amplitude is called the amplitude equation and is typically a partial differential equation (pde). Amplitude equations capture three basic ingredients of pattern formation: the growth of a perturbation about the spatially uniform state, the saturation of the growth by nonlinearity, and

---

[1] More precisely, each locally stripe-like region of a pattern will be described by a single amplitude. You will see in Chapters 7 and 8 examples of patterns that consist of superpositions of stripes or of two regions of stripes that have a common boundary. For such patterns, two or more amplitudes are needed to describe the dynamics, with each amplitude evolving according to its own equation. Carrying out the expansion to higher order can also lead to more than one amplitude field per stripe region.

what we will loosely call dispersion, namely the effect of spatial distortions. The interplay of these three effects lies at the heart of pattern formation and so amplitude equations have yielded many useful quantitative insights.

Amplitude equations also naturally extend into the weakly nonlinear regime the classification of pattern-forming systems based on the type of linear instability that was discussed in Section 2.5. Rather remarkably, it turns out that the form of the amplitude equation is dictated by the linear classification (type I-s, type III-o, etc.) together with the symmetries of the system and some simple assumptions about the effects of nonlinearity. The amplitude equations corresponding to the different linear transition types therefore contain certain behaviors that are characteristic of all such systems. Behaviors that are common to a class of diverse systems are often called universal.[2] The remarkably similar pattern formation that is observed in diverse systems can partially be understood as a consequence of the universal forms of the amplitude equations.

The increased generality of the states that can be investigated within the amplitude equation formalism comes with the penalty that amplitude equations have a restricted range of validity. First, amplitude equations are quantitatively accurate only sufficiently close to threshold since they are derived as expansions about threshold in the small reduced parameter

$$\varepsilon = \frac{p - p_c}{p_c}, \qquad (6.1)$$

where p is the control parameter and p_c is its critical value above which the uniform state becomes unstable in the ideal infinite system. Second, the distortions that can be studied are only those modulations of ideal patterns (stripes, squares, hexagons, etc.) that vary slowly in space and time compared to the basic length and time scales of the dynamical equations. Third, only lateral boundary conditions that vary sufficiently slowly in time and in space can be treated.

To keep the discussion manageable, in this chapter we discuss only amplitude equations for the type-I-s instability and postpone until Chapter 10 the discussion of amplitude equations for oscillatory instabilities (type I-o and III-o). We will also restrict our attention to the one-dimensional case of a single extended coordinate $x_\perp = x$. This assumption includes stripe states that vary in a direction

---

[2] The word *universal* is not meant to imply that the behavior applies to every system, but rather to a whole class of systems characterized by broad similarities, such as symmetries and instability type. The behavior for systems in different "universality classes" may be totally different. The idea of universal behavior is borrowed from the study of second-order equilibrium phase transitions, for which certain features such as critical exponents were found to be the same for all systems that had the same symmetry and spatial dimensionality, independently of the details of their atomic composition.

along the stripe normal when there are two extended directions. The next chapter will discuss stripes that can vary in transverse as well as longitudinal extended directions.

The rest of this chapter is organized as follows. Section 6.1 introduces the complex amplitude A that describes slow modulations of a stripe state near threshold. (This section generalizes the discussion of Section 4.1.1 by allowing the amplitude to vary spatially.) The form of the corresponding evolution equation is derived in Section 6.2 using symmetry arguments, and we explain how the coefficients in the resulting amplitude equation can be deduced from some simple calculations. (A systematic but more technical derivation of the amplitude equation that uses the method of multiple-scale perturbation theory is given in Appendix 2 and can be skipped upon a first reading.) Since the amplitude equation is a pde, boundary conditions must be prescribed for a unique solution to be obtained. These conditions can be derived from the boundary conditions on the physical fields, either heuristically or again by the more systematic method described in Appendix 2.

We then discuss in Section 6.3 several general properties of the amplitude equation. With an appropriate scaling of time, space, and magnitude, we display the universality of the amplitude equation and then discuss some physical implications of universality. We next show that, for certain boundary conditions, the one-dimensional type-I-s amplitude equation has potential dynamics, just like the Swift–Hohenberg equation that we discussed in the previous chapter (see Section 5.1.2). The potential associated with the lowest-order amplitude equation can provide a more intuitive understanding of the dynamics and can greatly simplify calculations such as deducing the velocity of a climbing dislocation (Fig. 4.11). On the other hand, potential dynamics can be misleading since the existence of the potential relies on ignoring higher-order terms in the perturbation expansion that yields the amplitude equation. Because all potential dynamics eventually relax to a time-independent pattern and since experiments and simulations show that a sustained time dependence is sometimes observed just above the onset of type-I-s instabilities, the lowest-order amplitude equation is not able to encompass the full richness of possible behaviors.

We conclude this chapter on the one-dimensional amplitude equation with three applications that illustrate the power of the method: how boundaries affect a nonlinear pattern, the structure of the stability balloon near threshold, and the dynamics of slowly varying compressions and dilations of a stripe pattern. The latter can be analyzed in terms of how the phase of the complex amplitude evolves in time and so illustrates the value of studying phase dynamics. More sophisticated applications of the amplitude equation will be described in Chapter 8 while phase dynamics is studied more generally in Section 9.1.1 and Section 9.1.2.

6.1 Origin and meaning of the amplitude

The idea of a slowly varying amplitude function that modulates some stationary periodic pattern of interest is a generalization to a continuum of modes of the beating that occurs between two sinusoidal modes that have almost the same periods. If you have ever tried to tune (or listened to someone tune) a stringed instrument like a guitar using a tuning fork, you will have heard a beating effect in which the combined sound of the tuning fork and of a vibrating string that is almost in tune increases and decreases slowly in intensity. A similar beating occurs for the sum of two spatially varying modes $\cos(q_1 x)$ and $\cos(q_2 x)$ with nearly identical amplitudes and nearly identical wave numbers $q_1 \approx q_2$. If we denote the mean wave number by $q_c = (q_1 + q_2)/2$, we can write

$$\cos(q_1 x) + \cos(q_2 x) = \left[2\cos\left(\frac{q_1 - q_2}{2}\right) x \right] \cos(q_c x). \tag{6.2}$$

The quantity in brackets can be identified as an amplitude $A(x)$ that varies slowly in space in that, for $|q_1 - q_2| \ll q_c$, the spatial position x has to change by a large amount $\Delta x \gg q_c^{-1}$ for A to change by a substantial amount.

A similar beating effect, with a slowly varying multiplicative modulation $A(x, t)$ of a more rapidly varying basic pattern, occurs for time-dependent patterns near onset that are close to one of the stationary periodic patterns described in Fig. 4.6. The key insight comes from Fig. 2.7. Just above the onset of instability, corresponding to the curve labeled $p > p_c$ in the figure, only modes whose wave numbers q lie in a narrow band $q_N^- < q < q_N^+$ are available to construct some nonlinear state. There will then be a slow spatial beating because the nonlinear state will be a superposition of modes with almost identical wave numbers, centered about the critical wave number q_c. We also expect the amplitude $A(x, t)$ to vary slowly in time because, sufficiently close to onset, all growth rates are small by continuity (since the growth rate vanishes at onset).

On the basis of these observations, we introduce an amplitude $A(\mathbf{x}, t)$ intentionally as a strategy to investigate pattern formation near onset, and then explore its properties and confirm its value through explicit calculations and by comparisons of the calculations with experiments and simulations. Assuming the simple case of a single extended direction $\mathbf{x}_\perp = x$, we define a spatially dependent complex amplitude $A(x, t)$ in terms of a perturbation $\mathbf{u}_p = \mathbf{u}(x, \mathbf{x}_\parallel, t) - \mathbf{u}_b(\mathbf{x}_\parallel)$ of the uniform base state $\mathbf{u}_b$ by the equation[3]

$$\mathbf{u}_p(x, \mathbf{x}_\parallel, t) = A(x, t)\mathbf{u}_c(\mathbf{x}_\parallel) e^{i q_c x} + \text{c.c.} + \text{h.o.t.} \tag{6.3}$$

---

[3] For a type-III-o instability, the same ansatz is used except that the "fast" stripe structure $e^{i q_c x}$ is replaced by a fast temporal variation $e^{-i\omega_c t}$, see Eq. (10.10). Similarly, for a type-I-o instability, a traveling wave $e^{i(q_c x - \omega_c t)}$ is used instead, see Eq. (10.62).

The notation "h.o.t." denotes higher-order terms that are smaller in magnitude than the displayed terms in the limit that the reduced bifurcation parameter Eq. (6.1) becomes sufficiently small. For a technical reason that simplifies the multiple-scale perturbation calculation described in Appendix 2, we choose to base our expansion around the critical onset mode $\mathbf{u}_c(\mathbf{x}_\|)e^{iq_c x}$, where q_c is the critical wave number that maximizes the growth rate $\sigma(q, p_c)$ at onset $p = p_c$ when σ first becomes zero.[4] (The growth rate σ is real since we assume a type-I-s instability.) The function $\mathbf{u}_c(\mathbf{x}_\|)$, which corresponds to $\mathbf{u}_{q_c}(\mathbf{x}_\|)$ in Eq. (2.22), is the shape of the critical unstable mode in the confined directions, and can be determined explicitly from a prior linear stability calculation of the uniform state.

It is not obvious that time-dependent solutions of the full evolution equations can be expressed in the form Eq. (6.3). The multiple-scales method of Appendix 2 confirms that this is possible as well as shows how to calculate the higher-order terms in Eq. (6.3) explicitly. For example, to lowest order in the expansion about a stripe solution, the amplitude A is found to depend on space x and time t as $A(\sqrt{\varepsilon}x, \varepsilon t)$. Sufficiently close to onset, positive powers of ε are small quantities and the amplitude then indeed changes slowly as x or t are varied by amounts of order one. Because the multiple-scales perturbation method requires a lengthy amount of algebra for even the simplest pattern-forming systems, for most of this book we will avoid the technical details of the systematic calculations and proceed more phenomenologically, with an emphasis on understanding the implications of the amplitude-equation formalism.

Figure 6.1 gives some intuition about the effects of a slowly varying amplitude modulation $A(x)$ on a rapidly varying critical stripe state $e^{iq_c x}$. We write a representative field $u(x)$ in the form

$$u(x) = A(x)e^{iq_c x} + \text{c.c.}, \qquad (6.4)$$

with base wave number $q_c = 1$ and with a modulation function

$$A(x) = [0.5 + 0.1\cos(0.1x)]e^{i\cos(0.2x)}, \qquad (6.5)$$

that is constructed ad hoc to produce a slow modulation in space. (We ignore the time dependence of A for this example.) The expression $|(1/A)dA/dx|$ is the effective local wave number in the Fourier expansion of A and you can verify graphically that its maximum is about 0.2. Since this is small compared to the critical wave number $q_c = 1$, the amplitude A indeed slowly modulates the periodic state $e^{iq_c x}$. An obvious feature of Fig. 6.1 is the modulation of the magnitude of the sinusoid. But if you look carefully, you will also see that the local periodicity of u (for example

---

[4] Alternatively, q_c is the critical wave number that minimizes the neutral stability curve $\text{Re}[\sigma(q)] = 0$, compare Fig. 2.2 with Fig. 2.3.

6.1 Origin and meaning of the amplitude

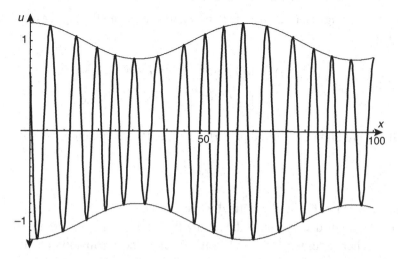

Fig. 6.1 Illustration of how a slowly varying amplitude $A(x,t)$ modulates a sinusoidal stripe-like behavior. The thick curve is the function $u = A(x)e^{iq_c x} + $ c.c. (with $q_c = 1$ and amplitude $A(x) = (0.5 + 0.1\cos(0.1x))e^{i\cos(0.2x)}$) that represents a stripe-like physical field close to onset. The magnitude modulation is shown by the two light curves $\pm|A(x)|$ that smoothly pass respectively through the local maxima and the local minima. The phase modulation causes the local periodicity to also vary slowly.

measured by the distance between two adjacent zero crossings) is no longer uniform so that the wave number also has a slow spatial modulation.

As was the case in Section 4.1.1, it is again useful to express the amplitude $A(x,t)$ in magnitude–phase form

$$A = ae^{i\Phi}, \qquad (6.6)$$

where $a(x,t)$ is its real-valued magnitude and $\Phi(x,t)$ is its real-valued phase. The magnitude a gives the size of the perturbation $\mathbf{u}_p$ near onset and typically evolves quickly, often decaying exponentially rapidly to a steady value. The phase Φ sets the *position* of the growing stripes, e.g. a change $\Delta\Phi$ in the phase translates the field $\mathbf{u}_p$ rigidly by a distance $-q_c^{-1}\Delta\Phi$ in the x-direction (see Section 4.1.1). Because of its link to translational and rotational symmetries of the system, the phase generally evolves more slowly than the magnitude and its dynamics can often be isolated and studied separately as we discuss in Section 6.4.3 and more generally in Section 9.1.2.

A slow variation in the amplitude's phase corresponds to a stretching of the wave number of the critical state (see Fig. 6.1). We can see this by examining the effect of the lowest-order non-constant terms of a Taylor-expansion of the phase $\Phi = \Phi(x_0) + k_x(x - x_0) + \cdots$ in the vicinity of some point x_0. The phase will vary slowly near x_0, provided that $|k_x| \ll q_c$, which we assume to be the case.

Then, on combining Eq. (6.3) with Eq. (6.6) and neglecting the higher-order terms, we see that

$$\mathbf{u}_p(x, \mathbf{x}_\parallel, t) \approx |a| e^{i(\Phi_0 + k_x(x-x_0))} \mathbf{u}_c e^{iq_c x} + \text{c.c.} \tag{6.7a}$$

$$= |a| e^{i(\Phi_0 - k_x x_0)} \mathbf{u}_c e^{i(q_c + k_x)x} + \text{c.c.}, \tag{6.7b}$$

so that the wave number q of the perturbation Eq. (6.3) is given by a value

$$q \approx q_c + k_x, \tag{6.8}$$

that is slightly shifted from the critical wave number.

We conclude our discussion of the ansatz Eq. (6.3) with a few comments about where the higher-order terms "h.o.t." come from. If the expansion in ε of Eq. (6.1) is formally carried out as discussed in Appendix 2, corrections indeed arise that are proportional to higher and higher powers of ε. Some of the corrections come from spatial harmonics that are generated by the nonlinearities, for example cubing $\cos qx$ creates a harmonic $\cos(3qx)$. But there are also corrections that arise at the linear level since, for a spatially varying amplitude, the structure $\mathbf{u}_c(\mathbf{x}_\parallel)$ will not give the precise solution to the evolution equations. For example, a variation corresponding to a shift of wave number will change $\mathbf{u}_c$ to $\mathbf{u}_q$ in the exponentially growing solution. In addition, the mode structure is perturbed if the control parameter is not exactly equal to its threshold value, which also leads to higher-order terms in Eq. (6.3).

6.2 Derivation of the amplitude equation

6.2.1 Phenomenological derivation

The amplitude equation for the amplitude A can be derived in a systematic way by substituting Eq. (6.3) into the evolution equations for the physical field $\mathbf{u}(\mathbf{x}, t)$ and then by using a formal expansion technique of the sort discussed in Appendix 2. Instead, we will proceed phenomenologically to deduce directly the form of the amplitude equation. This involves writing down terms that are low order in the various small quantities and then considering how various symmetries restrict the possible form. While this phenomenological approach suffices for the simple case of the lowest-order one-dimensional amplitude equation, ultimately a formal expansion is needed to understand the regime of validity of the amplitude equation, to obtain higher-order corrections that may be needed to understand particular experiments, to deduce appropriate boundary conditions, and to extend the method to more complicated situations such as degenerate bifurcations.

6.2 Derivation of the amplitude equation

We argue that the one-dimensional amplitude equation for a modulated stripe state near a type-I-s instability takes the form

$$\tau_0 \, \partial_t A(x,t) = \varepsilon A + \xi_0^2 \, \partial_x^2 A - g_0 |A|^2 A. \tag{6.9}$$

(The forms for type-III-o and type-I-o amplitude equations are discussed in Chapter 10, see Eq. (10.11) and Eq. (10.63).) Here ε is the reduced bifurcation parameter Eq. (6.1) while the parameters τ_0, ξ_0, and g_0 are constants that depend on details of the physical system and can be calculated from the known evolution equations. *However, the mathematical form of Eq. (6.9) does not depend on details of the physical system undergoing a type-I-s transition.* Its form is dictated completely by symmetry arguments, by a smoothness assumption that constrains which derivatives can appear, and by the fact that we are expanding about a base solution that minimizes the neutral stability curve.

The symmetry requirements that constrain the possible form of an amplitude equation arise from the need for Eq. (6.9) to be consistent with the symmetries that leave invariant the evolution equations for the physical field $\mathbf{u}$, with the correspondence given by Eq. (6.3). Thus we require that Eq. (6.9) be invariant under:

(i) translation symmetry, which means that the amplitude equation is unchanged after the substitution $A \to A e^{i\Delta}$ with Δ a constant. This corresponds to a translation of the pattern $\mathbf{u}_p$ through a distance $-\Delta/q_c$ in the x-direction;

(ii) parity symmetry, which means that the amplitude equation is unchanged after the double substitution $A \to A^*$ followed by $x \to -x$. This corresponds to an inversion of the horizontal coordinates in the original system, $\mathbf{x}_\perp \to -\mathbf{x}_\perp$.

When we extend the discussion to two extended coordinates in the next chapter, we will add a third symmetry, namely invariance under rotations.

The correspondence of these operations to a symmetry of the physical system can be seen from Eq. (6.3). As an example, under the substitution $A \to A e^{i\Delta}$ for some constant Δ, the solution $\mathbf{u}_p$ becomes (ignoring higher-order terms)

$$\mathbf{u}_p(\mathbf{x}_\perp, \mathbf{x}_\parallel, t) = A e^{i\Delta} \mathbf{u}_c(\mathbf{x}_\parallel) e^{iq_c x} + \text{c.c.} = A \mathbf{u}_c(\mathbf{x}_\parallel) e^{iq_c (x + \Delta/q_c)} + \text{c.c.}, \tag{6.10}$$

which corresponds to a translation of the field $\mathbf{u}_p$ by the amount $-(\Delta/q_c)$.

The required invariance of the amplitude equation under the symmetries of translation and parity restricts the possible terms in the amplitude equation in the following ways.[5] First, we observe that algebraic products of A and of its complex conjugate A^* that lead to odd powers such as A, $|A|^2 A \, (= A^* A^2)$, $|A|^4 A$, and so on are invariant under all the symmetries and so can appear in the amplitude equation.

---

[5] The argument for the possible nonlinear terms parallels the one on page 132 in Section 4.1.1.

Invariance under the substitution $A \to Ae^{i\Delta}$ rules out even powers such as A^2, $(A^*)^2$, $|A|^2$, and $|A|^2 A^2$ as well as some odd powers such as A^3, $(A^*)^3$, and $|A|^2 A^3$. The terms A and $|A|^2 A$ are the simplest ones that lead to growth and saturation. Although $|A|^2 A$ is of higher order than the linear term, the coefficient of the linear term is small near onset. (A subcritical transition would require an even higher-order term such as $|A|^4 A$ to saturate the exponential growth.) We will discuss in a moment why we do not include nonlinear terms such as $|A|^2 \partial_x^2 A$ that are allowed by symmetry but that contain partial derivatives.

Let us next consider what kinds of derivatives of A can appear in the amplitude equation. There must be some kind of time derivative since this is an evolution equation and the simplest guess would be that a first-order derivative $\partial_t A$ is sufficient. This is allowed by the above symmetries but is also the simplest choice consistent with a symmetry not mentioned above but implicit in all driven-dissipative pattern-forming systems, namely that the dynamics is *not* invariant under the time-reversal symmetry $t \to -t$.[6]

We next observe that a first-order spatial derivative of the form $i\,\partial_x A$ is allowed by the above symmetries, for example it is consistent with the parity symmetry $A(x) \to A^*(-x)$. However, such a term can be eliminated by a redefinition $A \to \bar{A} e^{i\Delta x}$ for a suitable constant Δ and so would play no essential role in the dynamics.[7] In fact, our choice of the critical wave number q_c as the reference in Eq. (6.3), i.e. the wave number that minimizes the neutral stability curve $p_c(q)$, already implies the absence of the $i\,\partial_x A$ term. We therefore assume that no such term appears in the amplitude equation.

A second-order derivative term $\partial_x^2 A$ is consistent with all the symmetries, and will occur in the amplitude equation. For an amplitude $A(x, t)$ that describes slow spatial modulations, higher-order spatial derivatives will be correspondingly smaller (roughly by the ratio of the basic wavelength of the pattern to the length scale of the modulation). We will therefore truncate the expansion at second order in the derivatives. For the same reason, we ignore nonlinear terms with spatial derivatives such as $|A|^2 \partial_x A$ since such a term is smaller than the existing cubic term $|A|^2 A$.[8]

---

[6] Given the correspondence Eq. (6.3), we need the amplitude equation to be dissipative if the evolution equation for **u** is also dissipative. For a single scalar field A, this is most easily achieved by having the amplitude equation resemble a diffusion equation, with a first-order time derivative and second-order spatial derivatives. Equation (6.9) has this property.

[7] Note that the second-order derivative term which we *will* include, when acting on this product produces terms including $2i\Delta\,\partial_x \bar{A}$. The value of Δ can then be chosen to cancel any first-order derivative term.

[8] Note that we are assuming that the amplitude equation is smooth so that we may expand in successive integral order derivatives $\partial_x^n A$. Rather surprisingly, there are some pattern-forming systems for which the assumption of smoothness does not hold near onset. An example is Rayleigh–Bénard convection of a fluid between so-called free-slip plates (an example would be a convecting fluid layer like water between denser and less dense liquid layers like mercury and oil) although the difficulties only appear in the two-dimensional amplitude equation. Something called a mean flow appears (see Section 9.1.3) that depends nonlocally on the physical fields. The lowest-order amplitude equation then turns out to involve two coupled fields whose dynamics can not be reduced to a single amplitude equation with simple derivative terms.

6.2.2 Deduction of the amplitude-equation parameters

Once we have deduced the form of the amplitude equation Eq. (6.9), the unknown parameters τ_0, ξ_0, and g_0 can be deduced from calculations already described in Chapters 2 and 3. Thus if we consider a small-amplitude disturbance

$$A = \delta A(t) e^{ikx}, \quad (6.11)$$

and linearize the amplitude equation about the zero solution $A = 0$ (which corresponds to the uniform base state), we see that the time dependence is exponential with a growth rate $\tau_0^{-1}(\varepsilon - \xi_0^2 k^2)$. But by the correspondence Eq. (6.3), this is the growth rate of a small physical perturbation $\mathbf{u}_p$ at wave vector $q = q_c + k$ and so must correspond to the growth rate $\sigma(q)$ of the linear stability analysis for the uniform base state $\mathbf{u}_b$. Thus we have

$$\sigma(q) = \tau_0^{-1}\left[\varepsilon - \xi_0^2(q - q_c)^2\right] + \cdots, \quad (6.12)$$

for small ε and for small $q - q_c$. The parameters τ_0 and ξ_0 are then identified as the ones introduced in the expansion of the linear growth rate about threshold, see Eq. (2.27) in Section 2.5.1. Alternatively, we can split the calculation into two pieces in which we first compare the amplitude growth rate with the dependence on ε of the growth rate σ_q at the critical wave number

$$\sigma(q_c) = \tau_0^{-1}\varepsilon + \cdots, \quad (6.13)$$

and then compare with the dependence of the critical control parameter value on wave numbers near q_c

$$\varepsilon_c(q) = \xi_0^2(q - q_c)^2 + \cdots. \quad (6.14)$$

The coefficient g_0 determines the saturation amplitude of the critical mode

$$|A| \to (\varepsilon/g_0)^{1/2}, \quad (6.15)$$

and so g_0 can be found from a Galerkin expansion calculation for the nonlinear saturation of the critical mode, as described in the example in Section 4.1.3.

Although the constants τ_0, ξ_0, and g_0 are needed to compare predictions of the amplitude equation with experiments, the qualitative dynamical behavior of the solutions to Eq. (6.9) does not depend on their values. We can see this by rescaling the variables in Eq. (6.9) as follows:

$$\tilde{A} = g_0^{1/2} A, \quad \tilde{x} = x/\xi_0, \quad \tilde{t} = t/\tau_0, \quad (6.16)$$

to obtain an equation in which only the parameter ε remains,

$$\partial_{\tilde{t}}\tilde{A} = \varepsilon\tilde{A} + \partial_{\tilde{x}}^2\tilde{A} - |\tilde{A}|^2\tilde{A}. \tag{6.17}$$

Solutions of Eq. (6.17) can be compared with experiment by transforming back to the more physical variables A, x, and t via Eq. (6.16). From the scaling, we see that the parameters τ_0, ξ_0, and g_0 serve to set the time, length, and magnitude scales for the problem.

6.2.3 Method of multiple scales

The method of multiple scales described in Appendix 2 formalizes the expansion procedure about threshold by tying together the various small effects: the small distance from threshold ε, the slow time dependence, the weak nonlinearity represented through saturation at a small magnitude $|A|$, and the slow spatial modulation. The details are technical and are not needed at the level of this chapter since, in the end, they only justify the phenomenological derivation in Section 6.2.1 of the lowest-order amplitude equation Eq. (6.9). However, in more complicated situations, the method of multiple scales becomes necessary since it may not be possible to derive the amplitude equation by symmetry arguments and by matching coefficients to simpler calculations. Even if you choose to skip the discussion of multiple scales in Appendix 2, you should appreciate that this method is widely used in the theory of pattern formation and in many other fields such as applied mathematics, engineering, plasma physics, and oceanography.

The phenomenological approach also is usually inadequate if we need to extend the calculation to higher order in the expansion in ε, because there are then too many terms to be pinned down by simple arguments. The extension to higher order may be necessary not just for quantitative accuracy, but because the results from the lowest-order calculation can be qualitatively misleading. An example of this is a finite one-dimensional system with realistic boundaries. Here the lowest-order amplitude equation suggests that a continuum of nonlinear stationary states exist, corresponding to an arbitrary translation of the stripes relative to the ends (see Section 6.4.1). Only by extending the expansion to the next order is the correct result recovered, that there is a discrete set of states such that the stripes have a preferred position relative to the ends. Another example where a higher-order calculation is required is to find the dependence of the zigzag instability boundary on the control parameter near threshold for a system with two extended dimensions. As we will discuss in Section 7.1.4, the zigzag instability boundary does not depend on ε according to the lowest-order amplitude equation, and so a higher-order calculation is needed to find the coefficient of the correct linear dependence $q_Z - q_c \propto \varepsilon$.

6.2.4 Boundary conditions for the amplitude equation

Since the one-dimensional amplitude equation Eq. (6.9) is a pde with second-order spatial derivatives, to obtain a unique solution a boundary condition must be specified at each point on the boundary. An easy case, often used in simple theoretical analyses, would be a periodic domain with the boundary condition $A(x + l, t) = A(x, t)$ for all x, where l is the domain size. But to make contact with experiment, we want to use more realistic boundary conditions on a finite domain, which we will assume in the following discussion to be the interval $-l/2 \leq x \leq l/2$ with boundaries at $x = \pm l/2$.

Physical boundaries can be divided into three types: those that tend to suppress the pattern formation, those that enhance pattern formation, and those that are neutral with respect to pattern formation (with all three cases being compared to patterns that occur in an ideal infinite system). The neutral case only occurs in rather special cases such as periodic conditions and no-flux boundary conditions for a reaction–diffusion system (see Exercise 3.11 for this latter case).

The case of suppressing boundaries is straightforward. Since these boundaries inhibit the onset of the pattern, the lowest-order boundary conditions for $A(x, t)$ take the form[9]

$$A(\pm l/2, t) = 0. \tag{6.18}$$

Note that these boundary conditions are independent of much of the underlying physics leading to the pattern formation and so are universal in a sense discussed in the next section, Section 6.3.1. Using Eq. (6.18) with the amplitude equation Eq. (6.9), one can show generally that pattern formation is suppressed in the sense that a larger value of the reduced bifurcation parameter, shifted upwards by an amount $O(1/l^2)$ above $\varepsilon = 0$, is needed to initiate pattern formation from the uniform state. This shift decreases rapidly with increasing system size so is a small but still observable effect for experiments that are many stripes wide. (Exercise 2.9 in Chapter 2 demonstrated this result in the specific case of the Swift–Hohenberg model on a finite domain.)

On the other hand, for enhancing boundaries the local driving by the boundaries is usually not small compared with the driving associated with the expansion parameter ε. As a result, a partial pattern typically forms near the boundaries even well below the threshold $p = p_c$ for an infinite system, and just above threshold the amplitude of the pattern near the boundary will be much larger than in the bulk. It

---

[9] Remember that the magnitude of the amplitude is expected to scale as $\varepsilon^{1/2}$ near threshold. Equation (6.18) should be interpreted in terms of the amplitude going to zero on this scale. There may be $O(\varepsilon)$ corrections to the zero on the right-hand side.

can then be argued that the boundary conditions take the form[10]

$$A(x \to \pm l/2) = \frac{\sqrt{2}\xi_0 e^{i\Phi_\pm}}{|x \mp l/2|}. \tag{6.19}$$

Here the phase constants $\Phi_\pm$ will depend on the details of the boundaries, and the values must be calculated by matching to a complete solution of the strongly driven region near the ends. The divergence of $|A|$ approaching the boundary on the large length scales of the amplitude equation corresponds to the physical statement that the disturbance becomes large near the boundary. Of course the amplitude equation description breaks down very close to the end (on length scales of order the pattern wavelength), so there is no actual divergence of physical quantities.

An experimental example of an enhancing boundary condition would be a heated wire that is wrapped around the sidewalls of a Rayleigh–Bénard convection system, which then causes a small horizontal temperature gradient to appear.[11] This horizontal gradient drives a convection flow at all Rayleigh numbers, even well below the threshold R_c of the instability in the ideal infinite system. What is observed with such a heated wire is that, well below threshold ($R < R_c$), a few convection rolls form adjacent to the sidewalls, and parallel to the sidewalls. (So a heated wire wrapped along a cylindrical convection cell will induce cylindrical rolls to form near the sidewall.) As the bifurcation parameter R is increased in small increments, the convecting region near the walls expands (more convection rolls appear) until, just above threshold, convection rolls fill the entire system. Another example of an enhancing boundary condition would be a rigid non-rotating end wall in the Taylor–Couette system of Fig. 1.11. Such an end wall is observed to drive a localized circulating vortex (called an Ekman vortex) for rotation rates below the onset of Taylor vortices, which is the stripe state for this system.

Since there is no sharp onset of pattern formation in most systems with an enhancing boundary, the onset of pattern formation is described as an imperfect bifurcation (see Appendix 1). However, imperfect bifurcations can also arise even for periodic or infinite systems so it is not necessarily the case that an imperfect bifurcation implies that the boundary conditions enhance pattern formation, although visual observation of the pattern below onset can directly settle this issue.

---

[10] This is the solution to the equation $\xi_0^2 \, d^2A/dx^2 = |A|^2 A$, which are the terms that dominate in the amplitude equation Eq. (6.9) near the boundary.

[11] This enhancing boundary condition can be roughly modeled via the Swift–Hohenberg equation, Eq. (5.9), by imposing a nonzero boundary condition $u = c$ with c a constant of order one, together with the usual second boundary condition $\partial_x u = 0$. Since Eq. (4.24) tells us that a Swift–Hohenberg stripe pattern has a small amplitude of order $\sqrt{r}$ near onset, there has to be a rapid variation near the boundaries for the stripes to have a value $c = O(1)$ on the boundaries.

6.3 Properties of the amplitude equation

6.3.1 Universality and scales

In our discussion of Eq. (6.17) above, we found that we could eliminate the scale factors τ_0, ξ_0, and g_0 from the amplitude equation by transforming time, space, and magnitude variables. If instead we scale the variables as follows:

$$\bar{A} = \left|\frac{g_0}{\varepsilon}\right|^{1/2} A, \quad X = \frac{|\varepsilon|^{1/2}}{\xi_0} x, \quad T = \frac{\varepsilon}{\tau_0} t, \quad (6.20)$$

we obtain a fully scaled amplitude equation from which *all* the parameters have been removed,

$$\partial_T \bar{A} = \pm \bar{A} + \partial_X^2 \bar{A} - |\bar{A}|^2 \bar{A}. \quad (6.21)$$

(The positive sign for the first term on the right-hand side corresponds to above threshold $\varepsilon > 0$, and the negative sign to below threshold $\varepsilon < 0$.) The absence of parameters in Eq. (6.21) dramatically demonstrates the *universality* of pattern forming phenomena near onset, when the amplitude equation is a good description, since we can analyze the behavior of Eq. (6.21) without referring back to the physical nature of the system. Actually, there is one parameter that characterizes different physical systems, namely the scaled system size

$$L = \frac{|\varepsilon|^{1/2}}{\xi_0} l, \quad (6.22)$$

since the boundary conditions on the scaled amplitude $\bar{A}$ must be applied at the scaled boundary positions $X = \pm L/2$. We deduce that, sufficiently close to onset, all one-dimensional stripe states in systems that have the same size after the scaling Eq. (6.22) will have the same properties.[12]

The absence of any explicit dependence on the small parameter ε in the scaled amplitude equation (6.21) immediately tells us the scaling behavior with small ε of the physical length, time, and pattern intensity, namely[13]

$$x = O\left(\varepsilon^{-1/2}\right), \quad t = O\left(\varepsilon^{-1}\right), \quad |A|^2 = O\left(\varepsilon^1\right), \quad (6.23)$$

in the limit $\varepsilon \to 0^+$. For example, since all the coefficients in Eq. (6.21) have magnitude one, any change in the value of the solution $\bar{A}$ of $O(1)$ will require an

---

[12] A caveat is that no other bifurcations occur for the same parameter value (a so-called degenerate bifurcation). Such a degeneracy would imply the existence of other amplitudes that vary slowly in space and time that could couple to A and so change its dynamics. Such degeneracies are unlikely in most experiments and simulations since they occur only when two separate system parameters are carefully tuned.

[13] These power-law dependencies are analogous to the divergences associated with the so-called "critical slowing down" that occurs at an equilibrium second-order phase transition, and indeed time, space, and intensity have the same scaling exponents when a so-called mean-field approximation is used to describe such a transition.

amount of time of $O(1)$ to pass for the scaled time variable T, which by Eq. (6.20) corresponds to the passage of an amount of laboratory time $t = (\tau_0/\varepsilon)T$ that diverges as ε^{-1} near threshold. Similarly, the length scale in the physical variable x over which the intensity of the stripe pattern grows from a small value near a suppressing boundary (or near a topological defect) to its $O(1)$ value in the bulk will diverge as $\varepsilon^{-1/2}$ near threshold. (For the case that we discuss in the next chapter of two extended coordinates, there turn out to be two different scalings of lengths for stripe systems, namely an $O(\varepsilon^{-1/2})$ scaling in the direction perpendicular to the stripes, and a slower $O(\varepsilon^{-1/4})$ scaling in the direction parallel to the stripes.) Finally, the amplitude of the pattern will have the characteristic square root dependence proportional to $\sqrt{\varepsilon}$, so that the intensity of the pattern, which is proportional to $|A|^2$, will increase linearly with ε.

Numerous experiments have confirmed the scalings Eq. (6.23). For example, Figures 6.2 and 6.3 show measurements from a so-called electroconvection experiment that consists of a thin freely suspended two-dimensional rectangular film of a smectic-A liquid crystal[14] that is driven to convect by a static electric field that is created by the two horizontal support wires (see Fig. 6.2(a)). The electric field causes charged impurities in the liquid crystal to move, and this motion couples to the smectic molecules and causes a net fluid motion. A one-dimensional amplitude equation can accurately describe the dynamics near onset since, to high accuracy, the suspended film has just one confined direction and one extended direction (which are respectively the vertical and horizontal directions in Fig. 6.2(a)). The film is so thin in the third direction, less than a micron, that this coordinate can be ignored. The lateral supports for the film, which lie to the left and right of the image shown in Fig. 6.2(a), turn out to act as suppressing boundaries so that the fluid velocity continuously decreases to a zero value at these boundaries. The corresponding boundary condition to use with the amplitude equation is therefore Eq. (6.18).

The velocity of the convective flow was measured optically from the motion of dust particles that adhere to the surface of the film without perturbing its behavior. To a reasonable approximation, the local maxima of the speeds in Fig. 6.2(b) can be used to estimate the magnitude of the amplitude function at those points so that we can interpret the velocity magnitudes plotted in panels (a) and (b) of Fig. 6.3 as the magnitude of the amplitude function. The spatial variation of the amplitude near the side boundary, where the flow velocity is suppressed, is shown in Figure 6.3(a). The variation of the length scale over which the amplitude recovers is

---

[14] A liquid crystal is a substance consisting of long rod-like molecules that can flow freely like a liquid but for which the orientation of its molecules are highly correlated like a crystal. A smectic A is a liquid crystal whose molecules align perpendicular to the planar layers. A smectic-A liquid crystal acts like an isotropic fluid for motion within a planar layer, which is the case in Fig. 6.2.

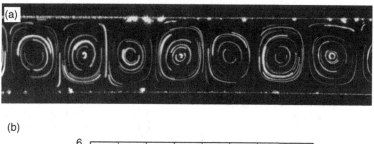

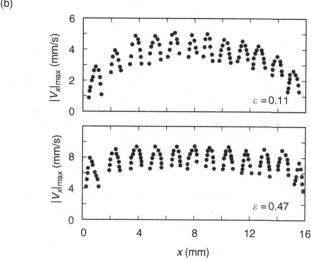

Fig. 6.2 Experimental verification of the amplitude equation scalings Eq. (6.23) just above the onset of electroconvection in a two-dimensional film consisting of a smectic-A liquid crystal. (a) The experiment consists of a rectangular vertical film of dimensions 20 mm by 2 mm supported between two long horizontal wires (they appear white in this photograph). The wires exert a static electric field vertically across the film which drives the liquid crystal out of equilibrium. Seven convection rolls out of the many are shown; the approximately circular velocity fields were made visible by the motion of dust particles that adhered to the film. (b) Spatial variation of the velocity field for values 0.11 and 0.47 of the reduced bifurcation parameter ε. The velocity goes to zero beyond the left and right sides of the plot. (From Morris *et al.* [76] and Mao *et al.* [67])

consistent with the expected scaling of $\varepsilon^{-1/2}$. Figure 6.3(b) shows the time dependence of the maximum flow velocity (which corresponds to the amplitude away from the boundaries) from a small initial magnitude until saturation. The increase near threshold of the time for this process is consistent with the scaling of ε^{-1}. In both panels of Fig. 6.3, the value of the saturated amplitude for a given value of ε can be read off from the large distance or large time value. The experimentalists also verified quantitatively that the saturated value increases as $\varepsilon^{1/2}$ near onset, consistent with Eq. (6.23).

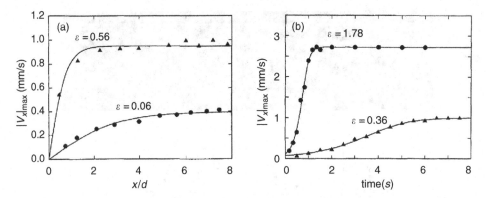

Fig. 6.3 Solutions Eq. (6.29) of the one-dimensional amplitude equation Eq. (6.21) give a quantitatively accurate fit to the experimental electroconvection data of Fig. 6.2. (a) Envelopes of the spatially varying velocity data of Fig. 6.2(b) for parameter values $\varepsilon = 0.06$ and 0.56. The "healing distance" (see Eq. (6.32)), over which the fluid recovers its bulk value from its zero value at the left boundary, increases closer to onset. An analysis (not shown) gives quantitative agreement with the expected scaling of $\varepsilon^{-1/2}$. The smooth curves are the analytical solution Eq. (6.31). (b) Temporal variation of the maximum velocity component $|v_x|_{\max}(t)$ for values $\varepsilon = 0.36$ and $\varepsilon = 0.78$. The longer time scale occurs for the smaller ε value, and a fit of the data for different ε values (not shown) confirms the expected ε^{-1} scaling for time. (Sources as in Fig. 6.2.)

6.3.2 *Potential dynamics*

In Section 5.1.2, we discussed how the Swift–Hohenberg equation has potential dynamics such that, for appropriate boundary conditions, there is a potential V that decreases monotonically for all initial conditions. This potential V greatly simplifies the analysis and understanding of solutions. For example, the existence of V implies that all nontransient states must be time-independent (so periodic, quasiperiodic, or chaotic behavior is not possible), and the final asymptotic state corresponds to a local minimum of V. The potential also provides a straightforward way to study the competition between two spatially extended patterns that are separated by a domain wall (see Fig. 5.1) and to analyze the wave number selected by some dynamical mechanism such as the climbing motion of a dislocation.

It turns out that the lowest-order amplitude equation Eq. (6.9) also has potential dynamics for periodic boundaries and for the suppressing boundary condition Eq. (6.18), and again explicit knowledge of the potential greatly helps to understand the properties of the amplitude equation. The existence of a potential at lowest order is surprising because it is not a property that we would generally expect for a system far from equilibrium. Indeed, a more careful analysis shows that the existence of a potential relies on the neglect of higher-order terms in the expansions leading to the amplitude equation and that retaining these terms leads to evolution equations that are no longer potential. However, some properties of the lowest-order amplitude equation are expected to be qualitatively correct on slow time scales of order ϵ^{-1}, with corrections to the lowest-order equation showing up on even slower time scales of order ϵ^{-2}.

We briefly discuss here the potential for the lowest-order one-dimensional fully scaled amplitude equation Eq. (6.21) and refer you to our earlier discussion in Section 5.1.2 for more details. See also Section 7.1.3, where the potential is given for the two-dimensional amplitude equation.

Consider a one-dimensional system defined on the domain $[a, b]$ in the (scaled) extended coordinate X. Then for appropriate boundary conditions on the amplitude $\bar{A}$ and assuming that $\bar{A}$ evolves according to Eq. (6.21), we claim that the expression

$$V[\bar{A}] = \int_a^b dX \left[-|\bar{A}|^2 + \frac{1}{2}|\bar{A}|^4 + |\partial_X \bar{A}|^2 \right], \tag{6.24}$$

is a potential since it evolves according to the equation

$$\frac{dV}{dt} = -2 \int_a^b dX \, |\partial_T \bar{A}|^2, \tag{6.25}$$

which implies that V decreases whenever $\partial_T \bar{A} \neq 0$. We can verify this claim with the same steps that we used for the Swift–Hohenberg equation: we take the time derivative of both sides of Eq. (6.24) (with respect to the scaled time variable T) to obtain

$$\frac{dV}{dT} = \int_a^b dX \left\{ \left(-\bar{A} + |\bar{A}|^2 \bar{A} \right) \partial_T \bar{A}^* + \partial_X \bar{A} \, \partial_X(\partial_T \bar{A}^*) + \text{c.c.} \right\}, \tag{6.26}$$

and then integrate by parts once to transfer the spatial derivative acting on the term $\partial_T \bar{A}^*$ to another term. The integration by parts leads to "surface" terms that are evaluated at the boundaries

$$\partial_X \bar{A} \, \partial_T \bar{A}^* \big|_b - \partial_X \bar{A} \, \partial_T \bar{A}^* \big|_a, \tag{6.27}$$

and these vanish for periodic boundaries or for the suppressing boundary condition Eq. (6.18). Assuming that the surface terms vanish, the remaining integral is easily seen to reduce to Eq. (6.25) provided that $\bar{A}$ evolves according to the fully scaled amplitude equation, Eq. (6.21).

6.4 Applications of the amplitude equation

6.4.1 Lateral boundaries

An early vexing question in understanding pattern formation was to determine the degree to which the ideal states of theory, based on laterally infinite systems or systems with periodic boundary conditions, had anything to do with the states seen in experiments on necessarily finite systems. Further, there was the question of how the properties of a laterally large system approached those of the infinite system. For systems that are large compared with the pattern periodicity, and for control parameter values close to onset, the amplitude equation formalism can readily address these issues. The approach is particularly well suited to systems with one extended coordinate x, or to systems with two extended directions with a pattern of stripes parallel to the boundary, since, in these situations, the formalism simplifies to a one-dimensional amplitude equation. An interesting experimental application is to the Taylor–Couette system, Fig. 1.11, where in the roll state the azimuthal symmetry renders the problem one-dimensional.[15] The general situation in a two-dimensional system is harder since the boundaries often tend to reorient the stripes, which leads to a pattern with large reorientations of stripes that cannot be treated within the amplitude equation description.

We study here the case of boundaries that tend to inhibit the pattern formation. For steady states with a one-dimensional spatial variation and suppressing boundaries, we want to solve the (fully scaled) amplitude equation Eq. (6.21) with no time variation

$$0 = \bar{A} + \partial_X^2 \bar{A} - |\bar{A}|^2 \bar{A}, \tag{6.28}$$

with the condition Eq. (6.18)

$$\bar{A} = 0, \tag{6.29}$$

at the boundaries. The amplitude equation then allows us to determine how the intensity of the pattern grows with distance away from the boundary, to approach the bulk saturated value far from the boundaries. There are also dramatic effects on the range of possible wave numbers of the pattern far away from the boundary, expressed through restrictions on the phase variation of the complex amplitude. We will see that the stationary solutions to Eqs. (6.28) and (6.29) in fact have a constant phase so that, within the accuracy of the lowest-order amplitude equations, the wave number of the stripes is unique and equal to the critical wave number. In contrast, for an infinite or periodic system, there is a band of stable stationary solutions of different wave numbers as represented by the stability balloon.

---

[15] Most physical end conditions for the Taylor–Couette apparatus correspond to the more difficult case of enhancing boundaries which we will not address in this section.

6.4 Applications of the amplitude equation

We illustrate how the amplitude equation can be used to understand the effect of lateral boundaries by considering first the case of a semi-infinite one-dimensional system $X \geq 0$, with a suppressing boundary at $X = 0$. You can verify by direct substitution that a solution to the amplitude equation Eq. (6.28) with boundary condition Eq. (6.29) is

$$\bar{A} = e^{i\Phi} \tanh\left(\frac{X}{\sqrt{2}}\right), \tag{6.30}$$

where the phase Φ is an arbitrary real constant. In the unscaled variables, this expression becomes

$$A = e^{i\Phi} \sqrt{\frac{\varepsilon}{g_0}} \tanh\left(\frac{x}{\xi}\right), \tag{6.31}$$

with

$$\xi = \sqrt{2}\xi_0 \varepsilon^{-1/2}. \tag{6.32}$$

The form of the magnitude $|A|$ of the solution is shown in Fig. 6.4. This simple solution demonstrates two important features of how a boundary or defect suppresses the amplitude of a stripe solution: the suppression of the bulk value extends over a characteristic length ξ called the healing length or coherence length, and this length diverges as $\varepsilon^{-1/2}$ toward threshold. The hyperbolic tangent form of $|A|$ and the scaling of the healing length Eq. (6.32) near threshold are predictions that have been amply confirmed by experiments, e.g. in the electroconvection liquid-crystal experiment of Figs. 6.2 and 6.3.

The solution Eq. (6.31) contains an arbitrary constant phase factor $e^{i\Phi}$ which means that stripes can have an arbitrary position relative to the boundary. On the other hand, there are no solutions with a spatially varying phase, which would

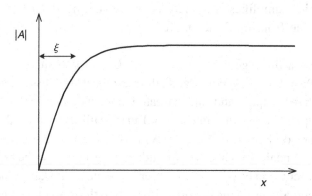

Fig. 6.4 Plot of the stationary amplitude magnitude $|A|$ from Eq. (6.31) as a function of position x near a suppressing boundary at $x = 0$ where $A = 0$. The arrow shows the length of the healing length $\xi = \sqrt{2}\xi_0 \varepsilon^{-1/2}$, which sets the length scale for the variation of the magnitude.

correspond to a deviation of the wave number of the stripes from the critical wave number. Thus, far from the boundary where the magnitude has saturated, the wave number of the stripe state is uniquely determined up to order $\varepsilon^{1/2}$ and has the value of the critical wave number:

$$q = q_c + 0 \times \varepsilon^{1/2} + O(\varepsilon). \tag{6.33}$$

(The notation $0 \times \varepsilon^{1/2}$ means that the lowest-order expected correction has a zero coefficient.) In contrast, for a laterally infinite or periodic system, the stripe wave number is not necessarily unique and has values that lie within a band that grows as $\varepsilon^{1/2}$ above threshold.

These conclusions are modified when the calculation is extended to higher order in ε. The phase Φ is then found to vary in space but slowly, in a manner consistent with Eq. (6.33). The solution far from the side wall again attains a constant magnitude $|A|$ that corresponds to saturated stripes but the wave number of the saturated stripes is not the unique value q_c of Eq. (6.33) but can take on values that lie within a band whose width increases linearly as ε near threshold. The band of observed wave numbers is therefore narrower than the width $\varepsilon^{1/2}$ for a periodic or infinite system. In addition, the stripe positions relative to the boundary become restricted to a discrete set of values.

The net result of this amplitude equation analysis is rather surprising: in a semi-infinite domain, a suppressing boundary strongly influences the possible wave numbers of the stripe state far from the boundary, causing the possible stripe wave numbers to lie in a narrow band (narrow compared with the band observed in a periodic or infinite domain) and the stripe positions to be discretized with respect to the boundary. These conclusions would be difficult to deduce directly from the basic evolution equations and indicate the power of the amplitude-equation formalism, which simplifies the analysis by separating dynamics that is slow in space and time from the faster more complicated dynamics of the evolution equations.

Now consider a finite geometry $0 \leq X \leq L$ with two suppressing boundary conditions $\bar{A}(0) = \bar{A}(L) = 0$. For large L, the regions of suppressed magnitude near the boundaries are far apart, and can be treated independently, so that $|\bar{A}|$ has a "top hat"-type X-dependence, saturating at $\bar{A} = 1$ in the bulk away from the boundaries, as shown by the solid curve in Fig. 6.5. As L is reduced, the suppression regions begin to overlap, and the maximum amplitude is reduced below the bulk saturation, as for the dashed curve in Fig. 6.5. For smaller L, the maximum amplitude decreases, and we can eventually use a linear approximation to the amplitude equation, which yields the following linear onset solution in the finite geometry:

$$\bar{A} = \bar{a} e^{i\Phi} \sin X, \tag{6.34}$$

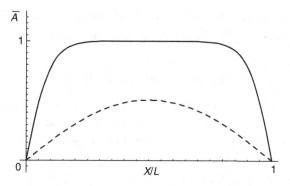

Fig. 6.5 Solution of the fully scaled amplitude equation Eq. (6.28) in a finite geometry of size L with boundary conditions $\bar{A} = 0$ at $X = 0, L$, plotted as a function of X/L. The full curve is for $L = 15$ (physical size $l = 15\varepsilon^{-1/2}\xi_0$), and the dashed curve is for $L = 3.5$ (physical size $l = 3.5\varepsilon^{-1/2}\xi_0$). Note that for system sizes large compared to the healing length, the amplitude away from the boundaries saturates at the bulk saturated value, whereas for sizes comparable to the healing length, the amplitude does not reach this value. For $L < \pi$ (corresponding to $\varepsilon < \pi^2 \xi_0^2/l^2$ in unscaled units) there is no nonzero solution. For L slightly larger than π, the solution is proportional to $\sin(\pi X/L)$, which is the solution to the linearized amplitude equation.

where Φ is an arbitrary real phase. The magnitude prefactor $\bar{a}$ is not determined by the linear equation,[16] but the solution only satisfies the boundary conditions if $L = n\pi$ with $n = 1, 2, \ldots$ Translating to the unscaled units in which the system size is l with $L = \varepsilon^{1/2} l/\xi_0$, we see that the $n = 1$ mode (the first mode that begins to grow as ε is increased) occurs not at the onset value $\varepsilon = 0$ for a periodic or infinite domain, but for a *shifted* larger value

$$\varepsilon_c = \pi^2 \left(\frac{\xi_0}{l}\right)^2. \tag{6.35}$$

This is an explicit general calculation of the suppression of the onset by finite size effects in the case of suppressing boundaries. The solution Eq. (6.34) again contains an arbitrary constant phase factor which corresponds to a continuum of onset solutions with different stripe positions. If the amplitude equation analysis is extended to higher order, this degeneracy is removed and a discrete set of onset solutions are obtained with onset values ε_c that are given by Eq. (6.35) with small corrections of order $(\xi_0/l)^4$. It is interesting to compare these general results with Exercise 2.9 of Chapter 2, which calculates the effects of suppressing boundaries on

---

[16] The magnitude $\bar{a}$ can be determined by substituting Eq. (6.34) into the nonlinear amplitude equation Eq. (6.21), followed by collecting terms in $\sin X$ while ignoring the higher harmonic terms like $\sin(3X)$ that are generated by the nonlinearity, see Exercise 6.7.

one-dimensional stripe solutions of the Swift–Hohenberg model, but using methods that are arguably less insightful and that are specific to that model.

6.4.2 Eckhaus instability

The amplitude equation provides a direct way to investigate the instability of stripe states with respect to spatially dependent perturbations, and so to construct the stability balloon near onset. The universal form of the equation implies that the stability balloon too will have universal features near onset. In addition, we can learn much more about the instabilities, for example how the wave vector of the fastest growing perturbation varies as the wave vector of the stationary nonlinear base state moves into the unstable region. Further, numerical simulations of the amplitude equation can be used to follow the growth of the perturbation to large amplitudes, so that dynamics that cause the wave number to change, for example by the elimination or creation of stripes, can be followed to completion.

The strategy for the linear stability analysis is the standard one: first construct an unperturbed base state (here the nonlinear saturated steady solution with a wave vector deviating from critical), and then investigate the dynamics of small perturbations by linearizing about the base state. With the one-dimensional amplitude equation, we can study the stability of stripe states to longitudinal perturbations. The stability of stripes to transverse perturbations is studied in Section 7.1.4 of the next chapter. The calculations for the lattice states are more involved, but follow the same ideas.

The stability balloon is obtained by testing the stability of the base states as a function of their wave number. The stripe state with wave vector differing slightly from the critical value q_c is given by the amplitude (in the scaled representation)

$$\bar{A}_K(X) = a_K e^{iKX}, \tag{6.36}$$

where the phase factor gives the wave number shift of the stripes

$$q = q_c + \xi_0^{-1} \varepsilon^{1/2} K, \tag{6.37}$$

and the magnitude prefactor is obtained as a simple result of substitution into the amplitude equation Eq. (6.21)

$$a_K^2 = 1 - K^2. \tag{6.38}$$

The existence band

$$-1 \leq K \leq 1 \quad \text{or} \quad q_c - \xi_0^{-1} \varepsilon^{1/2} \leq q \leq q_c + \xi_0^{-1} \varepsilon^{1/2}, \tag{6.39}$$

is the width of the band of wave numbers between the neutrally stable wave numbers.

6.4 Applications of the amplitude equation

The stability of these states is tested by adding to $\bar{A}_K$ a small perturbation $\delta\bar{A}$

$$\bar{A}(X,T) = \bar{A}_K(X) + \delta\bar{A}(X,T). \tag{6.40}$$

By linearizing the amplitude equation in $\delta\bar{A}$, we see that the perturbation evolves according to the following linear evolution equation:

$$\partial_T \delta\bar{A} = \delta\bar{A} + \partial_X^2 \delta\bar{A} - 2|\bar{A}_K|^2 \delta\bar{A} - \bar{A}_K^2 \delta\bar{A}^*. \tag{6.41}$$

The solution to Eq. (6.41) for $\delta\bar{A}$ turns out to be messy because of the spatial dependence of the coefficient $\bar{A}_K^2$ of the last term. Fortunately, since we are looking at the perturbation of a spatially periodic state A_K, a version of Bloch's theorem applies so that the stability eigenvalues and eigenvectors can be labeled by a Bloch wave number Q. (We will see that the perturbation $\delta\bar{A}$ actually has components with wave numbers $K \pm Q$.) The task is then to calculate the exponential growth rate $\sigma_K(Q)$, which will depend on the wave number K of the base state, and on the Bloch wave number Q that characterizes the perturbation. This result of this procedure is the growth rate

$$\sigma_K(Q) = -(1 - K^2) - Q^2 + \sqrt{(1 - K^2)^2 + 4K^2 Q^2}. \tag{6.42}$$

as shown in the following Etude.

Etude 6.1 Linear stability analysis of a stripe state using the amplitude equation
The usual form of Bloch's theorem introduced in Section 4.2.1 addresses the properties of a perturbation to a real solution. To study Eq. (6.41), we need to generalize Bloch's theorem for a complex base state. The form of the generalization can be discovered by trying an ansatz for the perturbation in the form $\delta\bar{A} \sim e^{iKX} e^{iQX}$. Substitution into Eq. (6.41) gives several terms with the same spatial dependence, but also generates a term $e^{iKX} e^{-iQX}$. Thus we try the more general ansatz

$$\delta\bar{A} = e^{iKX}[\delta a_+(T) e^{iQX} + \delta a_-^*(T) e^{-iQX}], \tag{6.43}$$

where we use the complex conjugate on δa_-^* for later convenience. Substituting this expression into Eq. (6.41), linearizing in $\delta a_\pm$, and collecting the coefficients of the two linearly independent functions $e^{i(KX \pm QX)}$ gives the pair of equations

$$d_T \delta a_+ = -(P^2 + U_+)\delta a_+ - P^2 \delta a_-, \tag{6.44a}$$

$$d_T \delta a_- = -P^2 \delta a_+ - (P^2 + U_-)\delta a_-, \tag{6.44b}$$

with

$$P^2 = 1 - K^2 \quad \text{and} \quad U_\pm = [K \pm Q]^2 - K^2. \tag{6.45}$$

The growth rate $\sigma_K(Q)$, defined by $\delta a_\pm \sim \exp[\sigma_K(Q)T]$, can be obtained by a standard eigenvalue calculation for a 2×2 matrix. One finds that the most positive growth rate about the state at wave number K is

$$\sigma_K(Q) = -P^2 - \frac{1}{2}(U_+ + U_-) + \left[P^4 + \frac{1}{4}(U_+ - U_-)^2\right]^{1/2}. \tag{6.46}$$

which simplifies to Eq. (6.42).

Given any expression for a growth rate like Eq. (6.42), it is usually interesting and easiest to explore the behavior at long wavelengths since this involves calculating the first few terms in a Taylor series about the small-wave-number limit $Q = 0$. Expanding Eq. (6.42) in the small quantity Q^2 and retaining the two lowest-order terms gives

$$\sigma_K(Q) = -\left(\frac{1-3K^2}{1-K^2}\right)Q^2 - \left(\frac{2K^4}{(1-K^2)^3}\right)Q^4 + O(Q^6), \tag{6.47}$$

which also shows that the instability is of type II since the growth rate is always zero for $Q = 0$. The coefficient of the Q^4 term is always negative within the existence band $|K| < 1$ (recall Eq. (6.39)) and so the parabola Eq. (6.47) (in Q^2) can give a positive growth rate only when the coefficient of the Q^2 term just becomes positive, which corresponds to the following inequality for the base-state wave number K:

$$|K| > \frac{1}{\sqrt{3}}. \tag{6.48}$$

Further, when instability first occurs (K lies just outside this band), it occurs with Q arbitrarily close to zero so the most unstable mode is the mode with the longest wavelength that fits into the domain.

Of course, Eq. (6.47) gives us information only about the long-wavelength longitudinal perturbations. We need to return to the general growth rate Eq. (6.42) and determine whether other instabilities might occur, say a finite-wavelength instability with $Q = O(1)$, in which case the band of stable wave numbers might be even smaller. This is found not to be the case. The two-dimensional amplitude equation discussed in the next chapter will in addition allow us to investigate further whether the stripe state is stable near onset with respect to wave vectors of arbitrary orientation relative to the stripe direction.

Since from Eq. (6.39) we know that the band of wave numbers for the existence of nonlinear stationary states satisfies $|K| < 1$, Eq. (6.48) tells us that, near threshold, the band of wave numbers that are stable with respect to a longitudinal long-wavelength instability has a universal form in that its width is $1/\sqrt{3}$ times the

6.4 Applications of the amplitude equation

width of the existence band, independent of any details of the system. Returning to physical units, the longitudinal instability occurs at the boundaries

$$q_c \pm \frac{1}{\sqrt{3}} \xi_0^{-1} \varepsilon^{1/2}. \tag{6.49}$$

The width of this band, $(2/\sqrt{3})\xi_0^{-1}\varepsilon^{1/2}$, grows as the square root of the distance $p - p_c$ above onset.

The instability we have just investigated is called the Eckhaus instability in honor of Wiktor Eckhaus, who first studied the instability in 1965. The form of the perturbation is a sinusoidal spatial modulation of the wave number of the pattern, with regions of compression and stretching as shown in Fig. 6.6. We will see in the next section (Section 6.4.3) that the result Eq. (6.48) for the boundary of the stability balloon can be obtained by a simpler calculation using the phase equation. However, the present calculation gives us additional insights into the instability. For example, you use Eq. (6.42) in Exercise 6.9 to show that, for wave numbers K unstable to the Eckhaus instability, the maximum growth rate occurs for a perturbation whose wave number Q_{max} is given by

$$Q^2_{max}(K) = 3\frac{(K^2 + 1)(3K^2 - 1)}{4K^2}, \tag{6.50}$$

with growth rate

$$\sigma_{max}(K) = \frac{(3K^2 - 1)^2}{4K^2}. \tag{6.51}$$

These results for the instabilities teach us several general lessons. We have learned that the stationary nonlinear stripe states near onset for any type-I-s system are

Fig. 6.6 Sketch of a stripe system undergoing an Eckhaus instability which is a longitudinal long-wavelength type-II modulation of the wavelength of the stripes.

unstable to a longitudinal long-wavelength Eckhaus instability whose boundaries in the stability balloon have a universal quantitative form near threshold. This is an important insight since the stability balloon gives us our basic understanding of the periodicities available for pattern formation. Ideas similar to those used in the primary instability calculation of the uniform state are often useful at these secondary instabilities too. For example, we found via Eq. (6.47) that the Eckhaus instability is itself a type-II-s instability. As in the analysis about the uniform state, the linear stability analysis of a stripe state leaves us with an exponentially growing perturbation in the unstable regions. We need to study effects nonlinear in the perturbation of the stripe state to understand the subsequent fate of the stripes, such as the important question of whether the perturbation saturates at small amplitude, or grows so large that it may catastrophically change the pattern, for example by eliminating a stripe pair. Such an analysis cannot be done analytically, but it is usually straightforward to simulate the amplitude equations numerically, and such calculations are often much easier than direct simulations near onset of fluids, liquid crystals, chemical systems, and plasmas based on their fundamental evolution equations.

6.4.3 Phase dynamics

The magnitude a and phase Φ of the complex amplitude $A = ae^{i\Phi}$ play different dynamical roles in the description of pattern formation. In particular, a perturbation of a will tend to relax to the value determined by the nonlinear terms in the amplitude equation on the time scale $\varepsilon^{-1}\tau_0$. On the other hand, a phase perturbation that is independent of position is simply a spatial translation of the whole pattern, and does not relax at all. Consequently, the relaxation of a phase perturbation on a length scale l will relax on a time scale that diverges with l (as l^2 or longer as we will see). We can therefore imagine situations where the phase relaxes much more slowly than the magnitude so that the magnitude can be evaluated as the value consistent with the instantaneous local phase field, as if the phase were time independent. (Mathematically, we neglect expressions involving time derivatives of the magnitude like $\partial a/\partial t$ since these are assumed small compared with other terms in the dynamical equation for the magnitude, such as εa.) This approximation method is known as adiabatic elimination and with this approximation the magnitude is said to adiabatically follow the phase variation. Adiabatic elimination allows us to derive a simple dynamical equation for the slow phase variation which is called the phase diffusion equation. Since a change in the phase at some position corresponds to a translation of the pattern at that point, the phase diffusion equation captures some of the essential features of pattern dynamics.

For simplicity of notation, we consider the amplitude equation in its fully scaled form Eq. (6.21). Our goal is to find a dynamical equation for the slow variation of Φ

6.4 Applications of the amplitude equation

that arises from a long-wavelength perturbation of a stripe solution. Although the calculation can be done more generally, we choose to look at small perturbations of a uniform stripe state with a wave number q shifted from critical $q = q_c + \xi_0^{-1} \varepsilon^{1/2} K$ that is described by the time-independent amplitude as in Eq. (6.36)

$$\bar{A} = a_K e^{iKX}, \qquad (6.52)$$

with the value of the constant $a_K = \sqrt{1-K^2}$ as in Eq. (6.38). We now write for the perturbed amplitude

$$\bar{A}(X,T) = a e^{i\Phi} e^{iKX}, \qquad (6.53)$$

with $a = a_K + \delta a(X,T)$, then linearize in the small-amplitude perturbation $\delta a(X,T)$ *and* linearize in low-order spatial derivatives of the phase $\Phi(X,T)$. Since the length scale of the perturbation is supposed long, we neglect higher-order spatial derivatives of the same quantity, i.e. we assume for example that $\partial_X^2 a \ll \partial_X a \ll a$. We also only keep those expressions that lead to terms in the final phase evolution equation that have up to second-order spatial derivatives of Φ.

The formal scheme is to insert Eq. (6.53) into the amplitude equation Eq. (6.21), to use the expression

$$\partial_T \bar{A} = (\partial_T a + ia \, \partial_T \Phi) e^{i\Phi} e^{iKX}, \qquad (6.54)$$

and a corresponding expression for the spatial derivative $\partial_X^2 \bar{A}$, to multiply through by $e^{-i\Phi} e^{-iKX}$ and finally to collect real and imaginary parts. When the dust clears, the real and imaginary parts of the amplitude equation are found to be

$$\partial_T a = (1 - K^2)a - a^3 + \partial_X^2 a - 2Ka \, \partial_X \Phi - a (\partial_X \Phi)^2, \qquad (6.55a)$$

$$a \, \partial_T \Phi = 2(K + \partial_X \Phi) \partial_X a + a \, \partial_X^2 \Phi, \qquad (6.55b)$$

which are evolution equations for a and Φ. Let us first consider Eq. (6.55a) and substitute $a = a_K + \delta a(X,T)$, retain only terms that are linear in δa, neglect the term $(\partial_X \Phi)^2$ as smaller than the term $K \partial_X \Phi$, and use the fact that $a_K^2 = 1 - K^2$. This yields the following linearized evolution equation for δa (see Exercise 6.10)

$$\partial_T \delta a = -2a_K^2 \, \delta a - 2Ka_K \, \partial_X \Phi + \partial_X^2 \delta a. \qquad (6.56)$$

Note that, for a spatially uniform perturbation, Eq. (6.56) shows that the magnitude perturbation δa relaxes exponentially as $\exp(-2a_K^2 T)$. Since a_K^2 is of order unity, this is a rapid decay of magnitude perturbation as was mentioned in the introduction to this section.

The phase variation $\partial_X \Phi$ in Eq. (6.56) drives a nonzero value of δa. In comparing the size of the terms in Eq. (6.56) that involve δa, we see that the dominant term is the first one on the right-hand side since all the other terms involve spatial derivatives or time derivatives of δa that are small for slow variations. Thus

$$a_K \, \delta a \simeq -K \, \partial_X \Phi, \quad (6.57)$$

and δa adiabatically follows the perturbations of the phase gradient. (Note that this expression is just the equation for the change δa in magnitude given by Eq. (6.38) arising from a change in the wave number K by $\delta K = \partial_X \Phi$.)

We now continue with the Φ-evolution equation Eq. (6.55b). We substitute $a = a_K + \delta a$, keep only terms linear in δa, and neglect $\partial_X \Phi$ as small compared with K to get

$$a_K \, \partial_T \Phi \simeq 2K \, \partial_X \delta a + a_K \, \partial_X^2 \Phi. \quad (6.58)$$

Eliminating δa with the adiabatic approximation Eq. (6.57) and using Eq. (6.38) leads to the following evolution equation for slowly varying phase perturbations

$$\partial_T \Phi = \frac{1 - 3K^2}{1 - K^2} \, \partial_X^2 \Phi. \quad (6.59)$$

This is a diffusion equation for the phase, with a diffusion constant $D_\parallel$ for variations along the stripe normal. Transforming back to the unscaled space and time variables, the equation becomes

$$\partial_t \Phi = D_\parallel \, \partial_x^2 \Phi, \quad (6.60)$$

with diffusion constant

$$D_\parallel = (\xi_0^2 \tau_0^{-1}) \frac{\varepsilon - 3\xi_0^2 k^2}{\varepsilon - \xi_0^2 k^2}. \quad (6.61)$$

for phase perturbations about the stripe state with wave number $q = q_c + k$, where $k = \xi_0^{-1} \varepsilon^{1/2} K$.

The phase equation Eq. (6.60) is a powerful tool, and many important results can be derived from it. For example, we know that diffusion equations lead to exponentially growing solutions if the diffusion constant is negative which here would signal the onset of an instability. Equation (6.61) therefore implies that a stripe state with number $q_c + k$ is unstable to long-wavelength longitudinal phase perturbations when $D_\parallel < 0$ or $|\xi_0 k| > \varepsilon^{1/2}/\sqrt{3}$. This is just the Eckhaus instability described previously in Section 6.4.2. (But, again, we cannot deduce detailed facts about the growth rate $\sigma_K(Q)$ of the stripe solution from the phase diffusion equation, which not surprisingly does not provide as much information as a full perturbation analysis.)

The topic of phase dynamics is one that will recur later in the book. Although the phase dynamics is easily derived from the amplitude equation formalism as we have just done, we will see in Chapter 9 that phase dynamics has a wider validity. We will argue that even away from threshold, the symmetry aspects of the pattern are captured by an appropriately defined phase variable. Again slow spatial variations of the phase necessarily evolve slowly in time, and this slow variation can be isolated mathematically from the faster dynamics of other degrees of freedom in the form of a phase diffusion equation, which is now nonlinear and so more difficult to analyze. The phase dynamics provides a simple way to investigate some important questions such as what are some of the instabilities that bound the stability balloon of the finite-amplitude nonlinear stripe states or parameter regimes or situations where the amplitude equation is not valid.

6.5 Limitations of the amplitude-equation formalism

Although many interesting questions can be addressed within the amplitude equation, it is important to bear in mind the limitations of the formalism. One limitation is that the amplitude equation is derived by perturbation expansion and truncation, and so is only a good approximation over a restricted range of parameters, in particular near onset and for long wavelength and temporally slow modulations of the ideal pattern.

There are limitations on the nature of the patterns that can be treated. For example, as we will see in the next chapter where we generalize the approach to two dimensions, because it is not possible to derive a rotationally invariant amplitude equation that respects the rotational invariance of the physical system, the only patterns that can be calculated quantitatively are those that are close to a single set of parallel stripes, or close to a superposition of stripes such as squares and hexagons. Patterns in which the orientation of stripes or lattices vary through large angles over large distances cannot be treated even though the rate of variation can be slow.

The way in which the amplitude equation approximation is physically relevant can be subtle. Indeed the answers to qualitative questions may be quite wrong! For example if we ask the question "Can system ABC show chaos near onset?" the lowest-order amplitude equation may immediately lead us to the answer "No," because of the existence of the potential. However, since the equation is derived as an approximation, we should not be so definite in any physical statement. Indeed, the correct answer might be, "The relaxational dynamics predicted by the amplitude equation should be a good coarse description of what happens." However, at long times, there may be slow persistent dynamics at a time scale beyond the $O(\varepsilon^{-1})$ time scale of the dynamics controlled by the amplitude equation, or there may be small-magnitude persistent dynamics, perhaps on a fast scale, that is again beyond what the amplitude equation can approximate. Alternatively, the amplitude equation may predict dynamics that is quenched by residual effects not captured by the perturbation formalism. An example of this is the propagation of fronts (mentioned briefly in Section 6.3.2 and to be discussed in Section 8.3) where the motion predicted by the amplitude equation may be quenched by the pinning of the fronts to the stripes themselves.

We have said that the amplitude equation is good "near onset." We might also question this phrase more carefully: does this mean asymptotically near onset, or just at some small but finite distance from onset? Finally, as a worst case, if the solutions predicted by the amplitude equation are actually sensitive to slight changes in the equations (such solutions are called structurally unstable), the predicted behavior might be misleading and bear no resemblance to the actual physical behavior. While these concerns are all important and some continue to be research issues, the fortunate fact remains that the amplitude equation formalism does succeed in explaining many experimental phenomena, and provides insights that are difficult to obtain when working with the full evolution equations.

6.6 Conclusions

In this chapter, we have introduced a powerful tool known as the amplitude equation that allows us to understand many aspects of pattern formation. This equation describes the slow space and time modulation of the critical onset mode, for states that are "close" to a stripe state, and provides a natural extension of the linear stability analysis into the weakly nonlinear regime for control parameter values near threshold. The amplitude equation captures three basic ingredients of pattern formation: the growth of the perturbation about the spatially uniform state, the saturation of the growth by nonlinearity, and the effect of spatial distortions of the pattern. As a consequence, it provides a useful starting point for thinking about conceptual issues even though the quantitative applicability is valid only near threshold.

It is worthwhile to note that the amplitude equation Eq. (6.9) is actually well known in other contexts. For example, it is the Ginzburg–Landau equation that describes the evolution of the complex-valued order parameter used to describe a superconductor or superfluid in a mean field theory of these thermodynamic phases. In those contexts, the phase of the order parameter is related to the supercurrents that flow in these systems. Thus stripe patterns with wave numbers away from the critical value q_c (i.e. having a phase gradient of the amplitude) correspond in this mathematical description to a supercurrent in a superconductor. Applying intuition gained from the study of one of these physical systems to the other can be quite productive. Given this analogy with an equilibrium system, it is not surprising that the amplitude equation has potential dynamics Eq. (6.25), although this results from the truncated perturbation expansion and will not be true if the expansion is continued to higher order.

We have shown that the amplitude equation can be reduced by a rescaling of variables to a universal parameter-free form, and that the only parameter of the description is the scaled system size where boundary conditions are applied. This explains why diverse physical systems that yield patterns through a type-I-s instability – be they fluid, chemical or biological systems – can show *the same* behavior in the regime of validity of the amplitude equation. Again, what we learn from the study of one system, may be transferred to the others.

We have so far discussed the amplitude equation in the simplest context of a type-I-s instability with spatial variations along a single extended direction, leading to the one-dimensional amplitude equation, Eq. (6.9). In the next chapter, we will study the generalization to two extended directions, particularly the case where the physical system has a rotational symmetry with respect to the extended directions. This will especially allow us to study the dynamics and competitions of lattice states, and further understand the stability balloon and the effect of lateral boundaries. We will discuss the amplitude equation formalism further in Chapter 8, which concerns localized structures that are observed in many experiments, and complete our discussion in Chapter 10, where the ideas are extended to oscillatory media whose uniform state undergoes a type-o instability.

6.7 Further reading

(i) The amplitude equation description of one-dimensional pattern formation was introduced by Newell and Whitehead [80] and Segel [93].
(ii) A general introduction to the technique of multiple scales perturbation theory can be found in *Advanced Mathematical Methods for Scientists and Engineers* by Bender and Orszag [11].
(iii) *Statistical Physics, Part 2* by Lifshitz and Pitaevskii [65]. Chapter V provides an introduction to the Ginzburg–Landau equation in the context of superconductivity.

Exercises

6.1 **Elimination of the linear derivative term in the amplitude equation:** Assuming that a term $\xi_1 i \, \partial_x A$ appears on the right-hand, side of Eq. (6.9), where ξ_1 is a real constant, find the value of the real constant Δ such that the transformation $A \to \bar{A} e^{i\Delta x}$ yields an amplitude equation for the new amplitude $\bar{A}$ that lacks this term.

6.2 **Coefficients in the amplitude equation for the Swift–Hohenberg equation:** Following the phenomenological approach discussed in Section 6.2.2, deduce the values of the amplitude equation constants τ_0, ξ_0, and g_0 for the one-dimensional Swift–Hohenberg model Eq. (2.4). For the purpose of this exercise, define the small parameter ε to be the Swift–Hohenberg parameter r, rather than using the experimentally appropriate relation Eq. (5.21).

6.3 **One-dimensional amplitude equation for system without $u \to -u$ symmetry:** Show that for the one-dimensional Swift–Hohenberg equation without $u \to -u$ symmetry

$$\partial_t u(x,t) = ru - (\partial_x^2 + 1)^2 u - g_2 u^2 - u^3, \tag{E6.1}$$

the *form* of the amplitude equation is unchanged but the value of the nonlinear coefficient g_0 is different.

6.4 **Amplitude equation for the Brusselator:** Use the phenomenological method of Section 6.2.2 to derive the coefficients τ_0 and ξ_0 in the one-dimensional amplitude equation for the stripe state of the Brusselator reaction diffusion system, Eqs. (3.23). You should treat b as the control parameter so $\varepsilon = (b - b_c)/b_c$.

6.5 **Method of multiple scales:** Use the method of multiple scales discussed in Appendix 2 to derive the lowest-order amplitude equation for the generalized Swift–Hohenberg equation in one dimension

$$\partial_t u(x,t) = ru - (\partial_x^2 + 1)^2 u + (\partial_x u)^2 \partial_x^2 u. \tag{E6.2}$$

6.6 **Numerical solutions of the amplitude equation in a finite geometry:** Use a computational environment like Mathematica, Maple, or Matlab to calculate the positive nonlinear steady solution $\bar{A}(X)$ to the amplitude equation Eq. (6.21) with boundary conditions $\bar{A}(0) = \bar{A}(L) = 0$ for various scaled system sizes L. (Note that varying L is equivalent to varying ε for fixed physical system size.)

(a) Show graphically how the shape of $\bar{A}$ evolves from sinusoidal for L just greater than π to a top-hat-like profile for larger L.
(b) Plot how the maximum amplitude, and (more work) the integrated intensity (both in unscaled units) vary with ε for some fixed (unscaled) length.

Although one can directly solve for steady state solutions of Eq. (6.21) with boundary conditions $\bar{A}(0) = \bar{A}(L) = 0$ as a time-independent boundary-value problem, it is easier to use the fact that the dynamics is potential to find a stationary state as the asymptotic behavior of some time-dependent state. So first find a high-level integrator for one-dimensional partial differential equations (e.g. the function NDSolve in Mathematica). Then specify the boundary conditions $\bar{A}(0) = \bar{A}(L) = 0$ and the amplitude equation Eq. (6.21), specify some positive initial condition that has a small overall magnitude (e.g. $\bar{A}(X,0) = 0.1\sin(\pi X/L)$, then integrate the equations forward in time until a stationary state is attained, say $F = 30$ time units. You will have to explore a bit how accurately you can approach a stationary solution for a given integration time T, but 10% accuracy is plenty for the purpose of this problem. Since you are interested only in the asymptotic state, you also can take large time steps (or specify a large integration error) since it is only the asymptotic time-independent state you are interested in, and this is determined by the spatial resolution.

As an example, the following Mathematica code will integrate the scaled one-dimensionl amplitude equation with the specified boundary conditions for an integration time of $F = 30$ units, then plot the asymptotic amplitude form $A(X,F)$, and also plot the entire space-time evolution over the region $[0,L] \times [0,F]$ so that you can examine the convergence to the asymptotic stationary state.

```
L = 4;     (* length of domain, should be bigger than Pi *)
F = 30;    (* integration time of 30 time units *)
solution = A /. First[
    NDSolve[              (* solve 1d pde with ics and bcs *)
    {
        D[A[X,T],T] == A[X,T] + D[A[X,T],{X,2}] - A[X,T]^3 ,
        A[X,0] == 0.1*Sin[Pi X/L] ,  (* the initial condition *)
        A[0,T] == 0,       (* boundary condition A(0,t)=0 *)
        A[L,T] == 0        (* boundary condition A(L,t)=0 *)
    },
    A,                     (* variable to solve for *)
    {X, 0, L},             (* range of X, namely [0,L] *)
    {T, 0, F}              (* range of T, namely [0,F] *)
    ]
] ;
Plot[ solution[X,F], {X, 0, L} ]
Plot3D[ solution[X,T], {X, 0, L}, {T, 0, F}, PlotRange -> All ]
```

6.7 **Magnitude of the solution near onset in a finite geometry:** Sufficiently close to onset, a sinusoidal function as in the linear expression for the amplitude Eq. (6.34) remains a good approximation to the shape of the solution, with the amplitude $a_1(\varepsilon)$ now fixed by the nonlinearity. More specifically we have

$$\bar{A} = a_1 \sin(\pi X/L) + a_3 \sin(3\pi X/L) + \cdots \qquad (E6.3)$$

with $a_3 \ll a_1$ for ε close enough to ε_c. The spatial period of the sine function is chosen so that the amplitude satisfies the boundary conditions Eq. (6.29). Remember $L = \varepsilon^{1/2} l/\xi_0$ with $L > L_c = \pi$ for $\varepsilon > \varepsilon_c$. By substituting this expression into the amplitude equation Eq. (6.21) and collecting coefficients of the orthogonal functions $\sin(n\pi X/L)$, show that a_1^2 grows linearly with ε

according to
$$a_1^2 \simeq \frac{4}{3}\frac{\varepsilon - \varepsilon_c}{\varepsilon_c} \tag{E6.4}$$
for small enough $\varepsilon - \varepsilon_c$. Find the ratio of the slope of the mean squared (unscaled) amplitude of the pattern against ε to the value for an infinite system (cf. the discussion of the effects of boundaries in Chapter 2). Discuss over what range of ε this expression might be expected to be valid, and how the sinusoidal shape of the amplitude changes to the top-hat shape as in Fig. 6.5 for larger ε.

6.8 **Growth rate curve for the Eckhaus instability:**

(a) Derive Eq. (6.46), which is the final result for the calculation of the growth rate for a long-wavelength perturbation to a stripe state.
(b) Plot the dependence of the growth rate $\sigma_K(Q)$ on the wave number Q of small perturbations for the Eckhaus instability for different values of the background wave number K and so verify the statement that the instability occurs first at long wavelengths ($Q \to 0$).

6.9 **Wave number for maximum growth rate beyond the Eckhaus instability:** For stripes with wave numbers K that are unstable to the Eckhaus instability, show that the maximum growth rate occurs for a perturbation of wave number
$$Q_{X\,\text{max}}^2(K) = \frac{(K^2+1)(3K^2-1)}{4K^2}, \tag{E6.5}$$
and that the corresponding growth rate is
$$\sigma_{\text{max}}(K) = \frac{(3K^2-1)^2}{4K^2}. \tag{E6.6}$$

6.10 **Derivation of the phase equation:** Derive with full attention to details the perturbation equations Eqs. (6.56) and (6.58), keeping only terms linear in δa and Φ and up to second order in spatial gradients.

6.11 **Invariants in the time-independent amplitude equation:** Show that the quantities Q and E defined by
$$Q = a^2 \partial_X \Phi \quad \text{and} \quad E = \frac{1}{2}(\partial_X a)^2 + \frac{Q^2}{2a^2} + \frac{1}{2}a^2 - \frac{1}{4}a^4 \tag{E6.7}$$
are invariants in that they are independent of X for $\bar{A} = ae^{i\Phi}$ satisfying the time-independent one-dimensional amplitude equation, Eq. (6.21).

Use this fact to show that if the amplitude magnitude a is zero at any point, then the phase must be a constant throughout the system, in which case the wave number of the pattern cannot deviate from the critical value q_c.

6.12 **Wave-number band with spatially dependent magnitude of the amplitude:** Assume that in some one-dimensional system the magnitude a of the amplitude varies spatially and has a maximum value denoted by $a_M = O(1)$ and a smallest value denoted by a_m. Using the spatial invariants Eq. (E6.7) introduced in Exercise 6.11, show that the local wave-number perturbation $q_M = \partial_X \Phi$ at the maximum of a, $a = a_M$, is given by the expression

$$q_M^2 \leq a_m^2(1 - a_m^2/2), \tag{E6.8}$$

and so is limited by the smallest value a_m that a takes anywhere in the system (e.g. at the boundary where the amplitude is suppressed).

Use this result to compute the band of possible wave numbers in regions where the amplitude saturates at the bulk value far away from the region of suppressed amplitude, and compare with the ideal system where the magnitude is saturated everywhere.

6.13 **Nonlinear phase diffusion equation:** When the longitudinal diffusion coefficient $D_\parallel$ passes through zero in the phase diffusion equation, Eq. (6.59), phase perturbations can grow exponentially so that a linearized evolution equation is no longer a good approximation. Also the shorter the wavelength of the perturbation, the more rapid the growth within the phase equation. We need to keep nonlinear terms, and higher-order derivative terms to control these effects.

(a) By considering only spatial variation in the X-direction but keeping nonlinear and higher-order derivative terms, show from the amplitude equation that the equation for X-dependent phase variations about the state $a_K e^{iKX}$ may be extended to

$$\partial_T \Phi = D_\parallel \partial_X^2 \Phi - \gamma \partial_X^4 \Phi + \beta (\partial_X \Phi) \partial_X^2 \Phi + \cdots. \tag{E6.9}$$

(b) Derive expressions for $\beta(K)$ and $\gamma(K)$.
(c) This phase equation has fourth-order spatial derivatives, but in the original amplitude equation terms in ∂_x^4 were ignored. Is this a consistent approximation?

7
Amplitude equations for two-dimensional patterns

In the previous chapter, we introduced and used a one-dimensional amplitude equation to study slow spatiotemporal modulations of a stripe pattern but with the restriction that the spatial variation could only be longitudinal (in the direction normal to the stripes). In the present chapter, we extend the previous chapter in two ways to study patterns that depend on *two* extended coordinates. The first generalization is the obvious one, which is to write down a two-dimensional amplitude equation that can treat a stripe pattern with modulations that vary along, as well as normal to, the stripes. The second generalization is to study superpositions of stripe states with different orientations, which will allow us to study quantitatively the periodic lattice states that we discussed in Section 4.3. It turns out that many useful insights about the stability of, and competition between, lattice states can be obtained by using just the zero-dimensional (no spatial dependence) amplitude equation similar to Eq. (4.6) to describe each stripe participating in the superposition. At the end of the chapter, we will discuss briefly the more general but also more difficult case of using a two-dimensional amplitude equation to describe general slow distortions of each stripe associated with lattice states and other stripe superpositions.

The two-dimensional amplitude equation that describes modulations along as well as normal to stripes turns out to have two different forms depending on whether the system is rotationally invariant (for example, Rayleigh–Bénard convection) or anisotropic (for example, a liquid crystal or a conducting fluid in the presence of a magnetic field). We first discuss the two-dimensional amplitude equation for rotationally invariant systems. This amplitude equation turns out to have a rather complex form, with spatial derivatives in the coordinate along the stripes occurring up to fourth order. The higher-order spatial derivatives in turn require additional boundary conditions to be determined which is a significant and subtle complication beyond what we discussed in Section 6.2.4 for the one-dimensional amplitude equation. The two-dimensional amplitude equation for the rotationally invariant case allows us to study analytically the transverse long-wavelength zigzag instability that we discussed briefly in Section 4.2.1 and calculated via a one-mode Galerkin approximation for the two-dimensional Swift–Hohenberg equation (see Etude 4.1). Our discussion in Section 7.1.4 will show that the zigzag instability is a universal instability of stripes near onset in rotationally invariant systems and is important since it bounds part of the stability balloon.

For systems without full rotational symmetry in the plane of the extended coordinates, the two-dimensional amplitude equation turns out to be simpler than for the rotationally symmetric case. In appropriately chosen scaled coordinates, the two-dimensional equation is obtained from the one-dimensional amplitude equation by simply replacing the second-order derivative ∂_X^2 with the symmetric combination $\partial_X^2 + \partial_Y^2$. A consequence of this simplicity is that results for the stability balloon near onset are easily obtained by generalizing our one-dimensional calculations.

We next discuss lattice states that can be formed from a collection of stripes with different orientations that go unstable at the same threshold (see Figures 2.11 and 2.13(b), and the discussion in Section 4.3). The existence of lattice states near threshold, and the competition between them and the stripe state, can be described in terms of coupled amplitude equations with one amplitude for each component stripe. (For example, two coupled amplitude equations are required to describe the perpendicular stripes that form a square state.) To illustrate the issues in the simplest context, we set up these amplitude equations without spatial derivatives, which corresponds to using uniform stripe states at the critical wave number q_c. Our discussion will identify a new quantity, a stripe coupling coefficient $G(\theta)$, that characterizes the nonlinear interaction between two sets of stripes at an angle θ to one another and whose values determine the relative stability of one lattice state compared to another. Hexagonal lattices, formed from three sets of stripes at an angle of $\pi/3$ to one another, are particularly important for systems which do not have the field inversion symmetry $\mathbf{u} \to -\mathbf{u}$, since there are then additional terms in the amplitude equation that favor hexagons near threshold.

Toward the end of this chapter, we include the spatial derivatives for the coupled amplitude equations. This allows us to investigate the cross-stripe instability introduced in Section 4.2.1. This instability is important in systems that are rotationally invariant in the plane, since, depending on the values of $G(\theta)$, the cross-stripe instability may bound part of the stability balloon close to threshold and preempt the Eckhaus instability as the wave number of the pattern is increased away from the critical value. Further applications of the coupled amplitude equations with spatial derivatives are postponed to Chapter 8, for example Section 8.2 where we discuss the dynamics of grain boundaries.

7.1 Stripes in rotationally invariant systems

We first consider the amplitude equation for stripe-forming systems that have two extended directions and that have rotational symmetry in this plane. (Rayleigh–Bénard convection and the Turing instability in a large aspect ratio planar geometry are examples of such a system.) The derivation of the amplitude equation follows along the same lines as for one-dimensional systems in Section 6.1 and Section 6.2 but now incorporates the two dimensions and the new rotational symmetry. Boundary conditions must again be specified to solve the equation in a specific geometry. We end this section by discussing the new features of the stability balloon near threshold for stripes in rotationally invariant two-dimensional systems.

The amplitude function is introduced in the same way as for the one-dimensional situation Eq. (6.3), but we now allow the amplitude to be a slowly varying function of the two coordinates in the extended directions, as well as of time

$$\mathbf{u}_p(\mathbf{x}_\perp, \mathbf{x}_\|, t) = A(\mathbf{x}_\perp, t)\mathbf{u}_c(\mathbf{x}_\|)e^{iq_c x} + \text{c.c.} + \text{h.o.t.}, \tag{7.1}$$

with $\mathbf{x}_\perp = (x, y)$. Although the physical system is rotationally invariant, we are forced by the amplitude equation formalism to introduce a reference set of parallel stripes at the critical wave number q_c, and to specify a particular reference direction normal to these stripes, which is the x-direction in Eq. (7.1). This stipulation of a reference direction in a rotationally invariant system is an undesirable but unfortunately unavoidable aspect of the formulation and limits the use of the amplitude equation. In particular, patterns in which the stripe orientation varies over large angles across the system, even if the rate of variation is slow, cannot be treated using the amplitude equation approach.

7.1.1 Amplitude equation

In Section 6.1, we used the correspondence of a slow variation in the amplitude's phase in the x-direction to a stretching of the wave vector of the critical state to understand the form of the gradient terms appearing in the one-dimensional amplitude equation. The first step in constructing the two-dimensional amplitude equation is to ask what is the significance of a slow variation of the phase in the y-direction along the stripes. This turns out to correspond to a small change in the orientation of the stripes. We can see this by examining a complex amplitude $A = ae^{i\Phi}$ with a non-constant phase of the form $\Phi = k_x x + k_y y$ with $|k_x|, |k_y| \ll q_c$ so that the phase is slowly varying. Substituting this variation into Eq. (7.1) leads to the spatial dependence

$$\mathbf{u}_p(\mathbf{x}_\perp, \mathbf{x}_\|, t) = 2a\mathbf{u}_c(\mathbf{x}_\|)\cos(\mathbf{q} \cdot \mathbf{x}_\perp) + \text{h.o.t.}, \tag{7.2}$$

7.1 Stripes in rotationally invariant systems

with wave vector

$$\mathbf{q} = (q_c + k_x)\hat{x} + k_y\hat{y}. \tag{7.3}$$

This wave vector corresponds to a rotation of the stripes through the small angle k_y/q_c to leading order in the small quantity $|k|/q_c$. Equation (7.3) also corresponds to a small stretching of the wave number by an amount that is first order in k_x and second order in k_y:

$$q \simeq q_c + k_x + \frac{k_y^2}{2q_c} + \cdots. \tag{7.4}$$

Now we establish the form of the two-dimensional amplitude equation for a rotationally invariant system by imposing the restriction that the form must be invariant under a small rotation. This is imposed by demanding that the amplitude equation be invariant under the substitution

$$A \to A \exp\left[i\left(\Delta y - \frac{\Delta^2}{2q_c}x\right)\right], \tag{7.5}$$

with Δ a small real constant, which corresponds to a small rotation through angle Δ/q_c.[1] It is only the derivative terms in the amplitude equation that will have a different form in the generalization to two dimensions, and these are determined as the lowest-order derivative terms that have the invariance under the transformation of Eq. (7.5). This leads to the result for the two-dimensional amplitude equation for a system that is rotationally invariant

$$\tau_0 \, \partial_t A(x, y, t) = \varepsilon A + \xi_0^2 \left(\partial_x - \frac{i}{2q_c}\partial_y^2\right)^2 A - g_0|A|^2 A. \tag{7.6}$$

The asymmetric way that the x- and y-derivatives appear in the equation results from the choice of a particular reference state, with stripes perpendicular to the x-direction. We can confirm that Eq. (7.6) has the correct form by looking at the growth of a small amplitude solution in the linear approximation

$$A = a e^{i(k_x x + k_y y)} e^{i\sigma_k t}. \tag{7.7}$$

Substituting into Eq. (7.6) gives for the growth rate

$$\sigma_\mathbf{k} = \tau_0^{-1} \xi_0^2 \left(k_x + \frac{k_y^2}{2q_c}\right)^2 \simeq \tau_0^{-1}\xi_0^2(q - q_c)^2, \tag{7.8}$$

where Eq. (7.4) has been used in the last approximate equality.

---

[1] The linear dependence of the phase on the y-coordinate causes a small rotation of the wave vector, and a second-order change Eq. (7.4) in the magnitude of the wave vector. This magnitude change is removed by the $O(\Delta^2)$ linear dependence of the phase on the x-coordinate, yielding in the end a pure rotation up to $O(\Delta^4)$.

As before, we can eliminate the constants τ_0, ξ_0, g_0, and now also q_c to obtain a scaled form of the amplitude equation that does not depend explicitly on physical parameters. We accomplish this by substituting the following scaled variables (which assumes g_0 is positive)

$$\tilde{A} = g_0^{1/2} A, \quad \tilde{x} = x/\xi_0, \quad \tilde{y} = y(q_c/\xi_0)^{1/2}, \quad \tilde{t} = t/\tau_0, \tag{7.9}$$

to obtain an equation in which only the parameter ε remains

$$\partial_{\tilde{t}} \tilde{A} = \varepsilon \tilde{A} + \left(\partial_{\tilde{x}} - \frac{i}{2} \partial_{\tilde{y}}^2\right)^2 \tilde{A} - |\tilde{A}|^2 \tilde{A}. \tag{7.10}$$

This generalizes Eq. (6.17) to the rotationally invariant two-dimensional case.

Alternatively, we can incorporate appropriate ε scales into the variables to eliminate this parameter from the equation:

$$\bar{A} = \left|\frac{g_0}{\varepsilon}\right|^{1/2} A, \quad X = \frac{|\varepsilon|^{1/2}}{\xi_0} x, \quad Y = |\varepsilon|^{1/4} \left(\frac{q_c}{\xi_0}\right)^{1/2} y, \quad T = \frac{|\varepsilon|}{\tau_0} t, \tag{7.11}$$

which yields the fully scaled equation

$$\partial_T \bar{A} = \pm \bar{A} + \left(\partial_X - \frac{i}{2} \partial_Y^2\right)^2 \bar{A} - |\bar{A}|^2 \bar{A}, \tag{7.12}$$

which generalizes Eq. (6.21). The positive sign for the first term on the right-hand side again corresponds to $\varepsilon > 0$, and the negative sign to $\varepsilon < 0$. This scaled version of the equation shows us the universality and scaling properties of the solutions. It also shows us that the spatial variation in directions parallel and perpendicular to the stripes will typically occur over lengths that scale in different ways with ε namely proportional to $\varepsilon^{-1/2}$ for the direction perpendicular to the stripes, but proportional to $\varepsilon^{-1/4}$ along the stripes. For example, the core of a dislocation defect (the region where the amplitude is suppressed from the bulk value by the rapid phase variation) will show this anisotropic structure.

7.1.2 Boundary conditions

The amplitude equations (7.6), (7.10) or (7.12) must be supplemented with boundary conditions. As in the one-dimensional case, periodic boundary conditions over a rectangular domain are often used in theoretical discussions as a convenient mathematical simplification. The question of the boundary conditions for realistic physical boundaries is more difficult. Part of the difficulty is that stripes tend to align along the normal to a boundary, which necessarily leads to large reorientations of the stripes over a simply connected region such as a rectangular or circular system. This situation cannot be treated within the amplitude-equation formalism, which

7.1 Stripes in rotationally invariant systems

requires stripes to be nearly parallel everywhere. On the other hand, the tendency toward a normal alignment does not seem to be a boundary *condition* and, particularly near threshold, patterns are often seen where the orientation of the stripes appears unaffected by the proximity of boundaries.

The simplest cases to analyze are boundaries parallel or perpendicular to the reference wave-vector direction used to construct the amplitude equation. For a boundary perpendicular to this direction, with normal along the x-direction in our conventional choice of axes, the problem reduces to the one-dimensional situation considered before, and the same conditions as before are sufficient. For a boundary that inhibits the instability, we have the same condition as Eq. (6.18)

$$A = 0 \quad \text{as } x \to \text{boundary}. \tag{7.13}$$

On the other hand, for a boundary normal to the y-direction, the variation of the amplitude induced by the boundary is in the y-direction. Since ∂_y appears up to fourth order ∂_y^4 in the amplitude equation, we might expect an additional condition to be required at each such boundary (for a total of two conditions specified at each boundary point). By matching to complete solutions of the basic equations near the boundary, it has been shown in specific cases that the "natural" conditions

$$A = \partial_y A = 0 \quad \text{as } y \to \text{boundary} \tag{7.14}$$

apply for boundaries that suppress the pattern formation. For stripes approaching a boundary at an arbitrary angle, a more complicated analysis of the region near the boundary seems to be necessary.

7.1.3 Potential

Using the same methods as in Section 6.3.2, you can show that, with appropriate boundary conditions (those of Eqs. (7.13) and (7.14) or periodic boundary conditions for example), the two-dimensional amplitude equation has potential dynamics just like the one-dimensional amplitude equation. Quoting the result for the scaled form of the equation (7.12), the potential is

$$\bar{V}[\bar{A}] = \iint dX\, dY \left[-|\bar{A}|^2 + \frac{1}{2}|\bar{A}|^4 + \left|\left(\partial_X - \frac{i}{2}\partial_Y^2\right)\bar{A}\right|^2 \right], \tag{7.15}$$

where the double integral goes over the interior of the domain. The potential evolves according to the equation

$$d_T \bar{V} = -2 \iint dX\, dY\, |\partial_T \bar{A}|^2, \tag{7.16}$$

and decreases if there is any dynamics of the amplitude.

7.1.4 Stability balloon

The extra flexibility of perturbations that have a spatial dependence transverse to the stripe wave vector allows new instabilities of the stripe state in addition to the Eckhaus instability discussed in Section 6.4.2. A transverse long-wavelength instability known as the zigzag instability is especially important since it limits the band of stable wave numbers of stripes in a rotationally invariant system. The two-dimensional amplitude equation allows us to calculate the boundary to the zigzag instability near threshold.

The procedure follows precisely the same path as in Section 6.4.2, but now we use the two-dimensional amplitude equation in its fully scaled form, Eq. (7.12). For the base state, we choose the amplitude $A_K(X)$ in Eq. (6.36), which corresponds to a stripe state whose wave vector differs slightly from q_c. We test the stability of the state by adding to $\bar{A}_K$ a small perturbation $\delta\bar{A}$ that now depends on X and Y:

$$\bar{A}(X,Y,T) = \bar{A}_K(X) + \delta\bar{A}(X,Y,T). \tag{7.17}$$

Linearizing the amplitude equation in $\delta\bar{A}$ gives the following evolution equation for the perturbation:

$$\partial_T \delta\bar{A} = \delta\bar{A} + \left(\partial_X - \frac{i}{2}\partial_Y^2\right)^2 \delta\bar{A} - 2|\bar{A}_K|^2 \delta\bar{A} - \bar{A}_K^2 \delta\bar{A}^*. \tag{7.18}$$

The analysis proceeds as in the Etude in Section 6.4.2 except that the Bloch ansatz Eq. (6.43) must be generalized to allow for a perturbation with an arbitrary wave vector $\mathbf{Q}$:

$$\delta\bar{A} = e^{iKX}\left[\delta a_+(T)e^{i\mathbf{Q}\cdot\mathbf{X}} + \delta a_-^*(T)e^{-i\mathbf{Q}\cdot\mathbf{X}}\right]. \tag{7.19}$$

The result is that the growth rate $\sigma_K(\mathbf{Q})$ has the same form as in Eq. (6.46),

$$\sigma_K(\mathbf{Q}) = -\left(1 - K^2\right) - \frac{1}{2}(U_+ + U_-) + \left[\left(1 - K^2\right)^2 + \frac{1}{4}(U_+ - U_-)^2\right]^{1/2}, \tag{7.20}$$

but with the quantities $U_\pm$ now defined as

$$U_\pm = \left[K \pm Q_X + Q_Y^2/2\right]^2 - K^2. \tag{7.21}$$

The stability boundaries are given by changing K from zero, which corresponds to moving the stripe wave number q away from q_c, and by asking when the maximum of $\sigma_K(\mathbf{Q})$ over all possible $\mathbf{Q}$ first becomes positive.

A number of useful results can be proved from the general expression Eq. (7.20). It can be shown that, as $|K|$ is increased from zero, the instability always occurs first

7.1 Stripes in rotationally invariant systems

Fig. 7.1 Schematic plot of the stripe distortion that arises from a zigzag instability, which suggests the origin of its name. This is a transverse instability whose perturbation wave vector $\mathbf{Q} = Q_Y \hat{\mathbf{Y}}$ (with $|Q_Y| \ll 1$) is perpendicular to the orientation $\hat{\mathbf{X}}$ of the stripe base state.

for either a purely longitudinal ($\mathbf{Q} = Q_X \hat{\mathbf{X}}$) or purely transverse ($\mathbf{Q} = Q_Y \hat{\mathbf{Y}}$) perturbation. Furthermore, the instability always occurs in the limit $Q \to 0$ which corresponds to long wavelengths. We have already discussed the longitudinal instability in Section 6.4.2, and so here we will give results for the transverse one.

Figure 7.1 illustrates why a transverse long-wavelength instability of stripes is described as "zigzag."[2] If we set $Q_X = 0$ in Eq. (7.20), expand in the small quantity Q_Y, and retain the two lowest-order terms, we find that

$$\sigma_K(Q_Y) = -K Q_Y^2 - \frac{1}{4} Q_Y^4. \tag{7.22}$$

Equation (7.22) shows that the zigzag instability is of type II, with a zero growth rate at $Q_Y = 0$. This occurrence of a type-II instability is a consequence of the rotational symmetry of the physical system, since a $Q_Y = 0$ perturbation corresponds to a small rotation of the stripes, which is a perturbation that neither grows nor decays. The instability develops as the deviation of the stripe wave number from critical K passes to negative values, and for a long-wavelength perturbation, $Q_Y \to 0$. In the unstable region $K < 0$, the maximum growth rate is at the wave vector

$$Q_Y = \sqrt{2(-K)}. \tag{7.23}$$

---

[2] This pattern is generated from the full amplitude $A + \delta A$ with the form of δA dictated by the linear stability analysis, with some choice of the wave vector of the perturbation Q_Y and size of perturbation. The figure is not intended to be quantitatively accurate. Indeed, as the perturbation increases in magnitude from a small value, nonlinear effects can cause a substantial change to such a pattern.

Returning to the physical variables, we see that the spatially periodic stripe solutions are unstable to the zigzag instability for $q < q_Z$ with

$$q_Z = q_c + 0 \times \varepsilon^{1/2} + O(\varepsilon), \qquad (7.24)$$

where the $O(\varepsilon)$ term is beyond the reach of the lowest-order amplitude equation. The zigzag instability has a universal form near threshold for a rotationally invariant type-I-s system (the zero at $O(\varepsilon^{1/2})$ in Eq. (7.24)). This behavior is sketched in Fig. 4.2(b). The $O(\varepsilon)$ term in Eq. (7.24), which determines the slope of the boundary in the εq plane as $\varepsilon \to 0$, is nonuniversal and depends on the system under consideration.

7.1.5 Phase dynamics

As we did for the one-dimensional amplitude equation in Section 6.4.3, we can derive a single equation for the dynamics of phase variations that are so slow that the magnitude adiabatically follows the local phase gradient (see Eq. (6.57)). We again choose to look at small perturbations from a uniform stripe state at a wave number q shifted from critical $q = q_c + \varepsilon^{1/2} K \xi_0^{-1}$, and keep only terms that are linear in the phase deviation $\Phi(X, Y, T)$ from this state (as defined in Eq. (6.53)), and that are up to second order in spatial derivatives of Φ. The derivation goes through as there except that there is an additional term involving Y-derivatives of the phase. For the fully scaled amplitude equation, Eq. (7.12), the linear equation for the phase dynamics is

$$\partial_T \Phi = \frac{1 - 3K^2}{1 - K^2} \partial_X^2 \Phi + K \partial_Y^2 \Phi. \qquad (7.25)$$

This is again a diffusion equation for the phase, now with different diffusion constants for variations parallel and perpendicular to the stripe wave vector. Returning to the unscaled units, Eq. (7.25) becomes

$$\partial_t \Phi = D_\parallel \partial_x^2 \Phi + D_\perp \partial_y^2 \Phi, \qquad (7.26)$$

with diffusion constants $D_\parallel$ and $D_\perp$ given by

$$D_\parallel = (\xi_0^2 \tau_0^{-1}) \frac{\varepsilon - 3\xi_0^2 k^2}{\varepsilon - \xi_0^2 k^2} \quad \text{and} \quad D_\perp = (\xi_0^2 \tau_0^{-1}) \frac{k}{q_c}, \qquad (7.27)$$

for phase perturbations about the stripe state with wave number $q = q_c + k$ (where $k = \xi_0^{-1} \varepsilon^{1/2} K$). Note that the transverse diffusion constant $D_\perp$ is zero for $k = 0$ corresponding to $q = q_c$. This signals the onset of instability for $q < q_c$, the zigzag instability discussed in the previous section.

7.2 Stripes in anisotropic systems

The extension of the amplitude equation into two dimensions is easier for a stripe-forming system that does not have full rotational symmetry, and the resulting equation is simpler (in appropriately chosen and scaled coordinates). We will consider the amplitude equation for the two different possibilities for uniaxial systems, where the linear instability may be as in panels (a) and (b) of Fig. 2.13. In the case of Fig. 2.13(b), stripes result from the growth of modes in just one pair of ellipses at $\pm \mathbf{q}$. The amplitude equation for the completely anisotropic case, with a single pair of ellipses of unstable modes as in Fig. 2.14, will have the same form as this latter case. The equations can be derived phenomenologically as in Section 6.2.1 or, for a given physical system, using the method of multiple scales perturbation theory.

7.2.1 Amplitude equation

The easiest case is for uniaxial symmetry with the critical wave vector along the preferred direction as in Fig. 2.13(a). We define the x-direction to be along the preferred axis, and the critical wave vector is then $q_c \hat{x}$. The amplitude equation is

$$\tau_0 \partial_t A = \varepsilon A + \xi_x^2 \partial_x^2 A + \xi_y^2 \partial_y^2 A - g_0 |A|^2 A. \tag{7.28}$$

There are no cross derivative terms $\partial_x \partial_y$ in Eq. (7.28) since the equations must be invariant separately under $x \to -x$ and $y \to -y$, but there are no other constraints arising from rotational invariance arguments, and so the lowest-order derivative terms consistent with these invariances appear. Note that the coefficients for variations along and perpendicular to the preferred direction are different in general. The constants ξ_x and ξ_y can be related to the expansion of the growth rate about $\mathbf{q} = q_c \hat{x}$

$$\sigma(\mathbf{q}) \simeq \tau_0^{-1}[\varepsilon - \xi_x^2(q_x - q_c)^2 - \xi_y^2 q_y^2]. \tag{7.29}$$

The quadratic terms in x and y wave vector changes with different coefficients correspond to the elliptical contours of Fig. 2.13(a) with principal axes that are aligned along the x and y coordinate axes.

For a uniaxial system in which the critical wave vector of the instability is not aligned with the preferred axis of the system as in Fig. 2.13(b), or for a system with no rotational symmetry as in Fig. 2.14, the amplitude equation (for a single set of stripes for Fig. 2.13(b)) takes the form

$$\tau_0 \partial_t A = \varepsilon A + \xi_x^2 \partial_x^2 A + 2\xi_{xy} \partial_x \partial_y A + \xi_y^2 \partial_y^2 A - g_0 |A|^2 A, \tag{7.30}$$

where again the quadratic derivatives correspond to elliptical contours of constant $\sigma(\mathbf{q})$ near threshold, but the principal axes of these ellipses do not have any particular relationship with the critical wave vector or the x and y axes.

By rescaling coordinates in Eq. (7.28), or by introducing scaled coordinates that are rotated to be along the principal axes of the constant-growth-rate elliptical contours for Eq. (7.30), both equations can be written in the parameter-free form

$$\partial_T \bar{A} = \pm \bar{A} + \partial_X^2 \bar{A} + \partial_Y^2 \bar{A} - |\bar{A}|^2 \bar{A}. \tag{7.31}$$

Thus the *anisotropic* system leads to an amplitude equation that is *isotropic* in appropriately chosen scaled coordinates, whereas the amplitude equation for the rotationally invariant system (but with stripes chosen with normals in the x-direction) has the x- and y-derivatives appearing in different ways. The simplicity of the derivative terms in Eq. (7.31) compared with Eq. (7.12), makes the analysis of the physical consequences considerably easier. An example is the structure of a dislocation defect, discussed in Section 8.1 in the next chapter.

For appropriate boundary conditions, the amplitude equation Eq. (7.31) has potential dynamics with potential

$$\bar{V}[\bar{A}] = \int\!\!\int dX\,dY \left[-|\bar{A}|^2 + \tfrac{1}{2}|\bar{A}|^4 + |\partial_X \bar{A}|^2 + |\partial_Y \bar{A}|^2 \right], \tag{7.32}$$

which evolves according to Eq. (7.16).

7.2.2 Stability balloon

The isotropy of the scaled amplitude equation (7.31) means that the calculation of the stability balloon is a straightforward extension of the one-dimensional calculation, Section 6.4.2. We start with the base solution

$$\bar{A}_{\mathbf{K}}(X, Y) = (1 - K^2)^{1/2} e^{i\mathbf{K} \cdot \mathbf{X}}, \tag{7.33}$$

with $\mathbf{K}$ in any direction. As before, the stability of the state is tested by adding to $\bar{A}_{\mathbf{K}}$ a small perturbation $\delta \bar{A}$

$$\bar{A}(X, Y, T) = \bar{A}_K(X, Y) + \delta \bar{A}(X, Y, T). \tag{7.34}$$

The equation of the perturbation given by linearizing the amplitude equation in $\delta \bar{A}$ is

$$\partial_T \delta \bar{A} = \delta \bar{A} + \left(\partial_X^2 + \partial_Y^2 \right) \delta \bar{A} - 2|\bar{A}_{\mathbf{K}}|^2 \delta \bar{A} - \bar{A}_{\mathbf{K}}^2 \delta \bar{A}^* \tag{7.35}$$

and the Bloch ansatz is

$$\delta \bar{A} = e^{i\mathbf{K} \cdot \mathbf{X}} [\delta a_+(t) e^{i\mathbf{Q} \cdot \mathbf{X}} + \delta a_-^*(t) e^{-i\mathbf{Q} \cdot \mathbf{X}}], \tag{7.36}$$

with $\mathbf{Q}$ the Bloch wave vector of the perturbation. Proceeding as in Section 6.4.2, you can show that the growth rate given by $\delta a_\pm(t) \propto e^{\sigma_{\mathbf{K}}(\mathbf{Q}) T}$ is

$$\sigma_{\mathbf{K}}(\mathbf{Q}) = (1 - K^2) - Q_\parallel^2 + \sqrt{(1 - K^2)^2 + 4K^2 Q_\parallel^2} - Q_\perp^2, \tag{7.37}$$

with $Q_\parallel$ the component of $\mathbf{Q}$ parallel to $\mathbf{K}$, and $Q_\perp$ the component perpendicular to this direction.

A nonzero $Q_\perp$ reduces the growth rate, and so the instability always occurs first for $\mathbf{Q}$ parallel to $\mathbf{K}$. This observation reduces the calculation to the same one as in Section 6.4.2, so that the instability occurs first for $Q \to 0$ and at a value of $K = 1/\sqrt{3}$, the same result as in Eq. (6.48).

7.2.3 Phase dynamics

The equation for small phase perturbations about a uniform stripe state is readily derived by following the methods of Section 6.4.3. Let us start with the rotated coordinates (if necessary) and the scaled variables leading to the simple form of the amplitude equation, Eq. (7.31). For a base state given by the amplitude

$$\bar{A}_K(X, Y) = (1 - K^2)^{1/2} e^{iKX}, \tag{7.38}$$

the linear equation for the phase dynamics is

$$\partial_T \Phi = \frac{1 - 3K^2}{1 - K^2} \partial_X^2 \Phi + \partial_Y^2 \Phi. \tag{7.39}$$

In agreement with the calculations of the previous section, we see that the long-wavelength instability, which is signaled by a negative diffusion constant, occurs for longitudinal perturbations (variation along the same direction as the phase of the base state amplitude) and for $K = 1/\sqrt{3}$. This result must be translated back to the original coordinates to get a physical description of the unstable mode.

7.3 Superimposed stripes

A question of fundamental importance in pattern formation is why some systems show stripe patterns, some hexagonal ones, and others squares or more exotic patterns. The qualitative aspects of this question were introduced in Section 4.3. The amplitude equation approach provides a systematic and quantitative way to answer this question near threshold where the weakly nonlinear theory applies.

A general and reliable way to investigate the nonlinear competition between different patterns is a linear stability analysis: first construct various nonlinear solutions, and then test the stability of each solution. If out of two states being compared, one state is stable, whereas the second is unstable, we can focus our attention on the stable state, and for most purposes ignore the unstable one.

On the other hand, both states may turn out to be stable, and then the stability test does not discriminate between the two solutions. For this case of bistability (or more generally, multistability), the potential associated with the amplitude dynamics can be used to assess the competition between the two states, see Fig. 5.1 and the related discussion in Section 5.1.2. Using a potential to compare two states is less general than the stability test. Firstly, it cannot be extended further away from threshold, where the dynamics is no longer potential, or even to situations described by amplitude equations that do not lead to a potential. Furthermore, although it is quite generally true that potential dynamics cannot evolve from the lower potential state to the higher one, it is not always true that a dynamical pathway exists to connect a higher potential state to a lower one. It is therefore not necessarily true that all experimental protocols will lead to the state with lower potential. For example, some initial conditions might favor the growth of a state that is a local minimum of the potential, but that minimum may have a higher potential than some other state. If a uniform configuration of the higher potential state develops, the competition between spatial domains of the higher and lower potential states via the motion of domain walls never arises, and there may be no other dynamical pathway that connects the two states. In this case, the higher potential state will persist.

The competition between the lattice and stripe states is straightforward to analyze within the amplitude equation approach. We first outline the general method, and then discuss two specific examples. For the first example, the competition between stripes and a general lattice state, the stability test leads to a sharp criterion for the competition, with either the stripe state or the lattice state being stable, depending on parameters, but not both together. For the second example, the competition between stripes and a hexagonal state for a system without field inversion symmetry $\mathbf{u} \to -\mathbf{u}$, there are parameter regions where bistability occurs. For these parameter values, an investigation of the potential provides further insights.

7.3.1 Amplitude equations

To study states that are the superposition of stripes that are oriented in different directions, we need to use a more general zeroth-order ansatz than Eq. (7.1). For example, for a lattice state based on stripes at wave vectors $\mathbf{q}_1$ and $\mathbf{q}_2$, both with magnitude q_c but with different directions, the ansatz takes the form

$$\mathbf{u}_p = A_1 e^{i\mathbf{q}_1 \cdot \mathbf{x}_\perp} \bar{\mathbf{u}}_c(\mathbf{x}_\parallel) + A_2 e^{i\mathbf{q}_2 \cdot \mathbf{x}_\perp} \bar{\mathbf{u}}_c(\mathbf{x}_\parallel) + \text{c.c.} + \text{h.o.t.} \qquad (7.40)$$

Note that the critical wave numbers of the two sets of stripes must be the same, so that both sets are weakly nonlinear together. For a rotationally invariant system this is no restriction since $\mathbf{q}_1$ and $\mathbf{q}_2$ may lie in any direction around the critical circle. However, for a uniaxial system, the wave vectors $\mathbf{q}_1$ and $\mathbf{q}_2$ must be the degenerate zig and zag directions of Fig. 2.13(b).

7.3 Superimposed stripes

Equation (7.40) introduces two amplitudes, A_1 and A_2. While these will generally vary slowly as functions of space and time, to simplify the discussion of the lattice states we will first assume there is no spatial dependence. This assumption corresponds to restricting the component stripe states to be uniform and at the critical wave number. The more general case is discussed in Section 7.3.4.

The rotational symmetry of the physical system implies that the evolution of the amplitudes A_1 and A_2 may depend on the angle θ between the wave vectors $\mathbf{q}_1$ and $\mathbf{q}_2$, but not on the individual directions. Also rotational symmetry tells us that the two sets of stripes must have the same z-dependence $\bar{\mathbf{u}}_c(\mathbf{x}_\|)$,[3] and that only the nonlinear terms involving both A_1 and A_2 may lead to differences from the single amplitude case. Then using the same type of symmetry arguments to restrict the possible terms as in Section 6.2.1, we argue phenomenologically that the lowest-order amplitude equations should take the form

$$\tau_0 \, d_t A_1 = \varepsilon A_1 - g_0 \Big(|A_1|^2 + G(\theta)|A_2|^2 \Big) A_1, \tag{7.41a}$$

$$\tau_0 \, d_t A_2 = \varepsilon A_2 - g_0 \Big(|A_2|^2 + G(\theta)|A_1|^2 \Big) A_2. \tag{7.41b}$$

(Etude 7.2 gives a more detailed justification for the more general case of many superimposed stripes.) The new parameter $G(\theta)$ gives the coupling between stripes at relative orientation θ. The function $G(\theta)$ must satisfy

$$G(\theta) = G(\pi - \theta), \tag{7.41c}$$

since these two angles define the same relative orientation. We have also assumed that the physical system is unchanged under reversing the sense of rotation, $\theta \to -\theta$ (which we call chiral symmetry), so that the same interaction coefficient appears in the two equations.

We shall see in Section 7.3.2 that the coefficient $G(\theta)$ determines the relative stability of stripe and lattice states, and in Section 7.3.5 that it is important in the cross-stripe instability. It can be shown quite generally to have the property

$$\lim_{\theta \to 0} G(\theta) = 2, \tag{7.42}$$

even though we might expect unity for this limit, since that would reproduce the coefficient of the one-amplitude nonlinearity. This difference arises because of the interference effects that occur if θ is identically zero. These disappear for any nonzero θ (actually for $\theta \gtrsim \varepsilon^{1/4}$).

---

[3] For a choice of the variables making up $\mathbf{u}$ that are themselves vectors in coordinate space, the onset solution $\mathbf{u}_c$ may in fact depend on the direction of the wave vector, and we would have to modify the argument slightly. For example, if we use fluid velocity as basic variables in Rayleigh–Bénard convection, the horizontal component will be along the direction normal to the rolls, i.e. along the wave vector. Components along fixed coordinate axes will then depend on q_{1x}, q_{1y} etc. Formulating the basic equations in terms of stream functions would eliminate this extra complication.

The value of $G(\theta)$ for a given physical system can be calculated from the calculation of the nonlinear saturation of the unmodulated lattice state that consists of two sets of stripes of equal amplitude A_L (the "lattice amplitude") at the critical wave number q_c and at a relative angle θ. From the amplitude equations Eqs. (7.41), we find that the lattice amplitude A_L has the following value

$$A_L = |A_1| = |A_2| = \sqrt{\frac{\varepsilon}{g_0}}(1 + G(\theta))^{-1/2}. \tag{7.43}$$

We can compare this value with the amplitude A_S of a saturated stripe state, which is a solution of Eq. (7.40) with one of the stripes eliminated, say by setting $A_2 = 0$. We find that

$$A_S = |A_1| = \sqrt{\frac{\varepsilon}{g_0}}, \quad |A_2| = 0. \tag{7.44}$$

Thus $G(\theta)$ can be deduced from the ratio of stripe to lattice intensities

$$G(\theta) = \frac{A_S^2}{A_L^2} - 1. \tag{7.45}$$

These intensities in turn can be calculated separately from a lowest-order Galerkin calculation as was discussed in Section 4.1.3 and as we discuss now in the following Etude. Alternatively, Eqs. (7.41) can be derived using the multiple-scales method of Appendix 2, which yields an explicit expression for $G(\theta)$.

Etude 7.1 Stripe coupling coefficient $G(\theta)$ for the Swift–Hohenberg equation
Our discussion here is brief since the calculation uses the same approach as Section 4.1.3, where most of the details have already been given. We are interested in stationary states $u(x, y)$ of the two-dimensional Swift–Hohenberg equation (see Eq. (5.9) in Section 5.1)

$$\partial_t u = ru - \left(\partial_x^2 + \partial_y^2 + 1\right)^2 u - u^3, \tag{7.46}$$

that are a superposition of two stripes with critical wave number $q_c = 1$. For such stationary states, Eq. (7.46) reduces to

$$0 = ru - u^3. \tag{7.47}$$

Now consider a lattice state made from equal amplitudes of stripes with wave vectors $\mathbf{q}_1$ and $\mathbf{q}_2$ (with $q_1 = q_2 = q_c = 1$)

$$u = a_L[\cos(\mathbf{q}_1 \cdot \mathbf{x}) + \cos(\mathbf{q}_2 \cdot \mathbf{x})] + \cdots. \tag{7.48}$$

For simplicity, we again use a cosine notation for the calculation rather than complex exponentials; to compare with Eq. (7.45), we would use $A_L = a_L/2$. The terms denoted by $\cdots$ include spatial harmonics at wave vectors $3\mathbf{q}_1$, $3\mathbf{q}_2$, $5\mathbf{q}_1$, and so on (see Section 4.1.3) as well as sinusoids whose wave vectors are sums and differences of $\mathbf{q}_1$ and $\mathbf{q}_2$, for example $2\mathbf{q}_1 \pm \mathbf{q}_2$. Sufficiently close to onset ($0 < r \ll 1$), the coefficients of these modes are all negligibly small compared with a_L.

The amplitude a_L can be obtained by substituting Eq. (7.48) into Eq. (7.47) and by collecting the coefficients of each sinusoidal mode separately and setting those coefficients to zero. The result we need is given by looking at (for example) the $\cos(\mathbf{q}_1 \cdot \mathbf{x})$ mode. The term u^3 generates

$$u^3 = a_L^3[\cos^3(\mathbf{q}_1 \cdot \mathbf{x}) + 3\cos^2(\mathbf{q}_1 \cdot \mathbf{x})\cos(\mathbf{q}_2 \cdot \mathbf{x})$$
$$+ 3\cos(\mathbf{q}_1 \cdot \mathbf{x})\cos^2(\mathbf{q}_2 \cdot \mathbf{x}) + \cos^3(\mathbf{q}_2 \cdot \mathbf{x})]. \tag{7.49}$$

Combining the products of cosines into cosines of sums and differences gives

$$u^3 = a_L^3 \left[\frac{9}{4}\cos(\mathbf{q}_1 \cdot \mathbf{x}) + \cdots \right], \tag{7.50}$$

where the $\cdots$ in the last expression contains sum and difference modes with wave vectors other than $\mathbf{q}_1$. Thus Eq. (7.47) for the $\cos(\mathbf{q}_1 \cdot \mathbf{x})$ mode gives

$$a_L = \frac{2}{3}\sqrt{r}. \tag{7.51}$$

Since we know from Eq. (4.24) that the amplitude of a stripe state is $a_S = \sqrt{4r/3}$, Eq. (7.45) tells us that

$$G(\theta) = 2. \tag{7.52}$$

Thus for the Swift–Hohenberg equation, $G(\theta)$ is independent of θ. This will not be the case for most evolution equations that describe pattern formation. For example, in Exercise 7.10 you can derive the more complicated expression, Eq. (E7.7), for the stripe coupling coefficient of the generalized Swift–Hohenberg model for rotating convection discussed in Section 5.2.4. Note that the result Eq. (E7.7) is consistent with the general statement Eq. (7.42).

Equations (7.41) can be generalized to states of any number of superimposed stripes[4]

$$\tau_0 d_t A_i = \varepsilon A_i - g_0 \left(|A_i|^2 + \sum_{j \neq i} G(\theta_{ij})|A_j|^2 \right) A_i, \tag{7.53}$$

in which each amplitude A_i is coupled to all the other amplitudes A_j, and the coupling coefficients $G(\theta_{ij})$ depend on the relative orientation ϑ_{ij} of the stripes

---

[4] Equations (7.53) break down if the angular separation between two sets of stripes becomes too close, namely when $\theta_{ij} \lesssim \varepsilon^{1/4}$ for some pair of stripes. Those stripes are then correctly described by a single spatially dependent amplitude.

(with $G(\theta) = G(-\theta)$ if there is chiral symmetry). A special case is three sets of stripes at equal relative angles of $\pi/3$. The nonlinear interaction of the disturbance given by two sets of stripes can then generate a disturbance along the third set of stripes. This corresponds to additional quadratic terms in the amplitude equations, for example

$$\tau_0 \, \partial_T A_1 = \cdots + \gamma A_2 A_3, \tag{7.54}$$

where the $\cdots$ represents the same terms as in Eq. (7.53), and the parameter γ is real-valued. There are corresponding equations for the amplitudes A_2 and A_3 which can be obtained by cyclic permutations of the indices.

As we show in the following Etude, the forms (7.53) and (7.54) of the amplitude equations follow from the translational invariance and parity symmetry, Note that if the system also has the field inversion symmetry $\mathbf{u} \to -\mathbf{u}$, so that the amplitude equations must be invariant under $A_i \to -A_i$, the coefficient γ of the quadratic nonlinear terms must be zero and there are only cubic nonlinear terms as in Eqs. (7.41).

Etude 7.2 Form of amplitude equation for coupled stripes

We require that the amplitude equations reflect the invariance of the physical system under an arbitrary translation $\mathbf{x} \to \mathbf{x} + \Delta \mathbf{x}$. *Under such a translation, each amplitude* A_i *appearing in the equation becomes multiplied by a different phase factor* $\exp(i\mathbf{q}_i \cdot \Delta \mathbf{x})$. *The amplitude equations must be invariant under this substitution, as discussed in Section 6.2.1. This immediately shows that the cubic terms must be of the form in Eqs. (7.41) or Eq. (7.53).*

Now consider possible quadratic terms in the equation for amplitude A_i

$$\tau_0 \, \partial_T A_i = \cdots + c_1 A_j A_k + c_2 A_j^* A_k + c_3 A_j A_k^* + c_4 A_j^* A_k^*, \tag{7.55}$$

where the four coefficients c_i *are complex numbers. Upon translation through* $\Delta \mathbf{x}$, *the left-hand side of Eq. (7.55) is multiplied by the phase factor* $\exp(i\mathbf{q}_i \cdot \Delta \mathbf{x})$, *and the first term* $c_1 A^{(j)} A^{(k)}$ *on the right-hand side is multiplied by* $\exp[i \, \Delta \mathbf{x} \cdot (\mathbf{q}_j + \mathbf{q}_k)]$. *The amplitude equation must be invariant under this symmetry operation. Therefore, for the coefficient* c_1 *to be nonzero these two phase factors must be equal for any* $\Delta \mathbf{x}$, *which requires that*

$$\mathbf{q}_i = \mathbf{q}_j + \mathbf{q}_k. \tag{7.56}$$

For the terms involving complex-conjugate amplitudes, the corresponding $\mathbf{q}_i$ *appears with a minus sign. This leads to the general condition*

$$\mathbf{q}_i \pm \mathbf{q}_j \pm \mathbf{q}_k = 0, \tag{7.57}$$

for a term to have nonzero coefficient, with the $+$ *sign for amplitudes appearing as the complex conjugate, the* $-$ *sign otherwise. Since we also know that* $q_i = q_c$

for all i, this equation can only be satisfied for wave vectors (or their negatives) equally spaced at angles of $\pi/3$. For three sets of stripes with wave vectors at $\pi/3$ so that $\mathbf{q}^{(1)} + \mathbf{q}^{(2)} + \mathbf{q}^{(3)} = 0$, the coefficient c_4 is the only nonzero one. The invariance under spatial inversion, corresponding to all $A_i \to A_i^*$, then shows that the coefficient c_4 must be real.

An important question is the existence of a potential for superimposed stripes, as in the discussion in Section 6.3.2. You can verify that the coupled amplitude equations Eq. (7.53) have a potential dynamics for the potential density[5]

$$v = \sum_i \left(-\varepsilon |A_i|^2 + \frac{g_0}{2} |A_i|^4 \right) + \frac{g_0}{2} \sum_{\substack{i,j \\ i \neq j}} G(\theta_{ij}) |A_i|^2 |A_j|^2, \qquad (7.58)$$

which decreases for any dynamics of the amplitudes A_i since

$$d_t v = -2 \sum_i |d_t A_i|^2. \qquad (7.59)$$

The quadratic terms in the amplitude equations for three stripes at angles of $\pi/3$ are also consistent with a potential, see Eq. (7.76) below.

If you differentiate both sides of Eq. (7.58) with respect to time and work through the algebra to derive Eq. (7.59), you will see that, at one crucial step, you will need to combine terms that are identical except that one set of terms has the coefficients $G(\theta_{ij})$ while the other set has the coefficients $G(\theta_{ji}) = G(-\theta_{ij})$ (since $\theta_{ij} = -\theta_{ji}$). So Eq. (7.58) is a potential *only* if the system has chiral symmetry for which $G(\theta) = G(-\theta)$. Although many physical systems have chiral symmetry, this is not always the case. For example, if a Rayleigh–Bénard system is rotated about a vertical axis, the amplitude equations still have the form Eq. (7.53) but now $G(\theta) \neq G(-\theta)$ due to the asymmetry induced by the rotation. This is sufficient to render the dynamics nonpotential, and indeed a chaotic state called domain chaos is observed in this system immediately at onset, as we mentioned in Chapter 1, Fig. 1.15. This system is discussed further in Exercise 7.20 and in Section 5.2.4 and Section 9.2.4.

7.3.2 Competition between stripes and lattices

In this section, we investigate the competition between stripe and lattice states for systems with field inversion symmetry so that there are no quadratic nonlinear

---

[5] For the general case in which the amplitudes A_i vary spatially, the potential V would be given by a double integral of a potential density over the domain, $V = \iint v \, dx \, dy$. Since the amplitudes do not depend on space here, we work directly with the potential density, which is also the appropriate quantity to compare for competing patterns that span different areas of some domain.

terms in the amplitude equations. The competition between states of superimposed stripes at different orientations is captured through the pairwise stripe interaction parameter $G(\theta)$. Introducing the scaled amplitudes $\bar{A}_i = \sqrt{g_0/\varepsilon} A_i$ and the scaled time $T = t/\tau_0$, Eq. (7.53) gives the evolution equation for each set of stripes with an amplitude $\bar{A}_i$

$$d_T \bar{A}_i = \bar{A}_i - \left(|\bar{A}_i|^2 + \sum_{j \neq i} G(\theta_{ij}) |\bar{A}_j|^2 \right) \bar{A}_i. \qquad (7.60)$$

The task now is to construct lattice state solutions of the sort discussed in Fig. 4.6 of Section 4.3, with N amplitudes $\bar{A}_i$ at orientations θ_i dictated by the choice of symmetry and by other assumptions. We then test the stability of the solutions with respect to small perturbations. Using Eq. (7.60), we can test stability within the space of the N chosen A_i, as well as to the growth of additional new modes at arbitrary new orientations. Although only spatially uniform perturbations can be studied using Eq. (7.60), this is often sufficient since the fastest growing mode is usually at the critical wave number, and can be described by a spatially uniform amplitude.

In the following Etude, we study the competition between a stripe state and a lattice state comprised of two superimposed sets of stripes. Note that Eq. (7.60) implies that the phases Φ_i of the different amplitudes are uncoupled, and each one can take an arbitrary constant value in the stationary solutions. For simplicity, we can choose the solutions of the amplitude equations to be real, and test stability with perturbations that are also real.

Etude 7.3 Competition between stripes and lattices
We consider the competition between the stripe state and a lattice state formed from the superposition of two sets of stripes at an angle θ. The case $\theta = \pi/2$ gives the square lattice, otherwise the lattice is orthorhombic (see Section 4.3). We study the competition by calculating the amplitudes in each stationary nonlinear state, and then the stability of these states.

First consider the stripe state. The solution to Eq. (7.60) for a single set of stripes is $\bar{A}_1 = \bar{A}_S$ with

$$\bar{A}_S^2 = 1. \qquad (7.61)$$

This simple statement contains the physical result that the amplitude of the stripes grows as the square root of the distance above onset, described by the unscaled amplitude

$$A_S \propto \sqrt{\varepsilon}. \qquad (7.62)$$

The linear stability to the growth of a second set of stripes of amplitude $\bar{A}_2$ at angle θ is tested by linearizing Eqs. (7.60) about the solution Eq. (7.61) for the

7.3 Superimposed stripes

perturbation to $\bar{A}_1$ and the small amplitude $\bar{A}_2$

$$\bar{A}_1 = \bar{A}_S + \delta\bar{A}_1, \qquad (7.63a)$$
$$\bar{A}_2 = \delta\bar{A}_2, \qquad (7.63b)$$

to give the equations

$$d_T \delta\bar{A}_1 = -2\,\delta\bar{A}_1, \qquad (7.64a)$$
$$d_T \delta\bar{A}_2 = [1 - G(\theta)]\delta\bar{A}_2. \qquad (7.64b)$$

From the second equation, we see that stripes are linearly stable for $G(\theta) > 1$, but for $G(\theta) < 1$, the stripes are unstable toward the superposition of a second set of stripes at the angle θ. In the latter case, we might expect that the perturbation would grow to a saturated lattice state consisting of superimposed stripes but the asymptotic nonlinear state cannot be determined by the linear analysis. The first equation shows that perturbation to $\bar{A}_1$ always decays at linear order. Notice that the calculations separate for linear instability toward a new stripe (the second equation), and for instability within the space of the already present stripe (the first equation). Also, the instability toward the addition of any number of new stripes separates into instability toward the individual stripes in the linear analysis.

Now consider the lattice state of two sets of superimposed stripes at the angle θ. The nonlinear saturated solution to Eq. (7.60) is $\bar{A}_1 = \bar{A}_2 = \bar{A}_L$ with

$$\bar{A}_L^2 = [1 + G(\theta)]^{-1}, \qquad (7.65)$$

The (unscaled) amplitude of the lattice pattern A_L grows as $\sqrt{\varepsilon}$, but with a different proportionality constant than for stripes.

The linear stability within the space of $\bar{A}_1$ and $\bar{A}_2$ is determined by writing $\bar{A}_i = \bar{A}_L + \delta\bar{A}_i$ and by linearizing

$$d_T \delta\bar{A}_1 = -2\bar{A}_L^2[\delta\bar{A}_1 + G(\theta)\delta\bar{A}_2], \qquad (7.66)$$
$$d_T \delta\bar{A}_2 = -2\bar{A}_L^2[\delta\bar{A}_2 + G(\theta)\delta\bar{A}_1]. \qquad (7.67)$$

The growth rates σ of the instability are then obtained by substituting $\delta\bar{A}_i(t) = \delta\bar{a}_i e^{\sigma t}$ and by carrying out an elementary eigenvalue calculation. The result is that

$$\sigma = -2\bar{A}_L^2[1 \pm G(\theta)] = -2\frac{1 \pm G(\theta)}{1 + G(\theta)}. \qquad (7.68)$$

The corresponding eigenvectors tell us the nature of the modes that grow or decay exponentially. We see that for $G(\theta) > -1$ the lattice state is stable to the symmetric mode ($\delta\bar{A}_1 = \delta\bar{A}_2$, $\sigma = -2$), and is stable with respect to one stripe growing and the other decaying ($\delta\bar{A}_1 = -\delta\bar{A}_2$, $\sigma = 2\left[G(\theta) - 1\right]/\left[G(\theta) + 1\right]$) *for* $|G(\theta)| < 1$, *and unstable for* $G(\theta) > 1$. *For* $G(\theta) < -1$ *the bifurcation to the lattice state is subcritical, and there are no stable small amplitude solutions. In the unstable case, we might expect the perturbation to grow until the amplitude of one set of stripes disappears, leading to the stripe state, but this can only be confirmed by a calculation that is nonlinear in the perturbation.*

Comparing the lattice and the stripe results in this Etude, we see that for $G(\theta) > 1$ stripes are stable with respect to the growth toward the lattice state (i.e. to the growth of an additional set of stripes at angle θ) and the corresponding lattice state is unstable with respect to the collapse toward stripes. For $|G(\theta)| < 1$ the reverse occurs and the stripes are unstable while the lattice state is stable. Thus the stability analysis gives a sharp criterion for the preferred state.

The case of squares is given by $\theta = \pi/2$, and is included in these general results. However, we expect the square lattice to be a particularly common occurrence since the simplest shape of $G(\theta)$ consistent with the stability of a lattice state gives a minimum at $\theta = \pi/2$. This is because the symmetry of $G(\theta)$ about $\pi/2$ means that the function has a minimum or maximum at $\theta = \pi/2$. In addition, the stability of a lattice state requires $G(\theta)$ to decrease from its value at $\theta = 0$. The simplest form of $G(\theta)$ consistent with these two facts is a monotonic decrease to a minimum at $\theta = \pi/2$.

The case $\theta = \pi/3$ is special since we can further ask the question of the stability of the two-stripe lattice to the addition of a third set of stripes at $\theta = 2\pi/3$ (equivalent to $\theta = -\pi/3$) which would lead to the hexagonal lattice. It can be shown that if stripes are unstable toward the lattice state of two superimposed stripes, i.e. $|G(\pi/3)| < 1$, then the two-stripe lattice state is itself unstable toward the addition of the third set of stripes.

7.3.3 Hexagons in the absence of field-inversion symmetry

As we saw at the end of the last section, the case of superimposed stripes at an angle $\theta = \pi/3$ is special, even for systems with $\mathbf{u} \to -\mathbf{u}$ symmetry, since a lattice made from this superposition is unstable to the addition of the third set of stripes forming the hexagonal lattice. The hexagonal lattice becomes even more important for systems without the $\mathbf{u} \to -\mathbf{u}$ symmetry, because there are then the additional quadratic terms Eq. (7.54) in the amplitude equations that enhance the growth of three sets of stripes at relative angles of $\pm\pi/3$.[6] This enhanced growth arises from the resonance in the nonlinear interaction between three modes with wave vectors

---

[6] Usually we think of the hexagonal lattice in this context. However, the same resonance terms may also favor more complicated states, such as quasicrystal states with 12-fold symmetry.

7.3 Superimposed stripes

on the critical circle $\mathbf{q}_i = q_c \hat{\mathbf{q}}_i$ that sum to zero

$$\mathbf{q}_1 + \mathbf{q}_2 + \mathbf{q}_3 = 0. \tag{7.69}$$

The analysis of this case leads to a general conclusion with wide applicability:

For a system where the $\mathbf{u} \to -\mathbf{u}$ symmetry is absent, the stripe solution that bifurcates continuously from the uniform state is *unstable*. Furthermore, there is a transcritical bifurcation[7] to a hexagonal state. If the breaking of the symmetry is small (given by the parameter γ in Eq. (7.54)), in which case the amplitude equation remains valid in the saturated nonlinear state, there is a *stable* finite amplitude hexagon solution near the critical value of the control parameter. A second hexagon solution, given by changing the sign of the amplitudes, bifurcates continuously from the uniform state, but is *unstable* near threshold.

We derive these results in the following Etude.

Etude 7.4 Competition between stripes and hexagons in systems with no $\mathbf{u} \to -\mathbf{u}$ symmetry

The starting point is the coupled amplitude equations in the absence of the $\mathbf{u} \to -\mathbf{u}$ symmetry, Eqs. (7.53) and (7.54). In the absence of any spatial dependence of the amplitudes, these equations take the form

$$\tau_0 \, d_t A_1 = \varepsilon A_1 + \gamma A_2^* A_3^* - g_0 \Big[|A_1|^2 + G_1 \big(|A_2|^2 + |A_3|^2 \big) \Big] A_1, \tag{7.70}$$

with similar equations for A_2 and A_3 obtained by cyclic permutation of the indices. We have written G_1 for $G(\pi/3)$. We stipulate that γ is positive. (This can always be arranged by a suitable definition of the amplitudes, since the redefinition $\mathbf{u} \to -\mathbf{u}$ will change the sign of γ.)

We can combine the parameters ε, γ, and g_0 in Eq. (7.70) into a single effective control parameter $\tilde{\varepsilon}$,

$$\tilde{\varepsilon} = \frac{g_0}{\gamma^2} \varepsilon, \tag{7.71}$$

by rescaling the amplitude and time

$$A = \frac{\gamma}{g_0} \tilde{A}, \quad t = \frac{g_0 \tau_0}{\gamma^2} \tilde{t} \tag{7.72}$$

to give

$$d_{\tilde{t}} \tilde{A}_1 = \tilde{\varepsilon} \tilde{A}_1 + \tilde{A}_2^* \tilde{A}_3^* - \Big[|\tilde{A}_1|^2 + G_1 \big(|\tilde{A}_2|^2 + |\tilde{A}_3|^2 \big) \Big] \tilde{A}_1, \tag{7.73}$$

together with similar equations for $\tilde{A}_2$ and $\tilde{A}_3$.

The stationary stripe solution to Eq. (7.73) is given by $\tilde{A}_1 = \tilde{a}_S e^{i\Phi}$, $\tilde{A}_2 = \tilde{A}_3 = 0$ (for example) with $\tilde{a}_S^2 = \tilde{\varepsilon}$. The phase Φ is arbitrary, and we will take it to be zero,

---
[7] See Appendix 1 for the nature of a transcritical bifurcation.

$\Phi = 0$. *The stability of the solutions can then be tested by linearizing Eqs. (7.73) about this solution.*

The hexagonal solution is found by setting the three magnitudes equal, $|\tilde{A}_1| = |\tilde{A}_2| = |\tilde{A}_3| = \tilde{a}_H$, with $\tilde{a}_H$ to be determined. A consequence of the evolution equations for $\tilde{A}_1$, $\tilde{A}_2$, and $\tilde{A}_3$ is that the relative phases of the amplitudes are no longer independent. Writing $\tilde{A}_i = \tilde{a}_H e^{i\Phi_i}$, two of the three phases Φ_i may be chosen arbitrarily by an appropriate translation of the coordinate origin, but there is an evolution equation that determines the sum of the phases

$$d_{\tilde{t}}(\Phi_1 + \Phi_2 + \Phi_3) = -3\tilde{a}_H \sin(\Phi_1 + \Phi_2 + \Phi_3). \tag{7.74}$$

This equation has two time-independent state solutions. Considering small deviations from the steady solutions shows that the stationary solution $\Phi_1 + \Phi_2 + \Phi_3 = 0$ is stable. The other stationary solution $\Phi_1 + \Phi_2 + \Phi_3 = \pi$ is unstable.

For the stable solution with $\Phi_1 + \Phi_2 + \Phi_3 = 0$, a suitable choice of the two free phases allows us to take all three $\tilde{A}_i$ to be real, so that $\tilde{A}_i = \tilde{a}_H$, and then the stationary solutions to Eq. (7.73) are

$$\tilde{\varepsilon} + \tilde{a}_H - (1 + 2G_1)\tilde{a}_H^2 = 0. \tag{7.75}$$

The solutions to this quadratic equation are easily found. The stability of the solutions can then be tested by linearizing Eqs. (7.73) about $\tilde{A}_1 = \tilde{A}_2 = \tilde{A}_3 = \tilde{a}_H$. The negative solutions for $\tilde{a}_H$ to this equation actually correspond to the second choice of the sum of the phases $\Phi_1 + \Phi_2 + \Phi_3 = \pi$, and are always unstable.

Figure 7.2 shows the full picture of the time-independent hexagon and stripe solutions to Eq. (7.73) and their stability properties. Note that the amplitude of the hexagon states increases linearly as $\tilde{\varepsilon}$ passes through zero so that this is a transcritical bifurcation from the uniform solution (see Appendix 1), although here the small-amplitude solution is unstable for both positive and negative $\tilde{\varepsilon}$. In addition to the small-amplitude solution near $\tilde{\varepsilon} = 0$, Eq. (7.73) predicts a finite-amplitude solution, which at $\tilde{\varepsilon} = 0$ has the magnitude $\tilde{a}_H = (1 + 2G_1)^{-1}$. Returning to the unscaled variables, we see that this corresponds to the magnitude $a_H = (\gamma/g_0)(1 + 2G_1)^{-1}$, which is proportional to the $\mathbf{u} \to -\mathbf{u}$ symmetry-breaking parameter γ. For the amplitude equation to retain its validity at this solution, the amplitude a_H must be small, which in turn requires the symmetry-breaking parameter γ to be small.

Equation (7.73) can be used to show that the finite-amplitude solution near $\tilde{\varepsilon} = 0$ is stable within the space of the three amplitudes $\tilde{A}_i$. As $\tilde{\varepsilon}$ is decreased from here, the stability of the solution switches at the saddle-node point $\tilde{\varepsilon} = \tilde{\varepsilon}_A < 0$ in Fig. 7.2. The stripe solution develops from the uniform state via a standard pitchfork bifurcation at $\tilde{\varepsilon} = 0$, but is unstable near onset toward the addition of stripes at angles $\pm \pi/3$. As $\tilde{\varepsilon}$ increases from zero, first the stripe solution becomes stable at $\tilde{\varepsilon}_S = 1/(G_1 - 1)^2$,

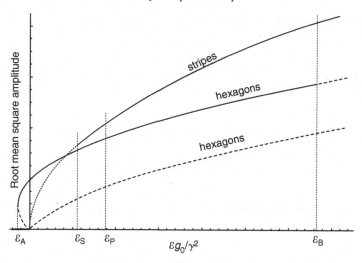

Fig. 7.2 Pattern intensity, as defined by the root-mean-square amplitude $\sqrt{\sum_i |\tilde{A}_i|^2}$, as a function of the scaled control parameter $\tilde{\varepsilon} = \varepsilon g_0/\gamma^2$ for the hexagonal and stripe solutions to the amplitude equations Eq. (7.73) for a system with no $\mathbf{u} \to -\mathbf{u}$ symmetry. The stable solutions are depicted by full lines, and the unstable ones by dashed lines. The control parameter labels indicate a saddle-node bifurcation (ε_A) when stable hexagons first appear, where the stripes become stable (ε_S), where the hexagons become unstable (ε_B), and where the stripes and hexagons have equal potentials (ε_P). There is bistability of stripes and hexagons over the range $[\varepsilon_S, \varepsilon_B]$: hexagons have a lower potential density and so are favored when $\varepsilon < \varepsilon_P$, and stripes have a lower potential density when $\varepsilon > \varepsilon_P$. The plot corresponds to a value of $G_1 = 4$.

and then at a larger value $\tilde{\varepsilon}_B = (G_1 + 2)/(G_1 - 1)^2$ the hexagon solution becomes unstable. Note that there is a range of control parameters for which both the stripe and hexagon solutions are stable. In addition to the stripe and hexagon solutions there are also "mixed solutions" with two magnitudes equal, but the third one taking on a different value (e.g. $|\tilde{A}_2| = |\tilde{A}_3| \neq |\tilde{A}_1|$). These solutions bifurcate from the stripe and hexagon solutions at $\tilde{\varepsilon}_S$ and $\tilde{\varepsilon}_B$ respectively. They are always unstable so we do not show them in Fig. 7.2.

Over a range of control parameters $\tilde{\varepsilon}_S < \tilde{\varepsilon} < \tilde{\varepsilon}_B$ the amplitude equation calculation shows that stripe and hexagon states are both stable so that the stability analysis does not lead to a result for the competition between the two states here. In such cases of bistability, we can use the potential nature of the amplitude equation dynamics to gain further insights into the competition. The comparison of the potentials of the hexagon and stripe states is straightforward but messy. If we continue to assume that the wave vectors are at critical so that there are no derivative terms in the amplitude equations, you can verify that the scaled amplitude equation Eq. (7.73) has potential dynamics for the potential density $\tilde{v}$ given by

$$\tilde{v} = \sum_{i=1}^{3}\left(-\tilde{\varepsilon}|\tilde{A}_i|^2 + \frac{1}{2}|\tilde{A}_i|^4 + \frac{1}{2}G_1\sum_{\substack{j=1\\j\neq i}}^{3}|\tilde{A}_i|^2|\tilde{A}_j|^2\right) - \left(\tilde{A}_1\tilde{A}_2\tilde{A}_3 + \text{c.c.}\right). \quad (7.76)$$

For the stripe state $\tilde{A}_1 = \sqrt{\tilde{\varepsilon}}, \tilde{A}_2 = \tilde{A}_3 = 0$ the potential density $\tilde{v}_S$ evaluates to

$$\tilde{v}_S = -\frac{1}{2}\tilde{\varepsilon}^2. \quad (7.77)$$

For the hexagon state, substituting $\tilde{A}_1 = \tilde{A}_2 = \tilde{A}_3 = \tilde{a}_H$ into Eq. (7.76) gives the potential density $\tilde{v}_H$

$$\tilde{v}_H = -3\tilde{\varepsilon}\tilde{a}_H^2 + \frac{3}{2}(1 + 2G_1)\tilde{a}_H^4 - 2\tilde{a}_H^3, \quad (7.78)$$

which we can evaluate further by using the fact that a_H is a root of Eq. (7.75). Some lengthy but straightforward algebra shows the value $\tilde{\varepsilon}_P$ for equal potentials $\tilde{v}_S = \tilde{v}_H$ to be

$$\tilde{\varepsilon}_P = \frac{1}{[2(1+G_1)]^{3/2} - 2(1+3G_1)}, \quad (7.79)$$

with $\tilde{v}_S < \tilde{v}_H$ for $\tilde{\varepsilon} > \tilde{\varepsilon}_P$, and vice versa.

Figure 7.3 shows an experiment for which bistability of hexagon and stripe states occurs in a chemical reaction–diffusion system. The experiment is performed on the chlorite–iodide–malonic acid system in a gel disk reactor as described

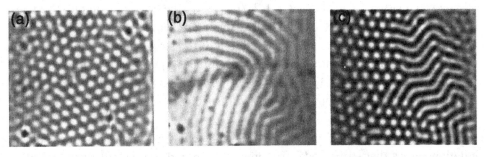

Fig. 7.3 Multistability of hexagon and stripe states in the chlorite–iodide–malonic acid (CIMA) chemical reaction–diffusion system in a gel disk reaction region (see Fig. 3.3). Panels (a) and (b) show hexagonal and stripe patterns at the same value of the control parameter (the malonic acid concentration) depending on whether the control parameter is swept up or down. Panel (c) shows hexagons and stripes separated by a grain boundary. The hexagonal domain is slowly invading the stripes at a speed of about two lattice sites per day. The concentration of malonic acid was 21 mM in (a) and (b) and 14 mM in (c). (From Ouyang et al. [85].)

in Section 3.2. The concentration of malonic acid in one of the reservoirs is used as the control parameter. Hexagons are found to be stable toward smaller values of the control parameter, and stripes toward larger values. There is a region of control parameter values where both states are seen, as shown in panels (a) and (b). Panel (c) shows a weakly dynamical state where stripes and hexagons are seen in different parts of the system. The hexagonal state is slowly invading the stripe state, so that eventually the system would fill with hexagons. However the dynamics is slow at the control parameter value shown, with the domain wall moving at a speed of about two lattice sites per day. A stationary domain wall would correspond to the equal potential point $\varepsilon = \varepsilon_P$ in the amplitude equation description. The patterns in these chemistry experiments do not show the regular, ordered states of the theoretical analysis: local regions show stripes or a lattice state, but the pattern is disordered on larger scales. Nevertheless, we expect the conclusions from the theoretical analysis to remain applicable, since they can be deduced by comparing the average potential densities which are not much affected by the disorder. Only with extraordinary care can perfectly regular states be formed, such as the hexagonal state in panel (b) of Fig. 1.14.

7.3.4 Spatial variations

So far we have insisted that the component stripes in a superposition state be spatially uniform and have wave vectors on the critical circle $q = q_c$. We can remove these assumptions by allowing the complex amplitudes A_i to vary spatially. It is straightforward to write down the corresponding amplitude equations by combining the ideas of Section 7.1 and Section 7.3. (In the following, we only consider the case of a system that is rotationally invariant in the plane.) In fact, at the lowest order of the amplitude equations, the spatial derivative terms and nonlinear terms are independent. Thus for coupled sets of stripes, the amplitude equations that generalize Eqs. (7.41) to include spatial variations are

$$\tau_0 \, \partial_t A_i = \varepsilon A_i + \xi_0^2 \left(\partial_{x_i} - \frac{i}{2q_c} \partial_{y_i}^2 \right)^2 A_i - g_0 \left(|A_i|^2 + \sum_{j \neq i} G(\theta_{ij}) |A_j|^2 \right) A_i. \quad (7.80)$$

Note that the coordinates (x_i, y_i), which appear with different orders of spatial derivatives, are defined relative to the direction of the wave vector of each set of stripes, so that $x_1 = \hat{q}_1 \cdot \mathbf{x}$, and y_1 is perpendicular to $\hat{q}_1$, etc.

Introducing scaled variables as

$$\bar{A}_i = \left| \frac{g_0}{\varepsilon} \right|^{1/2} A_i, \quad X_i = \frac{|\varepsilon|^{1/2}}{\xi_0} x_i, \quad Y_i = |\varepsilon|^{1/4} \left(\frac{q_c}{\xi_0} \right)^{1/2} y_i, \quad T = \frac{\varepsilon}{\tau_0} t, \quad (7.81)$$

yields the scaled form of the coupled equations

$$\partial_T \bar{A}_i = \pm \bar{A}_i + \left(\partial_{X_i} - \frac{i}{2}\partial_{Y_i}^2\right)^2 \bar{A}_i - \left(|\bar{A}_i|^2 + \sum_{j \neq i} G(\theta_{ij})|\bar{A}_j|^2\right)\bar{A}_i. \quad (7.82)$$

Of course the variables (X_1, Y_1) and (X_2, Y_2), etc., are not independent, and the correct way to treat the derivative terms requires some care. This is because we cannot introduce the different $\varepsilon^{1/2}$ and $\varepsilon^{1/4}$ scalings of the physical space variables along a single pair of orthogonal directions to eliminate the ε dependence from the derivative terms appearing in all the equations. Actually, we would expect the set of stripes with the slowest spatial variation along some direction (given by nonzero second-order derivatives with respect to displacements in this direction) to limit the rate of spatial variation. For a general situation then, the correct procedure will be to introduce a *single* scaling of the space variable $\mathbf{X} = \varepsilon^{-1/2}\mathbf{x}/\xi_0$ which eliminates all the derivative terms beyond second order. The resulting scaled amplitude equations for superimposed stripes are

$$\partial_T \bar{A}_i = \pm \bar{A}_i + \partial_{X_i}^2 \bar{A}_i - \left(|\bar{A}_i|^2 + \sum_{j \neq i} G(\theta_{ij})|\bar{A}_j|^2\right)\bar{A}_i. \quad (7.83)$$

with X_i the scaled coordinate along the roll normal. Section 8.2 discusses an interesting example where this reduction is not sufficient, namely the boundary between two perpendicular sets of stripes.

7.3.5 Cross-stripe instability

The amplitude equations for superimposed stripes in a rotationally invariant system, Eqs. (7.82), allow us to test a general stripe state for a cross-stripe instability, in which a second set of stripes begins to grow at some other orientation. (As mentioned earlier, this analysis does not allow us to deduce what the saturated nonlinear state will be, for example whether it will be a superposition of both stripes or whether the original stripe pattern will die out and be replaced by the new stripe pattern.) The cross-stripe instability generalizes the stability calculation of stripe states in Section 7.3.2, where we just looked at stripes at the critical wave number and asked if they were stable. Here we are moving the wave number away from critical and asking when the state first becomes unstable toward a second set of stripes. The resulting stability boundary often bounds part of the stability balloon.

Thus we start with a saturated set of stripes with wave vector arbitrarily chosen to lie in the x-direction and with a reduced wave number K given by the scaled

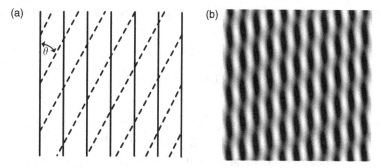

Fig. 7.4 Linear instability of a stripe state to growth of a second set of stripes with a different orientation. (a) Heavy lines denote the saturated stripe state with amplitude $\bar{A}_1 = \bar{A}_S$. To test the stability, we add a small amplitude δA_2 of a second set of stripes, denoted by the dashed lines, oriented at an angle θ to the first set, as well as a perturbation $\delta \bar{A}_1$ to $\bar{A}_1$. (b) Gray-scale plot of stripe state with superimposed stripes at angle 30° of relative intensity 0.16.

amplitude $\bar{A}_1(X)$ as in Eq. (6.36)

$$\bar{A}_{1K} = (1 - K^2)^{1/2} e^{iKX}. \tag{7.84}$$

Next, we test the stability of this solution to the growth of a second set of stripes at an angle θ as in Fig. 7.4. This set of stripes is described by the amplitude $\bar{A}_2$, and the coupled equations for $\bar{A}_1$ and $\bar{A}_2$ are Eqs. (7.82). We can also consider the second set of stripes with wave number away from critical, so that they are described by the amplitude

$$\delta \bar{A}_2 = \delta \bar{a}_2(T) e^{i\mathbf{K}_2 \cdot \mathbf{X}} = \delta \bar{a}_2(T) e^{iK_2 X_2}, \tag{7.85}$$

with $\delta \bar{a}_2$ small and $\mathbf{K}_2$ parallel to the wave vector of the second set of stripes with the magnitude K_2 giving the shift of the wave number. Substituting into Eq. (7.82) and linearizing in $\delta \bar{a}_2$ gives the evolution equation for the size a_2 of the perturbation δA_2:

$$d_T \bar{a}_2 = \left[1 - K_2^2 - G(\theta)\left(1 - K^2\right)\right] \bar{a}_2 \tag{7.86}$$

This expression differs from Eq. (7.64b) by the terms $1 - K^2$ and $1 - K_2^2$. The equation for $\delta \bar{A}_1$ is not needed in the calculation since, as you can verify, this perturbation always decays for the range of K for which stripes exist, $|K| < 1$.

Equation (7.86) shows that the most rapidly growing mode is always at $K_2 = 0$, which corresponds to the critical wave number. Instability occurs when the wave number of the first set of stripes reaches $K = \pm K_{CR}$ with

$$K_{CR} = \sqrt{1 - \frac{1}{G(\theta)}}. \tag{7.87}$$

In unscaled units, this becomes instability outside the cross-stripe stability boundaries $q = q_c \pm k_{CR}$ with

$$k_{CR} = \sqrt{1 - \frac{1}{G(\theta)} \xi_0^{-1} \varepsilon^{1/2}}. \tag{7.88}$$

The proportionality of $|q - q_c|$ on $\xi_0^{-1} \varepsilon^{1/2}$ is the same as found for the Eckhaus instability in Section 6.4.2, except there the proportionality constant is $1/\sqrt{3}$. The cross-roll instability will occur at a smaller value of $|q - q_c|$ than the Eckhaus instability, $k_{CR} < k_E$, if for any θ

$$G(\theta) < \frac{3}{2}. \tag{7.89}$$

If this relationship is satisfied, the cross-roll instability, rather than the Eckhaus instability, bounds the stability balloon near threshold on the large wave number side. For stripes in a rotationally invariant system, the zigzag instability, with a boundary $q_Z(\varepsilon) - q_c$ growing proportionately with ε as in Eq. (7.24), always forms the instability boundary near threshold on the small wave number side. The full stability balloon near threshold is shown in Fig. 4.2 in Chapter 4.

7.4 Conclusions

In this chapter, we have discussed with examples how the one-dimensional amplitude equation formalism of the previous chapter can be extended to describe patterns near onset that depend on two extended directions. The generalization allows us to treat two important aspects of pattern formation. The first is stripe states with spatial modulations along the stripe direction. One type of modulation (a linear variation of the phase of the complex amplitude) gives a rotation of the stripes. This means we can understand phenomena in which there is a spatial variation of the stripe orientation, for example the zigzag instability of stripes which bounds the stability balloon near threshold for systems that have rotational symmetry in the plane of two extended directions. Unfortunately, the formalism is limited to small reorientations from a parallel stripe reference state, and so cannot treat patterns in which the stripes are oriented in quite different directions in different parts of the system. Such patterns are commonly observed in experiment, for example see Fig. 4.8. We return to the discussion of this type of pattern in Section 9.1.2.

The second aspect we can now understand quantitatively near onset is the existence of various lattice states, and the nonlinear competition between stripes and these lattices. We discussed the simple lattice states formed from the superposition of stripes at two different orientations and, for the case of a hexagonal lattice, three sets of stripes. The discussion can be extended to the superlattice or quasiperiodic states, such as those shown in Fig. 4.7.

The amplitude equations derive from the lowest-order terms in a perturbation expansion in the small parameter $\varepsilon = (p - p_c)/p_c$. Physically, this corresponds to assuming that the effects of nonlinearity are weak, and that the modulations of a basic reference state vary slowly in space and time. This means that out of the vast range of phenomena seen in pattern-forming systems, some of which are described in Chapter 1, the amplitude equation description is accurate only for a small fraction. Nevertheless, the approach does give us a qualitative understanding of many questions, including ones we have discussed in this chapter such as stability balloons and the competition between different basic patterns. In the next chapter, we use amplitude equations to discuss localized structures, such as topological defects and boundaries between different patterns. Again, the range of quantitative accuracy is small but the qualitative insights gained are invaluable. These insights are even more precious because scientists and mathematicians have not yet developed any formalism of comparable power and wide applicability that can explain pattern formation for strong nonlinearity, e.g. when the reduced parameter ε is of order one in magnitude. How to analyze strongly nonlinear patterns remains an exciting research frontier as of the time that this book is being written. We touch briefly on this topic in Chapter 9.

7.5 Further reading

(i) Symmetry considerations become useful in classifying solutions to coupled amplitude equations for more complicated lattice states, such as the superlattices shown in Fig. 4.7(a) and (b), as well as quasiperiodic states as in Fig. 4.7(c). For an introductory account of these methods see *Pattern Formation: An Introduction to Methods* by Hoyle [47] and *The Symmetry Perspective: from Equilibrium to Chaos in Phase Space and Physical Space* by Golubitsky and Stewart [39]. A more technical reference is *Singularities and Groups in Bifurcation Theory: Volume 2* by Golubitsky, Stewart, and Schaeffer [40].

(ii) An interesting paper connecting the ideas of Section 7.3.2 and Chapter 3 is "Simple and superlattice Turing patterns in reaction-diffusion systems: bifurcation, bistability, and parameter collapse" by Judd and Silber [51].

Exercises

7.1 **Long-wavelength instability growth rate:** Derive with careful discussion of details and of assumptions Eq. (7.20) for the growth rate $\sigma_K(\mathbf{Q})$ of a stripe state in a rotationally symmetric system that is described by the perturbation equation (7.18).

7.2 **Growth-rate curve for the zigzag instability:** From the expression Eq. (7.20), plot the dependence of the growth rate $\sigma_K(Q_Y \hat{Y})$ on the wave number Q_Y of small perturbations for the zigzag instability for different values of the background wave number K and so verify the statement that the instability occurs first at long wavelengths ($Q_Y \to 0$).

7.3 **Instability of a stripe state for arbitrary directions of the perturbation wave vector:** Plot the growth rate $\sigma_K(Q_X, Q_Y)$ for perturbations about a stripe state in a rotationally invariant system given in Eq. (7.20) as a function of the perturbation wave vector Q_X, Q_Y (e.g. make a contour plot using Mathematica or another program): (a) for an initial wave number that is Eckhaus unstable but zigzag stable; (b) for an initial wave number that is zigzag unstable but Eckhaus stable; and (c) for an initial wave number that is unstable to both Eckhaus and zigzag instabilities. Verify that the maximum growth rate always occurs for the perturbation wave vector along or perpendicular to the stripes ($Q_X = 0$ or $Q_Y = 0$), but that there may be positive growth rate for general Q_X, Q_Y.

7.4 **Amplitude for stripes normal to a boundary:** Stationary stripes at the critical wave number approaching a side wall perpendicularly are described by the fully scaled amplitude equation, cf. Eq. (7.12),

$$0 = \bar{A} - \frac{1}{4}\partial_Y^4 \bar{A} - \bar{A}^3, \qquad (E7.1)$$

and the boundary conditions at $Y = \pm L/2$,

$$\bar{A} = \partial_Y \bar{A} = 0, \qquad (E7.2)$$

taking Y as the coordinate normal to the wall and $\bar{A}$ to be real. Here L is the appropriately scaled width of the system.

(a) Show that the linear onset in the finite system occurs at $L = L_c \simeq 3.34$ and interpret this in terms of the dependence of ε_c, the value of the reduced control parameter ε at onset on the system width l in unscaled units. Compare this with the expression Eq. (6.35) for stripes parallel to the side wall.

(b) Solve Eqs. (E7.1) and (E7.2) numerically for several values of the system size L with $L > L_c$, and interpret the solution in the physical unscaled variables. This is fairly straightforward to do with the NDSolve function of Mathematica, which requires only specification of equations and variables.

7.5 **Derivation of phase diffusion equations:** Derive the linear phase equations, Eq. (7.25) and Eq. (7.39), for small deviations from a state of parallel stripes for the isotropic and anisotropic systems.

7.6 **Amplitude equation for the Swift–Hohenberg equation in two dimensions:** Including spatial variations, derive the amplitude equation for stripes in two dimensions using the method of multiple scales of Appendix 2 for the two-dimensional Swift–Hohenberg equation

$$\partial_t u = ru - (\nabla^2 + 1)^2 u - u^3, \qquad (E7.3)$$

that is discussed in Section 5.1. You should generalize the method of Section A2.3.2 using the scalings introduced in Eq. (A2.6).

7.7 **A different rescaling of space variables:** For the Swift–Hohenberg equation in Exercise 7.6, derive the amplitude equation for stripes normal to the x direction using the following different choice of scalings

$$T = rt, \quad X = r^{1/2}x, \quad Y = r^{1/2}y. \qquad (E7.4)$$

This differs from the usual scalings for a rotationally invariant equation suggested in Appendix 2 in that the y variable now has the same scaling as the x variable. Compare the form of the amplitude equation to the one derived in Section 7.1.1 or Exercise 7.6. Is there anything to argue against the choice of scaling in Eq. (E7.4)?

7.8 **Scaled amplitude equation in anisotropic systems:** Show how to derive Eq. (7.31) from both of Eqs. (7.28) and (7.30) by an appropriate rotation (if necessary) and scaling of the variables.

7.9 **Stripe interaction coefficient $G(\theta)$ for a generalized Swift–Hohenberg equation:** By calculating the stripe interaction coefficient $G(\theta)$ for the two-dimensional generalized Swift–Hohenberg equation (see Section 5.2)

$$\partial_t u = ru - (\nabla^2 + 1)^2 u + \nabla \cdot \left[(\nabla u)^2 \nabla u\right], \qquad (E7.5)$$

show that squares, but not stripes, are stable near threshold for this equation and so should be observed.

(More advanced) Test your prediction by a numerical time integration of this evolution equation in a periodic domain of aspect ratio $\Gamma \geq 10$. One way is to start with random noise of small magnitude as an initial condition and verify that the noise grows into a square lattice state. Alternatively, you can start near onset with an analytical stripe solution and examine numerically how small-amplitude perturbations of different kinds evolve: noise, or a low-amplitude stripe oriented with angle θ with respect to the initial stripe, and so on.

7.10 **Stripe interaction coefficient $G(\theta)$ for a Swift–Hohenberg model of rotating convection:** As discussed in Section 5.2.4, the generalized Swift–Hohenberg equation

$$\partial_t u(x, y, t) = ru - \left(\nabla^2 + 1\right)^2 u - g_1 u^3$$
$$+ g_2 \hat{z} \cdot \nabla \times [(\nabla u)^2 \nabla u] + g_3 \nabla \cdot [(\nabla u)^2 \nabla u], \quad \text{(E7.6)}$$

can be used to model a convection system that is rotated with a constant angular frequency about the vertical axis. Here $\hat{z}$ is a unit vector perpendicular to the x- and y-axes, and g_1, g_2, and g_3 are real-valued parameters. Using Eq. (7.45), show that the stripe interaction coefficient $G(\theta)$ for this evolution equation is given by the expression

$$G(\theta) = \frac{6g_1 + 2g_2 \sin(2\theta) + 2g_3(2 + \cos(2\theta))}{3(g_1 + g_3)}. \quad \text{(E7.7)}$$

Why is the relationship $G(-\theta) = G(\theta)$ no longer satisfied? Does this expression for $G(\theta)$ satisfy the condition Eq. (7.42)?

7.11 **Patterns in the Brusselator model:** For the Brusselator model introduced in Etude 3.1, what pattern (stripes, squares, hexagons . . .) would you expect to see just above the Turing instability based on the symmetries of the equations? Verify your prediction by numerical simulations of the model for the parameters $a = 1.5, D_1 = 2.8, D_2 = 22.4$, and b just above b_c.

7.12 **Superlattice state:** Suppose that the amplitudes of the eight component modes for the square superlattice state shown in Fig. 4.7(a) can be described near threshold by the amplitudes in terms of four amplitudes (the amplitudes of the component modes with positive q_x) and their complex conjugates. Derive the condition on the interaction parameter $G(\theta)$ for the superlattice state described by four equal real amplitudes to be stable toward perturbations within the space of the four amplitudes.

7.13 **Stability of hexagonal state:** For Eqs. (7.60) and for a stripe interaction coefficient satisfying $G(\pi/3) < 1$, the hexagon state $\bar{A}_1 = \bar{A}_2 = \bar{A}_3 = \bar{A}_H$ is stable with respect to decay toward a stripe solution (e.g. $\bar{A}_1 = \bar{A}_S, \bar{A}_2 = \bar{A}_3 = 0$) and that stripes are unstable toward hexagons. Show that the opposite is true for the complementary inequality $G(\pi/3) > 1$.

7.14 **Potential for stripes and hexagons:** Write down the potential density v for Eq. (7.60). Using your expression, show that the values of the potential densities for the stripe and hexagon states are equal for $G(\pi/3) = 1$ so that using the potential density to study the competition between stripes and hexagons gives the same result as the stability analysis of Exercise 7.13.

7.15 **Instability of hexagons to stripes at new orientation:** Show that instability of the hexagon state toward growth of a fourth stripe, which makes an angle θ with respect to one of the three sets of hexagon stripes, occurs when

$$1 + 2G(\pi/3) > G(\theta) + G(\pi/3 - \theta) + G(\pi/3 + \theta), \qquad \text{(E7.8)}$$

where $G(\theta)$ is the stripe interaction coefficient defined in Eqs. (7.41).

7.16 **Potential and pattern intensity:** Consider the total intensity of the pattern

$$I = \sum_i |A_i|^2, \qquad \text{(E7.9)}$$

in a stripe and square state. (In convection, for example, the heat convected by the flow is proportional to this quantity.) Within the amplitude equation approximation, does the more stable state correspond to the larger value of I? Do you expect this result to be true in general?

7.17 **Potential for stripes and hexagons in a system without field inversion symmetry:** For the potential Eq. (7.76), verify the expression Eq. (7.79) for $\tilde{\varepsilon}_P$, the value of $\tilde{\varepsilon}$ for which the potential densities of the stripe and hexagon states are equal. Show (for example graphically) that

$$\tilde{\varepsilon}_S \leq \tilde{\varepsilon}_P \leq \tilde{\varepsilon}_B, \qquad \text{(E7.10)}$$

i.e. the value $\tilde{\varepsilon}_P$ lies between the values for the saddle-node bifurcation and the instability of hexagons defined in Section 7.3.3.

7.18 **Stability balloon and the cross-stripe instability:** Sketch the stability balloons in the εq plane for small ε calculated within the amplitude equation for two hypothetical systems, one system with $G(\theta) > 3/2$ for all θ, and the other with $G(\theta)$ taking on a least value of $5/4$ at $\theta = \pi/2$. Discuss what would happen for a third system with $G(\pi/2) = 3/4$.

7.19 **Cross-stripe instability for a stripe state in a uniaxial system:** Can a stripe phase in a uniaxial system show the cross-stripe instability within the amplitude equation description?

7.20 **Rotating Rayleigh–Bénard convection:** If a Rayleigh–Bénard system is rotated sufficiently rapidly about the vertical axis, a set of convection rolls becomes unstable by the so-called Kuppers–Lortz instability to the growth of another set of rolls at an angle of about $\pi/3$ in the direction of the rotation. When these rolls saturate, they in turn become unstable to a third set of rolls at $\pi/3$ ... and so on; there turn out to be no stable steady states. As a first approach to explore this situation, we can use the amplitude equation formalism for sets of stripes with the spatial dependence ignored (so all stripes are uniform and have wave vectors on the critical circle $q = q_c$). This situation can then be modeled near onset by the following set of three scaled amplitude equations:

$$d_T \bar{A}_1 = \bar{A}_1 - (\bar{A}_1^2 + g_+ \bar{A}_2^2 + g_- \bar{A}_3^2)\bar{A}_1,$$
$$d_T \bar{A}_2 = \bar{A}_2 - (\bar{A}_2^2 + g_+ \bar{A}_3^2 + g_- \bar{A}_1^2)\bar{A}_2,$$
$$d_T \bar{A}_3 = \bar{A}_3 - (\bar{A}_3^2 + g_+ \bar{A}_1^2 + g_- \bar{A}_2^2)\bar{A}_3,$$

where we have chosen amplitudes to be real, and the nonlinear coupling coefficients g_+ and g_- for stripes at $\pm\pi/3$ are no longer equal.

(a) Give brief phenomenological arguments justifying the form of the amplitude equations. How would you expect g_+ and g_- to depend on the angular rotation rate Ω for small rotation rates?

(b) What are the conditions on the parameters g_+ and g_- for stripe and hexagon states to be unstable?

(c) In the three-dimensional dynamical phase space $\bar{A}_1, \bar{A}_2, \bar{A}_3$, show the three fixed points corresponding to the stationary single stripe states, and the directions of the stable and unstable eigenvectors at these fixed points.

(d) Describe qualitatively what happens in this model if the dynamics is started with an initial condition that is slightly perturbed from the $\bar{A}_1 = 1, \bar{A}_2 = \bar{A}_3 = 0$ fixed point in the case where all the fixed points are unstable. Do you think that this predicted dynamics would be seen in an actual experiment?

The mathematics of this convection problem has appeared in numerous places outside the context of pattern formation. For example, a related problem was independently discussed earlier by May and Leonard [71], who were interested in asymmetric cyclic arrangements of predators and prey such that species A ate species B, species B ate species C, and species C ate species A. Related models have also been used to study voting patterns.

8
Defects and fronts

In this chapter, we will use the amplitude equations introduced in the previous two chapters to gain an understanding of defects and fronts in stripe patterns near onset. Our discussion also serves to illustrate more advanced applications of the amplitude equations, in particular the ability to treat patterns that vary spatially.

The importance of defects in patterns was introduced in Section 4.4. A defect can be thought of as a local imperfection in an otherwise perfect pattern, and experimental and natural patterns often contain many defects, either due to the effect of boundaries[1] or due to a spatially inhomogeneous initial condition. To approach the complicated realistic case of many defects, it is useful first to study single isolated defects and then attempt to build an understanding of natural patterns from these elementary ingredients. In Section 4.4.2, we introduced the notion of a topological defect, namely one that can be identified from properties of the pattern far from the location of the defect (such as the winding number of the phase, Eq. (4.58)), where the pattern approaches the ideal state. These defects are particularly important because they can persist for long times and so are robust. We will focus on two types of topological defects in stripe states, dislocations, and grain boundaries.

We first discuss the structure of stationary defects. Since defects represent deviations from the ideal pattern, the intensity of the pattern is reduced in their vicinity. The region over which the intensity is reduced, and where the pattern strongly deviates from the ideal one, is called the core of the defect. For stripe states in an isotropic system near threshold, the suppression of the pattern intensity occurs anisotropically, over distances proportional to $\varepsilon^{-1/2}$ along the normal to the stripes, and proportional to $\varepsilon^{-1/4}$ parallel to the stripes (see Section 7.1).

We then discuss how a defect can constrain the pattern far from the defect's position. For example, as mentioned in Section 4.4.2, a stationary dislocation solution

---

[1] Near onset, convection rolls, for example, are observed to be approximately perpendicular to the boundaries, as shown in Fig. 4.8. In a cylindrical domain, this causes defects to form since it is not possible for nearly equally spaced stripes of wavelength approximately $2\pi/q_c$ to also be locally perpendicular to a circular boundary.

only occurs for a particular value q_d of the wave number of the stripe pattern far from the dislocation. A related question that we discuss is to determine the dynamics of a defect when the distant pattern deviates from the constraints consistent with a stationary defect. For example, a dislocation in a pattern with a background wave number that is different from the value q_d is found to move along the stripes. This motion, called climb, will eventually add or subtract a stripe pair and so provides a way to change the average wave number of the pattern. We discuss the dynamics of dislocations and grain boundaries in Section 8.1 and Section 8.2 respectively.

We next turn to a discussion of fronts. Their importance in pattern formation can be appreciated by considering the evolution of a pattern from the perturbed uniform state. The linear instability analysis of the uniform state in Chapter 2 looked at the growth of a single Fourier mode perturbation that is delocalized over the whole system. In experiment or the natural world, a physical perturbation that initiates the growth from the uniform state is more likely to be limited to some localized region of the system, for example at or near a boundary. The perturbation may then saturate with a finite amplitude in the vicinity of its initial position before the perturbation has had a chance to spread through the remainder of the system. As time advances, the saturated region can grow in extent by invading the remaining region that contains the unstable uniform state. The invasion occurs through the propagation of an interface which, to a good approximation, has a constant profile and propagates at a constant speed. Such an interface is called a front. A constant profile propagating with a constant speed is perhaps surprising in a driven dissipative system since diffusion alone typically causes a localized structure to smear out over time but the instability from the uniform state is capable of countering the effects of diffusion.

More generally, a front is a boundary or domain wall between two regions or domains of different states of the system.[2] Mathematically, a front is a solution connecting two simpler solutions that hold far away from the front, and so is analogous to a heteroclinic orbit connecting two fixed points in dynamical systems theory. (The analogy of a homoclinic orbit in pattern formation would be a pulse, in which a local region of one kind of pattern connects two regions consisting of some other kind of pattern.) The analogy of a front to a heteroclinic orbit will be useful when we investigate the properties of front solutions.

Our discussion of fronts divides into two issues: existence and selection. In many cases (the criterion will be introduced later), it is found mathematically that a family of steadily propagating front solutions exist for each fixed set of the equation parameters, with a continuous range of propagation speeds. However, numerical simulations and experiments show that fronts approach a speed that, once initial

---

[2] A grain boundary can also be considered a type of front. In this section, we will be concerned with the situation where one of the domains is the uniform state, rather than a different orientation of the same pattern as for a grain boundary.

8.1 Dislocations

transients have died out, is *uniquely* fixed by the system parameters. This leads to the second question of selection, namely what is the physical mechanism (or possibly mechanisms) that selects a particular speed out of the continuum of possibilities? A related question is whether selection of some type occurs for the pattern that is laid down behind the moving front, for example whether a unique stripe wave vector is observed in the region where the unstable uniform state previously existed. We discuss what is known at a fairly qualitative level since the mathematical analysis is rather subtle and, in some cases, still not fully understood.

8.1 Dislocations

Dislocations in stripe states were introduced in Section 4.4.2. The presence of a dislocation is prescribed by the winding number condition, Eq. (4.58). In terms of the phase Φ of the complex amplitude introduced in Section 4.1.1, the phase winding condition is

$$\frac{1}{2\pi} \oint \nabla \Phi \cdot d\mathbf{l} = \pm 1, \tag{8.1}$$

since there is no phase winding due to the $e^{iq_c x}$ dependence. We will make our usual choice of coordinate system so that the stripes far from the dislocation are parallel to the y-axis, and the stripe wave vector is in the $\hat{\mathbf{x}}$- direction. Referring to Fig. 8.1, the plus sign in Eq. (8.1) corresponds to the stripe pair extending to negative y, and the minus sign to the extra stripe pair extending to positive y. (Remember that the

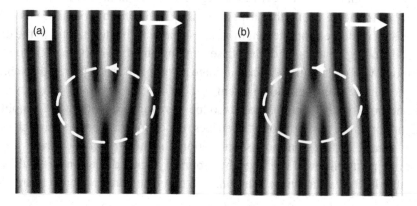

Fig. 8.1 The two signs of a dislocation in a stripe pattern. Once the sign of the background wave vector is arbitrarily chosen (indicated by the white arrow in the upper right part of each panel), accumulating the phase winding around the dashed contours in the direction shown indicates that the dislocations in panels (a) and (b) have respectively winding numbers of -1 and $+1$.

integration contour in Eq. (8.1) is traversed in the anticlockwise direction.[3]) The straight stripe pattern far away gives a boundary condition on the phase dependence of the amplitude, given in terms of $\mathbf{k} = \nabla \Phi$ by

$$\mathbf{k} \to (k_b, 0) + O(r^{-2}), \quad r \to \infty, \tag{8.2}$$

with r the distance from the dislocation. This corresponds to a stripe wave number far from the dislocation that we call the background wave number, $q_b = q_c + k_b$.

Since the length of the contour in Eq. (8.1) grows in proportion to the distance from the dislocation, the perturbation to the straight stripe pattern from the phase winding becomes small at large distances. (The $O(r^{-2})$ correction in Eq. (8.2) is the perturbation to the wave vector from the phase winding.) On the other hand, as the contour surrounding the dislocation shrinks to a small radius, the phase variation becomes more rapid. This causes the magnitude of the complex amplitude to be suppressed, which defines the core region of the defect. In fact, as the contour shrinks to zero radius, the phase variation becomes increasingly rapid. Thus the position of the dislocation is a singularity in the phase variable (the phase variable becomes undefined at this point). However, since the magnitude goes smoothly to zero here, the singularity in the phase variable becomes a smooth variation in the amplitude function, and therefore in the basic fields $\mathbf{u}$ as well.

An important result that we can derive from the amplitude equation, and that is also true more generally even away from onset, is that a stationary dislocation solution only exists for a particular value of the background wave number q_b. This wave number is called the dislocation selected wave number and is labeled q_d. The wave number q_d constrains stationary patterns in experiment and in nature, where dislocations commonly occur. If the background wave number deviates from q_d, the dislocation moves along the stripe direction.[4] This climb motion has the effect of increasing or decreasing the length of the stripe pair that terminates at the dislocation, depending on the sign of the phase winding and the direction of the motion. This motion will eventually add or eliminate a complete stripe pair and so change the wave number over the whole system.

The motion of dislocations can also be investigated within the amplitude equation. For climb motion along the stripes, the dislocation passes through states that are related by the continuous translational symmetry of the stripes in this direction. This motion will be driven by arbitrarily small deviations from the conditions that lead to a stationary solution. The requisite perturbation is a stretching

---

[3] There is actually no unambiguous way of assigning signs to the dislocations in a stripe pattern, since the sign of the wave vector $\mathbf{q}$ is itself not a-priori defined. Once we choose a convention for the sense of the wave vector, the sign of the dislocation can be determined by accumulating the phase winding from $\nabla \phi = \mathbf{q}$.

[4] We focus this qualitative discussion on the case of stripe states in isotropic systems or for uniaxial systems with the stripes along or perpendicular to the anisotropy axis, so that the (x, y) coordinates of the amplitude equation coincide with the stripe normal and direction respectively.

or compression of the wave number q_b of the stripes away from the selected value q_d. Motion perpendicular to the stripes, called glide, involves successive breaking and reconnection of stripes, and will not occur for infinitesimal perturbations from the stationary solution. However, these pinning effects are not captured in the perturbation theory leading to the amplitude equation, even if the theory is extended to higher orders, since they depend on terms of the form $\exp(-\alpha/\varepsilon^p)$, with α and p constants. Such effects are known as nonadiabatic effects.[5] Thus the amplitude equation does in fact predict glide motion for arbitrarily small perturbations from the stationary situation, even though a finite perturbation would actually be needed to overcome the pinning effect. Dislocation glide is often seen in experiment and in numerical simulations, presumably because the deviation from the stationary situation is large enough to overcome the pinning effects.

8.1.1 Stationary dislocation

Because of the asymmetric way in which the x- and y-derivatives appear in the amplitude equation Eq. (7.12) for stripes in an isotropic system, it is difficult to derive a dislocation solution. The full behavior of the amplitude in the core region requires numerical solution of the pde for the x- and y-spatial variation and even the phase variation far from the core must be obtained numerically (as the solution of an ode in the similarity variable $y/\sqrt{x}$). Some features of the solution are apparent from more general arguments. For example, the anisotropic scaling of x and y coordinates leads to an anisotropic core, with extent of order $\varepsilon^{-1/2}$ in the x-direction, and $\varepsilon^{-1/4}$ in the y-direction. The calculation of the core structure is simpler for the amplitude equation for the anisotropic system Eq. (7.31), and we describe this in the following Etude.

Etude 8.1 Dislocation solution in the amplitude equation for an anisotropic system

In scaled coordinates, the amplitude equation Eq. (7.31) is

$$\partial_T \bar{A} = \bar{A} + \partial_X^2 \bar{A} + \partial_Y^2 \bar{A} - |\bar{A}|^2 \bar{A}. \qquad (8.3)$$

We look for a stationary solution of the form $\bar{A} = \bar{a} e^{i\Phi}$, with $\nabla \Phi \to 0$ at large distances, which corresponds to a background wave number at critical. Since the equation and boundary conditions at infinity are then isotropic in (X, Y), the phase variation will be uniform in the polar angle Θ. The solution consistent with the 2π phase winding condition is, up to an arbitrary additive constant that just

---

[5] Nonadiabatic effects are studied quantitatively in Etude 9.1.

corresponds to a shift of the whole pattern,

$$\Phi = \Theta = \tan^{-1}(Y/X). \tag{8.4}$$

The magnitude $\bar{a} = |\bar{A}|$ will depend only on the radial coordinate $r = \sqrt{X^2 + Y^2}$. Expressing the Laplacian ∇^2 in polar coordinates

$$\nabla^2(\bar{a}(r)e^{i\Theta}) = \left[r^{-1}\partial_r(r\,\partial_r\bar{a}) - r^{-2}\bar{a}\right]e^{i\Theta}, \tag{8.5}$$

yields the following ode for the magnitude $\bar{a}$

$$d_r(r\,d_r\bar{a}) - r^{-1}\bar{a} + r\bar{a}(1 - \bar{a}^2) = 0. \tag{8.6}$$

The boundary conditions are that $\bar{a} \to 0$ linearly as $r \to 0$ (so that the complex amplitude is smooth here) and that $\bar{a} \to 1$ at large r (saturation). Since there are no parameters in the equation, we expect the amplitude to grow to saturation over a distance r of order unity which defines the core radius r_c. Equation (8.6) may be solved numerically to give the magnitude $\bar{a}$ and the core radius r_c, as shown in Fig. 8.2.

The stationary solution for the dislocation was calculated by assuming that the wave number far away from the dislocation was the critical value. In the next section, we verify using the potential for the amplitude equation that a wave number different from critical does indeed produce dynamics. Thus, within the amplitude

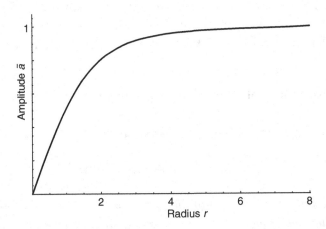

Fig. 8.2 Numerical solution of the ode Eq. (8.6) for the radial profile of the amplitude $\bar{a}(r)$ near a dislocation in an anisotropic stripe system. The amplitude is zero at $r = 0$ where the phase singularity occurs, and recovers toward saturation over a distance of about 2 in the scaled units which define the radius of the dislocation's core region. At large distances, $\bar{a} \simeq 1 - r^{-2}$, and is slightly depressed from unity by the phase winding.

equation description, the dislocation selected wave number q_d is

$$q_\mathrm{d} = q_\mathrm{c} + 0 \times \varepsilon^{1/2} + O(\varepsilon), \tag{8.7}$$

where the $O(\varepsilon)$ corrections may be found by methods that go beyond the lowest-order amplitude equation.

8.1.2 Dislocation dynamics

A powerful way to understand defect dynamics in the amplitude equation is to use the decrease in the potential governing the amplitude equation dynamics. To illustrate the method, we will look at the case of dislocation climb in a stripe state in an anisotropic system or in a uniaxial system with the stripes perpendicular to the special direction. The motion is driven by a deviation of the wave number of the stripes from the critical value. We will use the scaled space and time coordinates and amplitude, and so we look for steady climb in the $\pm Y$ direction along the stripes at (scaled) speed $\bar{v}$. The key to the method is the identity Eq. (7.16)

$$d_T \bar{V} = -2 \iint dX\, dY\, |\partial_T \bar{A}|^2. \tag{8.8}$$

Each side of this equation is evaluated assuming the steady motion of the dislocation. For climb, this gives the replacement

$$\partial_T \to -\bar{v}\, \partial_Y, \tag{8.9}$$

with $\bar{v}$ the scaled velocity. The left-hand side of Eq. (8.8) is evaluated from the change in the potential for a small displacement of the defect

$$d_T \bar{V} = -\bar{v}\, \partial_Y \bar{V}. \tag{8.10}$$

This is analogous to the work done by a force on a defect in an equilibrium system.[6] The right-hand side becomes

$$-2\bar{v}^2 \iint dX\, dY\, |\partial_Y \bar{A}|^2, \tag{8.11}$$

and is proportional to the square of the defect climb speed. This is analogous to a dissipation or drag term in an equilibrium system. Equating the two expressions gives the climb speed

$$\bar{v} = \frac{\partial_Y \bar{V}}{2 \iint dX\, dY\, |\partial_Y \bar{A}|^2}. \tag{8.12}$$

---

[6] In the context of dislocation motion in a crystal, this is known as the Peach–Koehler force.

The details of evaluating this expression for stripes in a uniaxial or isotropic system are described in the following two Etudes.

Etude 8.2 Dislocation motion for stripes in a uniaxial system

We first look at the case of dislocation climb in a stripe state in a uniaxial system with the stripes perpendicular to the special direction (the case of Fig. 2.13(a)) driven by a change in the wave number of the stripes from the critical value. The potential for this system is given by Eq. (7.32).

We look for steady climb in the $\pm Y$ direction along the stripes at (scaled) speed $\bar{v}$. The solution far away from the dislocation is taken to be uniform stripes at a changed wave number given by the scaled amplitude wave vector $(K_b, 0)$

$$\bar{A}_{K_b} \to \bar{a}_{K_b}^2 e^{iK_b X} \quad \text{with} \quad \bar{a}_{K_b}^2 = 1 - K_b^2. \tag{8.13}$$

The integral in the denominator of Eq. (8.12) is dominated by the region far away from the dislocation, where the important variation of the complex amplitude is from the phase, so that

$$|\partial_Y \bar{A}|^2 \simeq \bar{a}_{K_b}^2 (\partial_Y \Phi)^2, \tag{8.14}$$

and

$$\iint dX \, dY \, |\partial_Y \bar{A}|^2 \simeq \bar{a}_{K_b}^2 \iint dX \, dY (\partial_Y \Phi)^2. \tag{8.15}$$

The numerator of Eq. (8.12) is evaluated from the change in potential due to a shift δY of the dislocation. For a background wave-number deviation given by K_b, the wave-numbers for large positive and large negative Y are slightly changed to $K_b \pm \delta K$ by the winding of the phase around the dislocation. A translation δY of the dislocation has the effect of increasing the range of Y values below the dislocation with the wave number $K_b - \delta K$ by δY, and decreasing the range of Y values above the defect at wave number $K_b + \delta K$ by the same amount. The complicated, distorted region near the dislocation is just translated, so that the potential contribution from this region is unchanged. The potential per area $\bar{V}(K)/S$ of the stripe state with wave number K is given by Eq. (7.32) with $|\bar{A}| = \bar{a}_K e^{iKX}$ and is

$$\bar{V}(K)/S = -\tfrac{1}{2}\bar{a}_K^4 = -\tfrac{1}{2}(1 - K^2)^2. \tag{8.16}$$

The change in potential due to the dislocation climb through δY (an area $L_x \delta Y$ changes wave number from $K_b - \delta K$ to $K_b + \delta K$, where L_x is the extent of the system in the X-direction) is then

$$\delta \bar{V} = -4\bar{a}_{K_b}^2 K_b \, \delta K \, L_x \, \delta Y. \tag{8.17}$$

But the phase winding around the dislocation evaluates to $-2L_x \delta K$, and this is 2π for a $+1$ dislocation. Thus, simplifying and taking the Y-derivative, we have

$$\partial_Y \bar{V} = 4\pi \bar{a}_{K_b}^2 K_b. \tag{8.18}$$

8.1 Dislocations

Combining numerator and denominator gives the result for the climb velocity:

$$\bar{v} = -2\pi \left(\int\int dX\, dY (\partial_Y \Phi)^2 \right)^{-1} \times K_b. \tag{8.19}$$

Equation (8.19) gives us the result that the dislocation is stationary only if the background wave number is critical, $K_b = 0$. If K_b is positive so that the stripes are compressed and the background wave number is greater than critical $q_b > q_c$, the climb motion is in the $-Y$ direction, which eventually will eliminate a stripe pair and slightly reduce the wave number (remember we are looking at the $+1$ dislocation, Fig. 8.1(b)). Similarly, if K_b is negative, $q_b < q_c$, the climb motion will tend to increase the wave number. The equation also appears to show that the climb velocity is proportional to K_b, i.e. to the deviation of the background wave number from critical. This is actually not the case, since the integral appearing in Eq. (8.19) diverges for $K_b = \bar{v} = 0$, and so we need to keep these quantities nonzero in the evaluation.

To evaluate Eq. (8.19) to leading order in K_b, we need to solve the phase equation for a uniformly climbing dislocation. The equation for small phase variations was derived in Chapter 7, see Eq. (7.39). Putting $\partial_T \to -\bar{v}\,\partial_Y$ and ignoring the $O(K_b^2)$ terms in the parallel diffusion constant gives the equation

$$-\bar{v}\,\partial_Y \Phi = \partial_X^2 \Phi + \partial_Y^2 \Phi. \tag{8.20}$$

We might first try the $\bar{v} = 0$ solution of Section 8.1.1, where the solution Φ is just the polar angle Θ as in Eq. (8.4). Then writing $r = \sqrt{X^2 + Y^2}$ we have

$$\int\int (\partial_Y \Phi)^2\, dX\, dY = \int\int r^{-1} \cos^2\Theta\, dr\, d\Theta = \pi \int r^{-1}\, dr. \tag{8.21}$$

The integral diverges logarithmically,

$$\int\int (\partial_Y \Phi)^2\, dX\, dY = \pi \ln(r_m/r_c), \tag{8.22}$$

where r_m is a large-distance cutoff and r_c a constant of order unity coming from the core region, where the magnitude begins to change, so that the phase variation is not sufficient to evaluate $|\partial_Y \bar{A}|^2$ as done in Eq. (8.14). In a realistic physical system r_m might be defined by boundaries or by the presence of other defects in the stripe structure. But in the ideal mathematical situation of a dislocation in an infinite array of otherwise undistorted stripes, the divergence is eliminated by returning to Eq. (8.20) and including the velocity term. The phase variation solving the full equation can be shown to yield

$$\int\int (\partial_Y \Phi)^2\, dX\, dY = \pi \ln(\bar{v}_0/\bar{v}), \tag{8.23}$$

where $\bar{v}_0$ is a velocity of order unity that reflects the core cutoff. Thus finally, for dislocation climb in a uniaxial system the speed $\bar{v}$ is given by

$$\bar{v} \ln(\bar{v}_0/\bar{v}) = -2K_b, \qquad (8.24)$$

showing the logarithmic corrections to the simple result $\bar{v} \propto K_b$.

In setting up the discussion, we took the background wave vector deviation $\mathbf{K}_b$ to be in the $\hat{\mathbf{X}}$-direction. However, since for the uniaxial system, the scaled amplitude equation is isotropic, the calculation is independent of the direction of $\mathbf{K}_b$, and the dislocation will move at the speed Eq. (8.24) and perpendicular to $\mathbf{K}_b$ for any direction of $\mathbf{K}_b$. A small K_{by} corresponds to a rotation of the stripes and produces glide motion along $\hat{\mathbf{X}}$ in this calculation. It should be remembered, however, that glide motion does not take the dislocation through equivalent positions, so that there will be nonadiabatic pinning effects to the stripes that may resist the motion not captured by the amplitude equation.

The same scaled amplitude equation applies to the anisotropic system, and so our calculation also gives results for the dislocation dynamics in this case. We leave the reader to uncover the direction of the motion for various types of distortions to the stripes for this example.

Etude 8.3 Dislocation climb for stripes in an isotropic system

The main difference in the calculation for the isotropic system is that the phase diffusion constant for the Y derivatives is itself proportional to K_b, so that Eq. (8.20) is modified to (see Eq. (7.25))[7]

$$-\bar{v} \, \partial_Y \Phi = \partial_X^2 \Phi + K_b \, \partial_Y^2 \Phi. \qquad (8.25)$$

Again ignoring the term proportional to $\bar{v}$ on the left-hand side, a rescaling of the Y variable, $Y \to \tilde{Y} = Y/\sqrt{K_b}$ can be used to show that the integral appearing in Eq. (8.19) becomes

$$\iint (\partial_Y \Phi)^2 \, dX \, dY = \frac{\pi}{\sqrt{K_b}} \ln\left(\frac{r_m}{r_c}\right). \qquad (8.26)$$

So for the isotropic system, the climb velocity $\bar{v}$ is

$$\bar{v} \ln(v_0/\bar{v}) = -2K_b^{3/2}, \qquad (8.27)$$

and scales as $(q - q_d)^{3/2}$ (together with slowly varying logarithmic corrections). Notice that the difference between this result and Eq. (8.24) derives from the zero

---

[7] Actually, we should also include the fourth-order derivative terms, and also nonlinear terms in the phase equation, to be completely consistent. However, the basic idea of the result is given by this simpler equation.

transverse diffusion constant at the wave number q_d for which a dislocation is stationary (here $q_d = q_c$).

Results analogous to Eqs. (8.24) and (8.27) actually continue to hold away from threshold, with the climb velocity being proportional to $(q_b - q_d)$ (with logarithmic corrections) in the general case for both uniaxial and isotropic systems, but as $(q_b - q_d)^{3/2}$ (with logarithmic corrections) for any special cases where the dislocation-selected wave number coincides with the wave number at which the transverse diffusion constant is zero. This latter case turns out to occur when the equations governing the system are themselves potential, such as the Swift–Hohenberg equation.

8.1.3 Interaction of dislocations

If more than one dislocation is present in the system, the long-range phase distortion from one dislocation at the location of a second dislocation leads to an interaction between the defects. The dynamics induced by the interaction are given by combining the results from the previous two sections. For two dislocations separated along the stripe direction for example, the local wave-number change at the second defect, given by the gradient of the phase distortion from the first defect, will induce climb motion. You can check that defects of the same sign climb away from one another (repulsion), whereas defects of the opposite sign move toward one another (attraction).

Fig. 8.3 Dislocations in sand ripples on a sand dune in Death Valley. According to the theory developed in these sections, two dislocations aligned on the same stripe, such as in the picture, attract and tend to move together. These predictions are based on the amplitude theory valid near threshold, but are expected to hold qualitatively far from threshold, although it is not known whether the predictions apply to sand ripples.

290 Defects and fronts

8.2 Grain boundaries

Grain boundaries were introduced in Section 4.4.2, see Fig. 4.12. They are extended line defects that separate two half-spaces of differently oriented patterns, for example stripes with different wave vectors, which is the case that we will consider. Within the amplitude equation description, the boundary between two stripe regions can be constructed from two amplitudes, and the core of the grain boundary is defined by the region where the two amplitudes overlap. Such grain boundaries are called amplitude grain boundaries. The calculations for the two cases in Fig. 4.12 are a little different, as shown later in this section. If the angle between the stripes on opposite sides of the grain boundary is sufficiently small, the structure of the grain boundary changes in that the stripes may bend from one orientation to the other without terminating. Near threshold, such a small angle grain boundary can be described by an amplitude equation for a single set of stripes. Since most of the action is in the phase component of the complex amplitude, with relatively small changes in the magnitude passing through the boundary, these are also known as phase grain boundaries. In the core region, the stripe orientation bends continuously and there are no stripe terminations. This case is the subject of Exercise 8.3.

We first look at the structure of a grain boundary with general orientations of the two stripe states as in Fig. 8.5, using the coupled equations for the amplitudes of the two sets of stripes Eqs. (7.80). Let us define the coordinate system so that the grain boundary is along the infinite line $x = 0$, with a pattern of stripes at wave vector $\mathbf{q}_1$ for $x \to \infty$, and $\mathbf{q}_2$ for $x \to -\infty$. Let us also denote by θ_1 the angle between $\mathbf{q}_1$

Fig. 8.4 A phase grain boundary reconstructed from the amplitude equation.

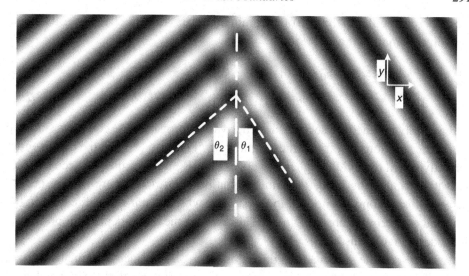

Fig. 8.5 Geometry of a grain boundary. The stripes on either side make an angle θ_1 and θ_2 with the grain boundary.

and the x-axis (equal to the angle between the stripes and the line of the boundary), and by θ_2 the angle between $\mathbf{q}_2$ and the x-axis. For the amplitude equation to be valid, the magnitudes of the two wave vectors must lie close to q_c and the control parameter must be near the threshold value. The amplitudes A_1 and A_2 of the stripes on either side of the boundary are coupled in the overlap region of the boundary through the interaction parameter $G(\theta_1 + \theta_2)$, which we will write from now on as G_{12}.

It is useful to simplify the full complexity of the coupled amplitude equations using some simple physical arguments. Since the wave number of each set of stripes can relax by translation of the stripes into the boundary, the wave numbers will evolve to approach the critical value far from the grain boundary to reduce the potential. Thus we have the wave number selection principle that the grain boundary structure is time independent only if the wave numbers of both sets of stripes are at critical, $|\mathbf{q}_1| = |\mathbf{q}_2| = q_c$. This condition corresponds to the boundary conditions

$$\begin{aligned} A_1(x \to -\infty) &= 0, & A_1(x \to \infty) &= \sqrt{\varepsilon/g_0}, \\ A_2(x \to -\infty) &= \sqrt{\varepsilon/g_0}, & A_2(x \to \infty) &= 0. \end{aligned} \quad (8.28)$$

(The asymptotic values can be chosen real by an appropriate choice of the coordinate origin[8]). For the general case of θ_1 and θ_2 not close to $\pi/2$, i.e. either set of stripes

---

[8] Since the system with the grain boundary along $x = 0$ remains translationally invariant in the y-direction, we might allow a y-dependent phase $A_i(x, y) \propto e^{ik_i y}$. However, this can be absorbed into a redefinition of the orientation of the stripes, and so we may set $k_i = 0$.

nearly perpendicular to the line of the grain boundary, Eqs. (7.80) reduce at leading order in the small parameter ε to

$$\tau_0\, \partial_t A_1 = \varepsilon A_1 + \xi_0^2 \cos^2\theta_1\, \partial_x^2 A_1 - g_0[|A_1|^2 + G_{12}|A_2|^2]A_1, \quad (8.29\text{a})$$

$$\tau_0\, \partial_t A_2 = \varepsilon A_2 + \xi_0^2 \cos^2\theta_2\, \partial_x^2 A_2 - g_0[|A_2|^2 + G_{12}|A_1|^2]A_2. \quad (8.29\text{b})$$

We have only retained the second-order derivative terms in these equation, since we expect the solutions to vary on the length scale of order $\varepsilon^{-1/2}$ and the higher-order derivatives will then yield higher powers of $\varepsilon^{1/2}$. Remember that the second-order derivatives appearing in the amplitude equations, Eqs. (7.80), are with respect to the coordinate along the normal to the stripes, which is *not* the x-direction in the coordinate system we are using here. This leads to the angle-dependent prefactors in the derivative terms in Eqs. (8.29).

We now seek the stationary solutions of Eqs. (8.29). Since the equations have real coefficients and boundary conditions, we can seek real solutions in which case Eqs. (8.29) reduce to the two equations

$$0 = \varepsilon A_1 + \xi_0^2 \cos^2\theta_1\, \partial_x^2 A_1 - g_0[A_1^2 + G_{12}A_2^2]A_1, \quad (8.30\text{a})$$

$$0 = \varepsilon A_2 + \xi_0^2 \cos^2\theta_2\, \partial_x^2 A_2 - g_0[A_2^2 + G_{12}A_1^2]A_2, \quad (8.30\text{b})$$

with $A_1(x)$ and $A_2(x)$ real functions that tell us how the magnitudes of the two sets of stripes vary with position. A full solution of these equations can be obtained numerically.[9] However, we can guess the nature of the solutions from the form of the equations. Let us put the position of the grain boundary at $x = 0$. Then, since for $G_{12} > 1$ the nonlinear coupling suppresses A_1 where A_2 is large, and suppresses A_2 where A_1 is large, we expect that A_1 will grow from zero to its saturated value $\sqrt{\varepsilon/g_0}$ over some healing region around $x = 0$, where also A_2 decreases from $\sqrt{\varepsilon/g_0}$ to zero. The width of the healing region can be estimated as about $\varepsilon^{-1/2}\xi_0$, provided that θ_1 and θ_2 are not too close to $\pi/2$. This is verified by the numerical solution of these equations shown in Fig. 8.6.

The grain boundary formed with one set of stripes perpendicular to the line of the boundary Fig. 4.12(b) is a special case. (Let us take this to be the second set of stripes, so that $\theta_1 = 0$ and $\theta_2 = \pi/2$.) This case is special since the coefficient of the ∂_x^2 in the amplitude equation for the perpendicular stripes disappears, and the higher-order fourth derivative terms must be introduced to get a smooth solution. We can also no longer use the phase unwinding argument to show that A_2 must be real. However, we can arrive at the same result by the following argument. The phase

---

[9] This is easy to do in a symbolic package such as Mathematica. A useful hint if you try this is to solve the time-dependent amplitude equation and look at the x-dependence after a sufficient time for the dynamics to effectively cease. Of course, the evolution must be done in a finite domain, large enough that the amplitudes saturate on either side of the grain boundary.

8.2 Grain boundaries

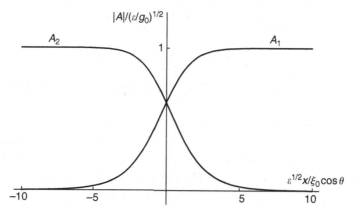

Fig. 8.6 Plot of the magnitudes of the amplitudes $|A_1|$ and $|A_2|$ in a grain boundary. We have chosen to plot the symmetric case for which the stripes make angles $\pm\theta$ with respect to the boundary. When the axes are scaled as in the figure, the plot only depends on the interaction parameter G_{12}. We have used the value $G_{12} = 1.5$ to plot the figure.

winding argument on the amplitude equation for A_1 still shows that $q_1 = q_c$. We use the potential to argue that the grain boundary will only be stationary for $q_2 = q_1$ so that, again, both wave numbers must be at the critical value for a stationary solution. The solution for the amplitudes in this case encounters some delicacies which we describe in the following Etude.[10]

Etude 8.4 Perpendicular grain boundary

We take the setup as in Fig. 8.5 but with $\theta_1 = 0$ and $\theta_2 = \pi/2$. We have argued that the grain boundary is stationary when the wave numbers of the states on either side are at critical. This means that the grain boundary may be described in terms of real amplitudes A_1 and A_2 of the stripes parallel and perpendicular to the boundary, respectively. The stationary grain boundary is then described by the two coupled amplitude equations

$$\varepsilon A_1 + \xi_0^2 \partial_x^2 A_1 - g_0(A_1^2 + GA_2^2)A_1 = 0, \qquad (8.31a)$$

$$\varepsilon A_2 - \frac{\xi_0^2}{4q_0^2} \partial_x^4 A_2 - g_0(A_2^2 + GA_1^2)A_2 = 0, \qquad (8.31b)$$

---

[10] We learned this illustrative application of the amplitude equation from Boris Shraiman.

with $G = G(\pi/2)$. We introduce the usual scalings, $X = \varepsilon^{1/2}x/\xi_0$ and $\bar{A}_i = (\varepsilon/g_0)^{-1/2}A_i$, which reduce Eqs. (8.31) to the form

$$\bar{A}_1 + \partial_X^2 \bar{A}_1 - (\bar{A}_1^2 + G\bar{A}_2^2)\bar{A}_1 = 0, \quad (8.32a)$$

$$\bar{A}_2 - \alpha \partial_X^4 \bar{A}_2 - (\bar{A}_2^2 + G\bar{A}_1^2)\bar{A}_2 = 0, \quad (8.32b)$$

where $\alpha = \varepsilon/(4q_0^2)$. Since α is a small number for ε small and all the other coefficients in the scaled amplitude equations (8.32) are of order unity, it might appear at first sight that we should be able to ignore the $\partial_X^4 \bar{A}_2$-term in Eq. (8.32b). Then $\bar{A}_2$ may be calculated algebraically in terms of $\bar{A}_1$ (this is an example in which one field $\bar{A}_2$ adiabatically follows another field $\bar{A}_1$) to give

$$\bar{A}_2^2 = \begin{cases} 1 - G\bar{A}_1^2, & \text{for } \bar{A}_1^2 \leq 1/G, \\ = 0, & \text{for } \bar{A}_1^2 > 1/G. \end{cases} \quad (8.33a)$$

Let us set the point where $\bar{A}_2$ first goes to zero and $\bar{A}_1^2 = 1/G$ as the origin $X = 0$. The equation for $\bar{A}_1$ is given by substituting these expressions into Eq. (8.32a) to yield

$$\partial_X^2 \bar{A}_1 + (1 - G)\bar{A}_1 - (1 - G^2)\bar{A}_1^3 = 0 \quad \text{for } X \leq 0, \quad (8.34)$$

$$\partial_X^2 \bar{A}_1 + \bar{A}_1 - \bar{A}_1^3 = 0 \quad \text{for } X \geq 0. \quad (8.35)$$

The solutions for $\bar{A}_1$ in the two regions must satisfy the matching conditions that $\bar{A}_1$ and $\partial_X \bar{A}_1$ are continuous at $X = 0$ since otherwise the second derivative is infinite and Eq. (8.32a) will not be satisfied at this point. The solutions can be found analytically

$$\bar{A}_1 = \begin{cases} \sqrt{\frac{2}{1+G}} \, \text{sech}\left[\sqrt{G-1}(X - X_1)\right], & X \leq 0, \\ \tanh\left[(X - X_2)/\sqrt{2}\right], & X \geq 0. \end{cases} \quad (8.36)$$

The integration constants X_1 and X_2 are fixed by the two continuity conditions at $X = 0$

$$\bar{A}_1(X = 0^-) = \bar{A}_1(X = 0^+), \quad (8.37)$$

$$\partial_X \bar{A}_1(X = 0^-) = \partial_X \bar{A}_1(X = 0^+). \quad (8.38)$$

This gives the solution for $\bar{A}_1$ and then Eqs. (8.33) give $\bar{A}_2$.

The solution for $\bar{A}_1$ and $\bar{A}_2$ are sketched in Fig. 8.7 with $\bar{A}_2$ going to zero at $X = 0$ as shown by the dashed line. This solution is not entirely satisfactory since the derivative of $\bar{A}_2$ is discontinuous at the origin, and Eq. (8.32b) is not satisfied here. In the language of boundary layer theory, such a solution is called the outer

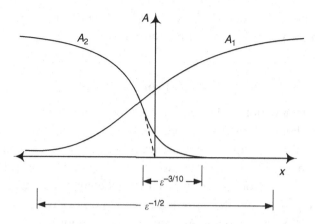

Fig. 8.7 Amplitudes A_1 (parallel stripes) and A_2 (perpendicular stripes) in the perpendicular grain boundary as a function of the coordinate x perpendicular to the grain boundary. The dashed line shows the singular behavior of A_2 on the "outer" scale of $\varepsilon^{-1/2}$. The full curve for A_2 shows the smooth behavior on the shorter "inner" scale $\varepsilon^{-3/10}$.

solution. The discontinuity must be resolved on an inner scale that is shorter than the $\varepsilon^{-1/2}\xi_0$ length scale of the X scaling. This is done by including the ∂_X^4 term that, so far, has been neglected from Eq. (8.32b) but which is clearly large near $X = 0$ where $\bar{A}_2$ is varyingly rapidly. On the other hand, the terms nonlinear in $\bar{A}_2$ are small here and can be neglected. Also, in the vicinity of the origin we can expand the coefficient of $\bar{A}_1$ in the last term of Eq. (8.32b) as

$$1 - G\bar{A}_1^2 = -\beta X + \cdots, \tag{8.39}$$

with β a coefficient of order unity that we can obtain from the $\bar{A}_1$ solution. Then for small X, Eq. (8.32b) becomes

$$\alpha \partial_X^4 \bar{A}_2 + \beta X \bar{A}_2 = 0. \tag{8.40}$$

Solution of this equation now shows that the kink in $\bar{A}_2$ is smoothed out on an X-length scale $(\alpha/\beta)^{1/5}$, as can be seen by a simple rescaling of the variables in the equation. Translating back to the original, unscaled coordinates, this means that the initial growth of A_2 away from zero occurs on a length scale that varies as $\varepsilon^{-3/10}$ near threshold. It is surprising to see the power law of 3/10 appearing when the obvious scalings give $\varepsilon^{-1/4}$ or $\varepsilon^{-1/2}$. This provides us with a useful lesson, that the actual variation of the solution might not follow our naive expectations!

We could now go on to look at the dynamics of grain boundaries when the criteria on the two wave vectors for time independence are not satisfied. For example, if the wave numbers differ across the grain boundary we could use the potential

296   *Defects and fronts*

argument analogous to Section 8.1.2 to study motion of the front. We will leave this calculation for you to explore in Exercise 8.5.

8.3 Fronts

We now turn our attention to fronts. We take the simplest case of a stripe state, with stripes parallel to the y-axis propagating in the x-direction into the unstable uniform state, which we suppose exists toward positive x. We suppose a (diffuse) boundary between stripe and uniform states that is a straight line also parallel to the y-axis – this is the front. Near threshold, the behavior can be analyzed using the amplitude equation and the amplitude $A(x, t)$ corresponding to this situation is shown in Fig. 8.8. In the amplitude equation description, there are solutions in which the front propagates without change of shape and at constant speed.[11] For steady propagation at fixed shape, we look for solutions that depend on x and t only through a comoving coordinate

$$\xi = x - ct, \tag{8.41}$$

with c the propagation speed, which is an unknown quantity to be determined from the equations. Introducing the comoving coordinate reduces a pde in x and t (with no dependence on y assumed) to an ode in ξ.

8.3.1 Existence of front solutions

Much of our understanding of front propagation derives from rigorous mathematical work in the 1970s on a class of nonlinear diffusion equations. Since the amplitude

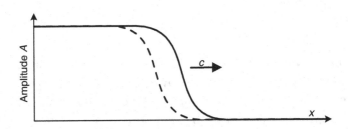

Fig. 8.8 A propagating front in the amplitude equation. For large positive x, the system is in the unstable $A = 0$ state. If an initial disturbance is localized at some large negative x, it will lead after some transient to a front as shown that propagates at a speed c. The dashed line shows the front at an earlier time.

---

[11] More generally, the front is moving in the periodic field of the stripes it establishes and so the motion will be jerky. This is another nonadiabatic effect that is not captured by the perturbation expansion in ε that leads to the amplitude equation.

equation restricted to real values of the amplitude falls within the class of equations considered in that work, it is useful first to understand this restricted class of real solutions. Note that if the $A = 0$ state, unstable for $\varepsilon > 0$, is perturbed by a real perturbation, the solution will remain real for all times. Thus this initial discussion tells us about the propagation of the $A \neq 0$ state from real localized initial conditions. This initial condition seems simple to engineer within the amplitude equation but physically corresponds to a small region of stripes precisely at the critical wave number, which would be hard to arrange experimentally. Thus later we will also want to investigate the front behavior when the initial data for A are complex-valued.

Fronts in the nonlinear diffusion equation

We start by considering the mathematical properties of front propagation in the one-dimensional amplitude equation restricted to real values, which we write in the form

$$\partial_t u = \partial_x^2 u + F(u). \tag{8.42}$$

For convenience, we have included the linear growth and nonlinear saturation terms in the real function $F(u)$, and we have chosen length and time scales to eliminate unnecessary constants, as in Eq. (7.10). We write u rather than the scaled amplitude $\tilde{A}$ to remind ourselves that we are dealing with a real variable. Equation (8.42) arises in many other contexts as well, and so our discussion of fronts here will also be useful in other situations, for example in Chapter 11 when we discuss signal propagation in excitable media.

To illustrate the general properties of fronts, we take the function $F(u)$ to be

$$F(u) = \varepsilon u - g u^3 - h u^5, \tag{8.43}$$

with $g = \pm 1$. The uniform $u = 0$ state is stable for $\varepsilon < 0$ and is unstable for $\varepsilon > 0$. For $g = 1$, the equation corresponds to the scaled amplitude equation for a supercritical bifurcation, the term in u^5 plays an inessential role, and we may put the coefficient of this term to zero, $h = 0$. On the other hand, for $g = -1$, the u^3 nonlinearity enhances the growth, the equation corresponds to a subcritical bifurcation at $\varepsilon = 0$, and we include a quintic term $-hu^5$ with $h > 0$ to saturate the growth at a finite value. The nonlinear states and their stability are shown in Fig. 8.9.

We seek the solution $u(x, t)$ for a front connecting the saturated nonlinear state $u = u_s$ (where u_s satisfies $F(u_s) = 0$) to the $u = 0$ state. For a steadily moving front solution, u is a function of the single comoving coordinate given by Eq. (8.41)

$$u(x, t) = u(\xi), \quad \xi = x - ct. \tag{8.44}$$

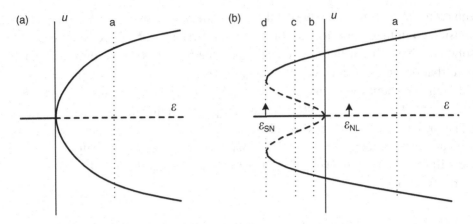

Fig. 8.9 Steady state solutions of the real amplitude equation, Eqs. (8.42) and (8.43), for (a) a supercritical bifurcation ($g = 1$), and for (b) a subcritical bifurcation ($g = -1$). Solid and dashed lines denote respectively stable and unstable solutions. The dotted vertical lines indicate values of ε for which the fronts are constructed in Fig. 8.10. Also shown is the value of $\varepsilon = \varepsilon_{NL}$ for which the front selection changes from "pulled" to "pushed," see Section 8.3.2.

Equation (8.42) then reduces to the ode

$$d_\xi^2 u + c\, d_\xi u + F(u) = 0. \tag{8.45}$$

Front solutions with the saturated nonlinear state toward large negative x and with the uniform state toward large positive x correspond to the boundary conditions

$$u(\xi \to -\infty) = u_s \quad \text{and} \quad u(\xi \to +\infty) = 0. \tag{8.46}$$

For positive c, the domain containing the nonlinear state $u = u_s$ grows at the expense of the $u = 0$ state.

Although Eq. (8.45) is a complicated nonlinear differential equation, it turns out we can get an almost complete qualitative understanding in terms of a simple "ball with friction in a potential" analogy. Consider the substitutions $\xi \to T$, $u \to X$, and $F(u) \to d\Phi/du$. Equation (8.45) then maps into the dynamical equation for the "position" X of a unit mass fictitious particle as a function of "time" T moving in a "potential" $\Phi(X)$ and acted on by a "frictional damping" with strength c. The form of Φ is sketched in Fig. 8.10 for the values of ε labeled a–d in Fig. 8.9(b). Please remember that this mapping of the front structure as a function of ξ into particle dynamics as a function of T is a mathematical analogy that helps us to solve for $u(\xi)$. The "dynamics" $X(T)$ does *not* describe the actual dynamics of the front $u(x, t)$ with physical time t.

Consider first the case $g = 1$ of the supercritical bifurcation, for which the nonlinear state exists only for $\varepsilon > 0$ (see Fig. 8.9(a)). Then the front is constructed

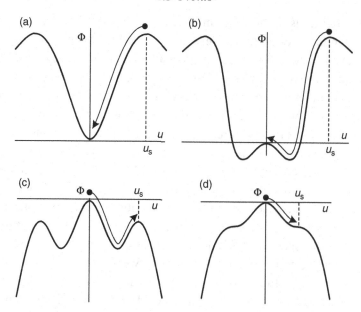

Fig. 8.10 Ball-in-potential analog for constructing front solutions. Panels (a)–(d) correspond to values of ε indicated in Fig. 8.9. In (a) and (b), the starting point is the stable stationary nonlinear state u_s, corresponding to a front propagating into the unstable $u = 0$ state. In (c) and (d) the ball trajectory is from $u = 0$ to u_s, and corresponds to a front propagating in the reverse direction.

as the dynamics of the fictitious particle starting ($T \to -\infty$) at the potential maximum at $u = u_s$, and ending ($T \to +\infty$) at the potential minimum at $u = 0$, as in Fig. 8.10(a). Our knowledge of particle motion tells us that we can construct a solution for any positive damping constant c. (We could construct a mathematical proof by using an energy argument on our fictitious particle.) This translates into the statement that a front solution can be constructed for any positive translational velocity. We can also learn about the approach of the front solution to $u = 0$ from the damped harmonic dynamics of the fictitious particle in the parabolic minimum. It is easily checked by linearizing for small u that the fictitious particle dynamics is over damped for $c > 2\sqrt{\varepsilon}$, giving a monotonic approach of the front to $u = 0$, but is under damped for $c < 2\sqrt{\varepsilon}$, translating into an oscillatory approach to $u = 0$. The borderline value $c = 2\sqrt{\varepsilon}$ turns out to play a special role in the subsequent theory.

The conclusions for the subcritical bifurcation $g = -1$ are the same for $\varepsilon > 0$. However for $\varepsilon_{SN} < \varepsilon < 0$, with ε_{SN} the value of ε at the saddle-node (see Fig. 8.9), the front is propagating into the stable $u = 0$ state, which corresponds to a maximum of the particle potential function $\Phi(u)$ (see Fig. 8.10(b)). It is now apparent that there is a unique value of the damping c (which can be found by a numerical integration) for which the fictitious particle starting at the maximum of Φ at $u = u_s$ will just climb the potential to end up precisely at the second maximum of Φ at $u = 0$. It is also clear that the fictitious particle must travel from the higher to the lower potential if c is positive. As we reduce ε from zero, we reach a point at which the relative heights of the maxima reverse. Now we get a trajectory with $c > 0$ starting from the $u = 0$ maximum and ending at the $u = u_s$ maximum, corresponding to a front propagating in the reverse direction, from the uniform state into the saturated state, Fig. 8.10(c). A special case occurs at the saddle-node point $\varepsilon = \varepsilon_{SN}$, where the stable and unstable nonlinear solutions collide and disappear. This leaves an inflection point in the function $\Phi(u)$, as in Fig. 8.10(d). With a little more effort, it can be shown that the trajectory from the maximum of Φ to this inflection point exists for sufficiently large damping $c > c_{SN}$, where c_{SN} is the limit of the unique front speed as ε approaches the saddle-node point from above $\varepsilon \to \varepsilon_{SN}$.

As a final remark, notice that for the front between two stable states, as in the case of the subcritical bifurcation for $\varepsilon < 0$, the direction of the propagation can be used as a way to label which state is selected or "preferred." This preference is only relevant in the context of the particular dynamics of front propagation, and other dynamics (e.g. the growth from small initial conditions spread over all space or survival in the face of random noise) may lead in general to a different selection.

Fronts in more general equations

If you review the preceding results, you will find that they are consistent with the following general statement. For a front propagating into an unstable state, uniformly propagating solutions may be constructed for a continuum of possible speeds (in the case we worked out, the possible speeds range over all positive values). On the other hand, for a front propagating into a stable state, there is a discrete set of propagation speeds fixed by the parameters (in our case a single value). The results stated in this way are found to extend to a wide range of equations far beyond our simple demonstration, for example to the amplitude equation with a complex amplitude, to generalized Swift–Hohenberg models that allow a subcritical bifurcation, and to many other examples.

To investigate the results for more general equations, we use the same technique of introducing the comoving coordinate $\xi = x - ct$. For equations such as the Swift–Hohenberg equation, in which the front lays down a spatially periodic state, the ansatz for the front solution must be generalized to be periodic in time in the moving frame, and the front motion itself will have a periodic component leading to a "jerky" motion. Mapping the equation for the ξ dependence of u onto a fictitious dynamical system remains useful, although the resulting dynamical system is, in general, more complicated than the ball-in-potential model since it will involve higher derivatives or more dynamical variables,[12] and the behavior for $\xi \to \pm\infty$ may correspond to a limit cycle of the dynamical system rather than to a fixed point.

Nevertheless, tools from dynamical systems theory that focus on the qualitative structure of the dynamical trajectories allow the same type of results to be derived. The fronts as a function of ξ correspond to heteroclinic dynamical trajectories that join the fixed points or limit cycles, and the task is then to understand when such heteroclinic orbits exist, for example for what ranges or values of the speed c. The key part of the argument that derives restrictions on values of c is to count the degrees of freedom of the front trajectory $\mathbf{u}(\xi)$ as this trajectory leaves or approaches the fixed points or limit cycles in the limit $\xi \to \pm\infty$. The behavior near the asymptotic values is given by linearizing about these values, which provides a tractable calculation. We illustrate the idea of this type of calculation in the following Etude, where we look for fronts in an amplitude equation with complex-valued solutions, although the analysis is too complicated for us to take it to completion (see also Exercise 8.6).

Etude 8.5 Complex fronts in the amplitude equation

We use the fully scaled form of the amplitude equation, Eq. (6.21). For the case of complex solutions, the ansatz for the steadily moving front requires some careful consideration. Although we expect the boundary between zero and the saturated magnitude to propagate steadily with speed c, the stripes far behind the front should be stationary and so the phase Φ of the complex amplitude should be constant there. Thus it is not correct to look for solutions in which the phase is just a function of the comoving coordinate ξ. Instead, we look for solutions for the scale amplitude in the form

$$\bar{A} = e^{iK_0 cT} a(\xi) e^{i\Phi(\xi)}, \tag{8.47}$$

with K_0 the (scaled) deviation of the wave vector of the stripes from critical for $\xi \to -\infty$ and $c > 0$. Notice that the solution ansatz is now periodic in the moving frame. The time dependence of the phase will be eliminated for $\xi \to -\infty$ by the boundary condition $d\Phi/d\xi|_{\xi \to -\infty} = K_0$. The evolution equation for $\bar{A}$ is reduced to a dynamical system of three first-order equations by introducing the fictitious time ξ,[13] and by defining variables

$$a = |\bar{A}|, \quad \kappa = |\bar{A}|^{-1} d|\bar{A}|/d\xi, \quad \text{and} \quad K = d\Phi/d\xi, \tag{8.48}$$

---

[12] You may remember from a course on dynamical systems that an equation of motion involving derivatives with respect to t up to order n can be replaced by a set of n equations that involve only first-order derivatives, by introducing supplementary variables such as $v_1 = du/dt$, $v_2 = dv_1/dt$, etc.
[13] We do not introduce the notation T for the fictitious time, since this symbol is already used for the scaled time of the amplitude equation.

with Φ the phase of $\bar{A}$. After some algebra, this gives the three odes

$$\dot{a} = \kappa a, \tag{8.49}$$

$$\dot{\kappa} = -c\kappa - 1 - \kappa^2 + K^2 + a^2, \tag{8.50}$$

$$\dot{K} = -c(K - K_0) - 2\kappa K, \tag{8.51}$$

where the dot denotes the derivative $d/d\xi$ with respect to the fictitious time. Note that the phase does not appear in these equations. Setting the time derivatives to zero gives several fixed point solutions. The fixed point

$$\kappa = 0, \quad a^2 = 1 - K^2, \quad K = K_0, \tag{8.52}$$

corresponds to the nonlinear saturated solution of the amplitude equation, while the two fixed points

$$a = 0, \quad K = K_\pm(c, K_0), \quad \kappa = \kappa_\pm(c, K_0), \tag{8.53}$$

correspond to the unstable uniform state, which we call the linear fixed points. Here $K_\pm$ and $\kappa_\pm$ are complicated functions of c and K_0 that can be written down analytically, but are too long to include here. To apply to a front that decays as $\xi \to \infty$, we require that the solutions for a, K, and κ be real, and that κ be negative.

The possibility of a heteroclinic connection between the two fixed points is analyzed by studying how the trajectory leaves the fixed point Eq. (8.52) as ξ increases from $-\infty$, and approaches one of the two fixed points Eq. (8.53) for $\xi \to \infty$. A linear stability analysis of the fictitious dynamical equations Eqs. (8.49) about the fixed point Eq. (8.52) shows that there is one unstable direction with positive eigenvalue, and two stable directions with negative eigenvalues. The heteroclinic trajectory departs from the fixed point along the unique unstable direction. Given this initial condition, a unique trajectory is defined by integrating forwards in time. Now we must address whether this trajectory will eventually arrive at one of the fixed points given by Eq. (8.53).

Suppose a linear stability analysis about the linear fixed point were to give one unstable direction, and two stable ones. If the trajectory from the nonlinear fixed point arrives in the vicinity of the linear fixed point, there will generally be some component of the deviation from the fixed point along the unstable direction, and this will grow as time increases. This means the trajectory will not approach the fixed point at long times, unless the component along the unstable direction can be eliminated by tuning some available parameter. Since there is a single trajectory leaving the nonlinear fixed point, we would have to tune c or K_0 to construct a heteroclinic orbit. This would lead to a single parameter family of front solutions, where we could vary say K_0 but then $c(K_0)$ is fixed by the tuning requirement.

In fact, the linear stability analysis about the fixed points Eq. (8.53) shows that all the eigenvalues at one of the two linear fixed points are stable.[14] *This means that if the trajectory leaving the nonlinear fixed point arrives in the vicinity of this linear fixed point, subsequent evolution will be an approach to the fixed point. In particular, if we know of a heteroclinic connection for some value of c and K_0, then in general we can change these values gently without eliminating the connection. The argument cannot be used to predict whether a front solution exists at all – we cannot guarantee that the trajectory leaving the nonlinear state in fact ever gets near the linear fixed point. However, if a front solution exists then it is necessarily a member of a two parameter family of fronts, in the sense that fronts should exist over continuous ranges of two parameters, namely c and K_0. The "ball-in-potential" argument for the special case of Eq. (8.42) provides more complete, global information.*

8.3.2 Front selection

For propagation into an unstable state, seeking a uniformly moving front solution does not specify the front velocity c since we are left with a continuous range of possible values, with no obvious reason to prefer one value over the other. Over the past few decades, the front selection problem has aroused much interest and the analysis has turned out to be unexpectedly subtle. Rigorous results are known mainly for the real, nonlinear diffusion equation, Eq. (8.42), for which it has been shown that, for sufficiently localized positive initial conditions and with some restrictions on the form of $F(u)$ in Eq. (8.43), the front velocity approaches the value $c_0 = 2\sqrt{\varepsilon}$ in the long-time limit. Thus although it is possible to set up by hand a family of uniformly propagating fronts, only a single one is accessible from localized initial conditions. The rigorous argument answers the question of velocity selection for fronts in the amplitude equation for the supercritical case, and for initial perturbations that are real and positive everywhere, but says nothing about the subcritical case, where $F(u)$ does not satisfy the constraints of the theorem, nor about the propagation from more general initial conditions. The methods used in the rigorous approach, which involves bounding one solution by another, do not generalize to these more general situations or to many other equations encountered in the study of patterns. This motivates the search for a more phenomenological approach that can be generalized to these cases.

There is now a consensus that two types of front-velocity-selection mechanisms operate. In the first case, which we will call pulled fronts,[15] the speed of the front

---

[14] At least, this is the result for small K_0, where the behavior of the fixed points can be analyzed completely.
[15] We are following the terminology of van Saarloos [109]

is completely determined by the behavior of the leading edge where the amplitudes are small. In this case, the properties of the front are given by a linear analysis and the selection mechanism is also called linear selection or linear marginal selection. We think of the complete nonlinear front as being "pulled" by the small-amplitude leading edge. For the second case that we will call pushed fronts, the full behavior of the nonlinear equations is needed to understand the front, i.e. the front is "pushed" by the large-amplitude region at and behind the front.

Pulled fronts

There have been many alternative rationalizations to justify the expressions for the speed of pulled fronts.[16] The simplest is a stationary phase analysis that goes back to Kolmogorov, Petrovsky, and Piskunov in 1937, and a related method known as the pinch-point analysis due to Lifshitz and Pittaevskii, that was developed in the context of plasma physics.

In the stationary phase approach, we assume – without very good justification – a completely linear analysis. In this case, the growth from any small initial condition $u_0(x)$ is given by a superposition of the linear Fourier modes with wave numbers q growing at the complex rates $\sigma(q)$, and with initial amplitudes given by the Fourier transform of the initial condition. This argument gives

$$u(x,t) = \frac{1}{2\pi} \int_{-\infty}^{\infty} dq \, e^{iqx+\sigma(q)t} \int_{-\infty}^{\infty} dx' \, u_0(x')e^{-iqx'}. \tag{8.54}$$

Here, the second integral gives the strength $\tilde{u}_0(q)$ of the component at wave number q in the original state, and then this strength is evolved with time and resummed in the first integral to give the field at a later time. This expression is certainly not correct over the whole space-time domain since we know that the nonlinear saturated solution occupies a growing region. However, the expression may be adequate if the solution far in the leading edge is dominated by the linear behavior.

We now evaluate $u(x,t)$ in Eq. (8.54) at a moving point $x = vt$, where v is a velocity as yet unspecified:

$$u(x=vt,t) = \frac{1}{2\pi} \int_{-\infty}^{\infty} dq \, e^{[iqv+\sigma(q)]t} \int_{-\infty}^{\infty} dx' \, u_0(x')e^{-iqx'}. \tag{8.55}$$

The integration over q can be evaluated asymptotically by considering the integral along the real q-axis to be a contour integral in the complex-q plane and then by moving the integration contour into the upper complex-q plane. For large time, the integral can be estimated from the contribution at the stationary phase point of the

---

[16] This section requires a knowledge of contour integration of complex functions and may be skipped if you are not familiar with that idea.

integrand, which is at the complex wave number $q = q_s$ given by the solution of

$$\frac{d}{dq}[iqv + \sigma(q)] = 0, \tag{8.56}$$

or

$$v = i\left.\frac{d\sigma}{dq}\right|_{q=q_s}. \tag{8.57}$$

This complex equation (two real equations) fixes the stationary value q_s (which is also complex in general) in terms of the real parameter v. Estimating the integral from the value of the integrand at the stationary phase point gives

$$u(x = vt, t) \sim \exp[(iq_s v + \sigma(q_s))\, t]. \tag{8.58}$$

The propagation speed of the front is then given by finding the value v of the moving frame in which the magnitude is constant, i.e. by setting the real part of the exponent to zero. Hence the front speed c is given by the value of v satisfying

$$v = \frac{\operatorname{Re}\sigma(q_s)}{\operatorname{Im} q_s}. \tag{8.59}$$

Equations (8.57) and (8.59) together fix the front speed and the asymptotic spatial dependence given by q_s and Eq. (8.58).

Let us look at the application of these expressions to the amplitude equation Eq. (7.10). The growth rate is

$$\sigma(q) = \varepsilon - q^2. \tag{8.60}$$

The two conditions to be solved together for the propagation speed c, Eqs. (8.57) and (8.59), become

$$c = -2iq_s \quad \text{and} \quad c = i\frac{\varepsilon - q_s^2}{q_s}. \tag{8.61}$$

Solving these yields $q_s = i\sqrt{\varepsilon}$ and then $c = 2\sqrt{\varepsilon}$. The speed agrees with the rigorous result for the real nonlinear diffusion equation.

Historically, other arguments have been made that lead to the same results in many cases. For example, the idea of "marginal stability" posits that the selected speed is the one that leads to the front that is on the edge of becoming unstable in the comoving frame, while the idea of "structural stability" demands that the selected front be insensitive to delicate changes to the equation that give a tiny jump as the final approach to $u = 0$. We refer the interested reader to the research literature to learn more.

The selected front speed is reached only for sufficiently localized initial conditions. We can see this from the derivation by noting that the second integral in

Eq. (8.55), the transform of the initial condition, will only converge if the fall off in the initial condition $u_0(x' \to \infty)$ overcomes the exponential growth $e^{\mathrm{Im}(q_s x')}$ of the last term. Thus the initial condition must fall off at least as fast as the exponential decay of the leading edge of the selected front. It is easy to see that the exponential growth of an initial condition with a shallower decay will appear as a faster propagating front.

An interesting extension of the analysis is to study the approach of the front speed $v(t)$ to its asymptotic value c for a localized initial condition. It is found that v approaches c from below with a slow $1/t$ decay

$$v(t) = c - \frac{3}{2 \,\mathrm{Im}\, q_s} \frac{1}{t} + O\left(\frac{1}{t^{3/2}}\right). \tag{8.62}$$

This result helps to explain some early experimental results on front propagation in Taylor–Couette rolls where velocities less than the expected value c were measured.

Pushed fronts

The linear selection principle is not valid for all equations and there is a second selection principle which sometimes takes over. Since the full nonlinear structure is needed to understand the selected front, we can think of these fronts as being "pushed" from behind. The mechanism is also known as nonlinear selection.

We can motivate the idea that the linear selection mechanism might break down from what we have so far learned by studying the subcritical real amplitude equation, Eqs. (8.42) and (8.43), with $g = +1$ as shown in Fig. 8.9(b). For $\varepsilon > 0$, a front propagates into the unstable $u = 0$ state and the linear selection principle suggests the speed $c = 2\sqrt{\varepsilon}$, which tends to zero as $\varepsilon \to 0$. On the other hand, for $\varepsilon < 0$, there is a unique front speed that remains nonzero as $\varepsilon \to 0$ (consult the particle-in-potential analogy). A discontinuity of behavior at $\varepsilon = 0$, where the stability properties of the $u = 0$ solution are changing smoothly, would be rather surprising. Numerical experiments in fact show a continuous behavior, with the front velocity deviating from the linearly selected value for values of ε below some particular value $\varepsilon_{\mathrm{NL}}$, as in Fig. 8.11.

The nonlinear selection works as follows. Since the amplitude equation in the comoving coordinates involves second derivatives in ξ, the front solution will approach $u = 0$ as the sum of two exponentially decaying pieces. For large ξ, the exponential with the slowest decay rate will dominate. This agrees with the prediction of the asymptotic dependence of the selected front given by the stationary phase argument. However, by tuning the speed c it may be possible to find a front solution that matches onto the small amplitude solution that is comprised of only

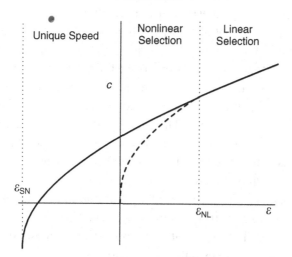

Fig. 8.11 Selected front speed c for the subcritical real amplitude equation (solid line). For $\varepsilon > 0$, linear selection gives a speed that goes to zero as $\varepsilon^{1/2}$ for small ε (dashed line). For $\varepsilon < 0$, there is a unique speed that remains nonzero as $\varepsilon \to 0$. A nonlinear selection mechanism gives a curve that interpolates between these two results for $0 < \varepsilon < \varepsilon_{NL}$.

the rapidly decaying exponential.[17] This will occur at a discrete value $c = c_{NL}$, if at all. Now we compare the speed of this front with the speed given by the linear selection c_L. It seems likely that if $c_{NL} > c_L$, then this is the front that will propagate from localized initial conditions, and numerical simulations confirm this. Thus the criterion for a pushed front is that both the decay rate of the special front is faster than that of the front given by linear selection *and* that its speed is faster. For the subcritical real amplitude equation the pushed front takes over as ε is decreased below a value of $\varepsilon_{NL} > 0$, and the pushed front speed c_{NL} is continuous at $\varepsilon = 0$ with the unique front speed that holds for $\varepsilon < 0$, Fig. 8.11. Pushed fronts are explored further in Exercises 8.6 and 8.8.

8.3.3 Wave-number selection

For the amplitude equation restricted to real values or for the nonlinear diffusion equation, the front velocity and shape are the only features to be determined. More general systems, such as the amplitude equation with complex solutions or the

---

[17] In terms of the type of analysis for heteroclinic connections in the fictitious dynamical system used in the Etude 8.5, this corresponds to the orbit approaching the alternative linear fixed point. Hints for a more complete analysis in this language are given in Exercise 8.6.

Swift–Hohenberg model, have a multiplicity of nonlinear states with different pattern wave numbers and the question of the selection between these states arises. In particular, we can ask what is the wave number of the patterned state laid down behind an advancing front from the propagation of the pattern solution into the unstable uniform state? Note that, for a front laying down a patterned state, the steady propagation ansatz is not adequate, and instead we must seek temporally periodic solutions in the moving frame, which leads to a more complicated analysis of the possible front solutions. The calculation for the fronts in the amplitude equation with complex solutions has already been given in Eq. (8.47). The stationary phase analysis can still be used to give the selected pulled-front speed, and pushed fronts can again be found in some cases.

A simple argument can be used to find the wave number of the pattern state laid down behind a pulled front. If the front moves with the velocity c, we see a time dependence Eq. (8.58). The exponential decay rate has been set to zero in deriving the front velocity, so this leaves only an oscillating phase

$$u \propto \text{Re} \exp[i(\text{Re}(q_s c) + \text{Im}\, \sigma(q_s))\, t]. \tag{8.63}$$

These oscillations establish a phase winding per unit length as the front advances, given by the temporal oscillation frequency divided by the front speed. If the phase winding is preserved through the growth to the saturated nonlinear solution, i.e. there are no additional nodes formed in u, this leads to the result for the wave number selected by front propagation q_{fp} in the nonlinear state behind the front

$$q_{\text{fp}} = \text{Re}\, q_s + c^{-1} \text{Im}\, \sigma(q_s). \tag{8.64}$$

For the amplitude equation with complex solutions, Eq. (8.64) leads to the rather uninteresting result of $q_{\text{fp}} = 0$. For other equations of pattern formation, such as the Swift–Hohenberg equation (see Exercises 8.9 and 8.10) or the complex Ginzburg–Landau equation introduced in Section 5.4 and discussed further in Section 10.2.2, the result Eq. (8.64) is nontrivial.

The validity of the node-conservation assumption is not a priori clear, but numerical experiments on many equations show that a well-defined wave number is indeed selected by front propagation, independently of the details of the initial conditions for example. The formation of patterns via front propagation is one of the few mechanisms of pattern formation that leads to a defect-free pattern with a precisely determined wave number. It is intriguing to wonder whether this mechanism is important in biological examples such as morphogenesis, where the pattern may have some functional significance.

8.4 Conclusions

A completely perfect pattern is a rigid object with little possibility of dynamics. Localized structures provide dynamical objects that can connect different patterns (e.g. ones of different wavelength) or that can create new patterns. In the present chapter, we have concentrated on some specific examples to illustrate the types of questions that arise, and have shown some of the answers to these questions that may be derived for systems near threshold using the amplitude equation approach developed in the previous two chapters. Another type of localized structure called a pulse – a small region of a pattern surrounded by the uniform state – is briefly described in the following chapter.

Localized structures are important far beyond the parameter regimes where the amplitude equation is valid. The existence of topological defects derives from the basic symmetry properties of the patterns and so these defects survive away from threshold. Pattern selection by defect dynamics is also to be expected away from threshold, although it is much harder to predict the selected state without the crutch of the amplitude equation. More general localized structures exist too. We have introduced fronts as boundaries between patterned and uniform states or between two patterned states. Fronts are important more generally as boundaries between any two nonequilibrium steady states that occupy different regions of space. In fact, patterns themselves can sometimes be built up from collections of fronts. For example, a stripe pattern can be constructed from a periodic array of fronts between a "black" and "white" state. We return to this idea in Chapter 11, where we construct periodic wave trains in reaction–diffusion systems as arrays of moving fronts, but the same method has also been used to develop theories of stationary stripe patterns in systems that are not near the linear instability of a uniform state.

8.5 Further reading

(i) A paper showing the anisotropic shape of a dislocation in a stripe state in an isotropic system is "The shape of stationary dislocations" by Meiron and Newell [72].
(ii) The full calculation of how dislocations move is complicated and you might want to refer to the original literature for the details. Bodenschatz *et al.* describe the uniaxial case in "Structure and dynamics of dislocations in anisotropic pattern-forming systems" [13], the potential method for climb in an isotropic system is discussed in "Dynamics of defects in Rayleigh–Bénard convection" by Siggia and Zippelius [96], and the paper "Climbing of dislocations in nonequilibrium patterns" by Tesauro and Cross [103] looks at the nonpotential situation away from threshold.
(iii) The study of front selection has a long history that is full of controversy. A recent review article "Front propagation into unstable states" by van Saarloos [109] provides an up-to-date account of the subject with references to the earlier work.

Exercises

8.1 **Structure of a dislocation:** Using a simple model for the dependence of the amplitude's magnitude $a(r)$ such as

$$a(r) = \tanh(r/\xi), \tag{E8.1}$$

plot in Mathematica, Maple or some other graphing program the reconstruction of the full pattern

$$U = \text{Re}[A(x, y)e^{ix}], \tag{E8.2}$$

for a dislocation in a uniaxial stripe system. (Note: since we have chosen the wave number of the base pattern to be 1, the healing length ξ should be chosen large compared to the wavelength 2π. Also, in using the $\tan^{-1}$ expression for ϕ, care must be used so that ϕ increases between 0 and 2π. The function ArcTan[x,y], which evaluates $\tan^{-1}(y/x)$, does this in Mathematica.)

8.2 **Motion of a dislocation:** By changing the amplitude dependence in Exercise 8.1 to $A(x - x_0, y - y_0)$ and by replotting for various choices of x_0 and y_0, study how the dislocation configuration changes as the center moves relative to the underlying stripes. Contrast the behavior for climb (keep $x_0 = 0$) with the behavior for glide (keep $y_0 = 0$).

8.3 **Phase grain boundary:** An example of a phase grain boundary is given by the exact solution of the amplitude equation Eq. (7.12)

$$\bar{A}(X, Y) = a(K_X, K_Y)e^{i\Phi(X,Y)}, \tag{E8.3}$$

where $\mathbf{K} = \nabla_X \Phi$ with

$$\Phi(X, Y) = -(Q^2/2)X + \ln[2\cosh(QY)], \tag{E8.4}$$

and

$$a(K_X, K_Y) = \sqrt{1 - (K_X + K_Y^2/2)^2}. \tag{E8.5}$$

(a) Find the wave vectors of the stripes as $X \to \pm\infty$ given by these equations, and hence argue that these equations do indeed describe a grain boundary.

(b) Show that the complex amplitude given by Eqs. (E8.3) and (E8.5) is a solution to the scaled amplitude equation Eq. (7.12).

(c) Use Mathematica or some other program to plot examples of phase grain boundaries given by Eqs. (E8.4) and (E8.5). You should reconstruct the full stripe solution by plotting $\text{Re}[A(X, Y)e^{iq_c x}]$ using various values of the parameter Q and some chosen value of ε to go between scaled and unscaled variables.

8.4 **Amplitude grain boundary:** For a symmetric amplitude grain boundary such as shown in Fig. 8.5 with $\theta_1 = \theta_2 = \theta$, determine how the width of the region of suppressed amplitude varies with θ. What do you predict for $\theta \to 0$ and $\theta \to \pi/2$? Do these results make sense physically for a grain boundary? If not, what do you expect the correct result to be?

8.5 **Motion of perpendicular grain boundary:** For the perpendicular grain boundary discussed in Section 8.2, set up the formalism analogous to the discussion of dislocation motion in Section 8.1.2. You should use the evolution of the potential to calculate the motion of the grain boundary if the stripes parallel to the boundary are at the critical wave number, but assume the perpendicular stripes have a wave number $q = q_c + k$ that deviate slightly from this value. Write down an expression for the drift velocity valid for small wave-number deviations, in terms of the solutions for the two amplitudes. For a small wave-number difference, the amplitudes can be approximated by the stationary solutions. How will the drift speed depend on $k \ll q_c$ and ε?

8.6 **Qualitative analysis of fronts in a nonlinear diffusion equation:** In this exercise you will investigate front propagation in the real nonlinear diffusion equation Eq. (8.42) with

$$F(u) = \varepsilon u + u^3 - u^5, \tag{E8.6}$$

using the qualitative methods discussed in Section 8.3.1. Note that this equation corresponds to an amplitude equation for a subcritical bifurcation at $\varepsilon = 0$ restricted to real solutions.

(a) Show that, for $\varepsilon > -1/4$, there is a fixed point $u = u_0 \neq 0$ and $v = 0$ that corresponds to the stable, saturated stationary nonlinear state of Eqs. (8.42) and (E8.6).

(b) Consider front solutions $u(\xi)$ with $\xi = x - ct$. Show that the equation for the front can be written as

$$u' = v, \tag{E8.7a}$$
$$v' = -F(u) - cv, \tag{E8.7b}$$

where primes denote $d/d\xi$. We will study Eqs. (E8.7) as a dynamical system with ξ acting as the fictitious time.

(c) Analyze the stability of the fixed point found in part (a) within the fictitious dynamical system Eqs. (E8.7), and show that it has one stable direction and one unstable direction for any positive value of c.

(d) For all values of c and ε, establish the stability of the $u = 0$ and $v = 0$ fixed point of the dynamical system Eqs. (E8.7) and the number of stable and unstable directions.

(e) For $\varepsilon > 0$ and $c > 2\sqrt{\varepsilon}$, sketch a trajectory in the uv plane corresponding to a "pulled" front of the nonlinear solution advancing into the $u = 0$ solution toward positive x, being careful to show the trajectory relative to the various eigenvectors near each fixed point. Use this to argue that if such a front exists, then it is a member of a continuous family.

(f) Sketch the trajectory corresponding to a "pushed front."

(g) By considering the trajectories linking the two fixed points for $-1/4 < \varepsilon < 0$, argue that the value of c must be tuned to find a possible front, i.e. there is a unique front propagation speed in this case.

(h) Can you show for $\varepsilon < 0$ that the speed of the pushed front for $\varepsilon > 0$ is continuous with the unique front speed?

8.7 **Complex fronts in the amplitude equation:** In this exercise, you will fill in some of the details for the Etude 8.5. To simplify the discussion, we will only consider the case of $c > 2$.

(a) Find the fixed point solutions of the dynamical system Eqs. (8.49) for the special case $k_0 = 0$.

(b) Show that the structure of the fixed points is as described in the Etude, namely that there is one unstable direction at the nonlinear fixed point, that there are two linear fixed points, and that all the directions are stable at one of these fixed points.

(c) Sketch the trajectory that corresponds to a front solution in the a, κ, k space.

We can now argue that the qualitative behavior is not going to change for small k_0 and hence that there is a *two-parameter family* of front solutions. (For $c < 2$, it turns out that some of the fixed points are marginal (real part of stability eigenvalue zero) for $k_0 = 0$ and so the stability of these fixed points at $k_0 \neq 0$ cannot be predicted from the $k_0 = 0$ results.)

8.8 **Analytic expression for a pushed front in the nonlinear diffusion equation:** It turns out that in some cases the pushed-front solutions can be obtained analytically. This exercise uses the analytic results for the real pushed fronts in the subcritical amplitude equation to illustrate the nature of pushed fronts. We will analyze the equation

$$\partial_t u = \partial_x^2 u + u + du^3 - u^5, \qquad (E8.8)$$

which is convenient for studying the behavior for both supercritical and subcritical regimes. The pulled-front solution takes the form

$$u_p(\xi) = \left(\exp\left(2\xi u_s^2/\sqrt{3}\right) + u_s^{-2}\right)^{-1/2}, \qquad (E8.9)$$

where $\xi = x - ct$ and u_s is the nonlinear saturated value satisfying

$$1 + du_s^2 - u_s^4 = 0. \tag{E8.10}$$

(a) Verify that $u_p(\xi)$ is a solution of Eq. (E8.8) with the propagation speed

$$c = \frac{2\sqrt{d^2 + 4} - d}{\sqrt{3}}. \tag{E8.11}$$

Note that c has a minimum value 2 at $d = \frac{2}{3}\sqrt{3}$.

(b) Show that the asymptotic decay of u_p for $\xi \to \infty$ is $e^{-\kappa x}$ with

$$\kappa = \frac{d + \sqrt{d^2 + 4}}{2\sqrt{3}}. \tag{E8.12}$$

(c) Show that a pulled front has a speed $c^* = 2$ and asymptotic decay rate given by $\kappa^* = 1$.

(d) Hence argue that the selected front will be a pushed front for $d > d_c = \frac{2}{3}\sqrt{3}$ because (a) the asymptotic fall off is faster than for the pulled front, and (b) the speed $c > c^*$.

(e) Plot the selected front speed as a function of d.

(f) For $d > 0$, show that Eq. (E8.8) can be mapped onto Eqs. (8.42) and (8.43) with rescaled space and time coordinates and $\varepsilon = d^{-2}$, $g = -1$, and $h = 1$. Hence construct the plot of the selected front speed as a function of ε for this system and compare with Fig. 8.11 for $\varepsilon > 0$.

(This problem was analyzed by van Saarloos [108].)

8.9 **Pulled fronts in the Swift–Hohenberg equation:** For the Swift–Hohenberg equation Eq. (2.4)

$$\partial_t u(x, t) = ru - (\partial_x^2 + 1)^2 u - u^3, \tag{E8.13}$$

the expressions

$$c = \frac{4}{3\sqrt{3}}(\sqrt{1 + 6r} + 2)(\sqrt{1 + 6r} - 1)^{1/2}, \tag{E8.14a}$$

$$q_r^* = \frac{1}{2}(\sqrt{1 + 6r} + 3)^{1/2}, \tag{E8.14b}$$

$$q_i^* = \frac{1}{2\sqrt{3}}(\sqrt{1 + 6r} - 1)^{1/2}, \tag{E8.14c}$$

give the velocity c of a pulled front and the complex wave number $q^* = q_r^* + iq_i^*$ that gives the stationary phase point.

(a) For small r, verify that these results agree with the results expected from the amplitude-equation approach.

(b) Show that the wave number q_{fp} of the pattern laid down by the front, using the conserved-nodes assumption is given by

$$q_{fp} = \frac{3(\sqrt{1+6r}+3)^{3/2}}{8(\sqrt{1+6r}+2)}. \qquad (E8.15)$$

(c) Expand q_{fp} in small r up to terms linear in r. Note that q_{fp} is *not* equal to the wave number that minimizes the potential for the Swift–Hohenberg equation (see Exercise 5.3).

8.10 **Derivation of the selection expressions for pulled fronts in the Swift–Hohenberg equation:** Derive equations Eqs. (E8.14) in the previous exercise.

9

Patterns far from threshold

We have approached the formation of patterns in nonequilibrium systems through the notion of states that develop via a supercritical linear instability and so saturate at small amplitudes near the threshold of the instability. The resulting patterns retain to some degree the features of the linearly growing mode and this allows many aspects of the pattern formation to be analyzed in a tractable way via the amplitude equation formalism. In nature, however, most nonequilibrium systems are not close to any threshold, and the amplitudes of their corresponding states cannot be considered small. Even near the threshold of a linear instability, structure can emerge via a subcritical bifurcation such that the exponential growth saturates with a large amplitude. What can be said about these strongly nonlinear patterns that are far from the linearly growing mode?

Experiments and simulations indicate that patterns far from threshold can be divided into two classes. One class qualitatively resembles patterns that, at least locally, take the form of stripes or lattices. The other class of patterns far from threshold involves novel states that do not correspond locally to lattice structures.

Far from threshold less can be said about stripe and lattice states with any generality. One question that can be addressed generally is the slow variation of the properties of the stripes or lattices over large distances in a sufficiently big domain. As we explain in Section 9.1.1 these slow dynamics are connected with symmetries of the system. Formally, the slow variation allows one to introduce a small parameter (the ratio of a lattice spacing to the length over which the structure varies slowly) which in turn allows one to carry out a perturbative analysis that leads to a "nonlinear phase diffusion equation" for the slowly varying phase of the lattice structure. The resulting formalism is not as universal as amplitude equations but still provides a valuable way to understand slowly varying spatial and temporal features of some states far from threshold. We discuss this formalism with some applications at a mainly qualitative level in Section 9.1.2, while Appendix A2.4 gives some of the mathematical details. In Section 9.1.3 we describe some extensions when the slow phase mode is coupled to other slowly varying dynamical variables.

The band of stable wave numbers of stripe and lattice states (the stability balloon, Section 4.2.1) can become large away from threshold, unlike near threshold where the band scales as $\varepsilon^{1/2}$. Experimentally, on the other hand, it is often found that the range of wave numbers in a pattern is significantly less than set by these stability considerations – the phenomenon of wave number selection. For applications to explain natural phenomena where the size of the structure plays a crucial role, such as morphogenesis in biology, the existence of wave number selection, the value of the wave number selected, and its robustness to changes in the system such as the size of the domain, are key issues in assessing the relevance of pattern-forming models. We have discussed wave number selection mechanisms at various places in previous chapters: in the present chapter, in Section 9.1.4, we collect these and other ideas together.

The other class of patterns far from threshold do not correspond locally to lattice structures and are less well understood. One possibility is localized structures rather than extended planforms, an example of which are the oscillons in Fig. 9.11, which are found over some parameter ranges when a thin layer of brass balls (a granular medium) is shaken vertically (see also the related Fig. 1.17). Another possibility is patterns whose spatial disorder is so strong that no local lattice structure can be usefully identified. (Fig. 9.9(b) later in this chapter is a numerical example.) Sufficiently far from the threshold of almost any linear instability, disordered structures are usually found to evolve chaotically, in which case the dynamics is called "spatiotemporal chaos."[1] Chaotic disordered states are perhaps the most challenging of nonequilibrium states to understand and represent an important frontier of nonequilibrium pattern formation.

As we discussed in Chapter 1 (see Fig. 1.20 and related text), most nonequilibrium systems cannot be driven arbitrarily far from threshold without something catastrophic happening to its physical properties so there are often natural limits to "far from threshold." One exception is fluids, which can be driven so strongly that the ratio of the size of the system to the finest spatial structure can be many orders of magnitude. This is the realm of fluid turbulence and is a hard problem indeed because of the strongly nonlinear and irregular dynamics. However, one simplification does occur in the limit of sufficiently large driving (sufficiently large Reynolds number) known as "Kolmogorov turbulence." In this regime, various quantities such as the energy have a scale-free power-law dependence on the wave

---

[1] There are a few rare examples in which spatiotemporal chaos bifurcates supercritically from a uniform state and so occurs arbitrarily close to a simple state. An experimental example is domain chaos, which is observed at the onset of convection in a convection chamber that is rotating sufficiently rapidly. But most spatiotemporal chaotic states are found far from threshold.

9.1 Stripe and lattice states

number k. We do not have space to describe fluid turbulence in this book but the topic is discussed in many fluid dynamics texts and is important for many branches of science and engineering.

9.1 Stripe and lattice states

The structure of ideal stripe and lattice structures far from onset can be understood following the principles of Chapter 4. Symmetry arguments can be used to guess what types of spatially periodic structures might exist, and a numerical Galerkin method can be devised to calculate nonlinear stationary structures of a particular symmetry and then test their linear stability. (Section 4.1, Section 4.3, and Chapter 12 explain some of the details of how a nonlinear stationary spatially periodic state can be found numerically, and Section 4.2.1 how its linear stability can be determined.) Far from threshold, there is no longer a restriction that the wave numbers associated with stable structures be close to the critical wave number q_c at threshold so that a wider class of lattices must be tested than in Section 4.3. Similarly, in a Galerkin analysis far from threshold, there is no particular advantage to use the linear onset modes as a basis. Instead, any convenient numerical basis can be used such as Fourier modes, Chebyshev polynomials, or modes that are suggested by experimental details.

To understand non-ideal and dynamic stripe and lattice patterns far from threshold where the amplitude expansion is no longer valid, new theoretical tools must be developed. The cost of the extension in the range of applicability is that the behavior will be less universal, since the unifying influence of the linear instability is less strong. In fact, there will be fast dynamics (on the characteristic time scale of the basic evolution equations) that are likely to be system specific, and are best understood using intuition developed for the phenomenology of each system, such as the wealth of understanding of fluid dynamics or of excitable media. On the other hand, there are also slow dynamics, that reflect the symmetries of the system and patterns. These dynamics will show a greater degree of universality, although the difference between systems is no longer simply accounted for by length and time scales (ξ_0 and τ_0) as in the amplitude equation.

We will concentrate the discussion on stripe states, and first introduce, based on rather general assumptions, an evolution equation for the phase variable that captures the symmetry-related dynamics. Examples of interesting results that can be derived from this equation are some, but not all, of the instabilities that limit the band of wave numbers for stable stripe states, and a rather sharp restriction on the wave number of stripes in disordered states with significant curvature of the stripes. We then briefly describe situations where the assumptions leading to the phase equation break down, and extended equations are needed. This usually occurs where there are additional slow degrees of freedom, to which the dynamics of the phase variable is coupled. One example, known as mean flow, is peculiar to fluid systems, and is connected with the incompressibility of the fluid on the dynamical scales of the pattern formation. In addition slow dynamics may occur near the onset of secondary instabilities, due to the slow growth or decay of the mode that goes unstable here.

The discussion of the dynamics of patterns far from onset eliminates one of the misleading lessons learned from the amplitude equation for stripes and lattices, namely the potential nature of the dynamics. Deductions that can be traced back to the potential nature of the amplitude equation may be quite misleading for systems far from threshold. Removing the constraint of a potential allows richer dynamics that include, for example, competition between wave number selection mechanisms operating at different spatial locations, and various types of persistent dynamics, including chaos.

9.1.1 Goldstone modes and phase dynamics

The breaking of the continuous symmetries of translation and rotation at the transition to a spatially periodic state has profound dynamical consequences for the ordered state. Based on the broken symmetry, we can predict that perturbations that are slow spatial modulations of the periodic pattern will have arbitrarily slow relaxation or growth rates. Positive growth rates correspond to instability of the spatially periodic state, and long-length-scale perturbations are often important in limiting the stability of stripe and lattice states. For stripe states, the instabilities are known as the Eckhaus and zigzag instabilities. We have already discussed these near threshold using the amplitude equation in Section 6.4.2 and Section 7.1.4, but these instabilities have a more general relevance because they have a degree of universality, applying to all stripe patterns in isotropic systems.

The qualitative argument for the slow dynamics of these long-length-scale perturbations goes as follows. If we take a particular spatially periodic state, any translation of this solution along the directions in which the equations have translational invariance will lead to a state that is also a stationary solution. A perturbation of the original state corresponding to such a translation will not relax. Now consider a perturbation that locally has the character of this translation, but with the size of the translation varying sinusoidally with a wavelength that is large compared with the wavelength of the pattern itself. Since locally the perturbation is just a translation that has no tendency to relax, the evolution depends on distant regions and the relaxation rate goes to zero as the wavelength increases. As we have seen in Section 4.1.1 and Section 4.4.1, a translation of a spatially periodic pattern can be represented as a change in the phase of the sinusoidal functions defining the periodic stripe or lattice states. Consequently, this low-frequency long-wavelength dynamics is known as phase dynamics, and its study provides one of the few theoretical tools available far from onset (see Section 9.1.2).

The argument for low-frequency long-wavelength modes invokes just symmetry ideas and so can be applied at other transitions that break continuous symmetries. In particle physics, the ideas were developed by Jeffrey Goldstone, and low-frequency dynamics associated with broken symmetries are often called Goldstone modes. The ideas are also important in the study of equilibrium phase transitions. An example is the existence of low-frequency transverse sound waves in crystalline solids. (Such transverse waves do not occur in liquids or gases since these support only the propagation of longitudinal waves.)

Near threshold, the phase was introduced as the phase of the complex amplitude. Away from threshold, the phase is conveniently introduced in terms of a local wave vector $\mathbf{q}$ as in Eq. (4.56)

$$\nabla \phi(\mathbf{x}_\perp, t) = \mathbf{q}(\mathbf{x}_\perp, t), \tag{9.1}$$

or

$$\phi(\mathbf{x}_\perp) = \int_0^{\mathbf{x}_\perp} \mathbf{q}(\mathbf{x}'_\perp) \cdot d\mathbf{x}'_\perp, \tag{9.2}$$

where the line integral is started at some convenient origin as reference point, here denoted by the vector $\mathbf{0}$. To investigate the slow dynamics implied by the Goldstone argument, we study the dynamics of this phase variable. This dynamics turns out to be diffusive as we will see in the following section, and so the equation is known as the phase diffusion equation, although the equation is now nonlinear since the diffusion constants depend on the wave number.[2]

The phase diffusion equation can describe much interesting pattern formation associated with stripe patterns far from threshold but cannot describe rapidly varying spatial structure associated with defects or near lateral boundaries. For example, a defect such as a dislocation involves perturbations to the stripe or lattice state that vary rapidly in space, and so are not completely described by the phase equation. In fact, a dislocation is a point where the phase variable is not defined and so is a singularity of the phase description. However, as we have seen in Section 8.1.2, there is a connection between topological defects and the phase variable: gradients of the phase in the vicinity of the defect drive the dynamics of the defect. For example, a phase gradient corresponding to a wave-number change drives dislocation climb.

We might hope to get a complete description of pattern dynamics, even far from threshold, in terms of two coupled descriptions, a phase dynamics that describes a slow spatial variation in the vicinity of a defect, and a fast local dynamics that

---

[2] Some readers may note that formulating the problem in terms of the phase integral Eq. (9.2) is analogous to the WKB formulation in linear wave problems such as quantum mechanics. The phase formulation for stripe dynamics is in fact sometimes called the nonlinear WKB method.

describes the response of a topological defect to the phase dynamics. Such a complete description has turned out to be difficult to derive and has not yet been convincingly carried out, even for the case of two interacting defects. Difficulties that arise are the need for rules defining how topological defects are created in regions of large pattern deformation, and how defects with opposite winding number may annihilate when they get close enough. In addition, the lateral boundary conditions to apply to the phase equation are not clear. Finally, in real systems, disordered regions are often encountered that cannot be described either as slowly modulated stripes or as stripes with topological defects, and it is not known how to incorporate such regions into the description.

9.1.2 Phase diffusion equation

In our discussion of the amplitude equation (Section 6.4.3), we saw that the dynamics of the complex amplitude's phase becomes increasingly slow as the length scale of the variation increases. The magnitude, on the other hand, evolves on the time scale $\tau_0 \varepsilon^{-1}$ set by the distance to threshold, and relaxes to the value consistent with the instantaneous and local wave number given by the spatial variation of the phase. Thus, for the evolution of large-length-scale phase variations, the magnitude need not be studied as an independent dynamical equation; we say that the magnitude adiabatically follows the phase variation. Further away from threshold, the phase dynamics continues to define the slow dynamics corresponding to large-scale spatial translations and reorientations of the pattern. The internal structure of the pattern will relax on a faster time scale, namely the intrinsic time scale of the basic evolution equations, so that it remains consistent with the local phase variation.

Derivation

Just as we did for the derivation of the amplitude equation, we can write down the expected form of the dynamical equations for the phase variable of a stripe phase phenomenologically, based on ideas of symmetry and of a smooth expansion in a small quantity. Alternatively, we can approach the problem more formally, using a method analogous to the multiple scales perturbation theory used to develop the amplitude equation. For simplicity, we will focus on the phenomenological approach and direct the more theoretically inclined readers to Appendix A2.4, which discusses the formal method.

A spatially uniform wave vector field $\mathbf{q}$ is a time-independent solution for a system that supports stationary stripe states. We therefore expect the time variation of the phase variable to depend on spatial derivatives of $\mathbf{q}$. For sufficiently slow spatial variations, we can ignore spatial derivatives that are higher than first order. A rotational transformation of the system rotates both the pattern wave vector, and the spatial gradient vector, and the phase equation must be invariant under these combined transformations in a rotationally invariant system. There are two possible first-order derivative terms of **q** that are consistent with this symmetry and so a lowest-order phase dynamics equation takes the general form

$$\partial_t \phi = f_1(q) \nabla \cdot \mathbf{q} + f_2(q) \mathbf{q} \cdot \nabla q, \tag{9.3}$$

with **q** related to ϕ by Eq. (9.1). Note that the wave number $q = \sqrt{\mathbf{q} \cdot \mathbf{q}}$ and the quantities $\nabla \cdot \mathbf{q}$ and $\mathbf{q} \cdot \nabla q$ are all rotationally invariant since they involve the scalar product of two vectors. The functions $f_1(q)$ and $f_2(q)$ will depend on the specifics of the system, and on system parameters. In contrast to the amplitude equation for stripes near threshold, for which the nature of the system appears only through three constants, the phase equation Eq. (9.3) requires two unknown *functions* to be determined. This reflects the lesser degree of universality away from threshold.

The phase equation is often written in a more condensed form that is obtained by using an integration factor $B(q)$ to make the right-hand side of Eq. (9.3) a total derivative. The phase equation then becomes

$$\tau(q) \partial_t \phi = \nabla \cdot [\mathbf{q} B(q)], \tag{9.4}$$

with

$$\tau(q) = \frac{B(q)}{f_1(q)} \quad \text{and} \quad \frac{d}{dq} \ln B(q) = \frac{f_2(q)}{f_1(q)}. \tag{9.5}$$

Instead of the two functions f_1 and f_2, the phase equation now depends on a q-dependent time-scale factor $\tau(q)$, and on the integration factor $B(q)$. We will use Eq. (9.4) as the phase equation of stripes far from threshold.

Since $\mathbf{q} = \nabla \phi$, the phase equation Eq. (9.4) involves second-order spatial derivatives of the phase, and the equation has the form of a nonlinear diffusion equation. The relation to a diffusion equation becomes clear if we linearize the equation to describe small perturbations about a state of undistorted stripes. For a base stripe state with wave vector $\mathbf{q}_b = q_b \hat{\mathbf{x}}$, we have $\phi = q_b x + \delta\phi$ and $\mathbf{q} = q_b \hat{\mathbf{x}} + \nabla \delta\phi$ with $\delta\phi$ the small phase perturbation. The phase diffusion equation linear in $\delta\phi$ is then

$$\partial_t \delta\phi = D_\parallel(q_b) \partial_x^2 \delta\phi + D_\perp(q_b) \partial_y^2 \delta\phi, \tag{9.6}$$

with diffusion constants $D_\perp$ and $D_\parallel$ respectively for spatial variations perpendicular and parallel to the stripes. Their mathematical forms are

$$D_\perp(q) = \frac{B(q)}{\tau(q)} \quad \text{and} \quad D_\parallel(q) = \frac{1}{\tau(q)} \frac{d(qB)}{dq}. \tag{9.7}$$

Even though the physical system is rotationally invariant, we get an anisotropic diffusion equation Eq. (9.6) with two diffusion constants since the stripe state about which we are perturbing defines a direction along $\mathbf{q}_b$. The diffusion of small displacements of the stripes will in general occur at different rates parallel and perpendicular to this direction.

Note that Eq. (9.6) is a diffusion equation only when the constants $D_\perp$ and $D_\parallel$ are both positive. This in turn is the case only for certain system parameters and for some values of the wave number q. In fact, as we show later in this section, the onset of the universal zigzag and Eckhaus instabilities can be determined mathematically by when these constants change sign, by taking on zero values.

The above phenomenological derivation of the phase equation suffices to tell us the structure of the phase equation, but to justify the result formally, and to evaluate the unknown functions $B(q)$ and $\tau(q)$ for any specific physical system, we must turn to a more systematic derivation. The details involve a lot of algebra so we direct the interested reader to Section A2.4.2 of the Appendix, where we outline the general method and then apply the method to the Swift–Hohenberg equation, for which the calculation is not too involved. Fig. 9.1 illustrates the type of result obtained by plotting the quantity $qB(q)$ versus the wave number q for the Swift–Hohenberg equation with the parameter value $r = 0.25$. (Plotting $qB(q)$ is useful for the discussion of instabilities, see Section 9.1.2.)

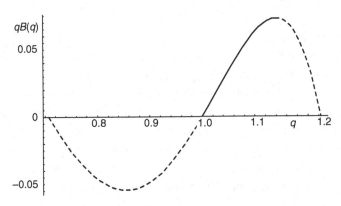

Fig. 9.1 Illustration of a nonlinear phase diffusion equation calculation for the Swift–Hohenberg model of pattern formation, Eq. (2.30), with parameter value $r = 0.25$. According to Eq. (9.7), the linearized phase diffusion equation Eq. (9.6) for a stripe state is diffusive when the quantities B and $(qB)'$ are both positive (the prime ' denotes differentiation with respect to q). Since q is non-negative, it suffices to plot qB versus q and identify where the curve qB and its slope are both positive. In this plot, the solid line denotes the only part of the curve where B and $(qB)'$ are both positive. The range of q corresponding to the solid curve turns out to be the wave number band for which stripes are linearly stable with respect to long-length-scale perturbations. The quantity $B(q)$ is calculated approximately in Section A2.4.2 of the Appendix, see Eq. (A2.75).

9.1 Stripe and lattice states

Application: diffusive dynamics

The phase equation tells us a remarkable fact about the stripe state: if we displace the stripes at some point, this local perturbation has long range and long-time-scale consequences, since the perturbation spreads diffusively. On the one hand, this is a phenomenon that is not immediately obvious just inspecting the microscopic equations, such as the Navier–Stokes equation for convection for example, and on the other hand the phenomenon is quite general, independent of the physics that produces the stripe state. It is a phenomenon that directly comes from the broken translational symmetry.

A direct experimental verification of the phase diffusion of the roll state in Rayleigh–Bénard convection was performed using the apparatus sketched in Fig. 9.2. The experiment was performed in a rectangular horizontal cell of width $5d$ and length $30d$, with $d = 0.6$ cm the depth of the cell. The fluid used was a silicon oil with Prandtl number 492, for which the basic time scale of the convection, the thermal diffusion time across the depth of the cell, was $\tau_d = 320$ seconds. The Rayleigh number was adjusted to have the value $R = 1.16R_c$, where R_c is the critical Rayleigh number for the onset of convection, so this experiment was not "far from threshold."

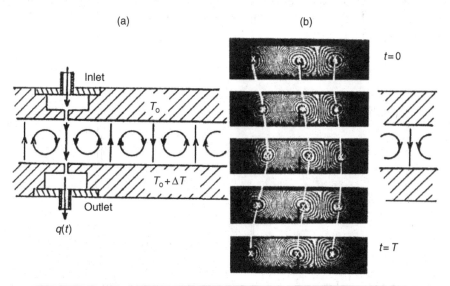

Fig. 9.2 Apparatus for experimental verification of phase diffusion in Rayleigh–Bénard convection. Panel (a) sketches the apparatus and the scheme for introducing a time-periodic perturbation to the rolls by injecting and withdrawing fluid periodically from the inlet and outlet respectively. Panel (b) shows the displacement of the rolls at some distance away over one period of the perturbation. (From Wesfried and Croquette [111].)

For this geometry and fluid parameters, the convection rolls align parallel to the short side to give a uniform stripe state. Midway along the long length, a localized perturbation was induced along a line parallel to the rolls by injecting periodically, with a frequency ω, fluid between slots in the top and bottom plates. (These slots are labeled Inlet and Outlet on the left side of Fig. 9.2.) This added flow perturbs the adjacent rolls in a complicated way, that includes a sideways displacement. Remember that a spatial displacement δ is equivalent to a phase perturbation $q\delta$ with q the wave number, so that for small perturbations the displacement and phase perturbation can be taken as proportional. For small frequencies $\omega\tau_d \lesssim 1$, the propagation of the phase perturbation can be understood from the phase equation, which predicts the space-time variation along the cell

$$\phi(x,t) = \phi_0 e^{-m_1|x|} \cos(m_2|x| - \omega t), \tag{9.8}$$

where x is measured from the point of disturbance. Note the exponential decay of the size of the phase perturbation, and also the oscillatory dependence on the distance x. The rate of decay m_1, and also the wave number of the oscillations of the disturbance m_2, are related to the parallel diffusion constant through $m_1 = m_2 = \sqrt{\omega/2D_\parallel}$ (see Exercise 9.3). The spatial dependence of Eq. (9.8), and the equality of m_1 and m_2, were quantitatively verified. The demonstration of the scaling of m_1, m_2 with $\omega^{1/2}$ is shown in Fig. 9.3.

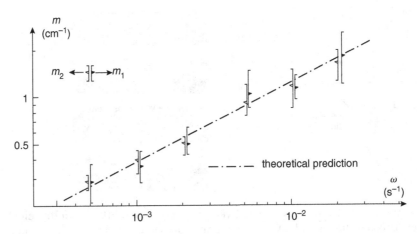

Fig. 9.3 Log-log plot of the decay rate m_1 and oscillation wave vector m_2 for the displacement perturbations in the Rayleigh–Bénard experiment shown in Fig. 9.2. The dashed line is the theoretical prediction, with a slope $1/2$, and a value of the diffusion constant predicted from theory. (Source as in Fig. 9.2.)

Application: instabilities

The diffusion equation (9.6) yields exponentially growing perturbations if either of the diffusion constants is negative. Equations (9.7) then show that the points where B and $(qB)'$ change sign identify the onset of instabilities. Since the phase equations describe slow spatial variations, these are long-wavelength instabilities. As was the case near threshold, we find the longitudinal or Eckhaus instability, which occurs for values of q such that $D_\|(q) < 0$ which implies $(qB)' < 0$. Similarly, the transverse or zigzag instability occurs when $D_\perp(q) < 0$ which implies $B(q) < 0$. Thus the stable wave numbers (the stability balloon) correspond to positive B and positive $(qB)'$. (This is the solid portion of the curve $qB(q)$ in Fig. 9.1, which was calculated in Appendix A2.4.2 for the Swift–Hohenberg equation.) The phase equation shows us that the Eckhaus and zigzag instabilities persist far from threshold, and are generic to stripe states in rotationally invariant systems. The locations of the instabilities are more system dependent than near threshold since they depend on a function $B(q)$ rather than on the two constants ξ_0 and τ_0 of the amplitude equation. Also system dependent is the issue of whether these instabilities will bound the region of stable wave numbers, or instead some other instability will occur at wave numbers inside the range stable to the two long-wavelength instabilities.

Application: wave number selection

The time-independent phase equation derived from Eq. (9.4),

$$\nabla \cdot [\mathbf{q}B(q)] = 0, \tag{9.9}$$

is reminiscent of Maxwell's equation $\nabla \cdot \mathbf{B} = 0$ for a static magnetic field $\mathbf{B}$, or for a static electric field in the absence of charges. In an introductory electromagnetism course, we learn about this equation in terms of field lines that do not end. If the geometry of the field is known, following a field line and inspecting how the separation of neighboring field lines changes allows us to relate the field strength at distant points. In the same way, Eq. (9.9) relates the wave number at distant points along curves drawn normal to the stripes, and this provides strong, nonlocal constraints on a stationary stripe pattern. In particular, if one of the points is chosen at the center of a focus or target singularity that exists in the pattern (see Section 4.4.2) we arrive at a powerful wave number selection principle, known as focus selection.

To illustrate the idea, first consider the case of axisymmetric stripes, Fig. 9.4(a). In polar coordinates (r, θ), the divergence-free equation $\nabla \cdot \mathbf{V} = 0$ becomes:

$$\frac{1}{r}\partial_r(rV_r) + \frac{1}{r}\partial_\theta V_\theta = 0, \tag{9.10}$$

where the general vector field $\mathbf{V} = V_r \mathbf{e}_r + V_\theta \mathbf{e}_\theta$ is resolved along the radial and azimuthal unit vectors $\mathbf{e}_r = (\cos\theta, \sin\theta))$ and $\mathbf{e}_\theta = (-\sin\theta, \cos\theta)$. An

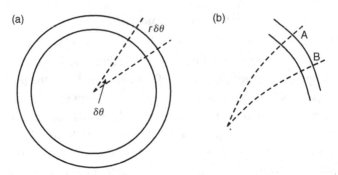

Fig. 9.4 Wave-number selection by centers of stripe curvature: (a) axisymmetric stripes; (b) generally curved stripes. The heavy lines are along the stripes and the dashed lines orthogonal to the stripes.

axisymmetric field does not depend on θ so the θ derivative in Eq. (9.10) vanishes. For the vector field $\mathbf{V} = \mathbf{q}B(q)$, the radial component is the magnitude of the field so $V_r = qB(q)$ and Eq. (9.10) reduces to

$$\frac{1}{r}\partial_r (rqB(q)) = 0, \tag{9.11}$$

which has the solution

$$qB(q) = \frac{C}{r}, \tag{9.12}$$

with r the distance from the center and C an integration constant. Even for r of order unity (i.e. approaching the core of the defect), we expect the wave number to remain of order the characteristic stripe wave number (which we take to set the unit of inverse length scale, and so by definition the typical wave number is of order unity), the integration constant C will also be of order unity. Thus for large distances r, we get the result $qB(q) \to 0$. Since we expect a patterned state with nonzero q, this implies

$$q \to q_\mathrm{f} \quad \text{where} \quad B(q_\mathrm{f}) = 0, \tag{9.13}$$

so that axisymmetric stripes approach a focus selected wave number q_f far from the core of the defect. Furthermore, for systems described by the phase equation (9.4), this selected wave number is right on the zigzag instability boundary $q = q_Z$ where the transverse diffusion constant $D_\perp = B(q)/\tau(q)$ is zero. The transverse diffusion constant $D_\perp$ plays the role of an elastic constant resisting the bending of the stripes. As a consequence, patterns in which the focus selection is operating to select the wave number, and which are described by the phase diffusion equation, are susceptible to the formation of sharp corners, rather than smoothly curved arcs.

The origin of the wave number selection is easily understood. Unless the perpendicular diffusion constant is zero, curved rolls tend to generate a stripe motion toward or away from the center of curvature, with the center providing a sink or source of the stripes. This argument for wave-number selection is not restricted to axisymmetric stripes. In general we can define an orthogonal pair of curvilinear coordinates along the stripes and along the normals, as shown in Fig. 9.4(b). (These coordinates are θ and r in the case of circular arcs.) We can think of Eq. (9.12) as telling us that the product $qB(q)$ is inversely proportional to the arc length $r\,\delta\theta$ between nearby radii as r varies along a constant θ trajectory. Stated in this form, the result is readily extended to general curvilinear coordinates, with the appropriate arc length AB replacing $r\,\delta\theta$.[3] Thus in patterns where the arc length between the nearby normal trajectories grows, the wave number at large distances will again approach q_f.

9.1.3 Beyond the phase equation

Mean flow

The assumptions motivating the general form of the phase equation Eq. (9.4), namely rotational symmetry and a smooth expansion in the phase gradients, seem mild. However the assumptions break down and the equation is *incorrect* for Rayleigh–Bénard convection at finite Prandtl numbers and for many other fluid systems because the smoothness assumption for the expansion in slow gradients in the phase breaks down. In the formal derivation of the phase equation (see Appendix A2.4), the breakdown can be traced to the existence of a slow mode. This second mode is associated with a horizontal flow with nonzero mean across the depth and so is called a mean flow $\mathbf{V}$. This flow carries the stripes along giving an extra advective term $\mathbf{V}\cdot\nabla\phi$ in the phase equation which now becomes

$$\tau(q)\,(\partial_t\phi + \mathbf{V}\cdot\nabla\phi) = \nabla\cdot[\mathbf{q}B(q)]. \qquad (9.14)$$

The mean flow velocity $\mathbf{V}(x, y, t)$ is in turn driven by distortions of the pattern described mathematically by gradients of the phase. Since a mean horizontal flow must necessarily be divergence free for an incompressible fluid, an attempt to eliminate the flow velocity $\mathbf{V}$ in favor of phase gradients can only be done at the expense of introducing a non-smooth dependence on the phase gradient. A divergence-free mean horizontal flow can also be accounted for in terms of a vertical vorticity Ω that quantifies the "stirring" of the fluid about the vertical axis. The vertical vorticity is defined in terms of the stream function ζ for the divergence-free

---

[3] This can be seen, for example, by applying Gauss's theorem to Eq. (9.9) integrated over an area between two nearby curves that are normal to the stripes.

two-dimensional horizontal velocity by

$$\Omega = -\left(\partial_x^2 + \partial_y^2\right)\zeta, \quad \text{where} \quad \mathbf{V} = (V_x, V_y) = (\partial_y\zeta, -\partial_x\zeta). \tag{9.15}$$

The equation describing the driving of the mean flow by phase gradients is in general quite complicated since it involves many functions of the wave number that must be calculated from the fluid equations. To give you an idea of the structure of the equation, we quote the result valid near threshold, where the driving term can be written in terms of the amplitude function

$$\nabla^2 \zeta = \gamma \hat{\mathbf{z}} \cdot \nabla \times \left[\mathbf{q}\nabla \cdot (\mathbf{q}|A|^2)\right]. \tag{9.16}$$

Here $\nabla = (\partial_x, \partial_y)$ is the horizontal gradient, $\nabla^2 = \partial_x^2 + \partial_y^2$ is the horizontal Laplacian, and γ is some constant that depends on the fluid Prandtl number. The magnitude $|A|$ depends on the local wave number q. The point to extract from this equation is that derivatives of the wave vector $\mathbf{q}$ drive the mean flow $\mathbf{V}$, which through the advective term in Eq. (9.14) gives an additional term in the phase dynamics. This is second order in phase derivatives, as are the regular terms in the phase equation, but cannot be expressed as a smooth expansion in derivatives of the phase.[4] A generalized Swift–Hohenberg model with mean flow was introduced in Section 5.2.3.

The mean flow introduces long-range effects in the pattern formation and singular terms in the phase diffusion equation. Some of the consequences are illustrated in Fig. 9.5. For example, the dynamics no longer just depends on the local curvature of the stripes but on the full geometry of the curved structures (compare panels (a) and (b) of Fig. 9.5). The curvature tends to drive a mean flow. This cannot occur in the axisymmetric geometry since a radial flow is not consistent with the incompressibility of the fluid. On the other hand, for portions of circular arcs in a corner of a cell this is not the case and mean flows can develop, as shown by the arrows in Fig. 9.5(b). This means that the wave number selection operating in (a), giving the focus selected wave number q_f, will not apply to (b). For the zigzag perturbed stripes in Fig. 9.5(c), again there is curvature driving a mean flow. However unlike panel (a), the flow is not suppressed by incompressibility conditions and may enhance or suppress the zigzag instability depending on the direction of the flow, which in turn is determined by the signs of various coefficients. For Rayleigh–Bénard convection, the sense is such that the mean flow tends to suppress the zigzag instability. Note that the effect of curvature in (a) and (c) is no longer the same and so the wave number for the zigzag instability q_Z will no longer coincide with the focus

---

[4] A simple example of a non-smooth quantity would be a ratio $f(x)/g(x)$ for which the numerator f is differentiable everywhere while the function $g(x)$ vanishes at one or more points, for example $x^2/\cos(x)$.

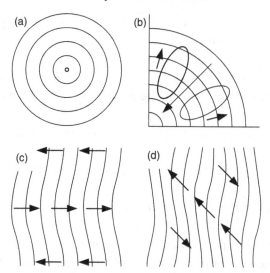

Fig. 9.5 Mean flow for a convecting fluid in various geometries. (a) A stripe curvature tends to induce a mean flow, but in an axisymmetric geometry this radial mean flow cannot develop because of the fluid incompressibility. (b) For incomplete circular arcs, the mean flow can now occur, in this case because of the suppression of the convection near the boundaries. (c) For a zigzag distortion, the roll curvature drives a mean flow that suppresses the instability. (d) A more complicated distortion leads to a mean flow that enhances the instability; this is what the so-called skew-varicose instability of straight parallel rolls looks like in the early linear stage. (Signs of the curvature-driven mean flow in the figure correspond to the case found in Rayleigh–Bénard convection.)

selected wave number q_f. In contrast to (c), a more complicated distortion of stripes in convection that mixes transverse and longitudinal distortions shown in (d) leads to a mean flow that enhances the distortion. This leads to a new long-wavelength instability called the skew-varicose instability that bounds the stability balloon of straight spatially periodic convection rolls at small Prandtl numbers.

Secondary bifurcations

The coupling of the phase dynamics to other slow degrees of freedom such as the mean flow **V** introduced in the previous section may occur in many other situations. The new slow mode may arise in various ways: conserved quantities, such as the total momentum in a fluid system with boundaries that do not damp this momentum; symmetries, such as the translation of a planar interface; or the slowly growing mode at secondary bifurcations. In many cases, because higher-order dynamical equations result, qualitatively new and unexpected behavior are found in these situations. We illustrate this phenomena in the context of an interesting secondary bifurcation that occurs in some experimental stripe systems, the parity-breaking secondary bifurcation of a stripe state.

The parity-breaking instability is an instability of a stripe state to a distorted state where the x and −x directions are no longer equivalent, for example due to the growth of a second harmonic of the spatial periodicity. This parity-breaking instability is quite common and has been seen in solidification fronts, convection in binary fluid mixtures, and other systems. If the stripes are no longer symmetric in the forward–backward directions, they will tend to drift along their normals with a speed that depends on the magnitude of the asymmetry, which goes to zero at the secondary bifurcation.

The slow degrees of freedom at the bifurcation are the amplitude of the parity-breaking distortion, and the translational degree of freedom of the stripes described by the phase ϕ. The basic approach for constructing appropriate amplitude equations for parameter values near the instability follows the approach near the original threshold of the uniform state. An amplitude of the perturbation B is defined so that the parity-breaking deviation from the parity-symmetric nonlinear stripe state is

$$\delta \mathbf{u} = B(\mathbf{x}_\perp, t) \mathbf{u}_1(\mathbf{x}_\perp, \mathbf{x}_\|) + \text{h.o.t.} \tag{9.17}$$

Here $\mathbf{u}_1$ is the Bloch eigenvector of the parity breaking instability. (See Section 4.2.1 for the terminology used in the stability analysis of a stripe state.) This eigenvector has the same period as the stripe state, and the parity-breaking instability is spatially uniform corresponding to a Bloch wave vector zero. The amplitude B is a real variable, which corresponds to the physics that the perturbation has the same spatial periodicity as the stripes. The coupled equations for B and ϕ must be invariant under the combined operations $x \to -x$, $\phi \to -\phi$, $B \to -B$, which corresponds to the parity symmetry of the original equations. This restricts the amplitude equation to the following form if we keep up to second-order derivatives in either variable and restrict our focus to spatial variations along the stripe normal:

$$\partial_T B = \mu B \pm B^3 + \xi_1^2 \partial_X^2 B + \xi_2 B \partial_X B + \xi_3 B \partial_X \phi + \xi_4^2 \partial_X^2 \phi, \tag{9.18a}$$

$$\partial_T \phi = B + \partial_X^2 \phi + \xi_5^2 \partial_X^2 B. \tag{9.18b}$$

Here the scaled coordinate X and the scaled time T have been chosen to make the coefficients of the first two terms on the right-hand side of Eq. (9.18b) equal to unity, and the ξ_i are constants. The parameter μ measures the distance from the onset of the parity-breaking instability.

Equation (9.18b) is particularly interesting since it directly shows that any nonzero amplitude B of the parity-breaking mode will cause a drift of the underlying pattern, in a direction determined by the sign of B. A number of other interesting predictions can be made from Eqs. (9.18). For example, the drifting wave pattern that forms immediately beyond the instability is actually unstable to a long-wavelength modulational instability. This in turn can lead to localized regions of drifting waves that propagate through a stationary background at a wave number that is not unstable to the parity-breaking instability. Such disturbances have been seen in the periodic cellular pattern formed in solidification patterns.

9.1.4 Wave-number selection

The linear stability analysis of the uniform state tells us the rough magnitude of the length scale to expect in the patterned state. However, above onset, spatially periodic nonlinear solutions exist within the stability balloon (see Section 4.2.1), which typically covers a rather broad range of possible periodicities around the length scale given by the linear analysis. On the other hand, experiments and simulations in realistic physical geometries often produce a rather narrow distribution of wave numbers in the pattern, corresponding to a quite well defined spatial periodicity. This raises the question of what physics might lead to a tighter restriction on the spatial periodicity and what is the precise wave number selected.

Wave number selection turns out to be a surprisingly difficult question to answer in systems far from equilibrium. It is worth contrasting this with the same question in an equilibrium system, for example the question of the precise periodicity of a crystal lattice. For the crystal, there is a simply stated principle: minimize the appropriate thermodynamic potential as a function of the lattice spacing a. In the absence of external stresses and at a given temperature, we minimize the free energy $F(a)$, where $F = E - TS$, with E the internal energy, S the entropy, and T the temperature. If equal normal stresses P are applied on all surfaces, we minimize the quantity $G(a)$ with $G = E - TS - PV$, with V the volume. Thus we only have to evaluate static properties at each lattice spacing, and then compare the different states through the thermodynamic potential. On the other hand, in a system far from equilibrium we have no general notion of a state function like F or G that can be minimized. We cannot compare two states with different wave numbers simply through the value of some function evaluated in these states. The only way in general we can "compare" the two states is to set up a dynamical situation that somehow connects the two states. The crux of the wave number selection problem is that, where they have been tested experimentally, theoretically or numerically, different dynamical processes that can change the wave number typically lead to different wave numbers, so that there is no consistency between the comparison between states produced by different connections, and there seems to be no particular choice of connection that is intrinsically more natural than other ones.

Let us focus on this issue a little more. It is again useful to think about the theoretical construction of an ideal straight stripe state parallel to one pair of edges of a rectangular cell with periodic boundary conditions. These are exactly the conditions assumed for the stability balloon analysis and we know from those calculations that, at each value of the control parameter, the stripe state is linearly stable for a nonzero range (a band) of wave numbers. Effectively, the number of spatial periods, and hence the wave number, is conserved for small perturbations about a state within the stability band. To allow for the wave number to evolve, we need to apply a finite disturbance or reorganization to the ideal straight stripe state. Although infinitesimal perturbations in a linear analysis can be classified and understood in terms of the Bloch modes, many different finite amplitude disturbances that might relax the constraint of a fixed number of periods can be imagined. In a system far from equilibrium, different finite perturbations often lead to different selected wave numbers.

Various dynamical mechanisms for defining a "preferred" wave number might be envisioned.

One choice that is natural from a theoretical perspective is the center of the wave number distribution found for the dynamical equations supplemented with noise, in the limit of small noise strength. For nonequilibrium systems that are near local equilibrium, thermal noise dictated by the principles of statistical mechanics (in particular the fluctuation–dissipation theorem) is appropriate. Even for small amplitude noise, occasionally a large enough fluctuation will initiate some sort of event that leads to a change in the number of stripes. Averaging over a long enough time, we expect to find a narrow distribution of wave numbers about some "noise selected" value. However, this idea is unlikely to be relevant to most experimental systems since the microscopic noise averages out to tiny values for patterns on a macroscopic scale.[5]

An alternative choice is the center of the distribution in the state developing at long times from a random initial condition (assuming that the final distribution is insensitive to the size and distribution of initial conditions under some constraints, and that the dynamics tends to relax the state to one with an adequately well defined wave number). In this protocol for forming the pattern, the growth from the random initial conditions produces a state at early times that is strongly disordered. As the dynamics proceeds, regions of ordered stripes develop, with different regions or domains showing different orientations. The ordered domains grow through competition between neighboring regions – a process known as coarsening. As the

---

[5] Carefully designed experiments on Rayleigh–Bénard convection using a fluid (SF_6) near its critical point to enhance the size of thermal fluctuations have measured the effect of thermal fluctuations on the convection pattern near threshold. The Swift–Hohenberg equation, was originally written down (supplemented by noise terms) to investigate exactly this question. Swift and Hohenberg predicted that the transition would actually be discontinuous, with a jump in properties such as the heat flow, rather than the continuous transition predicted by bifurcation theory for the noise-free situation. The expected size of the jump is tiny, but was verified by the experiments.

size of these ordered regions grows, the width of the wave number distribution decreases, and if the process continues to long times,[6] a well-defined wave number may result. Such quench protocols have been investigated numerically for model equations yielding stripe states such as the Swift–Hohenberg equation (an example is shown in Fig. 5.3), and also for the fluid equations describing Rayleigh–Bénard convection, although for shorter times and in smaller system sizes. In these simulations, the long time state is indeed found to be one with rather well defined wave number. Starting from small enough random initial conditions, at early times the fastest growing linear modes will dominate the wave number distribution. At later times, nonlinear effects take over and a different dominant wave number is found. The quench protocol has not been investigated in experiment, and it is by no means clear that it is relevant to most experiments or to natural patterns.

A different growth mechanism of the patterned state from the unstable uniform state is for the pattern to grow locally, perhaps at a boundary or an inhomogeneity, and then to spread into the rest of the system by a propagating front of the type discussed in Section 8.3. Such a propagating front tends to lay down a well-ordered state, with a wave number that can be predicted as in Section 8.3.3.

Rather than considering the growth of the pattern from the unstable uniform state, we can also consider specific dynamical mechanisms in the established pattern state and ask at which wave number the dynamics ceases. An important dynamical mechanism for wave-number relaxation is the dynamics of defects. This role of defects in selecting a wave number was introduced in Section 4.4.2. One example comes from dislocations: injecting a dislocation defect allows the wave number to increase or decrease by its climb motion. The direction and speed of climb depends on the wave number of the surrounding stripes, as discussed in Section 8.1.2. Successive injection of dislocations will eventually lead to the dislocation climb selected wave number q_d at which the climb velocity is zero. On the other hand the presence of focus defects allows the wave number to relax through the creation or destruction of the tiny stripes near the center, relaxing the system to a wave number q_f, the focus-selected wave number as discussed in Section 9.1.2. There is no reason to expect these two wave numbers to be equal in general, and experiment or calculation in a number of systems confirms this, for example experiments in Rayleigh–Bénard convection shown in Fig. 9.6. Grain boundaries also allow the wave number to relax, and again will in general lead to a stationary pattern at a different wave number q_g. (This was discussed near threshold in Section 8.2.) Thus the wave number selected may well depend in a complicated way on which defects happen to be present. In experiment, this often depends on the nature and geometry of the side walls and also on initial conditions. The lateral boundaries themselves

---

[6] Alternatively, the pattern may freeze into a disordered state, as in glass formation from a rapidly cooled liquid.

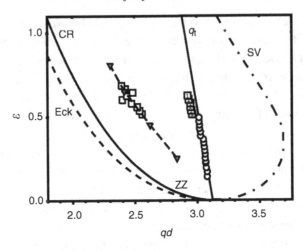

Fig. 9.6 Wave-number selection by targets and spirals (q_t, called q_f in the text) and by dislocations (q_d) in Rayleigh–Bénard convection. The plot shows various wave numbers q multiplied by the depth of the fluid d as the abscissa and the reduced control parameter ε as the ordinate. The wave numbers q_t and q_d would be equal for a system that has potential dynamics. For q_t, the open circles are experimental values for targets, the solid line is theory for targets, and the squares with a 1 inside are experimental wave numbers selected by 1-arm spirals. For q_d, the squares are experimental values while the triangles and dashed line are theoretical calculations. Also shown is the stability balloon. The experiments were performed in a cylindrical geometry of aspect ratio $\Gamma = 38.8$ using a fluid with Prandtl number $\sigma = 1.4$. (From the work of Plapp reported in the review by Bodenschatz et al. [12].)

may also provide regions where the stripe pattern is less well developed so that the wave number may relax (see Section 6.4.1). Rather than an abrupt termination of the pattern at a physical side wall, it is also possible to set up conditions so that the pattern gradually fades away over some long distance before the boundary is reached. This may be done by arranging for some control parameter to change slowly over space, with a supercritical value in one part of the cell and a subcritical value in some other part of the cell. Such a slow modulation of a parameter is called a control parameter ramp.

For example, in a Rayleigh–Bénard convection experiment with fixed plate temperatures, the Rayleigh number Eq. (1.2) varies as $R \propto d^3$ so that an R interpolating between values above and below R_c can be established by varying the depth slowly over space. (This in turn is accomplished in practice by machining the top or bottom plate to have a slowly varying thickness.) If the convection rolls orient parallel to the ramp (roll normal and direction of control parameter variation aligned) then the wave number in the bulk where the control parameter reaches its supercritical value may relax through the slow disappearance or appearance of new low amplitude rolls in the region of the ramp where R passes through R_c. It is rather straightforward to analyze this situation for exceedingly slowly varying ramps. The prediction, verified in a number of careful experiments, is that for a given ramped quantity (such as the depth in the Rayleigh–Bénard convection example) and for a given control parameter value in the bulk, the wave number in the stripe pattern takes on a unique value (rather than a band), so that the ramp acts as a wave number selection mechanism. However, if different quantities making up the dimensionless control parameter can be varied (for example the depth and the temperature difference in a convection system), these different spatial ramps will lead to different selected wave numbers. Again there seems to be no way to discover a unique "preferred" wave number independent of the detailed configuration.

An experimental example of wave number selection by ramps is shown in Fig. 9.7 for a Taylor–Couette experiment (see Fig. 1.11). The Taylor–Couette system is particularly convenient for studying wave number selection since the azimuthal symmetry of the rolls around the cylinders is maintained for some control parameter range above the threshold of the pattern so that the pattern-formation problem remains one dimensional. In contrast, for a convection cell with a ramp,

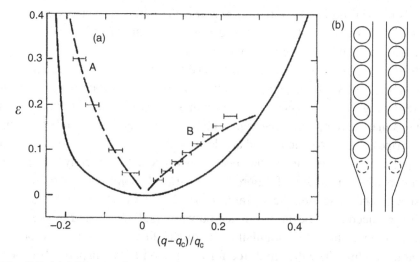

Fig. 9.7 (a) Wave numbers as a function of the reduced control parameter ε in Taylor–Couette systems with linear ramps in the cylinder radii. Bars: range of wave numbers observed in experiment; dashed lines: theoretical predictions of wave number selection. The solid lines show the Eckhaus instability for the infinite cylinder. The ramp angles for the inner and outer cylinders were (A) −0.0075 and 0 and (B) 0.0074 and 0.0151 (where a positive angle corresponds to cylinder radii decreasing away from the uniform region). (From Ning et al. [82].) (b) Sketch of the apparatus for the case of outer cylinder ramp with negative ramp angle, and no ramp on the inner cylinder.

the orientation of the rolls may also be affected by the ramps. The control parameter ramps in Fig. 9.7 were created by machining the radii of the inner or outer cylinder to vary slowly and linearly toward one end of the cylinders as shown in cross section in panel (b). The ramp is characterized by the geometrical angles of the walls of the cylinder. For small values of the angles, a narrow band of wave numbers is selected in the bulk of the system where the radii are uniform, as predicted by the general theory. Furthermore, the selected wave number is found to depend on the ratio of the ramp angles of the two walls in agreement with the general theoretical expectation, and the values measured are in good agreement with the theoretical predictions based on the fluid-dynamical equations.

Various extrema principles have been proposed to explain the observed narrow distribution of wave numbers such as maximizing some global property of the system, for example the total heat flow in convection or the rate of entropy production. However none of these seems to have any wide generality or any satisfying derivation from more basic principles. We can use the empirical fact that there appears to be no unique wave number selection, at least in the various systems far from equilibrium where this issue has been investigated, to argue against the existence of a useful minimization principle in these systems, and perhaps in general. (By the word "useful" we mean, for example, that the quantities to be minimized are integrals of local functions of the fields and their spatial derivatives over the system.) Opinions of pattern formation researchers continue to differ on whether such a minimization principle might ultimately be found.

If different dynamical processes can select different wave numbers, an interesting question arises of what happens where two or more such processes are operating in different regions of the same system. In these situations, a state with persistent dynamics may develop, with the region favoring a larger wave number tending to create stripes that propagate to, and are destroyed in, the region favoring a smaller wave number. Such dynamics has been constructed theoretically and observed experimentally in a number of different geometries. In some cases, the dynamics is periodic, in other cases chaotic. A simple example is shown in Fig. 9.8. Here stripes may be produced by the center, radiated outward, and annihilated by the climb of the dislocation around the azimuthal direction, leading to a continuous rotation of the spiral structure. The driving force for the motion is the disparity between the wave number selected by the central core, which is essentially the same as the focus selected wave number, and the dislocation selected wave number. (This mechanism is quite different than the one that produces rotating spirals in an excitable medium such as those discussed in Fig. 1.18(a) and Section 11.6. There the spiral core directly generates the dynamics, and there is no competition between two different selection mechanisms.) Two different control parameter ramps provide another

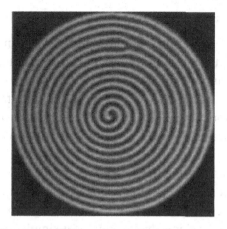 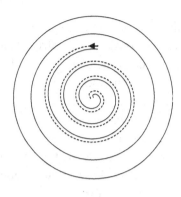

Fig. 9.8 Spiral pattern terminating in a dislocation defect. The left panel shows a shadowgraph from a Rayleigh–Bénard convection experiment in a cylinder of aspect ratio $\Gamma = 28.4$ for a fluid with Prandtl number $\sigma = 1.38$. The Rayleigh number is $1.72R_c$. (Source as in Fig. 9.6.) The right panel shows a schematic of the dynamics: the dashed line shows the phase front at later time and the arrow shows the motion of the dislocation as the spiral rotates in a counter-clockwise direction.

simple example where a dynamic state resulting from incompatible wave-number-selection mechanisms can be established and studied.

9.2 Novel patterns

For the second class of patterns far from onset – those which bear little or no resemblance to modes growing at a linear instability – much less can be said in general. Since strongly nonlinear systems are hard to understand in general, much of the understanding of the structure of these new types of patterns relies on experimental and numerical investigations, or on perturbative approaches based on other small parameters specific to particular systems. As yet, the understanding of patterns far from threshold is far from complete. Here too the question arises of "What is a pattern?" This is straightforward to answer for the regular or quasiregular states seen near threshold, but what should be included in patterns far from onset – or excluded so that some general truths can be stated – is by no means clear. In this section, we present a few of the new features encountered in patterns far from onset. Extending the methods and models we have discussed in previous chapters into more highly nonlinear regimes can give insights into patterns far from onset. In many cases, we are pushing the models beyond their range of validity and the results cannot be considered to be quantitatively reliable. However, the qualitative insights remain useful.

9.2.1 Pinning and disorder

Spatially disordered patterns are common far from onset. We can gain some insight into this by considering a stripe state taken far from onset. Near threshold, the patterns are constructed from slow modulations of a state that is locally stripes. The length scale of the modulation is much longer than the wavelength of the stripes. For example defects, such as dislocations or grain boundaries, involve a strongly distorted core that varies on length scales of order $\varepsilon^{-1/2}$ or $\varepsilon^{-1/4}$. Because of this separation of scales, the pattern variations evolve rather independently of the underlying stripe structure. Compare this with the situation far from threshold. Now there is no a priori separation of length scales. Pattern variations and defect cores occur at the scale of the stripes, and are easily pinned to the underlying structure. A simple illustration is given in Fig. 9.9, which shows numerical simulations of the Swift–Hohenberg equation from random initial conditions. Near threshold, the pattern steadily evolves from the short scale structure of the initial conditions, toward a final state that is close to one of straight parallel rolls, with perturbations from the effects of the boundaries. Far from threshold, on the other hand, evolution only continues a short time, before the pattern freezes into a highly disordered state that seems rather independent of the boundaries.

Some quantitative understanding of this pinning phenomenon, and why it is more important away from threshold, are established in the following Etude.

Etude 9.1 Nonadiabatic effects for fronts in the Swift–Hohenberg equation
Consider the Swift–Hohenberg equation in one spatial dimension

$$\partial_t u = ru - (\partial_x^2 + 1)^2 u - u^3. \tag{9.19}$$

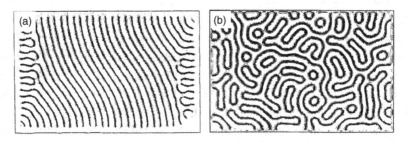

Fig. 9.9 State at long times in numerical integrations of the Swift–Hohenberg equation (5.9) in a rectangular box of dimensions 29.2×19.5, starting from small-amplitude random initial conditions: (a) For the control value $r = 0.1$, the random initial structure evolves into one of a few states consisting of slowly varying parallel rolls with defects occurring only along the shorter sides. (b) For $r = 2.0$, the random initial state evolves into a frozen highly disordered state whose details are sensitive to the initial state. (From Greenside and Coughran [43].)

9.2 Novel patterns

For $r > 0$, we can form fronts that propagate from the nonlinear stripe state into the uniform $u = 0$ state. In an amplitude equation description, the front is given by a real amplitude $\bar{A}(X)$ with $X = \varepsilon^{1/2}x$ and $\varepsilon = r$. The scaled amplitude $\bar{A}$ varies between 0 in the uniform state and 1 in the saturated nonlinear state at wave number $q = 1$. The full field is then

$$u = 2\varepsilon^{1/2}\bar{A}(\varepsilon^{1/2}x)\cos(x + \phi), \qquad (9.20)$$

where ϕ gives a shift of the position of the rolls. Remember the dynamics of the Swift–Hohenberg equation is potential, see Section 5.1.2. For the front solution, the potential Eq. (5.11) is

$$V = \varepsilon^2 \int dx \left\{ -2\bar{A}^2 \cos^2(x + \phi) + 4\bar{A}^4 \cos^4(x + \phi) + 8(\bar{A}')^2 \sin^2(x + \phi) \right\}, \qquad (9.21)$$

where $\bar{A}'$ denotes $d\bar{A}/dX$. For small ε, we would usually evaluate this expression by first ignoring the slow variation of A and then replacing the rapidly oscillating functions by their averages ($\cos^2 \to \frac{1}{2}$, $\sin^2 \to \frac{1}{2}$, $\cos^4 \to \frac{3}{8}$). This gives an integral over the slow variable

$$V = \varepsilon^2 \int dX \left\{ -\bar{A}^2 + 4[\bar{A}']^2 + \frac{3}{2}\bar{A}^4 \right\}, \qquad (9.22)$$

which is clearly independent of the front's position relative to the rolls since the dependence on ϕ has dropped out. However we have ignored terms such as

$$\Delta V = \varepsilon^2 \int dx \left\{ -\bar{A}^2(\varepsilon^{1/2}x)\cos[2(x + \phi)] \right\}, \qquad (9.23)$$

which comes from the $-2\bar{A}^2\cos^2(x + \phi)$ term. We can estimate how this correction scales with ε by noticing that the integral is the Fourier transform at wave number 2 of a function smoothly varying over a scale $\varepsilon^{-1/2}$. From the general properties of Fourier transforms, we know that the integral is exponentially small for small ε. In fact, the full correction to Eq. (9.22) varies as

$$\Delta V \sim \exp(-a/\sqrt{\varepsilon}), \qquad (9.24)$$

where the constant a and the prefactors depend on the details of $\bar{A}(X)$. However, most importantly, ΔV will depend periodically on ϕ, which determines the position of the front relative to the stripes. For small ε, the correction ΔV is negligible compared to the potential difference of the stripe and uniform state that drives the front forward. However, for large enough ε, the correction becomes sufficiently strong that the front position may become pinned in one of the preferred locations that minimize ΔV and the front motion will cease. These extra effects are known as nonadiabatic effects since they cannot be obtained through a perturbation expansion in a small parameter ε. Indeed, functions of the form $\exp(-a/\varepsilon^b)$ for $b > 0$ have the unusual property of being infinitely differentiable at $\varepsilon = 0$ yet not having a Taylor series about this value since, as you can verify, all derivatives of this function vanish at $\varepsilon = 0$.

We could also consider the weakly subcritical version

$$\partial_t u = ru - (\partial_x^2 + 1)^2 u + gu^3 - u^5, \qquad (9.25)$$

where g is a small and positive constant. The amplitude equation predicts a stationary front at an isolated value of $r < 0$, when the potentials of the stripe and uniform states are equal. Including the nonadiabatic corrections spreads this balance point into a range of r for which the front is stationary.

9.2.2 Localized structures

A second feature of patterns far from threshold is that they may be based on localized structures rather than on extended planforms of stripes and lattices. This can be partially motivated from the previous example. Since the fronts in the subcritical Swift–Hohenberg equation (9.25) can become pinned to the underlying stripes, we can construct one-dimensional localized pulse solutions, consisting of two fronts confining a small region of the nonlinear saturated state. In the absence of the nonadiabatic pinning effects, the saturated region would be expected to either grow or shrink, depending on the value of the control parameter. However because of the pinning to the underlying periodic stripe pattern, the fronts can be stationary over a range of control parameters, leading to a pulse solution in one dimension. This can also be described in terms of an interaction between the two fronts, which for the Swift–Hohenberg equation is a repulsive interaction at some distances and attractive at others. For this equation, for which a potential exists, the interaction can be understood in terms of an effective potential that varies non-monotonically with separation, although the same result is also expected for nonpotential modifications of the equation. Pulses are also found in numerical solutions of the subcritical Swift–Hohenberg equation in two and three dimensions. In general, it is not possible to calculate analytically the structure of localized solutions in the strongly nonlinear regime. The counting arguments used in the discussion of fronts in Section 8.3 may be useful in deciding whether pulses are likely to exist in one dimension. Such general methods do not exist for localized solutions in two and three dimensions.

Another mechanism for the formation of localized structures appears in the subcritical complex Ginzburg–Landau equation

$$\partial_t \bar{A} = \varepsilon \bar{A} + (1 + ic_1)\nabla^2 \bar{A} + (1 - ic_3)|\bar{A}|^2 \bar{A} - (1 - ic_4)|\bar{A}|^4 \bar{A}. \qquad (9.26)$$

(We discuss the importance of this equation for pattern formation in Chapter 5.) If the coefficients in the equation are real, $c_1 = c_3 = c_4 = 0$, Eq. (9.26) reduces to the one we used in Section 8.3 to analyze the properties of fronts. There we found that there is a particular (negative) value of ε at which a front between the stable finite magnitude solution and the $\bar{A} = 0$ solution is stationary. This value can again be identified as the one at which the two states have equal potentials since Eq. (9.26) is potential for $c_i = 0$. There are no stable pulse solutions, corresponding to a localized region of one state in a background of the other state, because of the interaction between the two fronts bounding the localized region, and the monotonic dependence of the interaction on position. However for the complex equation, $c_i \neq 0$, pulse solutions of localized regions of the finite amplitude state are found over a range of ε below the bifurcation, for both one and two space dimensions, as in Fig. 9.10. These pulses can be qualitatively understood as arising from the amplitude dependence of the frequency. This in turn affects the wave number k of the complex amplitude, which gives an effective parameter $\varepsilon - k^2$ that controls the dynamics of the fronts bounding the pulse solution. The result is that a self-consistent solution is found with stable pulses over a range of ε. The wave number changes as ε changes so that the combination $\varepsilon - k^2$ remains nearly constant.

A dramatic experimental example of a localized structure is an oscillon, which appears in shaken horizontal trays of granular material such as sand (see Fig. 9.11). Here the tray of granular material is shaken with a large enough amplitude so that the acceleration of the plate exceeds gravity and the grains become airborne over part of each cycle. For a given granular material, the parameters of the experiment are the shaking amplitude and frequency. For some ranges of these parameters, and in a layer much wider than the depth of the material, patterns familiar from previous

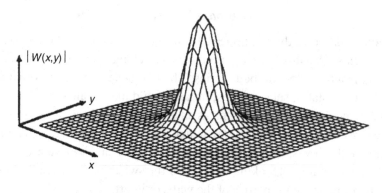

Fig. 9.10 Localized pulse solution in the subcritical complex Ginzburg–Landau equation (9.26), showing the magnitude of the complex amplitude (called W in the plot) as a function of two space dimensions. Parameters correspond to $\varepsilon = -0.192, c_1 = 4, c_3 = -0.30, c_4 = 0.40$. (From Thual and Fauve [104].)

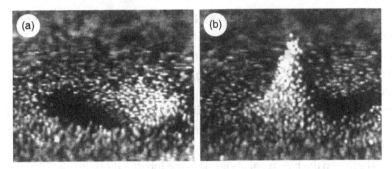

Fig. 9.11 Localized structures called oscillons are found in a thin layer of a sand-like granular medium consisting of small bronze balls (0.17 mm diameter) when the layer is shaken vertically with a suitable acceleration and frequency. The structure is a subharmonic response to the driving and the trough one period in (a) becomes a peak at the next period in (b) and vice versa. (From Umbanhowar *et al*. [107].)

chapters form such as stripes, squares, hexagons, and even spirals. But over other parameter ranges, the localized structure shown in the figure forms instead. These structures oscillate at half the drive frequency, so that at successive cycles of the drive the oscillon will manifest as a peak or a trough. There are therefore two different phases of oscillons, and both may be present in the same experiment. Multiple oscillon solutions are also seen. If the oscillons are separated by more than a few diameters they appear to be independent, and slowly diffuse around the system. When oscillons come closer together, they can form bound clusters of a pair of oscillons of opposite phase, and also more complicated structures as shown in Fig. 9.12. Similar localized structures are also seen in experiments on shaken fluid layers.

9.2.3 Patterns based on front properties

A common situation leading to nonequilibrium patterns arises where two states of a system coexist. A pattern can then form by switching spatially from one state to another via fronts or domain boundaries. The two states may be two equilibrium thermodynamics states, such as the solid and liquid in crystallization, may be two states of a dynamical description such as an amplitude equation, or an approximate notion such as the "products" and "reactants" in a chemical system. The patterns may form by the invasion of one state into the other and can take many forms, both static and dynamic. Thinking in terms of the switching between the two states focuses attention on the properties of the walls or interfaces, for example the direction of motion of a single front (which state grows at the expense of the other), the interaction between two fronts (do the fronts repel at short distances, or attract and annihilate), and possible instabilities of the fronts.

9.2 Novel patterns

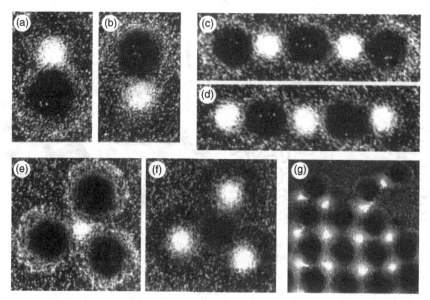

Fig. 9.12 Molecule-like localized structures formed from bound clusters of oscillons. (Source as in Fig. 9.11.)

Transitions in the properties of the walls may lead to changes in the morphology of the patterns. One such transition is called the Ising–Bloch transition. (The nomenclature comes from analogies with walls between magnetic domains in solid state physics.) This is a transition from a situation where there is a single type of wall between the two states that moves at a single parameter-dependent speed c (known as an Ising-like wall) to a situation where there are walls with two different speeds (Bloch-like walls) that, near the Ising–Bloch transition, will have values $c \pm \delta c$, with δc small. In the particular case for which the evolution equations are unchanged by interchanging the two states, the Ising wall will be stationary and the Bloch walls will move with nonzero speeds $\pm v$. The Ising–Bloch transition tends to produce more dynamic patterns.

The fronts themselves may be pattern-forming systems that may be analyzed using the methods introduced in previous chapters. For example, the stability of the fronts to transverse undulations at some wave number q can be investigated. This type of instability of a planar solid–liquid interface, known as the Mullin–Sekerka instability, is the starting point for understanding the rich pattern formation that occurs in solidification.

In numerical simulations, the combination of the Ising–Bloch transition and the transverse instability of a front does indeed appear to be associated with a change in the morphology of the pattern, from a stationary labyrinthine state to a

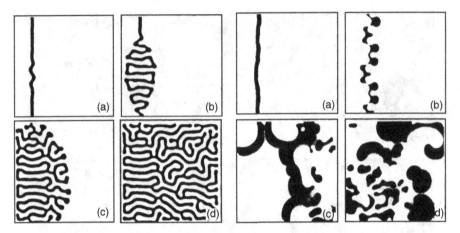

Fig. 9.13 Two kinds of pattern formation associated with an Ising–Bloch transition of the front between two stable uniform states that are denoted by black and white. The panels (a)–(d) on the left show the evolution from an initial condition of a single stripe of the black state in a background of the white state, for parameters such that the domain wall is of the Ising-like type and is unstable to transverse undulations. The pattern evolves into the stationary labyrinthine pattern of (d). For different parameter values, the wall is Bloch-like and the same initial condition evolves into a dynamic state consisting of disordered propagating waves with spiral sources, as in the panels (a)–(d) on the right. These results were obtained by numerical simulation of Eqs. (9.27) with parameters $a_0 = -0.1, a_1 = 2$, and $\varepsilon = 0.05, \delta = 4$ and $\varepsilon = 0.014, \delta = 2.8$ for the two cases. (From Hagberg and Meron [45].)

spatiotemporal state of dynamic spirals. For example, Fig. 9.13 shows the evolution of the chemical reaction–diffusion equations

$$\partial_t u = u - u^3 - v + \nabla^2 u, \tag{9.27a}$$

$$\partial_t v = \eta(u - a_0 - a_1 v) + D\nabla^2 v. \tag{9.27b}$$

Similar equations were introduced in Chapter 3 in the study of the Turing instability, and closely related ones will be studied in Chapter 11 in the context of propagating waves in excitable media. The results shown in the figure are obtained for parameter values for which Eqs. (9.27) have two stable stationary states (denoted by black and white in the figures). Starting from an initial condition of a single stripe of one state in the other leads to quite different final states depending on whether the walls are Ising-like or Bloch-like. (In both cases the walls are unstable to transverse undulations.) For the Ising-like, quasistationary, domain walls shown in panels (a)–(d) on the left, the transverse instability of the fronts leads eventually to a labyrinthine state, where the black state has invaded the white background as a tortuous stripe. The width of the stripe can be understood in terms of a repulsion between the two fronts at short distances, that prevents the invading white state from complete invasion. On the other hand, as shown in panels (a)–(d) on the right, the transverse instability of the Block-like dynamic fronts leads eventually to a dynamic state where the walls propagate, and form spiral structures. Both types of patterns – labyrinthine and dynamic spiral states – are often seen in highly nonlinear patterns. The labyrinthine state is also a feature of some equilibrium systems such as the magnetic domains in a ferrofluid. The dynamic spiral state is reminiscent of the pattern found in the complex Ginzburg–Landau equation that is the amplitude equation for oscillatory instabilities (see Fig. 10.4). We will discuss this further in Chapter 11.

The idea of a pattern in terms of two states separated by domain walls may be developed into a quantitative calculational scheme if the width of the boundary is small compared to the width of the domains. In this case, the ratio of widths provides a small parameter that can be exploited in the analysis. This approach is illustrated in Chapter 11 in the context of dynamical patterns in chemical reaction–diffusion equations, but has also been used to study static patterns.

9.2.4 Spatiotemporal chaos

Many pattern-forming systems driven far from onset show disorder in space and chaotic dynamics in time, together known as spatiotemporal chaos. The term "chaos" denotes persistent irregular behavior of a deterministic system. Much of the work on chaos of the last thirty years has dealt with systems that could be represented by a small number of degrees of freedom (see Section 1.4 for a qualitative discussion of low- and high-dimensional dynamical systems). Methods have been developed for analyzing chaotic behavior in such low-dimensional systems, for example measuring Lyapunov exponents and fractal dimensions of strange attractors by embedding time series into spaces of various dimensions and then measuring statistical properties of the resulting clouds of points. Pattern-forming systems cannot be described by a small number of degrees of freedom and so the disordered dynamics found in these systems represents a new type of chaos since the description appears to require a large number of chaotic elements distributed in space. Many of the tools developed for characterizing and understanding low-dimensional chaos have not been successfully generalized to spatiotemporal chaos, and it remains a challenge to develop appropriate methods and tools.

There are several qualitative ways to see why spatiotemporal chaos should occur. One is that, as any nonequilibrium system is driven further from onset, states with a particular symmetry like stripes or lattices eventually become unstable to states with a reduced symmetry so eventually there is spatial disorder. For example, a stripe state may become linearly unstable to zigzags as some parameters is varied. But as the infinitesimal zigzag perturbations grow in magnitude, they may eventually saturate because of nonlinearities, with a magnitude comparable to that of the stripes themselves and so can alter the structure of the stripes, perhaps causing neighboring stripes to connect and form defects (see Fig. 4.3). If the disordered state remains time dependent, there is then the possibility for nonperiodic behavior that arises from the spatial disorder.

Another way that spatiotemporal chaos can arise is through an uncontrolled initial condition in a large cell for which the influence of the lateral boundaries is weak. For example, as a control parameter is changed so that a stable uniform state becomes unstable, tiny random perturbations due to thermal noise or small geometric irregularities in the physical properties of the cell can grow into local regions of stripes or of lattices but these regions can have different orientations and different wave numbers in different parts of the experiment. These regions can then interact in such a way as to produce a time-dependent state, which can then be chaotic because of the spatial disorder. More precisely, random initial conditions can lead to a state with many defects such as focus singularities at boundaries and in the bulk grain boundaries, dislocations, disclinations, and perhaps clusters of these defects. The different wave number selection principles discussed in Section 9.1.4 can then be active simultaneously and the competition between the different selection mechanism will cause the system to be time dependent. In the absence of a potential, there is no reason for the system to settle into a stationary state so periodic and chaotic behaviors are possible.

A third reason why spatiotemporal chaos can be expected as a system is driven ever further from threshold is that strong driving tends to produce spatial structure at ever smaller length scales (a consequence of spatial harmonics arising from nonlinear terms, e.g. cubing a stripe solution of the form $\cos(qx)$ creates a term proportional to $\cos(3qx)$ on a smaller length scale than $2\pi/q$). Thus even in a system that is so small that just one or a few stripes or lattice units can fit in the cell, the dynamics can become high-dimensional and chaotic by the appearance of many modes over smaller and smaller length scales. For strongly driven large cells, different selection principles and fine spatial structure can occur at the same time and the dynamics is indeed complicated. The weather systems of planetary atmospheres would be examples of large systems with strong driving.

Rayleigh–Bénard convection provides some specific examples that illustrate these general ideas. Calculations based on the quantitatively accurate fluid dynamics equations (the Boussinesq equations) show that the stability balloon for straight parallel convection rolls ceases to exist above some maximum Rayleigh number. (This maximum value depends on the fluid's Prandtl number.) This is illustrated in Fig. 4.4 on page 150; the shaded regions indicate where stripe states are stable and these regions cease to exist above some maximum Rayleigh number. Thus even in

9.2 Novel patterns

the idealized case of a periodic geometry for which there is no forcing by lateral boundaries and starting with straight parallel rolls of constant wave number that is stable near onset, this stripe state must eventually become unstable to some new state for large enough Rayleigh numbers. Although the initial instability of the stripe state might evolve to a stationary pattern of some other geometry (say wavy rolls, a state whose stability is not described by the stability balloon) or to an oscillatory time dependence, a complex and dynamic spatial structure is usually found as the Rayleigh number is further increased. Similarly, convection experiments in small boxes (say of aspect ratio 2) show that a chaotic behavior with fine spatial structure is observed for sufficiently large Rayleigh numbers.

A convection example of how spatiotemporal chaos can arise through an uncontrolled initial condition is the spiral defect chaos state shown in Fig. 9.14. (We discussed this state briefly in Section 1.3.2, see Fig. 1.15.) The picture is a snapshot of a complex sustained dynamic pattern. In movies of this state, the eye picks out spiral structures which form and rotate a few times before being destroyed by the general disorder, and there are also dislocations that migrate around the spiral structures, sometimes annihilating, sometimes being created. It turns out that, for this particular Rayleigh number and Prandtl number, the stability balloon shows

Fig. 9.14 Spiral defect chaos in Rayleigh–Bénard convection. The convecting fluid is pressurized gaseous carbon dioxide at 33 bar with Prandtl number 0.96, the aspect ratio (ratio of radius to depth) is 44, and the Rayleigh number is $R = 1.4R_c$ where R_c is the critical number for onset. This state shows a sustained nonperiodic dynamics for as long as experimentalists had the patience to observe it (many horizontal diffusion times). (From Morris *et al.* [77].)

348   *Patterns far from threshold*

that stable stripe states exist. Indeed, numerical simulations of three-dimensional convection in a large rectangular geometry with periodic lateral boundary conditions for the same Rayleigh and Prandtl numbers show that either stable stationary stripes or spiral defect chaos can be found, depending on what initial condition was used to start the simulation. A further intriguing point is that the distribution of wave numbers found in the spiral defect chaos state (as obtained from a Fourier transform of the pattern) all lie within the stability balloon. The bistability of spiral defect chaos and stationary stripe states in a periodic domain demonstrates that the stability balloon, while valuable for understanding when stationary stripes may exist and how they become unstable, is not sufficient to predict the occurrence of spatiotemporal chaos.

Another interesting example of spatiotemporal chaos is given by rotating Rayleigh–Bénard convection in which the whole experimental system is placed on a platform rotating about the vertical axis. It turns out that if the angular frequency Ω of the rotation exceeds a critical rotation rate Ω_c, then a chaotic state is found arbitrarily close to the threshold of instability of the uniform structureless state. Unlike the spiral defect chaos of Fig. 9.14, this spatiotemporal state takes the appearance of a system of interacting domains with finite lifetimes and with a characteristic size that increases the closer the domain chaos state is to onset (see Figure 9.15). The domains are regions of stripe solutions whose orientations differ

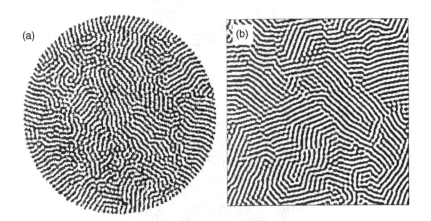

Fig. 9.15 Chaotic domain state: (a) in experiments on rotating convection in carbon dioxide at 20 atmospheres in a cylindrical cell with hot and cold fluid visualized using a digitally enhanced shadowgraph; and (b) from numerical simulation of a generalized Swift–Hohenberg model with positive values of the field shown black and negative values white. Both patterns continue to evolve in an irregular way. The similarity between the numerical results and the portion of the experimental cell away from the boundaries suggests a boundary-independent spatiotemporal chaotic state. (From Ning *et al.* [83] and Cross *et al.* [27].)

9.2 Novel patterns

by approximately 60°. Each domain grows into some other domain simultaneously so domain chaos is an example of a spatiotemporal chaos where each region is changing in time but one region of structure cannot grow at the expense of all the other regions to dominate the pattern.

The rotating convection system is particularly interesting for the study of spatiotemporal chaos because it is one of the very few experimental examples of complex dynamics that grows continuously from a uniform state. We can then hope that weakly nonlinear theories such as the amplitude equation and models related to the Swift–Hohenberg model might give good accounts of the behavior. Also, the chaotic dynamics can be directly associated with the rotation. In the absence of rotation, the system should be described by the amplitude equations for superimposed stripes introduced in Section 7.3. These equations are potential (ignoring any effects of the boundaries) and so cannot lead to chaotic dynamics. On the other hand, in the presence of rotation, the terms in the amplitude equations describing the nonlinear interaction of two sets of stripes at an angle of θ lose the $\theta \to -\theta$ symmetry which renders the dynamics nonpotential as discussed in Section 7.3 (see also Exercise 7.20). Similarly, the Swift–Hohenberg equation, which is a model of non-rotating convection, is potential and cannot show spatiotemporal chaos. A generalization of the model to include the effects of rotation, Eq. (5.32) in Chapter 5, is nonpotential and shows dynamic patterns remarkably similar to the experimental ones as shown in Fig. 9.15(b).

These examples of spatiotemporal chaos and many others raise a number of theoretical issues that are still poorly understood.

A basic question is what are appropriate ways to characterize the complex dynamical state, both qualitatively to formulate a compact way to define spatiotemporal chaos, and quantitatively, for example to compare theory, numerics, and experiment. In our present weak stage of understanding, it is natural to attempt to characterize spatiotemporal chaos in terms of representative times and lengths. For example, we might use the typical domain size in Fig. 9.15 to define a characteristic length of the system. Or we could measure the two-point correlation function of an elementary field $\langle u_1(x, y, t + \tau) u_1(x', y', t) \rangle_t$ (with $\langle \ \rangle_t$ denoting the average over time). Then the decay of the correlations with spatial separation of the two points and with time defines a correlation length and time in the way often used in systems near equilibrium. The relationship between times or lengths defined through different properties is unknown and could be quite complicated.

The conventional characterization techniques for chaos with a few degrees of freedom, such as Lyapunov exponents and fractal attractor dimensions, involve the geometry of the motion in the full phase space of the system. For a spatially distributed system, the dimension of phase space is enormous and it is not clear whether these traditional diagnostics can be suitably modified. One possible way to approach spatiotemporal chaos through dynamics is to adopt a thermodynamic style of thinking, which is to try to average over the fine dynamical structure (say defects and their complex interactions) and extract quantities that characterize bulk properties of the spatiotemporal chaos. It is not obvious this can be done because, when you look at how thermodynamics is justified, many key concepts like energy, entropy, and the chemical potential turn out to be associated with fundamental conservation laws and these laws do not apply to the open driven dissipative systems of this book. For example, the total energy of a convecting fluid between two plates is not conserved since the fluid can sometimes lose more energy to the cold plate than it gains from the hot plate, and similarly the total mass of chemical reagents in a gel like that of Fig. 3.3 need not be conserved since matter from reservoirs can flow into and out of the gel region with a possibly complicated time dependence.

An important observation in thermodynamics is that, for homogeneous states of matter such as a volume filled throughout with steam or ice, there are so-called extensive variables like the energy E, entropy S, and mass M that characterize the bulk property of the matter and whose values simply increase in proportion to the system volume V. This implies that numerical values of the quantities E, S, and M are not interesting since they just reflect the size of the system, and so it is better to work with intensive variables that involve ratios of extensive variables such as the mass density M/V or derivatives of extensive variables such as dS/dE which leads to the definition of a temperature. For sufficiently large volumes, intensive quantities are independent of the volume or surface area of the matter. For thermodynamic systems, intensive quantities are also often easier to measure than extensive quantities. For example, one readily measures the temperature of some sample with a thermometer, which is easier than measuring its extensive entropy (by measuring specific heats over some range of temperatures) and its extensive energy (say by burning the sample in a calorimeter).

Motivated by a suggestion of David Ruelle in the context of fluid turbulence, simulations on parallel computers of various pattern-formation models (and in a few cases of fundamental evolution equations like the Boussinesq equations) have shown that the fractal dimension D of a homogeneous spatiotemporal state acts like an extensive thermodynamic quantity in that, for a sufficiently large volume V of medium, D increases linearly with V.[7] Given this empirical observation, researchers have explored characterizing spatiotemporal chaos through an intensive dimension density defined by $\delta = \lim_{V \to \infty} D/V$ (if this limit exists), or perhaps in

---

[7] A technical aside: there is an infinity of fractal dimensions D_q associated with any attractor and the range of these dimensions can be greater than one so it is not obvious which of the many dimensions should be used to characterize a spatiotemporal chaotic state. For high-dimensional spatially extended systems, it turns out that only one fractal dimension, the so-called Lyapunov fractal dimension $D_\mathcal{L}$ can be readily computed and all dimension studies of spatiotemporal chaos have been based on computations of $D_\mathcal{L}$. However, each of the different fractal dimensions should themselves scale extensively.

a more practical way by a local derivative dD/dV for locally linear regions of a plot of D versus V. However, no one has yet succeeded in finding the equivalent of a nonequilibrium thermometer, that can determine the value of a dimension density or similar intensive dynamical quantity by studying small subsystems of the spatially extended system. For this reason, and because fractal dimensions greater than about 5 cannot be accurately estimated from experimental time series, a comparison of the fractal dimension for experimental examples of spatiotemporal chaos and for corresponding numerical calculations has not yet been possible.

This thermodynamic approach does lead to some interesting insights and questions. It is already not obvious why the fractal dimension of a spatially extended chaotic system should increases linearly with system volume. One might guess that this is a natural consequence of the spatial and temporal disorder of the spatiotemporal chaos; two regions that are far apart are presumably uncorrelated and so contribute their own internal properties in an additive way. But no one has yet succeeded in turning this into a rigorous argument, nor has anyone yet had a good scientific insight as to what features (if any) of the spatiotemporal chaos are local and so decorrelate from other features. Calculations and some theory also suggest that, unlike the fractal dimension, the largest Lyapunov exponent λ_1 is an intensive quantity that is independent of system volume or system shape for sufficiently large systems. Since the quantity $1/\lambda_1$ is the magnitude of the time with which accurate details of a dynamical system can be forecast into the future, this thermodynamic reasoning leads to the interesting conclusion that, for homogeneous spatiotemporal chaotic states, the ability to forecast details is independent of the system size or fractal dimension (for sufficiently large systems).

Another approach to characterize spatiotemporal chaos is to determine whether reduced descriptions of the complex dynamics can be found mathematically. One approach might be to look for collective coordinates, such as the coordinates of the spirals and dislocations in Fig. 9.14, or of the domain walls in Fig. 9.15. Rather than trying to model specific experiments, we might first seek simpler systems that show related phenomena. Just as the study of maps (discrete-time dynamical systems) were useful in advancing the understanding of low-dimensional chaos, a lattice of maps coupled between the lattice sites can be used to study spatiotemporal chaos numerically and, in some simple limits, analytically.

Once we have learned how to define and characterize spatiotemporal chaos, we might ask what broad classes of behavior exist in these systems. In analogy with equilibrium systems, which of course at the microscopic scale show strong dynamical fluctuations that we call thermal fluctuations, can different phases, separated by sharp transition points, be identified within the chaotic state? Are there any universal features at transitions to the chaotic states or at the transitions between the different chaotic states such as are found at second-order phase transitions in equilibrium systems? Do practical properties like the transport of energy or mass change in a systematic way upon the transition to spatiotemporal chaos, or as the properties of the chaos itself changes with parameters?

These are all difficult questions on which there is much ongoing work. A strong motivation for understanding these questions is the existence of far-from-threshold spatiotemporal chaotic states of great relevance to the human race such as predicting or controlling the weather, and predicting or controlling cardiac arrhythmias that often lead to death.

9.3 Conclusions

Pattern formation far from threshold is a difficult topic since there are far fewer general tools available compared to the tools available near a supercritical linear instability of a uniform state. We have divided our discussion into two parts: patterns that are qualitatively the same as ones near threshold and novel patterns.

In the former case, we can imagine steadily increasing the control parameter from the threshold value to a value of interest, even if the threshold value is not physically accessible. If there are no qualitative changes of behavior, represented mathematically as bifurcations, then the properties of the pattern will change continuously as the control parameter is increased. This means that much of the physical insight gained from the amplitude equation approach should continue to apply far from threshold. In particular, for a stripe state we are led to expect a range of possible wave numbers bounded by instabilities which perhaps have universal characteristics (Eckhaus and zigzag); long range and temporally slow effects represented by phase dynamics that ultimately derive from the translational symmetry of the physical system; and the importance of topological defects in the dynamics. Thus many of the same questions should be posed as near threshold, while answering the questions may be much harder and quite specific to each system. One feature of the behavior near threshold that does not extend further away is the potential nature of the dynamics found in the amplitude equation description. Eliminating the potential removes strong constraints, such as the consistency of different wave number selection mechanisms and the absence of chaos, which suggests that patterns in experiment and nature are likely to be more dynamic away from threshold. On the other hand, we have learned that pinning effects that tend to lock the modulations of the patterns to the underlying structure of the pattern tend to become stronger as the nonlinearity increases so that some types of motion such as dislocation glide might be quenched away from threshold.

The second type of patterns far from threshold we considered is ones that do not resemble those near threshold. These might occur through bifurcations from threshold-like patterns, such as occurs at the boundaries of the stability balloon, or as novel patterns that develop directly from the uniform state. Examples mentioned in the chapter are localized structures, that may then group to form regular structures, and patterns that are disordered in space and perhaps time as well. Disorder in both space and time is called spatiotemporal chaos.

As systems are driven ever further from equilibrium, the wealth of behavior becomes larger and the particular characteristics of each system, suppressed near threshold by the dominance of the few unstable linear modes, become more evident. Thus fluid systems become more dynamic since the dissipative viscous effects effectively compete with the strong driving only at small length scales; a large number of degrees of freedom then participate in the dynamics. For strong enough driving, this takes us into the realm of turbulence where statistical descriptions, rather than ones inspired by visual observation, seem more appropriate. In contrast, chemical reaction and diffusion systems, rather than becoming more dynamic when driven further from onset, seem to develop more complex spatial structures that might be thought of as interspersed regions of various product combinations with sharp interfaces between them. Far from threshold, we must therefore recognize the importance of both the issues of pattern formation, and the particular phenomenology of each physical, chemical, or biological system.

9.4 Further reading

(i) For a general discussion of Goldstone modes in equilibrium phases see *Principles of Condensed Matter Physics* by Chaikin and Lubensky [17], Chapter 8.

This chapter has briefly covered a number of more advanced topics. For further discussion, as well as the specific references already given in the figure captions consider the following:

(ii) The phase diffusion equation far from threshold has largely been worked on in the context of Rayleigh–Bénard convection. The comprehensive but difficult paper "The phase diffusion and mean drift equations for convection at finite Rayleigh numbers in large containers" by Newell *et al.* [81] describes this work and gives references to earlier papers.

(iii) For more on the analysis of the parity-breaking secondary instability mentioned in Section 9.1.3 see "Instabilities of one-dimensional cellular patterns" by Coullet and Iooss [49], and for the observation in experiments on cellular patterns on solidification fronts see "Dynamics of one-dimensional interfaces – an experimentalist's view" by Flesselles *et al.* [36].

(iv) You can find a wealth of information on wave number selection in Rayleigh–Bénard convection in the book *Rayleigh-Bénard Convection: Structures and Dynamics* by Getling [37]. We do not, however, endorse his conclusion that the evolution from random initial conditions is preferred over other mechanisms.

(v) A discussion of the stationary patterns based on front properties is given in "General theory of instabilities for patterns with sharp interfaces in reaction-diffusion systems" by Muratov and Osipov [78].

(vi) Our discussion of spatiotemporal chaos is based partly on the brief review "Spatiotemporal chaos" by Cross and Hohenberg [26].

Exercises

9.1 Goldstone modes: List four different Goldstone modes in equilibrium phases of matter. What are the corresponding symmetries that the phase transition to the state breaks?

9.2 Phase diffusion functions from amplitude equation: By comparing Eq. (9.6) with the phase equation derived from the amplitude equation in Section 7.1.5, show that near threshold we can use expressions for the parameters of the phase diffusion equation Eq. (9.4) of the form

$$B(q) = \xi_0^2(q-q_c)[\varepsilon - \xi_0^2(q-q_c)^2], \tag{E9.1}$$

$$\tau(q) = \tau_0[\varepsilon - \xi_0^2(q-q_c)^2]. \tag{E9.2}$$

Plot $qB(q)$ as a function of q and use this plot to discuss the stable range of q and the instabilities bounding this range.

9.3 Derivation of solution Eq. (9.8) to the linear phase diffusion equation: This exercise is related to the experimental confirmation of phase diffusion shown in Fig. 9.2.

(a) In applying the linear phase diffusion equation Eq. (9.6) to a convection cell that is long in the x coordinate but short in the y coordinate (where the vertical direction is the z coordinate), explain why it is fine to treat Eq. (9.6) as a one-dimensional diffusion equation $\partial_t \phi(x,t) = D_\| \partial_x^2 \phi$.

(b) Derive Eq. (9.8) by solving the linearized phase diffusion equation Eq. (9.6) in an infinite one-dimensional domain with coordinate x and with boundary condition $\phi(x=0,t) = \phi_0 \cos(\omega t)$. This boundary condition approximates the local periodic injection of fluid into a convection experiment, as described in Fig. 9.2.

(c) The solution Eq. (9.8) is correct for an infinitely long cell. Discuss qualitatively how the structure of this solution changes for a finite domain.

9.4 Eckhaus and zigzag instabilities for the Swift–Hohenberg model: Using the expression for $B(q)$ for the Swift–Hohenberg model derived in Appendix 2 Eq. (A2.75), find expressions for the wave numbers of the Eckhaus and zigzag instabilities as a function of the control parameter r for the two-dimensional Swift–Hohenberg equation Eq. (5.9), calculating $q-1$ to $O(r^{\frac{1}{2}})$.

Exercises

9.5 Mean flow: A simple form for the mean flow velocity driven by perturbations of the roll pattern in convection near threshold is given by Eqs. (9.15)–(9.16). In this exercise you will investigate the mean flow for small, long-wavelength perturbations to a stripe state at wave number q defined by the phase

$$\phi = qx + a\cos\mathbf{Q}\cdot\mathbf{x}, \tag{E9.3}$$

with a and $\mathbf{Q}$ very small, and taking the simple ansatz for the wave number dependence of the amplitude (actually corresponding to the Swift–Hohenberg model)

$$|A(q)|^2 = \frac{1}{3}\left[\varepsilon - (q^2 - 1)^2\right]. \tag{E9.4}$$

(a) Show that the phase Eq. (E9.3) corresponds to a displacement of the rolls in the normal direction by a distance Δ with

$$q\Delta \simeq -a\cos\mathbf{Q}\cdot\mathbf{x}. \tag{E9.5}$$

(b) The zigzag instability is described by a perturbation with $\mathbf{Q} = (0, Q)$. Show that for $a \to 0$ the x component of the mean flow for this perturbation is

$$V_x = \gamma\,|A(q)|^2 q Q^2 a \cos(Qy). \tag{E9.6}$$

Use this to argue that if $\gamma > 0$ the mean flow is *stabilizing* for the zigzag instability (cf. Fig. 9.5(c)).

(c) The skew-varicose instability is defined by a perturbation Eq. (E9.3) with a general orientation of the wave vector $\mathbf{Q} = Q(\cos\theta, \sin\theta)$. For $a \to 0$, show that V_x has the same spatial dependence as the stripe displacement $V_x = V_x^{(0)}\cos\mathbf{Q}\cdot\mathbf{x}$ and calculate the strength $V_x^{(0)}$.

(d) Verify the form of the stripe perturbation and the mean flow as sketched in Fig. 9.5(d) for a skew-varicose perturbed stripe pattern. For the plot use the parameters $\varepsilon = 1$, $q_0 = 5/4$, $\mathbf{Q} = (1/4, 1/8)$, and $a = 0.1$.

(e) For the skew-varicose perturbation in part (c) and $\varepsilon = 1/4$, plot the region in the $q\theta$ plane for which $V_x^{(0)} < 0$, so that the component of the mean flow normal to the rolls is in the same direction as the roll displacement and is therefore *destabilizing*. This destabilization leads to the skew-varicose instability in convection.

9.6 Mean flow and phase diffusion: Show that the growth rate σ of an infinitesimal phase perturbation defined by Eq. (E9.3) with the phase dynamics including mean flow given by Eqs. (9.14–9.16) is

$$\sigma = -(D_\parallel^{\mathrm{eff}} Q_X^2 + D_\perp^{\mathrm{eff}} Q_Y^2), \tag{E9.7}$$

with effective diffusion constants

$$D_\parallel^{\text{eff}} = D_\parallel + \gamma q^2 \frac{d}{dq}(qA^2)\sin^2\phi, \quad \text{(E9.8a)}$$

$$D_\perp^{\text{eff}} = D_\perp + \gamma q^2 A^2 \sin^2\phi, \quad \text{(E9.8b)}$$

with $\sin\phi = Q_Y/Q_X$ and $D_\parallel, D_\perp$ the values in the absence of mean flow as in Eqs. (9.7).

9.7 **Dynamics from control parameter ramps:** This exercise is based on Fig. 9.7. Describe qualitatively what you expect to happen in a long Taylor–Couette cylinder with ramp A at one end and ramp B at the other for reduced control parameter values (a) $\varepsilon = 0.1$ and (b) $\varepsilon = 0.25$. How would you make the answer to part (a) quantitative?

9.8 **Spiral rotation:** Derive the rotation frequency of the spiral in Fig. 9.8 if the dislocation is at radius r_d and supposing that the wave number dependence of the perpendicular diffusion constant $D_\perp$ and the climb velocity v_d of the dislocation can be taken to be linear

$$D_\perp = \alpha(q - q_f), \quad \text{(E9.9)}$$

$$v_d = \beta(q - q_d), \quad \text{(E9.10)}$$

with q_f the focus selected wave number and q_d the wave number at which dislocations are stationary. You may assume that $q_f - q_d$ is small compared with their mean to simplify the calculation. (You might want to look at the paper *Dynamics and Selection of Giant Spirals in Rayleigh-Bénard Convection* by Plapp *et al.* [87].)

9.9 **Spatiotemporal chaos in a coupled map lattice:** Exercise 2.8 of Chapter 2 introduced the concept of a coupled map lattice (CML), which is a dynamical system with continuous variables that evolves on a discrete space-time lattice. You should review that exercise in preparation for this one.

Consider a $N \times N$ two-dimensional periodic CML given by the evolution equation

$$u_{i,j}^{t+1} = f(u_{i,j}^t) + D\left[\frac{1}{4}\left(f(u_{i+1,j}^t) + f(u_{i-1,j}^t) \right.\right.$$
$$\left.\left. + f(u_{i,j+1}^t) + f(u_{i,j-1}^t)\right) - f(u_{i,j}^t)\right]. \quad \text{(E9.11)}$$

The real number D is a nearest-neighbor coupling constant, the variables i and j are lattice indices which run from 1 to N, the integer variable $t \geq 0$ indicates successive moments in time, and the function $f(x)$ is the quadratic function Eq. (E2.7) with parameter a that is associated with the logistic map.

The periodicity of the lattice is imposed by letting the lattice "wrap around" when the indices i,j in Eq. (E9.11) have the value $N+1$ or 0:

$$u^t_{N+1,j} = u^t_{1,j} \quad \text{and} \quad u^t_{0,j} = u^t_{N,j}, \quad 1 \le j \le N, \tag{E9.12}$$

$$u^t_{i,N+1} = u^t_{i,1} \quad \text{and} \quad u^t_{i,0} = u^t_{i,N}, \quad 1 \le i \le N. \tag{E9.13}$$

Using a computer mathematics environment like Mathematica or Matlab, write a program to evolve this system and display the results graphically. Investigate the behavior for choices of a in the range $3.57 \le a \le 4$ and D in the range $0 \le D \le 1$, using initial values $u^0_{i,j}$ that lie in the interval $[0, 1]$. (The range for a is motivated by the fact that the logistic map shows chaos for ranges of a with $a > 3.57$.) For some values you should find spatiotemporal chaos. What measures might you use to characterize the complex dynamics? Can you determine whether the chaotic dynamics changes quantitatively as you vary the logistic map parameter a and the lattice coupling constant D?

10

Oscillatory patterns

One of the fundamental ways that a stationary dynamical system can become time dependent as some parameter is varied is via a Hopf bifurcation (see Appendix 1). In the supercritical case, a fixed point becomes unstable at the same time that a stable periodic orbit grows smoothly out of the fixed point. In this chapter, we use amplitude equations and comparisons of calculations with experiments to discuss the universal dynamics that arise near the onset of a Hopf bifurcation in a spatially extended homogeneous nonequilibrium medium. Although many of the concepts and issues are similar to those already discussed in Chapters 6–8 for the type I-s instability (e.g. amplitude equations, stability balloons, defects, phase equations), a new feature of oscillatory media is the appearance of propagating waves. For media with one extended direction, there are typically right- and left-propagating waves that interact in a nonlinear way with each other, and these waves can also interact nonlinearly with waves generated by reflection from a lateral boundary. An intriguing one-dimensional example that we discuss later in this chapter is the blinking state, which can be observed when a binary fluid (e.g. a mixture of water and alcohol) convects in a narrow rectangular domain, see Fig. 10.11. In a two-dimensional oscillatory medium, the propagating waves most often take the form of rotating spirals (see Figures 1.9, 1.18(a), 10.3, and 10.4).

Propagating waves and spiral structures also are observed in so-called excitable media. Unlike an oscillatory medium, interesting dynamics arise from the uniform state only when the magnitude of a perturbation exceeds a finite threshold, in which case the medium responds locally with large-amplitude growth followed by decay back to the local stationary state. Because the spatiotemporal dynamics arises from finite-amplitude events, the dynamics can be strongly nonlinear and different mathematical tools are needed. We discuss the dynamics of excitable media separately in Chapter 11.

Historically, nonlinear oscillatory media were studied for a long time in the context of biological signaling and were also investigated experimentally in the context of oscillatory chemical reactions starting with the pioneering work of Belousov in the 1950s. (Indeed, the research by chemists provided a parallel line of development that, for a decade or so, was largely independent of the work on stationary structures.) The theoretical approach was largely based on approximation schemes that were tailored to the particular dynamical equations of the chemical systems, for which there are typically small parameters derived from the ratio of reaction rates. Perturbation techniques based on these small parameters yielded tractable methods even though the chemical states of interest were strongly nonlinear (not near a linear instability of a spatially uniform state). Fortunately, many of the key concepts and questions can be introduced by first considering the simpler limit of "close to onset." In this regime, the amplitude equation formalism provides a unifying explanation of many experimental details and predicts new ones.

This chapter is organized as follows. In Section 10.1, we introduce the concepts of absolute instability and convective instability, which are not encountered in the study of type-s stationary instabilities and that are important for understanding type-o (I-o and III-o) oscillatory instabilities. (Recall the classification scheme that we introduced in Section 2.5.) Convectively unstable states play an important role in solidification, the growth of a crystal in a supercooled melt, for example. The fact that macroscopic details of convectively unstable states can depend on molecular noise (an example from solidification would be the thermal noise at the tip of a growing dendrite) largely solves a question raised by Johannes Kepler 400 years ago, which was to explain the infinite variety of snowflake shapes (Fig. 1.8).

We discuss the type III-o instability first, in Section 10.2; this is the case when the uniform state becomes unstable to a spatially uniform oscillation. The lowest-order amplitude equation is none other than the complex Ginzburg–Landau equation which we discussed in Chapter 5 as a classic and much studied theoretical model of pattern formation. This equation, together with the associated phase equation that describes the slow evolution of long-wavelength phase variations, will allow us to determine the general structure of the stability balloon near onset and to analyze some of the defects that arise in oscillatory media with one and two extended directions. The theory has been tested in chemical reaction–diffusion experiments and in fluid experiments, although much remains to be understood, e.g. how spatiotemporal chaos emerges continuously from the uniform state and whether the resulting chaos agrees with the dynamics of the complex Ginzburg–Landau equation.

In the succeeding sections, we discuss the type-I-o oscillatory instability, which produces structure at a nonzero wave number. The nonlinear propagating states have different properties, depending on whether the system does or does not have a parity symmetry. For systems that lack this symmetry (Section 10.3), there is an asymmetry between the forward and backward directions so that the instability occurs first to waves propagating in a particular direction. This case has been studied experimentally in fluid systems such as Rayleigh–Bénard convection and the Taylor–Couette system, by imposing a flow along one extended direction. The nonlinear propagating wave with a characteristic wavelength then corresponds to circulating fluid rolls that drift along the extended direction.

For type I-o systems that have a parity symmetry, the instabilities to waves in the forward and backward direction coincide and we must consider the interaction of such waves, for example whether nonlinear traveling or standing waves are to be expected in ideal geometries. An experimental example is convection in a mixture of two fluids known as binary fluid convection. Near the onset of these instabilities, the growth rate is of course small but the propagation speed of the waves has some finite value. The dominant propagation effects near threshold make the treatment harder, and different approximations are appropriate depending on ratios of time scales such as the ratio of the doubling time of the growth to the propagation time over the system. Most experimental and theoretical work for parity-symmetric systems has been for the case of a single extended coordinate along which the waves propagate to and fro. We describe this case in Section 10.4. Even in this simple system, states with complex spatial structure and rich dynamics are observed near the onset of the instability, where an amplitude equation description provides at least a partial understanding. Our understanding of waves propagating in two-dimensional rotationally invariant systems on the other hand rests largely on experiment. As an example, we discuss briefly experiments on binary fluid convection in a two-dimensional cylindrical geometry in Section 10.5. This system has a rich spatiotemporal dynamics for which spirals do not play a role.

This chapter completes the discussion that we began in Chapter 6, of the fundamental supercritical transitions of a homogeneous nonequilibrium medium (type-s and type-o) and how details of these transitions and the resulting saturated nonlinear states can be understood near onset via the amplitude equation formalism.

10.1 Convective and absolute instability

Two new concepts associated with oscillatory instabilities are convective instability and absolute instability. These concepts are motivated by the need to understand how propagation of a perturbation affects the instability of the uniform state. We note that the effect of propagation is important for any oscillatory instability at nonzero wave number, whether for the primary instability of the uniform state or for a secondary instability of an existing pattern. For example, this effect is important for the oscillatory instability of stable straight convection rolls shown in Fig. 4.5. (We saw in an earlier chapter that, for Prandtl numbers of order one or smaller, the oscillatory instability bounds the stability balloon as the Rayleigh number is increased, see Fig. 4.4(a).) In the following, we will phrase our discussion

10.1 Convective and absolute instability

in terms of the uniform state. The case for instability about a nonlinear pattern is the same except that the simple Fourier modes would be replaced with Bloch waves of the form Eq. (4.31).

The conventional linear stability analysis of a spatially uniform state is carried out with respect to a single sinusoidal mode perturbation that extends over all space or over the whole of the physical domain. We can also ask about stability with respect to an infinitesimal disturbance that is localized over some region of space. (Such a disturbance might correspond better to practical situations.) For a stationary instability, the same results are recovered as for a delocalized perturbation since a localized initial condition will spread by diffusion and so will eventually cover a sufficient domain that the growth can be understood in terms of the single mode analysis.

However, for an instability at wave number q with nonzero frequency ω_q, we know from elementary courses on waves that wave packets travel at the group speed $s_q = d\omega_q/dq$. A localized initial disturbance may grow in amplitude but also propagate away at the group speed sufficiently rapidly that the magnitude of the disturbance at a fixed point decays, as in Fig. 10.1(b). In this case, the system is said to be convectively unstable. A system where a localized perturbation yields a growing disturbance at a fixed point, as in Fig. 10.1(a), is said to be absolutely unstable.

For stationary instabilities, the two criteria coincide since propagation is not an issue. However, for wave instabilities (type I-o with $q_c \neq 0$ and $\omega_c \neq 0$), the growth rate of the most unstable Fourier mode is zero at onset whereas the group speed is usually nonzero, and so propagation effects dominate. In these cases, the system remains absolutely stable at the onset of convective instability and it is necessary to drive the system a finite amount above this onset, so that the growth rate is

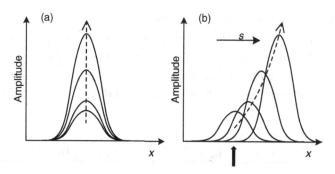

Fig. 10.1 Schematic illustration of absolute and convective instabilities. (a) For an absolute instability, a localized disturbance at a fixed point grows in time. (b) For a convective instability, a localized disturbance grows in time but also propagates away at the group speed s. As a result the disturbance at some fixed point of interest (e.g. at the solid arrow) decays at long times.

finite, for the absolute instability to develop. Absolute instability is not a necessary requirement for the observation of nonlinear wave states since, in the convectively unstable region, different mechanisms can sustain the state. Examples would be defects which act as persistent wave sources, boundaries or inhomogeneities that drive the system, and thermal noise.

The condition for convective instability is the same as the condition for instability to a Fourier mode perturbation. (If you imagine carrying out the linear stability analysis in a frame of reference moving with the pulse, the same arguments apply as for a stationary instability.) The condition for absolute instability is stricter, and is derived by balancing the growth rate against the propagation. Since the time dependence has exponential growth as well as oscillations, the criterion involves a generalization of the group speed to a complex-valued number. We arrive at the criterion for absolute instability using an argument that follows the growth of a localized initial condition by expressing the spatial profile as a superposition of Fourier modes and by using a stationary phase argument to evaluate the long-time asymptotic behavior. This is similar to the argument that we used in Section 8.3.2 that gave us the speed of pulled fronts, although here we are doing a linear instability analysis so the principle of superposition used in the method is better justified. As explained briefly in the following Etude, the result is that the system is absolutely unstable if

$$\operatorname{Re} \sigma_{q_s} = 0, \tag{10.1}$$

where σ_q is the growth rate and where q_s is a *complex* wave vector given by the solution of the stationary phase condition

$$\frac{d\sigma_q}{dq} = 0. \tag{10.2}$$

Etude 10.1 Criterion for absolute instability

In the linear regime, the disturbance growing from any given initial condition $u_p(\mathbf{x}, t = 0)$ can be expressed as the sum of Fourier modes growing at the complex rates σ_q, with initial amplitudes given by the Fourier transform of the initial condition. Restricting attention to one spatial dimension for simplicity, this argument gives

$$u_p(x, t) = \int_{-\infty}^{\infty} dq\, e^{iqx+\sigma_q t} \int_{-\infty}^{\infty} dx'\, u_p(x', 0) e^{-iqx'}. \tag{10.3}$$

If we rewrite the integral as

$$u_p(x, t) = \int_{-\infty}^{\infty} dx'\, u_p(x', 0) \int_{-\infty}^{\infty} dq\, e^{iq(x-x')+\sigma_q t}, \tag{10.4}$$

the integration over q can be evaluated by moving the integration contour into the complex q plane. For large time and at fixed distance from the support of the localized initial condition, the integral over q can be estimated from the contribution at the stationary phase point, i.e. the complex wave number $q = q_s$ given by the solution of the equation

$$\frac{d\sigma_q}{dq} = 0. \tag{10.5}$$

Estimating the integral from the value of the integrand at the stationary phase point gives

$$u_p(x = 0, t) \sim \exp(\sigma_{q_s} t). \tag{10.6}$$

Thus the system will be absolutely unstable for $\mathrm{Re}\,\sigma_{q_s} > 0$.

In the convectively unstable regime, the growing disturbance propagates away (unless there is some persistent source to maintain a local disturbance) and so the system returns to the unstable uniform state. Since the system remains unstable, the system is susceptible to the influence of remnant noise, either due to imperfections in the prepared state or ultimately due to the thermal noise arising from molecular fluctuations.

Solidification – the growth of a crystal into a supercooled liquid melt – is an interesting and important example where thermal fluctuations are believed to sustain a convective instability. The growth of a planar interface is unstable to sharp needles that shoot out into the melt (dendritic growth). A smooth needle shape with a parabolic shape at the tip that grows steadily into the melt is a solution of the equations but is itself unstable to undulations that develop into side branches. It turns out that this instability is convective rather than absolute, and experiments and theory suggest that the pattern of side branches is sustained by the amplification of molecular thermal noise in the tip region. The exquisite sensitivity of the needle growth of the ice crystals to noise and to tiny changes in the growth conditions associated with the convective nature of the side branching instability largely explains the seemingly infinite diversity of snowflake shapes.

10.2 States arising from a type-III-o instability

10.2.1 Phenomenology

Instability toward a uniform oscillation ($q_c = 0$, $\omega_c \neq 0$, type-III-o) can be understood by considering the growth or decay of the perturbation

$$\mathbf{u}_p(\mathbf{x}, t) = \mathbf{u}_0(\mathbf{x}_\|) e^{-i\omega_c t} e^{\mathrm{Re}\,\sigma_0 t} + \text{c.c.}, \tag{10.7}$$

where

$$\omega_c = -\mathrm{Im}\,\sigma_0, \tag{10.8}$$

is the frequency at the onset of the instability (the zero subscript refers to the zero wave number of the perturbation) and Re σ_0 gives the growth rate. This perturbation is uniform in the extended directions $\mathbf{x}_\perp$, but may depend on the confined directions $\mathbf{x}_\parallel$, as determined by the equations of motion and the boundary conditions. We can think of this perturbation as a coherent or synchronized state of spatially distributed oscillators.

Above the threshold to such an instability, the time-independent uniform state is also unstable to nearby wave states

$$\mathbf{u}_q(\mathbf{x}, t) = \mathbf{u}_q(\mathbf{x}_\parallel) e^{i(\mathbf{q} \cdot \mathbf{x}_\perp - \omega_q t)} e^{\operatorname{Re} \sigma_q t} + \text{c.c.}, \tag{10.9}$$

for a narrow band of wave numbers q with a width that grows with the distance from the onset, as in Fig. 2.9.[1] These are propagating dephasing disturbances of the oscillators. For a forward bifurcation, weakly nonlinear spatially uniform oscillations or wave states may be found above onset, although we will see that all such states may in fact be unstable for some parameter values. From the linear stability analysis alone, either traveling or standing waves might seem possible but the weakly nonlinear analysis we will study below shows that only traveling waves have the possibility of being stable in the weakly nonlinear regime.

Nonlinear waves behave strangely to those familiar with linear waves. Counter-propagating waves do not pass through each other as for linear waves, but annihilate each other at defects called sinks or shocks. At a physical boundary, although reflection may occur, the reflected wave is usually suppressed by the nonlinear interaction with the stronger approaching waves, so that the net effect is absorption of the incident wave. Because of these two effects, the initial field will tend to die out since an initial field of waves will tend to propagate away or dissipate at other waves or boundaries. Defects may provide a persistent source of waves, and are therefore particularly important in maintaining nonlinear wave patterns. In one spatial dimension, the defects are known as sources. These may also form as line defects in a two-dimensional system but here nonaxisymmetric point defects known as spirals and axisymmetric point defects known as targets are particularly important.

For some parameter values, the homogeneous oscillation may be unstable immediately above threshold via an instability known as the Benjamin–Feir instability. This instability was first discovered (in 1967) in the analysis of nonlinear waves, in which the energy of a wave with wave number q is transferred by the instability into new waves with almost equal wave numbers.[2] The new wave states are even

---

[1] For simplicity we will confine our remarks to a one-dimensional system such that $\mathbf{q} \cdot \mathbf{x}_\perp$ reduces to qx, or to a system that is rotationally invariant in two or three extended directions. In these cases, the growth rate depends only on the magnitude q of the wave vector $\mathbf{q}$.

[2] The Benjamin–Feir instability is often called a "sideband instability." The concept is borrowed from radio communication theory, in which a sideband of some carrier wave is a band of frequencies slightly higher or

more unstable in this situation so that there are then *no* stable weakly nonlinear states simply related to the linear instability modes. In some cases, this means that the system is chaotic immediately above threshold, a feature that has led to much theoretical and numerical study since supercritical transitions from a uniform state to spatiotemporal chaos are rare and are especially amenable to a theoretical analysis. (The domain chaos state – see Fig. 9.15 and the related discussion – is another example of a supercritical transition to spatiotemporal chaos but arises via a type-I-s instability.) However, the instability of the wave states may be convective (rather than absolute) so that even the "unstable" wave states may have physical relevance.

A useful way to begin to understand these properties of nonlinear waves is to investigate the weakly nonlinear states near a type-III-o instability using the same type of amplitude equation approach as in Chapter 6 for the stationary instability. Since the following discussion follows the one used there quite closely, we will focus on the new results and give only brief derivations. Although the results we will find apply quantitatively only in the weakly nonlinear regime near the threshold of instability, many of the insights are found to remain valid even in strongly nonlinear systems such as is typical of chemical reaction–diffusion systems. Many of the same phenomena will be discussed in this context in Chapter 11.

10.2.2 Amplitude equation

The weakly nonlinear states near threshold may be understood in terms of an evolution equation for the amplitude $A(\mathbf{x}_\perp, t)$. As was the case for the type-I-s instability (cf. Eq. (6.3) and also Eq. (10.7)), we define the amplitude in terms of the following perturbation of the uniform time-independent state

$$\mathbf{u}_p(\mathbf{x}, t) = A(\mathbf{x}_\perp, t)\mathbf{u}_0(\mathbf{x}_\parallel)e^{-i\omega_c t} + \text{c.c.} + \text{h.o.t.} \qquad (10.10)$$

The amplitude is again complex although its phase has a rather different significance than for the type-I-s system. Here the phase is the local phase of the temporal oscillation, and a change in phase corresponds to a shift of the time coordinate. The magnitude and phase of the amplitude A describe slowly varying spatial and temporal modulations of the spatially uniform "fast" oscillation $e^{-i\omega_c t}$. However, in the absence of a finite critical wave number q_c that sets an obvious length scale, the reference scale for defining the "slow" spatial variation must be deduced from the underlying physical equations.

The amplitude equation for A can be derived by symmetry arguments analogous to those used in Chapter 6, or by multiple scales perturbation methods for specific

lower than the frequency of the carrier wave. Radio sidebands are generated by modulation of the carrier wave, while here sidebands arise via an instability.

systems. The result is the following evolution equation:

$$\tau_0 \, \partial_t A = \varepsilon(1 + ic_0)A + (1 + ic_1)\xi_0^2 \, \nabla^2 A - (1 - ic_3)g_0|A|^2 A. \tag{10.11}$$

The derivative terms are simpler than for the type-I-s amplitude equation, Eq. (7.10), since we are describing slow modulations of a spatially uniform state in a rotationally invariant system. This implies that the lowest-order derivative terms have the symmetric form $\nabla^2 A = (\partial_x^2 + \partial_y^2)A$ for two extended directions.

The coefficients on the right-hand side of Eq. (10.11) are in general complex. These coefficients have been written in terms of real quantities with dimensions (ξ_0 a length, τ_0 a time, etc.), multiplied by complex numbers of the form $(1 \pm ic_i)$, where the real constants c_i give the imaginary parts of the coefficients. The numbers c_0, c_1, and c_3 are the main way that the amplitude equation Eq. (10.11) differs from the type-I-s equation. The choice of labeling – for example, that there is no c_2 and the various signs in front of the c_i – is a widely used convention. You will need to be alert for different choices of labels and signs in the research literature, however.

The presence of complex coefficients implies that the type-III-o amplitude equation Eq. (10.11) is not invariant under the transformation $A \to A^*$, which was the case for the type-I-s equation Eq. (7.10). This difference can be understood by reconsidering the symmetry argument used in Section 6.2.1 to derive the real amplitude equation for the stripe state. There we noted that the conjugate amplitude A^* describes the amplitude of the space-reversed component $e^{-iq_c x}$, whose properties were related to those of the $e^{iq_c x}$ component by space-inversion symmetry. In the present case, A^* gives the amplitude of the time-reversed component $e^{i\omega_c t}$, and dissipative systems do not have time-reversal symmetry that would relate the behavior of the $e^{\pm i\omega_c t}$ components. Thus the amplitude equation for a type-III-o instability is not required to be invariant under $A \to A^*$. Invariance under a phase change, $A \to e^{i\Delta}A$, is still required, although this now corresponds to the symmetry of the system under a time translation instead of under a spatial translation.

Some physical implications of the complex coefficients can be understood by considering a simple saturated nonlinear wave state

$$A = a_k e^{i\mathbf{k} \cdot \mathbf{x}} e^{-i\Omega_k t}. \tag{10.12}$$

Here $\mathbf{k}$ gives the wave vector of the wave state, and Ω_k the *change* in the frequency from ω_c. Substituting into Eq. (10.11) shows that the frequency difference Ω_k is given by

$$\tau_0 \Omega_k = -\varepsilon c_0 + c_1 \xi_0^2 k^2 - c_3 g_0 |a_k|^2. \tag{10.13}$$

Thus the coefficients c_0, c_1, and c_3 capture the dependence of the oscillation frequency on the control parameter ε, on the wave number k, and on the magnitude of the disturbance $|a_k|$ (i.e. c_3 gives the nonlinear frequency shift).

As usual, it is desirable for theoretical and numerical analysis to eliminate as many constants as possible in Eq. (10.11) by redefinitions and rescalings. The coefficient c_0 can be eliminated using a redefined amplitude

$$A \to A e^{-i c_0 \varepsilon t / \tau_0}. \tag{10.14}$$

This corresponds to using the amplitude that describes modulations of the oscillations at frequency ω_ε for the given value of the control parameter ε:

$$\omega_\varepsilon = \omega_c - c_0 \varepsilon \tau_0^{-1}. \tag{10.15}$$

The dimensional quantities τ_0, ξ_0, and g_0 can be eliminated as in Chapter 6 by introducing scaled space, time, and amplitude coordinates. Thus we write

$$\tilde{A} = g_0^{1/2} A e^{-i c_0 \varepsilon t / \tau_0}, \quad \tilde{\mathbf{x}} = \mathbf{x}/\xi_0, \quad \tilde{t} = t/\tau_0, \tag{10.16}$$

to obtain an equation in which only the parameters ε, c_1, and c_3 remain,

$$\partial_{\tilde{t}} \tilde{A} = \varepsilon \tilde{A} + (1 + i c_1) \nabla_{\tilde{\mathbf{x}}}^2 \tilde{A} - (1 - i c_3) |\tilde{A}|^2 \tilde{A}. \tag{10.17}$$

Finally, the control parameter ε can be eliminated to give a fully scaled amplitude equation by introducing the slow scales $\mathbf{X} = \varepsilon^{1/2} \mathbf{x}/\xi_0$, $T = \varepsilon t/\tau_0$, and $\bar{A} = (\varepsilon/g_0)^{-1/2} A e^{-i c_0 \varepsilon t/\tau_0}$. This leads to the canonical form

$$\partial_T \bar{A} = \bar{A} + (1 + i c_1) \nabla_X^2 \bar{A} - (1 - i c_3) |\bar{A}|^2 \bar{A}. \tag{10.18}$$

Note that the parameters c_1 and c_3 cannot be eliminated in general, and the physical behavior of the system depends crucially on their values. There are again different conventions used for the symbols c_1 and c_3 and for the signs in front of these numbers.

Equation (10.18) is known as the complex Ginzburg–Landau equation, which we will abbreviate as the CGLE in the rest of this chapter. It is a complex equation for a complex field $\bar{A}$, in contrast to the amplitude equation Eq. (6.21) derived in Chapter 6, which is a *real* equation (all the coefficients are real) for a complex field. Unlike the amplitude equation for type-I-s systems, which mimics equations that had appeared before in the context of equilibrium phase transitions, the CGLE had not appeared previously in the description of other physical systems. The equation provides a canonical model for the properties of spatially extended nonlinear oscillatory systems. For these reasons, the equation has been extensively studied as a new paradigm of dynamical systems, as was briefly mentioned in Chapter 5.

The properties of uniform oscillations of wave states in infinite or periodic geometries are easily calculated from the amplitude equation, Eq. (10.18). Spatially uniform nonlinear oscillations are given by

$$\bar{A}(X,T) = a_0 e^{-i\Omega_0 T}, \tag{10.19}$$

with

$$a_0^2 = 1 \quad \text{and} \quad \Omega_0 = -c_3. \tag{10.20}$$

Returning to physical units, this corresponds to an amplitude of oscillation growing as $\sqrt{\varepsilon} \propto \sqrt{p - p_c}$, and a frequency shift that is linear in $\varepsilon \propto p - p_c$. These results are the standard ones for a forward Hopf bifurcation (Appendix 1), which is not a surprise since the spatial aspects of the problem are not involved for uniform oscillations.

Equation (10.18) also admits traveling wave solutions

$$\bar{A}_K(X,T) = a_K e^{i(\mathbf{K} \cdot \mathbf{X} - \Omega_K T)}, \tag{10.21a}$$

$$a_K^2 = 1 - K^2, \tag{10.21b}$$

$$\Omega_K = -c_3 + (c_1 + c_3)K^2. \tag{10.21c}$$

The group speed of the waves in scaled units is

$$S = d\Omega_K/dK = 2(c_1 + c_3)K. \tag{10.22}$$

This corresponds in unscaled units to the group speed

$$s = \frac{2\xi_0}{\tau_0}(c_1 + c_3)k. \tag{10.23}$$

Standing waves can be constructed based on the addition of waves with wave vectors $\mathbf{K}$ and $-\mathbf{K}$ but they are unstable with respect to traveling waves for the positive sign of the cubic term that is needed in Eq. (10.18) for a forward bifurcation that saturates growth at small amplitudes near onset. You get to show this in Exercise 10.1.

10.2.3 Phase equation

The equation for the slow evolution of long-wavelength phase variations can be derived from the amplitude equation as in Section 6.4.3. Again, the idea is that, for variations over spatial scales long compared with $\varepsilon^{-1/2}\xi_0$, the phase relaxes slowly whereas the magnitude relaxes to a value consistent with the local wave number given by the gradient of the phase, on a time scale that does not increase with the length scale. The type-III-o amplitude equation is sufficiently simple that it is possible and informative to keep nonlinear terms in the phase equation. Using the fully scaled form Eq. (10.18) and keeping all terms up to second order in the spatial derivatives of the phase gives the nonlinear phase equation

$$\partial_T \Phi = -\Omega + \alpha \nabla_X^2 \Phi - \beta (\nabla_X \Phi)^2, \qquad (10.24)$$

with

$$\alpha = 1 - c_1 c_3, \quad \beta = c_1 + c_3, \quad \text{and} \quad \Omega = -c_3. \qquad (10.25)$$

These equations are derived in the following Etude.

Etude 10.2 Derivation of the nonlinear type-III-o phase equation
Writing $\bar{A} = a e^{i\Phi}$, we can calculate the following derivatives of A:

$$\partial_T \bar{A} = (\partial_T a + i a\, \partial_T \Phi) e^{i\Phi}, \qquad (10.26)$$

$$\nabla_X^2 \bar{A} = \left\{ \left[\nabla_X^2 a - a(\nabla_X \Phi)^2\right] + i\left[2(\nabla_X \Phi) \cdot (\nabla_X a) + a(\nabla_X^2 \Phi)\right] \right\} e^{i\Phi}. \qquad (10.27)$$

We now substitute these expressions into Eq. (10.18), multiply through by $e^{-i\Phi}$, and collect real and imaginary parts. The imaginary part gives

$$a\,\partial_T \Phi = \left[2(\nabla_X \Phi)\cdot(\nabla_X a) + a(\nabla_X^2 \Phi)\right] + c_1\left[\nabla_X^2 a - a(\nabla_X \Phi)^2\right] + c_3 a^3, \qquad (10.28)$$

and the real part gives

$$\partial_T a = a + \left[\nabla_X^2 a - a(\nabla_X \Phi)^2\right] - c_1\left[2(\nabla_X \Phi)\cdot(\nabla_X a) + a(\nabla_X^2 \Phi)\right] - a^3. \qquad (10.29)$$

These equations are exact deductions from the CGLE. We now assume slow spatial variation so that each partial derivative ∇_X introduces a power of the expansion parameter. We will keep all terms that are of up to second order in this expansion. Even on setting $\nabla_X = 0$ in Eq. (10.29), we would find that the magnitude a relaxes on a time scale of order unity to its steady state value. For the slow time variation induced by long-distance phase variations, this means that we can set a equal to its steady-state value and ignore $\partial_T a$ in Eq. (10.29). Furthermore, all terms involving ∇_x acting on a can be neglected since, as we will see self-consistently from Eq. (10.30) below, these lead to terms that are of third order or higher in ∇_x acting on the phase. These arguments lead to the simplified magnitude equation

$$a^2 = 1 - (\nabla_X \Phi)^2 - c_1(\nabla_X^2 \Phi). \qquad (10.30)$$

The second term on the right-hand side is the same K^2 correction to the magnitude as in Eq. (10.21b). The third term is a new one that depends on the derivative of the wave number K. Substituting Eq. (10.30) into Eq. (10.28) and ignoring terms involving more than two factors of ∂_X leads to Eq. (10.24).

The constant α in the phase equation, Eq. (10.24), plays the role of the diffusion constant. A negative diffusion constant indicates instability. This occurs when the Newell criterion is satisfied:

$$1 - c_1 c_3 < 0. \qquad (10.31)$$

This important criterion signals the instability of the homogeneous oscillations to the long wavelength Benjamin–Feir sideband instability. The nonlinear wave states described by Eq. (10.21) are also unstable if Eq. (10.31) is satisfied (see Eq. (10.36a) below) so that the Newell criterion indicates the instability of *all* the simple weakly nonlinear oscillation or wave states that exist near threshold. Numerical simulations of the CGLE in large domains with periodic boundaries show that the resulting dynamics is then often chaotic.

10.2.4 Stability balloon

We can use the amplitude equation to look at the full stability band of the nonlinear wave states near threshold. Although the complete analysis is involved algebraically, fortunately in most situations the stability boundaries occur for long-wavelength perturbations of the wave state that can be captured by a simpler phase equation approach, and so we will present this method. Deriving the stability of wave states from a phase equation analogous to Eq. (10.24) requires going to higher order in derivative terms. Instead, we directly derive the linear equation for small perturbations about a nonlinear wave state.

For a small perturbation about the nonlinear wave state Eqs. (10.21), we have $\bar{A} = ae^{i\Phi}$ with the phase variable taking the form

$$\Phi = KX - \Omega_k T + \delta\Phi, \qquad (10.32)$$

and the magnitude

$$a = a_K + \delta a. \qquad (10.33)$$

with $a_K^2 = 1 - K^2$. We will do a similar expansion to the one in the previous section but will now restrict our attention to terms linear in δa and $\delta\Phi$ as well as keeping only terms up to second order in derivatives. The result is

$$\partial_T \delta\Phi + S\, \partial_X \delta\Phi = D_\|(K)\partial_X^2 \delta\Phi + D_\perp(K)\partial_Y^2 \delta\Phi, \qquad (10.34)$$

with a new advection term on the left-hand side that arises from the wave propagation. The group speed S is given by

$$S = 2(c_1 + c_3)K. \qquad (10.35)$$

The constants $D_\|(K)$ and $D_\perp(K)$ are respectively the longitudinal and transverse diffusion constants and are given explicitly by

$$D_\| = (1 - c_1 c_3) \frac{1 - \nu K^2}{1 - K^2}, \tag{10.36a}$$

$$D_\perp = (1 - c_1 c_3), \tag{10.36b}$$

with

$$\nu = \frac{3 - c_1 c_3 + 2c_3^2}{1 - c_1 c_3}. \tag{10.37}$$

The following Etude outlines how to derive Eq. (10.34).

Etude 10.3 (Linear Phase Equation for Perturbations about a Wave State)
The derivatives $\partial_T \bar{A}$ and $\nabla_X^2 \bar{A}$ are generalizations of Eqs. (10.26) and (10.27):

$$\partial_T \bar{A} = \left\{ \partial_T \delta a + i \left[(-\Omega_K + \partial_T \delta \phi)(a_K + \delta a) \right] \right\} e^{i\Phi}, \tag{10.38}$$

$$\simeq \left\{ \partial_T \delta a + i \left[-\Omega_K (a_K + \delta a) + a_K \partial_T \delta \Phi \right] \right\} e^{i\Phi}, \tag{10.39}$$

$$\nabla_X^2 \bar{A} = \left\{ \left[\nabla_X^2 \delta a - (a_K + \delta a)(K + \partial_X \delta \Phi)^2 + (\partial_Y \Phi)^2 \right] \right.$$
$$\left. + i \left[2(K + \partial_X \delta \Phi)(\partial_X \delta a) + 2(\partial_Y \delta a)(\partial_Y \delta \phi) + a_K (\nabla_X^2 \delta \Phi) \right] \right\} e^{i\Phi}, \tag{10.40}$$

$$\simeq \left\{ \left[\nabla_X^2 \delta a - K^2 \delta a - a_K (K^2 + 2K \partial_X \delta \Phi) \right] + i \left[2K \partial_X \delta a + a_K (\nabla_X^2 \delta \Phi) \right] \right\} e^{i\Phi}. \tag{10.41}$$

In the approximate evaluations, we ignore all nonlinear terms in $\delta \phi$, δa and their derivatives. It is now not too hard to substitute into the CGLE, Eq. (10.18), multiply through by $e^{-i\Phi}$, collect real and imaginary terms, and proceed as before neglecting all terms leading to more than two spatial derivatives acting on $\delta \Phi$.

Equation (10.36a) is analogous to Eq. (6.59) for the longitudinal diffusion constant in type-I-s systems although the width of the stability band now depends on parameters through Eq. (10.37) and so is not universal near onset. The condition $D_\| = 0$ signals the onset of the longitudinal Benjamin–Feir instability, which takes the form of the growth of sidebands, i.e. wave components at wave numbers $K \pm Q$ with $Q \to 0$ at the onset of instability. The instability occurs for

$$|K| \geq \Lambda_B = \nu^{-1}, \tag{10.42}$$

which, for $1 - c_1 c_3 > 0$, leaves a stable band of wave numbers with a width that is a fraction ν^{-1} of the existence band $|K| < 1$. The analogy to the Eckhaus instability of Section 6.4.2 is evident and the longitudinal instability of the wave states is often called an Eckhaus instability rather than the Benjamin–Feir instability. Note that

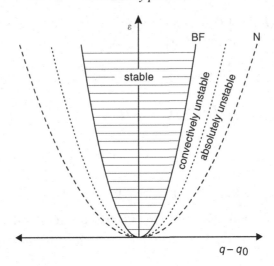

Fig. 10.2 Stability balloon of type-III-o system near threshold. The width of the stable region relative to the existence band depends on parameters c_1 and c_3, and for some values the instability does not occur first at long wavelengths. At the Benjamin–Feir instability boundary (BF: solid line) the instability is convective in nature – the absolute instability boundary is shown by the dotted line. The dashed line (N) is the linear instability of the uniform state.

the perpendicular diffusion constant Eq. (10.36b) is negative only if the Newell criterion Eq. (10.31) is satisfied so that the transverse instability does not limit the band of stable waves further.

An important consequence of the propagation term in Eq. (10.34) is that the criterion Eq. (10.42) is the one for convective instability. The wave states become absolutely unstable only for larger wave numbers

$$|K| \geq \Lambda_A > \nu^{-1}. \tag{10.43}$$

Exercise 10.3 gives you a chance to calculate the value of Λ_A.

More generally we could look at the instability of the wave states Eqs. (10.21) within the full amplitude equation using methods analogous to Section 6.4.2. For most values of c_1 and c_3, the stability limits Eq. (10.42) and Eq. (10.37) set by the long-wavelength phase calculation are sufficient. However, for some small regions of the $c_1 c_3$ parameter plane, a short-wavelength instability that lies beyond a phase equation description sets a smaller limit for Λ_B.

10.2.5 Defects: sources, sinks, shocks, and spirals

Defects may provide a persistent source of the propagating waves and so may have a profound influence on the pattern even far from the defect. In one spatial dimension, the important defects are sources, which produce outgoing wave disturbances, and sinks or shocks for incoming waves. Incoming waves do not simply pass through each another as in a linear system, but annihilate one another so that the shock acts as an absorber or sink of the two sets of waves. This is dramatically evident in pictures from chemical experiments as in Fig. 1.18(a), and is also apparent from simulations of the CGLE, as in Fig. 10.4.

One-dimensional defects

For wave numbers that are small enough, one-dimensional defects are nicely treated within the nonlinear phase equation (10.24). (Small enough means small compared with the typical wave number that induces magnitude variations, i.e. $q \ll \xi_0^{-1}\varepsilon^{1/2}$ in physical units.) In one dimension, this is an equation known as the Burgers equation.[3] Miraculously, this nonlinear equation can be transformed into a linear equation by a clever transformation known as the Cole–Hopf transformation. Interesting solutions such as sinks and shocks can then be formed by superimposing solutions within this linear equation. This transformation is described in the following Etude.

Etude 10.4 Cole–Hopf transformation of the Burgers equation

Let us first make the replacement

$$\Phi \to \Phi - \Omega T, \qquad (10.44)$$

to reduce the phase equation to

$$\partial_T \Phi = \alpha \nabla_X^2 \Phi - \beta (\nabla_X \Phi)^2. \qquad (10.45)$$

The Cole–Hopf transformation for Eq. (10.24) is

$$\chi(X, Y, T) = \exp[-\beta \Phi(X, Y, T)/\alpha]. \qquad (10.46)$$

This transforms the nonlinear phase equation into a linear *equation for* χ

$$\partial_T \chi = \alpha \nabla_X^2 \chi. \qquad (10.47)$$

Remember that α *must be positive for stability, and that the phase is a real variable. Simple solutions to Eq. (10.47) that are consistent with these restrictions are*

$$\chi = \exp\left[\left(\mp \beta K X + \beta^2 K^2 T\right)/\alpha\right]. \qquad (10.48)$$

---

[3] In one space dimension, differentiating the nonlinear phase equation Eq. (10.24) with respect to the spatial coordinate x gives a commonly written form of the Burgers equation, $u_t + uu_x = cu_{xx}$ with $u = \partial_x \Phi$. This looks like the Navier–Stokes equation for a single velocity field $u(x, t)$ in one space dimension, which helps to explain why the Burgers equation has been studied intensely in applied mathematics and fluid dynamics.

These correspond to the phase variations

$$\Phi = \pm KX - \Omega_K T, \tag{10.49}$$

with

$$\Omega_K = \beta K^2. \tag{10.50}$$

These solutions are traveling waves of wave number K in the $+X$ or $-X$ direction, and correspond to the results Eqs. (10.21).

Since the χ evolution equation is linear, we can superimpose a pair of solutions, for example ones with wave number K_+ for the right-moving wave and K_- for the left-moving wave (choosing $K_\pm$ to be positive)

$$\chi = \exp\left[\left(-\beta K_+ X + \beta^2 K_+^2 T\right)\big/\alpha\right] + \exp\left[\left(+\beta K_- X + \beta^2 K_-^2 T\right)\big/\alpha\right]. \tag{10.51}$$

For large positive X, the second term is exponentially larger than the first term (we present results for β positive), and so the solution reduces to a nonlinear wave with left-moving waves. Similarly, for large negative X, the solution reduces to right-moving waves. Thus Eq. (10.51) is the solution for a sink or shock, where a pair of incoming waves collide. For $K_+ \neq K_-$, a more detailed analysis (see Exercise 10.6) shows that the shock moves. We will look explicitly at the simpler stationary shock solution with $K_+ = K_- = K$. In this case, the expression for the phase is

$$\Phi = -\frac{\alpha}{\beta} \ln\left\{\exp\left[\left(-\beta KX + \beta^2 K^2 T\right)\big/\alpha\right] + \exp\left[\left(\beta KX + \beta^2 K^2 T\right)\big/\alpha\right]\right\} \tag{10.52a}$$

$$= -\beta K^2 T - \frac{\alpha}{\beta} \ln[2\cosh(\beta KX/\alpha)]. \tag{10.52b}$$

Since the Φ evolution equation is not linear, the superposition Eq. (10.51) in the χ variable does not, of course, mean that the Φ solution is the superposition of the individual wave solutions, as is clear from Eqs. (10.52). For large $|X|$, the phase is given by (again for βK positive)

$$\begin{aligned}\Phi \to \Phi_- &= -KX - \beta K^2 T, \quad \text{for } X \to +\infty, \\ \Phi \to \Phi_+ &= KX - \beta K^2 T, \quad \text{for } X \to -\infty.\end{aligned} \tag{10.53}$$

Furthermore, to look at the size of the right-moving waves on the "wrong" side of the shock $X > 0$, we can expand the right side of Eqs. (10.52) to one higher order in the exponentially small terms to find

$$\Phi(X > 0) \approx \Phi_- - \frac{\alpha}{\beta} \exp(-2\beta KX/\alpha). \tag{10.54}$$

We see that the effect of the right-moving wave decays exponentially to the right of the shock with the decay length $\alpha/(2\beta K)$.

A more complete amplitude equation treatment gives a complicated picture of sources and sinks. In general, there continues to be sink solutions of any velocity with incoming waves of equal but arbitrary group speed relative to the sink velocity. This is a two-parameter family of solutions since solutions exist over a continuous range of two parameters, for example the two wave numbers of the incoming waves. In addition, there exists a one-parameter family of exactly known *Nozaki–Bekki* holes which are sources of outgoing waves. These hole solutions exist over a continuous range of propagation velocities. However, these holes are a nongeneric feature of the CGLE in the sense that adding additional higher-order terms such as a fourth-order nonlinearity $|A|^4 A$ to the equation eliminates the continuous family of solutions, even if the coefficients of the new terms are small. Also, the Nozaki–Bekki holes are not found by the counting arguments of Section 8.3.1.

Two-dimensional defects

In two spatial dimensions, the most important defects are spiral defects, which are point defects that act as persistent sources of waves. The robustness of spiral defects is suggested by their nontrivial topological nature since they can be characterized by a nonzero winding number of the phase around any contour surrounding the core (cf. the discussion in Section 4.4.2)

$$\frac{1}{2\pi} \oint \nabla \Phi \cdot d\mathbf{l} = m, \tag{10.55}$$

with m a positive or negative integer. The defect with winding number m has the appearance of an m-armed spiral. In oscillatory systems one-armed spirals are typically seen, as shown in Fig. 10.3 and Fig. 10.4; a three-arm spiral appears in Fig. 1.14, although this one is not related to an oscillatory medium. At large distances from the core, an m-armed spiral takes the form

$$\bar{A} = a \exp[i(Kr + m\theta - \Omega T)], \tag{10.56}$$

in polar coordinates (r, θ), where $r = \sqrt{X^2 + Y^2}$ is the distance from the core and where θ is the polar angle around the core. The frequency $\Omega = \Omega(K)$ is given by the dispersion relation Eq. (10.21c) and $a = a_K = \sqrt{1 - K^2}$. Approaching the core, the magnitude a will go to zero[4] so that the complex amplitude is continuous here. The wave number K may also depend on radius.

---

[4] The vanishing of the amplitude at the core provides a practical way to locate spiral cores numerically, by looking for crossings of the zero contour of the amplitude's real part with the zero contour of the amplitude's imaginary part, see Fig. 10.4.

376   *Oscillatory patterns*

Fig. 10.3 Schematic drawing of a one-armed spiral with winding number $m = 1$, see Eq. (10.55). This is a time-dependent state that appears to rotate (here counter-clockwise) as the waves propagate outward from the central core.

Spiral sources have long been studied in chemical oscillations, where they are a dominant feature of experiments on oscillatory chemical reactions such as the Belousov–Zhabotinsky reaction in unstirred reactors, see Fig. 1.18. There are also experiments on heart muscle that suggest the importance of spiral sources in some pathological heart conditions, see Fig. 1.10 and Section 11.6.5.

A key question in the chemical experiments and in the CGLE is whether there is a family of spirals giving a continuous range of possible frequencies Ω (which then determines the wave number of the outgoing waves at large distances through the dispersion relation), or whether there is a unique frequency with a prescribed frequency that selects a particular wave number, or perhaps a discrete set of possible frequencies and spiral structures. From chemical experiments, difficult analytic calculations on models of chemical reaction systems in various tractable limits, perturbative treatments of the CGLE, and numerical simulation of various models, the evidence consistently points to a unique frequency rather than to a continuous family. In this case, the wave number K_s of the waves far from the core will be fixed by the unique spiral frequency Ω_s and by the dispersion relation

$$\Omega(K_s) = \Omega_s. \tag{10.57}$$

One argument for the unique spiral frequency follows from an analysis of the amplitude equation, Eq. (10.18). It can be shown that periodic solutions of the CGLE corresponding to different values of c_1 and c_3 but with the same value of $|c_1 + c_3|$

Fig. 10.4 Numerical simulation of the CGLE, Eq. (10.18), showing wavefronts arising from $m = 1$ spiral sources. The black and gray lines are respectively the zero contours of the real and imaginary parts of the complex amplitude field $\bar{A}(X, Y, T)$. These contours cross where $\mathrm{Re}(\bar{A}) = \mathrm{Im}(\bar{A}) = 0$ which defines a defect core where $\bar{A} = 0$. The wave fronts propagate out from the spirals, as shown by the arrows on the plot, and collide at shocks forming the boundaries between neighboring spirals. The calculation was performed in a periodic square domain of size $L = 256$ for the parameter values $c_1 = 2$ and $c_3 = 0.2$, for which plane waves are linearly stable (no Benjamin–Feir instability). (From Chaté and Manneville [19].)

can be transformed into one another. This means that solutions with $c_1 + c_3 = 0$ can be constructed from solutions to the real amplitude equation $c_1 = c_3 = 0$, which is the type-I-s amplitude equation, Eq. (7.31), for a uniaxial system. The phase-winding topological defects of this equation are the dislocations calculated in Section 8.1. It is then possible to perturb away from this solution in the small quantity $|c_1 + c_3|$ to construct the spiral solution for the complex equation. This calculation predicts a unique stable spiral structure, with a wave number K_s that varies as

$$K_s \to \frac{1.018}{|c_1 + c_3|} \exp\left[-\frac{\pi}{2|c_1 + c_3|}\right]. \tag{10.58}$$

The $m = 0$ version of Eq. (10.56) takes the form of an axisymmetric structure known as a target. Since there is no phase winding, there is no topological argument which suggests that such defects should be stable, and their existence and stability must be determined for each system individually. Target solutions are not believed to be stable within the amplitude description. Since targets are a prominent feature of many chemical experiments, it is generally believed that targets result from imperfections such as dust particles that anchor the core.

If spirals are the source of the waves in a system, then the wave numbers that will be observed in the bulk of the system are not determined by the stability balloon but are restricted to the unique wave number produced by the spiral sources. The stability limit of the wave state is then determined by the limit of stability of the particular waves at wave number K_s rather than by the Newell criterion, or perhaps by stability limits of the core regions of the spirals. Since waves with nonzero K are more unstable than the homogeneous oscillations, the waves produced by the spirals may be unstable, and complex dynamics may occur even when the Newell criterion is not satisfied. On the other hand, it is probably the absolute instability point of these waves that is the important criterion, rather than the convective instability of Eq. (10.42). The complicated interplay between these various instability criteria for the CGLE is shown in Fig. 10.5.

If there are several spirals present in the system, the waves from the different sources will eventually collide. This leads to a shock between the two spirals, as shown in Fig. 10.4. Since the waves decay exponentially as they pass through the shock, the shock effectively isolates one spiral from the influence of another. Calculations show that, because of the shocks, the interaction of one spiral with

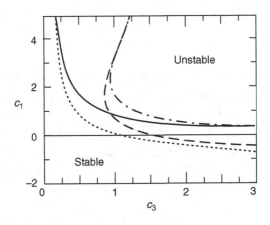

Fig. 10.5 Stability lines of the CGLE, Eq. (10.18), as a function of the parameters c_1 and c_3. Solid line: Newell criterion $c_1 c_3 = 1$; dotted line: (convective) Benjamin–Feir instability of spiral-selected wave number; dashed: absolute instability of spiral-selected wave number; dashed-dotted: absolute instability of whole wave number band. Unstable sides are toward larger positive $c_1 c_3$. (After Aranson et al. [5].)

10.3 Unidirectional waves in a type-I-o system

A type-I-o instability leads to a wave state of growing amplitude given by

$$\mathbf{u}_p(\mathbf{x}, t) = \mathbf{u}_\mathbf{q}(\mathbf{x}_\parallel) e^{i(\mathbf{q} \cdot \mathbf{x}_\perp - \omega_\mathbf{q} t)} e^{\text{Re}(\sigma_\mathbf{q} t)} + \text{c.c.} + \text{h.o.t.}, \tag{10.59}$$

where the maximum growth rate occurs at a nonzero wave number and where $\omega_\mathbf{q} = -\text{Im}\,\sigma_\mathbf{q}$ is the oscillation frequency. The threshold of instability occurs when the maximum of the real part of the growth rate, $\max \text{Re}\,\sigma_\mathbf{q}$, passes through zero, and this determines the critical wave vector $\mathbf{q}_c$ and the critical frequency at threshold

$$\omega_c = \omega_{\mathbf{q}_c}. \tag{10.60}$$

The conditions $\omega_c \neq 0$ and $q_c \neq 0$ define the type-I-o instability.

Unlike the stationary type-I-s case, it is not true in general that the $-\mathbf{q}$ mode goes unstable at the same control parameter value as the $+\mathbf{q}$ mode; this is a consequence of the fact that the complex conjugate term in Eq. (10.59) is not equivalent to changing the sign of $\mathbf{q}$ and so does not give a wave traveling in the reverse direction. An additional symmetry such as parity symmetry (equations invariant under $\mathbf{x}_\perp \to -\mathbf{x}_\perp$) in one dimension or rotational symmetry in more than one dimension is needed to guarantee that the wave propagating in the reverse direction

$$\bar{\mathbf{u}}_p = \mathbf{u}_{-\mathbf{q}}(\mathbf{x}_\parallel) e^{i(-\mathbf{q} \cdot \mathbf{x}_\perp - \omega_{-\mathbf{q}} t)} e^{\text{Re}\,\sigma_{-\mathbf{q}} t} + \text{c.c.} + \text{h.o.t.}, \tag{10.61}$$

will go unstable at the same control parameter value and with the same growth rate.

In the present section we look at the non-symmetric case, where there is only a wave propagating in one direction. Since there is a single propagation direction, we focus on one-dimensional systems. In the next section we discuss parity symmetric one-dimensional systems with counter-propagating waves. As in the discussion of stationary patterns, we will base much of the discussion in these two sections on the appropriate amplitude equations. Finally, in Section 10.5 we briefly describe some experiments on rotationally invariant two-dimensional systems where waves may propagate in any direction in the plane.

Equation (10.59) defines waves that propagate at the phase speed ω_c/q_c. A small amplitude wave packet made up of a superposition of modes near q_c will propagate at the group speed $s = \partial \omega_q / \partial q \big|_{q=q_c}$. In general, the group speed will be nonzero at the onset of the instability whereas the growth rate goes to zero here. This means that propagation effects dominate near threshold.

In an infinite system, or one with periodic boundary conditions (experimentally this would be implemented as a narrow annulus) we can simply transform to a frame moving with the group speed, the amplitude equation reduces to the same form as for a type-III-o instability, and we can take over many of the results from the previous section. If the bifurcation is supercritical, nonlinearity saturates the growth of the linear mode Eq. (10.59) to give a small amplitude traveling wave near threshold. As in the type-III-o waves, the Benjamin–Feir instability, which can be understood again in terms of a sideband instability of the nonlinear wave state, is important in limiting the band of wave numbers for stable wave states.

Because of the nonzero group speed, the initial instability to the wave state is convective so that near onset a small local disturbance will propagate away before it has a chance to grow locally. The instability only becomes absolute when the growth rate can overcome the propagation effect, as discussed in Section 10.3.2. End walls in a physical system absorb approaching waves, since there is no possibility of reflection in a system that does not support propagation in the reverse direction. As a result, we show in Section 10.3.3, a wave state only develops in such a finite system for control parameter values above the point of absolute instability. This means that the spatially uniform state for parameter values above the convective instability but below the absolute instability is very sensitive to small disturbances, and we discuss how noise or spatial inhomogeneity can lead to sustained wave states even when the ideal system does not support these.

10.3.1 Amplitude equation

Using the expression

$$\mathbf{u}_p = A(x,t)\mathbf{u}_0(\mathbf{x}_\parallel)e^{i(q_c x - \omega_c t)} + \text{c.c.} + \text{h.o.t.}, \tag{10.62}$$

for the perturbation $\mathbf{u}_p$ about the uniform state leads to the one-dimensional amplitude equation

$$\tau_0(\partial_t A + s\,\partial_x A) = \varepsilon(1 + ic_0)A + (1 + ic_1)\xi_0^2 \partial_x^2 A - (1 - ic_3)g_0|A|^2 A. \tag{10.63}$$

The coefficients s, ξ_0, τ_0, c_0, and c_1 can be obtained by matching to the growth rate of the linear instability calculation (see Section 6.2.2). For the parameters g_0 and c_3, a nonlinear calculation is needed in which case g_0 gives the saturation amplitude and then the parameter c_3 determines the nonlinear frequency shift.

The term $s\,\partial_x A$ on the left-hand side of Eq. (10.63) is a new term that does not appear in the type-III-o amplitude equation, Eq. (10.18). According to our standard procedure of neglecting "higher-order" terms, once this first-derivative term is present it might seem appropriate to ignore the second-order derivative terms. However, the physical roles of these two terms is quite different since one leads to propagation and the other to spreading and dispersion and so it is often necessary to include both terms to get a sensible description. For example, if we ignored the supposedly higher-order second-derivative term, an initial disturbance would propagate without change of shape in the linear description. However, if we include this term, there is now no single scaling of space, time, and magnitude with powers of ε that reduces Eq. (10.63) to a scaled form in which the small parameter ε disappears. This shows us that the equation does not then give a formally consistent approach.

In a formal multiple-scales derivation of the amplitude equation, *two* time scales are introduced. First, the reduced amplitude $\bar{A} = \varepsilon^{-1/2} A$ and the slow length scale $X = \varepsilon^{1/2} x$ are introduced in the usual way, guided by the expected magnitude of the nonlinear saturated solution and by the width of the wave-number band above threshold. In addition to the usual time scale $T = \varepsilon t$ on which spreading and dispersion occur, we introduce a new slow time scale $T_p = \varepsilon^{1/2} t$ that corresponds to the propagation time over the slow length scale. The scaled amplitude is written as a function of X and of the *two* time scales $\bar{A}(X, T_p, T)$. The equation derived at lowest order in the expansion in ε is a propagation effect

$$\partial_{T_p} \bar{A} + s \partial_X \bar{A} = 0, \tag{10.64}$$

which describes propagation of the envelope at the group speed. The solution to this equation is that $\bar{A}$ is a function of the reduced coordinate $\xi = X - sT_p$ rather than of the variables X and T_p separately

$$\bar{A}(X, T_p, T) = \bar{A}(\xi, T). \tag{10.65}$$

Physically, this simply corresponds to transforming to a frame moving at the group speed s. At the next order in the expansion, the dispersion, diffusion, and nonlinear saturation are found in this moving frame

$$\tau_0 \partial_T \bar{A} = (1 + ic_0)\bar{A} + (1 + ic_1)\xi_0^2 \partial_\xi^2 \bar{A} - (1 - ic_3) g_0 |\bar{A}|^2 \bar{A}. \tag{10.66}$$

Then by using

$$\xi = \varepsilon^{1/2}(x - st) \quad \text{and} \quad T = \varepsilon t, \tag{10.67}$$

or the inverse expressions

$$x = \varepsilon^{-1/2}\xi + \varepsilon^{-1} sT \quad \text{and} \quad t = \varepsilon^{-1} T, \tag{10.68}$$

so that

$$\partial_\xi \to \varepsilon^{-1/2} \partial_x \quad \text{and} \quad \partial_T \to \varepsilon^{-1}(\partial_t + s \partial_x), \tag{10.69}$$

we obtain Eq. (10.63).

In an annulus or other periodic geometry, Eq. (10.66) can be used to study growth, spreading, and saturation in the usual way. In the moving frame, the amplitude equation is the same as for the type-III-o instability and we take over the results calculated there. This approach is not valid if the system is inhomogeneous in space since, in the moving frame, the inhomogeneities lead to a fast time variation. An example where this occurs is in the lasing instability, where the population inversion leading to the instability may be inhomogeneous because of the pumping mechanism or because of the feedback of the wave itself. For one-dimensional geometries with fixed boundary conditions, such as a narrow rectangular box with rigid end walls, a transformation to the group speed frame is not useful since then the end walls are moving with this speed. Thus the interaction with the end walls is not completely given by the amplitude equation.

Although Eq. (10.63) may not be a consistent approximation in general, in the next sections we will use it to understand some of the behavior near a type-I-o instability. A formal justification for this is to restrict our attention to systems where the group speed is also small. In this case, for values of s of order $\varepsilon^{1/2}$, the terms in Eq. (10.63) are of consistent magnitude.

10.3.2 Criterion for absolute instability

Since the group speed is nonzero at the threshold of instability to a plane wave perturbation while the growth rate is zero, the instability is convective. Absolute instability (local growth) develops only at some finite distance above threshold. We can use the amplitude equation to find an approximate condition for absolute instability by using the method discussed in Section 10.1.

The linear dispersion relation from the amplitude equation

$$\sigma(q) = -isq + \tau_0^{-1}\varepsilon(1 + ic_0) - \tau_0^{-1}\xi_0^2(1 + ic_1)q^2, \qquad (10.70)$$

leads to the stationary phase point $\sigma'(q_s) = 0$ at the complex wave number

$$q_s = -\frac{is\tau_0}{2\xi_0^2(1 + ic_1)}. \qquad (10.71)$$

The criterion for absolute instability $\operatorname{Re} \sigma_{q_s} = 0$ is then $\varepsilon = \varepsilon_a$ with

$$\varepsilon_a = \left(\frac{s\tau_0}{2\xi_0}\right)^2 \frac{1}{1 + c_1^2}. \qquad (10.72)$$

If the group speed s is small then the parameter ε_a is also small and the calculation is within the range of validity of the amplitude equation. Since this approach involves just the linear dispersion relation, which is often easy to calculate for a system of

10.3.3 Absorbing boundaries

For a convectively unstable system, any disturbance in the system will eventually die out if there are no driving forces at the frequency of the waves. This is because the disturbance will propagate to one end, and waves moving in the opposite direction are not supported in the unidirectional case so that their amplitude will decrease exponentially away from the ends. This leads to the remarkable consequence that, although the conventional linear stability analysis predicts that the uniform state is unstable, in the absence of driving forces this unstable state survives in the finite geometry.

We can analyze this effect with the amplitude equation Eq. (10.63) for boundary conditions $A(x=0) = A(x=l) = 0$ that correspond to the absence of driving from the left boundary and to an absorbing right boundary. The linear onset solution of Eq. (10.63) with these boundary conditions is of the form[5]

$$A = a \sin(\pi x/l) e^{\kappa x} e^{-i\Omega t}, \tag{10.73}$$

where the frequency Ω is real but the spatial growth rate κ may be complex. Substituting this into the linearized version of Eq. (10.63) leads to the conditions

$$\kappa \xi_0 = \left(\frac{s\tau_0}{2\xi_0}\right) \frac{1 - ic_1}{1 + c_1^2}, \tag{10.74a}$$

$$\Omega = -\left(\frac{s\tau_0}{2\xi_0}\right)^2 \frac{c_0 + c_1}{1 + c_1^2} - (c_0 - c_1)\left(\frac{\pi \xi_0}{l}\right)^2, \tag{10.74b}$$

and fixes the threshold value of the control parameter $\varepsilon = \varepsilon_c$ to be

$$\varepsilon_c = \left(\frac{s\tau_0}{2\xi_0}\right)^2 \frac{1}{1 + c_1^2} + \left(\frac{\pi \xi_0}{l}\right)^2. \tag{10.75}$$

The magnitude a of the amplitude in Eq. (10.73) is of course undetermined by the linear equation.

The form of the amplitude A at ε_c is shown in Fig. 10.6. For any $\varepsilon < \varepsilon_c$, all solutions decay in time. The value of ε_c is close to the onset of absolute instability Eq. (10.72)

$$\varepsilon_c(l) = \varepsilon_a + O(l^{-2}), \tag{10.76}$$

---

[5] The $\sin(\pi x/l)$ factor may be replaced by $\sin(n\pi x/l)$ with n any integer but this will give a higher value of the onset ε.

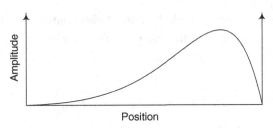

Fig. 10.6 Onset amplitude solution $A(x)$ of Eq. (10.73), for a unidirectional traveling wave in a finite geometry with boundary conditions $A(x=0) = A(x=l) = 0$.

with $O(l^{-2})$ corrections to that result. The solution we have found explicitly demonstrates the difficulty of treating an $O(1)$ group speed s since the length scale of the exponential dependence $(\text{Re}\,\kappa^{-1})$ is then of order unity and is not a slow length scale as required for the validity of the amplitude equation.

10.3.4 Noise-sustained structures

A prediction of the last section is that a finite system will remain in the convectively unstable uniform state for a range of ε above the instability to plane waves. Such a system is highly sensitive to experimental noise. For fluid experiments in which there is a net inflow of fluid, for example pipe flow, the noise might be due to experimental imperfections in the control of the nominally steady fluid flow at the inflow.[6] In other cases, such as growth of a solid dendrite into a supercooled melt during solidification, the noise might be thermal noise arising from the molecular nature of the system.

The convectively unstable state amplifies the noise to give a disturbance that grows and propagates downstream. If the initial disturbance is sufficiently strong or if the system size is sufficiently large to produce a big amplification, the magnitude of the disturbance may become large enough that nonlinearity eventually saturates the growth, leading to a noise sustained state. Taking the propagation direction to be to the right, the noise perturbations at the far left end of the system undergo the largest amplification and are the most important. The amplification factor over a distance l will be of order $e^{\sigma_m l/s}$ with σ_m the maximum growth rate of the waves and with s the group speed. For a noise disturbance of magnitude η, saturation will occur if $\eta e^{\sigma_m l/s}$ becomes of order $\varepsilon^{1/2}$. This will lead to the situation sketched in Fig. 10.7. The downstream magnitude saturates to a constant value but the position of the growth front (indicated by the double-headed arrow in the figure) and the phase of the waves depend on the driving force, and fluctuate with the noise.

---

[6] A common way to generate a steady inflow is to use a pump, whose mechanical vibrations can couple to the fluid and therefore act as a noise source.

10.3 Unidirectional waves in a type-I-o system

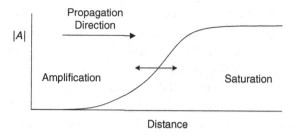

Fig. 10.7 Schematic drawing of the amplitude magnitude $|A(x,t)|$ for a noise-sustained state with nonlinear saturation. The double-headed arrow indicates the noise-induced fluctuations of the position of the growth front.

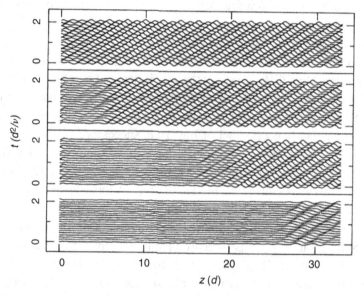

Fig. 10.8 Experimental example of noise-sustained dynamical states in Taylor–Couette flow between two concentric cylinders with an imposed axial flow (from left to right in each panel) and a fixed outer cylinder. The reduced inner Reynolds number ε has the values 0.1020, 0.0822, 0.0632, and 0.0347 for the top through bottom panels respectively. The uniform state for the conditions of the top panel is absolutely unstable and the nonlinear state is not sustained by the noise nor is it sensitive to noise. The fluid in the lower three panels is convectively unstable and the patterns observed are noise-sustained states. For each of the three lower panels, the growth-front position and the wave phases at successive times fluctuate because of the noise. (From Babcock et al. [8].)

An experimental example of a noise sustained state is shown in Fig. 10.8. The experiments used a long (98 cm) Taylor–Couette apparatus (see Fig. 1.11) with a rotating inner cylinder and fixed outer cylinder. The apparatus was modified to produce a steady azimuthally symmetric flow along the axis of the apparatus by injecting fluid at one end of the cylinders and by withdrawing fluid from the other end. Since the outer cylinder did not rotate, the experiment was characterized by two dimensionless parameters, the inner Reynolds number $\mathcal{R}_i \propto \omega_i$ of Eq. (1.4), which determines how rapidly the inner cylinder rotates, and an axial Reynolds number $\mathcal{R} = \langle w \rangle d / \nu$, which characterizes the average speed $\langle w \rangle$ of the axial flow, where $d = 0.68$ cm is the radial width of the fluid between the cylinders, and ν is the fluid's kinematic viscosity. In Fig. 10.8, the axial Reynolds number was $\mathcal{R} = 3.0$ and the four panels are labeled by the reduced inner Reynolds number $\varepsilon = (\mathcal{R}_i - \mathcal{R}_{ic})/\mathcal{R}_{ic}$, where $\mathcal{R}_{ic}$ is the critical inner Reynolds number for the onset of Couette rolls.

The axial flow converts the type-I-s instability of the uniform state to stationary Couette rolls into a type-I-o wave instability, in which a wave consists of Couette rolls that propagate down the length of the cylinders in the direction of the axial flow. The experimentalists quantified the behavior of the system by measuring how the pattern intensity[7] $I(z, t)$ varied spatially along the axis of the cylinder at successive moments in time. Here distance along the axis is denoted by a z-coordinate in units of the radial fluid depth d, and $z = 0$ corresponds to the end of the cylinder where fluid is injected to create the axial flow; the time t is measured in units of the viscous diffusion time d^2/ν. In each panel of Fig. 10.8, successive snapshots of the axial variation of $I(z, t)$ are displaced slightly upwards so time runs vertically in this figure. Each local spatial oscillation of I for a fixed t corresponds to a Couette roll, and the diagonals caused by the small shift of local features to the right at successive times indicate propagation with constant speed of the rolls along the axis. The figure shows about one quarter of the length of the apparatus.

As the control parameter ε is increased from the convectively unstable regime into the absolutely unstable regime (bottom to top in Fig. 10.8), the measurements show a transition from a fluctuating noise-sustained state in the lower three panels to a steady state where the noise is not significant in the top panel. The value of ε at which this transition occurs agrees well with the predictions Eq. (10.72) of the amplitude equation for the absolute instability. Note also that, as expected, the amplification length for the magnitude to reach saturation in the convectively unstable regime is larger for smaller ε.

10.3.5 Local modes

Another way that persistent waves can be sustained in a convectively unstable region of the system is if there is some local source of waves that feeds into this region. One way this can occur is if the control parameter varies spatially so that the

---

[7] The intensity $I(z, t)$ was determined by measuring with a charge-coupled-device camera, located outside the transparent glass outer cylinder, how much light was reflected (from small shiny particles suspended in the fluid) from a fluid region of coordinate z at time t.

system is absolutely unstable over some region. An instability to a local oscillating mode can then be a source of waves that travel into the convectively unstable region over the rest of the system. This type of situation occurs in many of the classic fluid flow instabilities such as the von Kármán vortex street produced behind a circular cylinder in a low Reynolds number flow.

The amplitude equation Eq. (10.63) can be used to illustrate the existence of local modes. A situation that can be worked out analytically is where the control parameter varies linearly in a one-dimensional space $0 \leq x < \infty$

$$\varepsilon(x) = \varepsilon_0 + \varepsilon_1 x, \quad \varepsilon_1 < 0, \qquad (10.77)$$

and the amplitude satisfies the boundary condition $A(x = 0) = 0$. The solution is discussed in the following Etude.

Etude 10.5 Local modes in the amplitude equation

The linearized version of Eq. (10.63) that gives the growth or decay of small amplitude disturbances can be solved analytically. For the analysis, it is convenient to introduce the frequency shift and scalings as in Eq. (10.15) and Eq. (6.16) and use the linearized equation in the form

$$(\partial_{\tilde{t}} + \tilde{s}\,\partial_{\tilde{x}})\tilde{A} = \varepsilon \tilde{A} + (1 + ic_1)\nabla^2_{\tilde{x}}\tilde{A}, \qquad (10.78)$$

with $\tilde{s} = s\tau_0/\xi_0$.

The solution can be written in the form $\tilde{A} = \chi_n(\tilde{x})e^{\sigma_n \tilde{t}}$ with

$$\chi_n(\tilde{x}) = e^{\tilde{s}\tilde{x}/2}\,\mathrm{Ai}\!\left([-\varepsilon_1/(1+ic_1)]^{1/3}\tilde{x} + \zeta_n\right), \qquad (10.79\mathrm{a})$$

$$\sigma_n = \varepsilon_0 - \tilde{s}^2/[4(1+ic_1)] + \left\{(1+ic_1)\varepsilon_1^2\right\}^{1/3}\zeta_n. \qquad (10.79\mathrm{b})$$

Here Ai is the Airy function and its zeros (which are all negative numbers) are denoted by ζ_n so that $\mathrm{Ai}(\zeta_n) = 0$. The largest growth rate $\mathrm{Re}\,\sigma_n$ occurs for the first zero $\zeta_1 \simeq -2.3381$ and we will concentrate on this mode. The growth rate becomes positive for $\varepsilon_0 > \varepsilon_\mathrm{L}$ with

$$\varepsilon_\mathrm{L} = \varepsilon_\mathrm{a} + (-\zeta_1)|\varepsilon_1|^{2/3}(1+c_1^2)^{1/6}\cos\!\left(\tfrac{1}{3}\tan^{-1} c_1\right), \qquad (10.80)$$

where ε_a is the constant value Eq. (10.72) for absolute instability in an infinite system. For $\varepsilon > \varepsilon_\mathrm{L}$ and for c_1 of order unity, you can see from Eq. (10.79a) that the amplitude is localized in the region $0 \lesssim \tilde{x} \lesssim (-\zeta_1^3/\varepsilon_1)^{1/3}$.[8] A finite amplitude oscillating solution will be established near $x = 0$ that sends waves out into the

---

[8] Perhaps surprisingly, this expression does not depend on ε_0 and therefore does not vary with the region of positive ε as ε_0 changes. In fact, as ε_0 changes, the growth rate increases but the linear modes are unchanged.

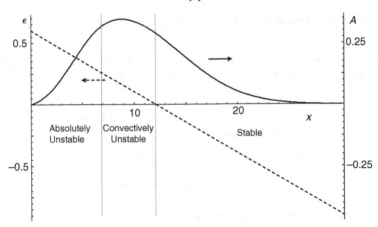

Fig. 10.9 Local mode for the linear amplitude equation Eq. (10.78) with $c_1 = 0$ for a linear variation of $\varepsilon(\tilde{x})$ given by Eq. (10.77). Solid line: amplitude $\tilde{A}(\tilde{x})$; dashed line: $\varepsilon(\tilde{x})$. Parameters used are $\varepsilon_0 = 0.6$, $\varepsilon_1 = -0.05$, and $\tilde{s} = 1$.

convectively unstable region $0 < \varepsilon(x) < \varepsilon_a$. The waves ultimately decay in the region $\varepsilon(x) < 0$. An example of such a solution is shown in Fig. 10.9. Parameters used in constructing the plot were $c_1 = 0$, $\varepsilon_0 = 0.6$, $\varepsilon_1 = -0.05$, $\tilde{s} = 1$. For these values, $\varepsilon_a = 0.25$ and $\varepsilon_L \simeq 0.567$. For further details and discussion, see the paper by Chomaz et al. [21].

10.4 Bidirectional waves in a type-I-o system

We now turn to the case of a one-dimensional system with parity symmetry so that the waves with wave number q and $-q$ go unstable together. At the linear level, any superposition of the waves propagating in the two directions may be formed. However the nonlinear competition will favor a particular combination, either a single wave traveling in either direction, or a standing wave made up of equal contributions of the two waves. In an infinite geometry, or one with periodic boundary conditions, this competition is easily analyzed using amplitude equations, as we do in Section 10.4.1. However in finite geometries, or systems with inhomogeneities, the analysis becomes much harder.

The difficulties of analyzing counter-propagating wave states even near onset can be illustrated with a simple example of counter-propagating waves in an annulus. Although calculating the properties of the simple weakly nonlinear plane wave states is easy enough by transforming the description into a frame moving with the group speed, more general situations are much less easy. For example, if the annulus contains a spatially inhomogeneous distribution of clockwise and counter-clockwise waves, then in the time scale ε^{-1} typical of the growth of intensity of a wave, any point on the clockwise-moving wave will feel the nonlinear influence of the counter-clockwise-moving wave averaged over a distance of order $s\varepsilon^{-1}$. This length scales in a different way than the $\varepsilon^{-1/2}$ length typical of diffusion and dispersion effects. In addition, the interaction between the waves is highly nonlocal and cannot be described by local pdes. The situation simplifies if the circumference L of the annulus satisfies $s\varepsilon^{-1} \gg L \gg \varepsilon^{-1/2}$ since then the clockwise wave travels many times around the annulus in the time scale ε^{-1} for its amplitude to respond. In this case, the influence of the counter-clockwise wave is experienced just through a spatial average.

A systematic and comprehensive amplitude equation is not available for spatially varying counter-propagating wave states. However, some important yet simple questions can be addressed using various approximations. In this section, we will first look at the competition between standing and traveling waves in the nonlinear state in a one-dimensional periodic geometry, which can be approximated in the laboratory by waves propagating around a narrow annulus. We will then discuss the nature of the onset state in a one-dimensional geometry with physical boundaries that act to reflect the waves. Without going into the technical details, we will then present some results of experimental and theoretical investigations that extend these results. In particular, we will describe some of the interesting oscillatory and chaotic states that can occur.

10.4.1 Traveling and standing waves

The question of the stability of a spatially uniform traveling wave versus the stability of a standing wave can be readily discussed near onset in a periodic geometry such as an annulus. The question is similar to our earlier discussion of the competition between stripes and lattice states in Chapter 7 since we want to compare the stability of nonlinear states that are constructed by adding growing modes that are known from a linear stability analysis. We follow the same methods but now use the amplitude equations for counter-propagating waves. The amplitudes are introduced through the equation

$$\mathbf{u}_p = \left[A_R(x,t) e^{i(q_c x - \omega_c t)} + A_L(x,t) e^{i(-q_c x - \omega_c t)} \right] \mathbf{u}_0(\mathbf{x}_\|) + \text{c.c.} + \text{h.o.t.} \quad (10.81)$$

For the easier case of spatially uniform amplitudes, the generalization of Eq. (10.11) to the counter-propagating case is

$$\tau_0 \, \partial_t A_R = \varepsilon(1 + ic_0) A_R - g_0[(1 - ic_3)|A_R|^2 + G(1 - ic_2)|A_L|^2] A_R, \quad (10.82)$$

for the right-moving wave amplitude, and

$$\tau_0 \, \partial_t A_L = \varepsilon(1 + ic_0) A_L - g_0[(1 - ic_3)|A_L|^2 + G(1 - ic_2)|A_R|^2] A_L, \quad (10.83)$$

for the left-moving wave amplitude. The new real parameter G quantifies the nonlinear interaction between right- and left-moving waves.

The competition between traveling and standing waves is now analyzed by constructing the various nonlinear solutions and then testing their stability using the amplitude equations Eqs. (10.82) and (10.83). The results of the analysis are that for $G < -1$ traveling waves are unstable and there is no saturation of standing waves; for $-1 < G < 1$ standing waves are stable and traveling waves unstable; and for $G > 1$ traveling waves are stable and standing waves are unstable. You may derive these results in Exercise 10.10.

10.4.2 Onset in finite geometries

For unidirectional waves in a finite geometry, we saw in Section 10.3.3 that the onset of instability is delayed until near the onset of absolute instability. The reason is that, in the convectively unstable regime, a perturbation from the uniform state is whisked away into the absorbing boundary. With counter-propagating waves, end walls can reflect one component into the other. This leads to the sustained growth of a perturbation as it propagates to and fro, and to a saturated nonlinear state even in the convectively unstable parameter regime. A simple estimate illustrates this.

In a system of size l, the propagation time for a wave with group speed s is l/s. Over this time, a disturbance, say a right-moving wave, will grow by a factor $\exp(\varepsilon l/s\tau_0)$. The disturbance is now reflected into a left-moving wave by the boundary at $x = l$, as in Fig. 10.10. Let us define the reflection coefficient r to be the complex-valued ratio of the amplitude of the right-moving wave to the amplitude of the reflected left-moving wave at the boundary, so that there is a reduction in the magnitude of the disturbance by a factor $|r|$. The threshold for instability in a finite geometry is given by the condition that the net growth be zero, which in turn requires that the amplification and reflection factors just balance. This occurs at

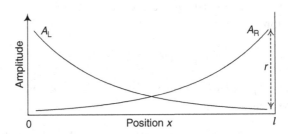

Fig. 10.10 Exponentially growing amplitudes of right- and left-moving waves in a finite geometry of length l. The ratio of amplitudes upon reflection is defined to be the complex-valued reflection coefficient r. Near the boundaries, the behavior will deviate from exponential and is not shown.

$\varepsilon = \varepsilon_c(l)$, where $\varepsilon_c(l)$, is given by

$$\varepsilon_c(l) \sim s\tau_0 l^{-1} \ln(|r|^{-1}). \tag{10.84}$$

For large systems, this is close to the threshold for convective instability $\varepsilon = 0$, rather than near the threshold for absolute instability ε_a as was found for the one wave case, Eq. (10.75). Clearly, reflection at boundaries is important in the case of counter-propagating waves.

These results can also be obtained by an analysis of the linearized amplitude equations for counter-propagating waves. To derive a first approximation for the threshold, we can look for exponentially growing solutions of the amplitude equations that include growth and propagation but that ignore dispersion and diffusion:

$$\tau_0(\partial_t A_R + s \partial_x A_R) = \varepsilon(1 + ic_0)A_R, \tag{10.85a}$$

$$\tau_0(\partial_t A_L - s \partial_x A_L) = \varepsilon(1 + ic_0)A_L. \tag{10.85b}$$

These equations are first order in spatial derivatives, and so require a *single* boundary condition at the endpoints $x = 0$ and $x = l$. They will take the form of empirical reflection conditions. If the two end walls are physically identical, the conditions are

$$A_R(x = 0) = rA_L(x = 0), \tag{10.86a}$$

$$A_L(x = l) = re^{2iq_c l} A_R(x = l), \tag{10.86b}$$

where the phase factor $e^{2iq_c l}$ in the second equation takes into account the phase variation of the underlying waves. If we seek oscillatory solutions with zero real growth rate $A_{R,L} \propto e^{-i\Omega t}$, the spatial dependence will be exponential $\propto e^{\pm \kappa x}$

$$A_R = a_R e^{ikx} e^{\kappa_r x}, \tag{10.87a}$$

$$A_L = a_L e^{-ikx} e^{-\kappa_r x}, \tag{10.87b}$$

with $\kappa_r = \operatorname{Re}\kappa$, $k = \operatorname{Im}\kappa$ given by

$$\kappa_r = \frac{\varepsilon}{\tau_0 s}, \quad k = s^{-1}\left(\varepsilon \tau_0^{-1} c_0 + \Omega\right). \tag{10.88}$$

The boundary conditions Eqs. (10.86) then give

$$1 = r^2 e^{2i(q_c+k)l} e^{2\kappa_r l}, \tag{10.89}$$

so that

$$\kappa_r = l^{-1} \ln\left(|r|^{-1}\right) \quad \text{and} \quad k = (n\pi - \Phi_r - q_c l)/l, \tag{10.90}$$

with n an integer and Φ_r the phase of r. The value of κ_r together with Eq. (10.88) reproduces the estimate for the onset value of the control parameter ε_c given in Eq. (10.84). The equation for k gives a discrete set of possible wave numbers that are fixed by a phase matching condition. In the approximation used, these different wave numbers all give the same threshold. Because of this degeneracy, a state with a complex initial condition that corresponds to a superposition of these different modes has a simple exponential growth rate in the approximation used, so that there is no convergence onto a fastest growing mode. We would expect there to be $O(k^2)$ corrections to ε_c coming from the second-order derivative terms left out in Eq. (10.85) that would tend to smooth out complex spatial structure removing this degeneracy. Alternatively, the nonlinearity may lead to mode competition and a unique solution even in the absence of second derivative terms.

10.4.3 Nonlinear waves with reflecting boundaries

The effect of nonlinearity on the linear state shown in Fig. 10.10 is interesting. As the amplitudes of the waves increase, the region of large amplitude right- or left-moving waves tends to further suppress the reflected counter-propagating waves. The interplay of the nonlinear suppression of one wave by the other over the bulk of the system and of the reflection of one wave into the other at the boundaries leads to the possibility of complex dynamics even close to threshold. The dynamics typically appears as a periodic or chaotic motion of the interfaces where either set of waves grows to large amplitudes. The excursions of the interface can be small, or may become large enough so that there are complete oscillations between right-moving waves that dominate, and then left-moving waves that dominate.

Figure 10.11 shows an experimental example of a state sustained by reflection called a blinking state. Here a binary fluid (a 0.3 percent by weight solution of ethanol in water) of mean temperature $21.4\,°C$ convects in a rectangular cell of relative dimensions $16.25 \times 4.9 \times 1$. Near onset ($\varepsilon = 3.2 \times 10^{-4}$ for the data in Fig. 10.11), a periodic state spontaneously evolves that consists of right- and left-moving waves. Using a data analysis technique called demodulation (see the brief description below in Section 10.5), the experimentalists were able to estimate the slowly varying right and left amplitudes, $A_R(x,t)$ and $A_L(x,t)$, which they then compared with calculations based on amplitude equations. In agreement with theory, the experimentalists found that the right and left amplitudes alternately become larger and then smaller on the right and left sides of the cell; these changes in amplitude correspond to changes in brightness in the shadowgraph images in Fig. 10.11, hence the moniker "blinking state." The experimentally extracted amplitudes also confirm a subtle prediction of the theory, that a pattern like this is not symmetric in that the maximum of the peak amplitude of the right-moving wave is larger than the

10.5 Waves in a two-dimensional type-I-o system

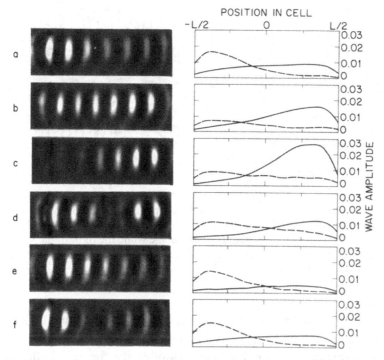

Fig. 10.11 Blinking state observed near onset in a binary fluid convection experiment whose uniform state undergoes a type-I-o instability. The panels on the left show shadowgraph images at equally spaced times of propagating convection rolls. The panels on the right show the corresponding amplitudes of right (full) and left (dashed) moving waves as extracted from the shadowgraph images by a demodulation method. The vertical panels span one period of the competition between the two sets of waves. (From Kolodner et al. [57].)

maximum of the peak amplitude of the left-going wave. The theoretical explanation of this dynamics is rather involved so we do not include a full discussion here.

10.5 Waves in a two-dimensional type-I-o system

Our knowledge of type-I-o waves in two extended dimensions rests primarily on experimental observations. Figure 10.12 shows a snapshot of a binary fluid convection experiment for which traveling waves emerge from a subcritical type-I-o bifurcation of the uniform state as the Rayleigh number is increased.[9] The first

---

[9] The fluid is an 8% by weight mixture of ethanol in water at an average temperature of 26 °C for which the Prandtl number is $\sigma = 12$ and the Lewis number, a dimensionless measure of the rate of concentration diffusion to the rate of heat diffusion, is $\mathcal{L} = 10^{-2}$. The vertical thermal diffusion time is $\tau = d^2/\kappa = 124$ s so the dynamics are actually rather slow on a human scale. The experimental cell has an aspect ratio of $\Gamma \approx 26$.

Fig. 10.12 Snapshot of a two-dimensional wave pattern in a cylindrical binary fluid convection experiment whose uniform state undergoes a subcritical type-I-o instability. The left frame shows a shadowgraph image while the right frame shows the phase field $\phi(x, y, t)$ of the waves, as obtained by demodulating the shadowgraph image. (From La Porta and Surko [88].)

panel is a shadowgraph of the convection rolls (see Fig. 1.13 and the related discussion in Chapter 1). The images as a function of time show that the convection rolls are propagating – the predominant motion in the figure is in the clockwise direction – so that the state is one of traveling convection rolls. The large-scale structure is one of domains of straight or weakly curved stripes. The sources of the waves are typically in corners of the domains, and the sinks are domain boundaries. The domain boundaries may be so-called zipper boundaries, for which the waves propagate largely parallel to the boundary but in opposite directions on opposite sides of the boundary, or may be perpendicular grain boundaries, for which one set of rolls approaches the boundary while the other set propagates parallel to the boundary.

The behavior of the waves is depicted more clearly in the second panel, where a demodulation technique was used to display the phase field of the waves $\phi(\mathbf{x}, t)$. The phase was extracted by assuming that shadowgraph intensity can be written in the form

$$I(\mathbf{x}, t) = a(\mathbf{x}, t) \exp[-i(\omega t - \phi(\mathbf{x}, t))] + \text{c.c.}, \tag{10.91}$$

where ω is the average frequency of the waves and where the amplitude a and phase ϕ are assumed to vary slowly in time. If we now multiply this expression by the Fourier mode $e^{i\omega t}$ and integrate to filter out the high-frequency rapidly oscillating component, we are left with the expression

$$I_\text{f}(\mathbf{x}, t) = a(\mathbf{x}, t) e^{i\phi(\mathbf{x}, t)}, \tag{10.92}$$

from which the phase ϕ is readily extracted. In Fig. 10.12, the phase is shown in a grey scale such that the phase increases in the order white to gray to black. The direction of propagation (which is evident from watching a movie of the dynamics) is in the direction of increasing phase.

The domain morphology of the patterns seen in these and other binary fluid convection experiments with two extended directions is quite different from the wave states in the CGLE that describes the behavior near a type-III-o instability, and differs from the patterns in chemical wave experiments, where spiral sources are typical. Whether this reflects a general difference between systems near a type-I-o instability and the propagating wave states in oscillatory or excitable media, or is simply a feature of the particular systems so far studied, is not yet clear and is a topic for future research.

10.6 Conclusions

In this chapter, we have studied the properties of collective nonlinear oscillator and wave states that emerge via respectively a type-III-o and type-I-o instability of a time-independent spatially uniform state. The methods we used generalized the amplitude-equation and phase equation methods that we developed for stationary patterns in Chapters 4 through 7.

As was the case for stationary patterns, theory currently can only address the weakly nonlinear region near threshold so many of our quantitative results are unlikely to apply to natural patterns nor to many experiments. Nevertheless, many interesting and novel properties of nonlinear collective oscillations and waves can be addressed near onset, and the qualitative insights gained should continue to apply for pattern formation in the more strongly nonlinear regime. For example, the feature that counter-propagating waves annihilate one another, which is surprising to those of us educated by elementary physics classes to understand the behavior of linear waves which can pass through one another, can be readily analyzed by the methods discussed in this chapter. As another example, the important question of whether spiral sources of waves have a unique frequency determined by the parameters of the system, or may have a range of frequencies, can be addressed within the amplitude-equation formalism (and this was done in 1982). The answer found – that indeed the frequency is unique so that the waves sustained by spiral sources have a well-determined frequency and wave number – was confirmed much later to apply to strongly nonlinear waves in reaction–diffusion systems, even in those that are excitable, for which a disturbance of finite magnitude is required in order to initiate wave propagation.

Many of the ideas introduced for stationary pattern-forming instabilities – such as the notion of a stability balloon, the use of a phase equation to capture the slow dynamics connected with the broken symmetries, and the importance of defects – continue to be important in nonlinear oscillator and wave systems. The essential difference of the latter systems is that *propagation* is a new ingredient. Not least, this means that a growing perturbation may propagate away before destroying the spatially uniform state, the idea of convective instability. This means that the presence of boundaries plays a particularly important role in systems supporting counter-propagating waves since the boundaries may reflect waves. The propagation introduces a new time scale l/s with l the length scale of pattern variation (e.g. the system size) and s the group speed. The interplay of this time scale with the growth time scale proportional to ε^{-1} means that for the wave instabilities (type-I-o rather than type-III-o) the amplitude equation approach has a restricted validity and more delicate theoretical methods may be needed. You can read about these from the list of further reading.

In Chapter 11, we return to the topic of propagating disturbances in nonequilibrium systems but in the context of excitable systems. You will see that many of the issues raised in the present chapter occur there as well.

10.7 Further reading

(i) An extensive review of the properties of the complex Ginzburg Landau equation is "The world of the complex Ginzburg-Landau equation" by Aranson and Kramer [6].

(ii) The first discussion of the uniqueness of spiral rotation frequencies can be found in the paper "Spiral waves in reaction diffusion equations" by Hagan and Cohen [44].

(iii) The book *Chemical Oscillations, Waves, and Turbulence* by Kuramoto [60] gives a more advanced discussion of nonlinear oscillators and waves.

(iv) To follow up on the question of the theory of counter-propagating waves and blinking states for wave states in finite geometries see: "Traveling and standing waves in binary-fluid convection in finite geometries" by Cross [24]; "Amplitude equations for travelling wave convection" by Knobloch and De Luca [56]; and "Finite size effects near the onset of the oscillatory instability" [69] and "Dynamics of a hyperbolic system that applies at the onset of the oscillatory instability" [70] by Martel and Vega.

Exercises

10.1 **Instability of standing waves near onset for a type-III-o transition:** Using the amplitude equation Eq. (10.17), show that standing waves are always unstable near the onset of a spatially uniform oscillatory instability.

10.2 **Stability balloon from the type-III-o amplitude equation:** Use the scaled type-III-o amplitude equation, Eq. (10.18) and the steps outlined below to derive the full expression for the complex growth rate σ for longitudinal perturbations about the plane wave base state of wave number K described by the amplitude $\bar{A}_K$ of Eq. (10.21a).

(a) Writing the scaled amplitude as $\bar{A} = \bar{A}_K + \delta \bar{A}$, derive the linear evolution equation satisfied by the perturbation $\delta \bar{A}(X, T)$.

(b) Show that the linear evolution equation may be satisfied using a perturbation at Bloch wave number Q of the form

$$\delta \bar{A}(X,T) = e^{i(KX - \Omega_K T)}[\delta a_+ e^{\sigma T} e^{iQX} + \delta a_-^* e^{\sigma^* T} e^{-iQX}], \quad \text{(E10.1)}$$

with $\sigma = \sigma(Q, K)$ the complex growth rate and with δa_+ and δa_-^* constants.

(c) Find the quadratic equation satisfied by the complex growth rate σ.

(d) By expanding the growth rate σ given by this quadratic equation in the small quantity Q up to $O(Q^2)$, verify the results for the stability balloon derived using the phase equation in Section 10.2.4. Note: The expression you have obtained can also be used to show that for some values of the CGLE parameters c_1 and c_3, a short-wavelength instability becomes important.

10.3 **Convective nature of the Benjamin–Feir instability:** Show that within the phase equation approximation, Eq. (10.34), the Benjamin–Feir instability that limits the wave number range of stable traveling waves near a type-III-o instability is always convective. Note: To find the quantity Λ_A in Eq. (10.43) that determines where the instability becomes absolute, you will need to use the full expression derived from the amplitude equation in Exercise 10.2. Why is this so?

10.4 **Phase and magnitude variation in a shock:** Use the solution for a shock in Etude 10.4 and the results of Section 10.2.3 to plot the phase variation of a stationary shock for the amplitude equation Eq. (10.18) with parameter values $c_1 = 0$ and $c_3 = 1.25$ for incoming waves with wave numbers $K = 0.1$. Also calculate from the amplitude equation how the *magnitude* of the complex amplitude varies near the shock by keeping terms up to second order in spatial derivatives of the phase. Based on a plot of the magnitude variation for the same parameter values, does the magnitude increase or decrease in the shock?

10.5 **Sources and shocks:** In the following, use the Cole–Hopf transformation discussed in Section 10.2.5 to analyze stationary shock defects.

(a) Show that for $\beta > 0$, there is no solution of the nonlinear phase equation (10.24) that corresponds to outgoing waves (a source).

(b) What is the large-distance behavior of the phase for the stationary shock solution for $\beta < 0$? Verify that the general rule is that the group velocity $d\Omega/dK$ is inwards, although the phase velocity Ω/K might be either inwards or outwards.

10.6 **Moving shocks:** Using the Cole–Hopf transformation Eq. (10.46), derive an expression for a shock solution to the nonlinear phase equation (10.24) for $\beta > 0$ in which the asymptotic wave numbers of the incoming waves are $K_L = K_0 + K_1$ on the left and $K_R = K_0 - K_1$ on the right. Show that the shock moves with a velocity $v = 2\beta K_1$ and that this velocity is consistent with the wave front conservation condition

$$\Omega_L - K_L v = \Omega_R - K_R v, \qquad (E10.2)$$

where $\Omega_{L,R}$ are the wave frequencies on the left and right sides of the shock.

10.7 **Targets in the nonlinear phase equation:** In two spatial dimensions, a pattern of outgoing circular waves from a point source is called a target pattern. Targets may be investigated using the enhanced nonlinear phase equation

$$\partial_T \Phi = -g(R) + \alpha \nabla_X^2 \Phi - \beta (\nabla_X \Phi)^2, \qquad (E10.3)$$

where R is a radial coordinate that starts from the center of the target, and $-g(R)$ is an additional radially dependent frequency term that models an imposed local inhomogeneity in the medium such as a speck of dust. For $g(R) > 0$, the inhomogeneity raises the local frequency of oscillation. (Remember that the frequency is $-\partial_T \Phi$.)

(a) Apply this version of the Cole–Hopf transformation

$$\Phi = -\Omega_T T - \frac{\alpha}{\beta} \ln \chi(\mathbf{X}) \qquad (E10.4)$$

to Eq. (E10.3) to obtain the following equation for χ

$$-\Omega_T \chi = -\frac{\alpha^2}{\beta} \nabla^2 \chi - g(R)\chi. \qquad (E10.5)$$

Notice that this equation has the same form as the time-independent Schrödinger equation for a particle in a potential, where χ corresponds to the particle wave function ψ, $g(R)$ corresponds to a potential energy V, and Ω_T corresponds to an "energy" eigenvalue E.

(b) Show that a solution for χ that is everywhere positive and that decreases exponentially at large radii corresponds to a target solution for Φ.

(c) Using the Schrödinger analogy or otherwise, show that there are no target solutions for the homogeneous case $g(R) = 0$.

(d) The Schrödinger equation in two dimensions with a localized attractive potential always has a bound state (state with negative energy) with a wave function that is positive and that decays exponentially for large R. Use this result to show that a target solution always exists for an inhomogeneity that raises the local frequency. As a next step, relate the frequency and wave number of the waves at large distances to the parameters of the solution to the corresponding Schrödinger problem. If there is more than one bound state in the Schrödinger problem, does this imply a family of targets?

(e) Show that there are no target solutions if the inhomogeneity *lowers* the local frequency.

10.8 **Local modes in a convectively unstable system:** Use Mathematica or a similar program to construct the nonlinear local mode solution for the parameters of Fig. 10.9 using the evolution equation Eq. (10.78) supplemented by the nonlinear term $-|\tilde{A}|^2\tilde{A}$. (Assume that the solution is real.) Compare the solution with the fastest growing linear onset mode, Eq. (10.79a) for $n = 1$. Study how the solution to the nonlinear equation changes as you vary ε_0, for example verifying the critical value for the onset of the local mode.

10.9 **Experimental study of a convectively unstable growth of a crystal:** Read the paper "Development of sidebranching in dendritic crystal growth," A. Dougherty, P. D. Kaplan, and J. P. Gollub, *Phys. Rev. Lett.* **58**, 1652 (1987). Then write a short summary of the paper in which you explain the experimental technique used and the evidence put forward by the authors to support their claim that molecular noise plays a role in the formation of side branches of a growing dendrite.

10.10 **Traveling and standing waves in a type-I-o system:** In this exercise you use the amplitude equations for a type-I-o instability with counter-propagating waves, Eqs. (10.82) and (10.83), to investigate the stability of traveling and standing waves near threshold.

(a) The solution for a right-moving traveling wave is given by $A_L = 0$, $A_R = a_T e^{-i\Omega_T t}$ where a_T may be chosen real. Find expressions for a_T and Ω_T by substituting into Eqs. (10.82) and (10.83).

(b) A standing wave is given by $A_R = A_L = a_S e^{-i\Omega_S t}$, where again a_S may be chosen real. Find a_S and Ω_S by substituting into Eqs. (10.82) and (10.83). Argue that, for $G < -1$, the bifurcation to standing waves is subcritical so that, within the equations used, there is no saturation of the standing wave solution for $\varepsilon > 0$.

(c) To test the stability of the traveling waves write

$$A_R = e^{-i\Omega_T t}(a_T + \delta a_R(t)), \quad A_L = e^{-i\Omega_T t}\delta a_L(t). \qquad \text{(E10.6)}$$

By substituting into Eqs. (10.82) and (10.83), keeping terms to linear order in δa_L and δa_R, and looking for solutions $\delta a_L, \delta a_R \propto e^{\sigma t}$ show that traveling waves are stable (all $\operatorname{Re}\sigma < 0$) for $G > 1$ but unstable for $G < 1$.

(d) To test the stability of standing waves write

$$A_R = e^{-i\Omega_s t}(a_S + \delta a_R(t)), \quad A_L = e^{-i\Omega_s t}(a_S + \delta a_L(t)). \qquad (E10.7)$$

We will guess that the most unstable eigenvector is for the perturbations δa_R and δa_L real. (You can show this at the cost of more algebra.) With this simplification and looking for solutions $\delta a_L, \delta a_R \propto e^{\sigma t}$ again, show that the growth rates σ are given by

$$\sigma = -\frac{2\varepsilon_0 \tau_0^{-1}}{1+G}[(1 - ic_3) \pm G(1 - ic_2)], \qquad (E10.8)$$

and therefore argue that standing waves are stable (both $\operatorname{Re}\sigma < 0$) for $-1 < G < 1$.

11
Excitable media

The waves that are most familiar in daily life are sound waves, light waves, electrical waves, water waves, and mechanical waves (say a standing wave on a piano string). These familiar waves have the property that their magnitude decreases as they propagate away from their source or, for standing waves, once their source is turned off. The decrease in magnitude is a result of dissipative effects in the medium such as fluid viscosity, electrical resistance, or friction that drain energy from the wave and that restore the medium to thermal equilibrium.

These familiar waves have the additional property of often being accurately described by a linear evolution equation such as the wave equation. Because the evolution equation is linear, one can superimpose sinusoidal waves to get localized pulses of arbitrary shape, and these pulses can also propagate. (For example, clapping your hands once loudly creates a localized sound pulse that propagates away.) Because each Fourier component in the superposition is itself damped in typical media, the propagating pulses also damp out and disappear over time. Even in the absence of damping, dispersive effects can cause the different Fourier components to travel at different speeds so, again, waves and pulses change their shape and decrease in magnitude during propagation.

We have seen in earlier chapters that sustained nonequilibrium systems allow many dynamical states that can propagate or exist in a local spatial region but are such that these states do not damp out over time or they preserve their shape and speed as they propagate away from a source. Examples include topological defects such as dislocations that can climb arbitrarily far in a stripe state, and fronts, pulses and waves, examples of which we discussed in the previous chapter in the context of states that arise from a type-o instability of a time-independent uniform state. Here the mechanisms that sustain a system out of equilibrium feed energy locally to a wave or pulse and so counter the effects of dissipation and dispersion. For sustained nonequilibrium systems, the evolution equations for waves and pulses are typically nonlinear (and can often be accurately described by amplitude equations near onset).

We have seen numerous times earlier in the book that nonlinearity has a profound effect in limiting the possible structures in terms of their shape, wavelength, and propagation speed.

In this chapter, we discuss the propagation of waves and localized objects in so-called excitable media. While many of the phenomena such as fronts, pulses, and spirals are similar to those discussed in Chapter 10, excitable media have the novel property that the uniform state is linearly stable so that interesting dynamics arise only when the medium is perturbed sufficiently strongly that the amplitude of the perturbation locally exceeds some finite threshold, which is a property of the excitable medium.

That a finite-amplitude perturbation of the uniform state is needed to initiate dynamics means that we cannot use a perturbative approach about the uniform state such as the amplitude equation formalism to study excitable media. But the situation is not then hopeless, and advances in theory and in experiment over the last few decades have shown that one can develop a broadly applicable and insightful framework to understand the phenomenology of excitable systems.

The dynamics of excitable media are also worth discussing because they play a central role in many fascinating and timely questions related to biology, physiology, and medicine (also in chemistry and chemical engineering, such as the propagation of chemical waves on the surface of a metal catalyst). Thus even single cell organisms such as a paramecium have evolved to create excitable electrical waves along their membrane and they use these waves to control the mechanical rotation of their cilia and hence their swimming motions. An excitable electrical wave plays an important role in the merging of a sperm with an egg; the wave triggers a rapid change in the egg's membrane that prevents a second sperm from entering, which would confound the development of the organism. Multicellular organisms have developed nervous systems in which information is transmitted from one neuron to another (or from neurons to muscles) in the form of brief pulses via specialized one-dimensional excitable connections called axons. In the case of a large animal like a blue whale, these pulses might propagate for tens of meters without a decrease in amplitude or speed, a remarkable evolutionary achievement.

Because over fifteen million people die each year worldwide from cardiovascular disease, there is immense interest in understanding and preventing such disease. The most deadly form of heart disease is ventricular fibrillation, in which the heart is no longer able to pump blood successfully and the body's tissues, including the heart itself, begin to die from lack of oxygen. From our point of view, the human heart is a ball of sustained nonequilibrium excitable muscle tissue that contains four blood-filled chambers (two small atria and two large ventricles). The excitable muscle supports nonlinear electrical waves that cause contraction of muscle cells as the waves sweep past the cells. But there are poorly understood dynamical transitions in which the usually coherent waves become disordered in space and time such that small regions of the heart contract out of phase with nearby regions, which reduces greatly the ability of the heart to pump blood. The study and prevention of heart arrhythmias provides a strong motivation to understand pattern formation in nonequilibrium excitable media.

To give you a physical sense of excitable dynamics and of its mathematical description, we discuss in the first half of this chapter how excitable pulses called action potentials arise in neurons and in hearts. Neurons and heart muscle are classic excitable systems with the property that a small-amplitude perturbation – corresponding say to injection of a small amount of current via a metal electrode – decays away while a perturbation that causes the voltage across a cell's membrane to exceed a certain threshold causes a large rapid amplified response, which leads to a pulse that can propagate long distances without decaying or changing shape.

As part of the discussion of neurons and hearts, we also introduce one of the most famous mathematical models in biology, the Hodgkin–Huxley equations, which describe quantitatively the propagation of electrical pulses along a one-dimensional nerve fiber (axon). We then discuss how this equation can be extended to describe the propagation of electrical waves inside heart muscle. Since heart tissue behaves approximately like a three-dimensional continuous (but admittedly inhomogeneous) excitable medium, the pattern-formation aspects of its dynamics play a more central role than for neurons. We can only outline some of what is known since neurons and hearts have been found experimentally to be extremely complicated. Models based on the accumulating experimental data are now quite complex and may have about 100 dynamical variables and numerous empirical functions whose many parameters must be determined by fits to experimental data.

An early theoretical advance in the study of biological excitable media was the discovery that the dynamics of the difficult four-variable Hodgkin–Huxley equations could be described qualitatively and insightfully by a simple two-variable model known as the FitzHugh–Nagumo model (see Section 11.1.3). This model, like many of the models that we discussed in Chapter 5, provides a way to understand general properties of excitable media without being burdened with too many details. This and other reduced models of excitable dynamics[1] have also provided a valuable way to explore by simulations the dynamics of different states in large two- and three-dimensional domains, including domains that accurately approximate the geometry of an adult human heart.

---

[1] We have seen in Chapter 3 that simple two-variable descriptions also provide a useful starting point for understanding reaction–diffusion chemical systems. For example, in that chapter we introduced the Brusselator and the Oregonator models for oscillatory chemical reactions. The reduction to two equations (from the many equations needed to represent a chemical system accurately) arises from the vastly different reaction rates that often characterize the different reactions.

In the second half of this chapter, we sketch a line of argument that leads from the two-component equations to many fascinating phenomena that are observed in diverse excitable systems. The thread connecting the ideas is the following. First, we discuss how the nullclines of a two-variable dynamical model in the absence of diffusion can be used to understand qualitatively whether the dynamics is excitable or oscillatory. The nullclines also help to explain some of the time scales associated with the dynamics, e.g. why there may be a rapid amplification of a perturbation followed by a slow decay back to the uniform state. Restoring diffusion to the two-variable models, we then show how to describe propagating disturbances in the form of fronts, pulses (which can be constructed from two fronts), and wave trains (which can be constructed from pulses).

Next, we discuss the properties of spiral structures since they are the most widely observed self-sustaining sources of propagating disturbances in an excitable medium with two or three extended coordinates. Analytical, numerical, and experimental studies indicate that spirals in excitable media select a unique frequency, a fact that helps to explain how multiple spirals interact with one another. Finally, we discuss instabilities of spiral structures and how such instabilities can lead to more exotic states with quasiperiodic and perhaps chaotic dynamics. Some of these states are believed to correspond to specific heart arrhythmias such as tachycardia (an elevated heart beat caused by the formation of an electrical spiral in the heart muscle), and ventricular fibrillation (clinically observed as an irregular weak heart beat) which may involve a chaotic state that has many small spirals which become unstable to new spirals, which themselves become unstable in a never-ending process.

11.1 Nerve fibers and heart muscle

11.1.1 Hodgkin–Huxley model of action potentials

Experimental studies in the 1950s of squid giant axons, which are about 1 mm in diameter[2] (see Fig. 11.1), led to the first quantitative understanding of how neurons transmit signals to each other. The resulting theoretical description known as the Hodgkin–Huxley equations continues to provide the most widely used way to model neurons and other biological excitable tissue like heart tissue.

The dynamic phenomenon involved with the transmission of information along an axon is the controlled transport of electrical charge across the axon's membrane.

---

[2] Theory shows that the speed of an action potential along an axon scales as a positive power of the axonal radius. Thus animal reflexes that need to be rapid, e.g. the escape reflex of a squid that involves squirting a jet of water to propel the animal away from danger, are typically mediated by axons with big diameters. In contrast, axons in the mammalian central nervous system have diameters that are one hundred to one thousand times smaller and are densely packed so are much more difficult to study.

11.1 Nerve fibers and heart muscle

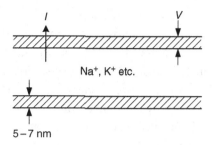

Fig. 11.1 Schematic of an axon, a one-dimensional excitable medium that carries information from one neuron to another neuron or to a muscle in the form of brief electrical pulses called action potentials. The axon can be thought of as a long hollow cylindrical tube whose wall consists of an extremely thin (5–7 nm thickness for the squid giant axon) impermeable lipid bilayer. The bilayer contains numerous complicated proteins called channels that can change their shape in response to voltage changes across the membrane. The change in channel shape can open up a hole through the membrane that selectively allows only certain ions to flow across the membrane, resulting in a current I across the membrane. The current can cause neighboring channels to change their shape which initiates a nearby current flow, and so on. As a result, an action potential of duration about 1 ms propagates along the axon with constant speed, amplitude, and shape. The variable $V(x,t)$ denotes the local transmembrane voltage.

This transport alters the voltage V across the axonal membrane, which in turn affects in a strongly nonlinear way the flow of the ion currents across the membrane, which leads to excitable behavior.

The membrane consists of a thin lipid bilayer that is impermeable. However, complicated proteins embedded in the membrane serve two roles that lead to a flow of charge across the membrane. First, proteins called ion pumps sustain the axon in a nonequilibrium state by using chemical energy (ATP) to pump potassium ions K^+ into the neuron and sodium ions Na^+ out of the neuron, resulting in two opposing concentration gradients.[3] Second, proteins called channels can change their shape in response to a change in membrane voltage (some channels also change their shape when certain chemicals bind to them) and open up holes through the membrane that selectively allow only one kind of ion (K^+ ions for potassium channels, Na^+ ions for sodium channels) to diffuse passively down its concentration gradient.

In the unexcited rest state of the axon, some of the K^+ channels are open and all of the Na^+ channels are closed so that only K^+ ions can diffuse across the membrane. This selective diffusion of K^+ ions down their concentration gradient to the outside

---

[3] There are other ions involved in ionic transport such as chloride Cl^- and calcium Ca^{2+} but we will ignore these to simplify the discussion. Currents caused by calcium ions are especially important in other functions of the nervous system but not for signal transmission along axons.

of the membrane creates a separation of charge (more positive ions outside than inside). The resulting electric field across the membrane drives a K$^+$ current in the opposite direction, back into the axon. The corresponding voltage V across the membrane (measured from the inside relative to the outside) changes until it reaches a value of about -68 mV at which point a steady equilibrium state is reached such that the outward current driven by the concentration gradient and inward current driven by the electrical field balance and there is no net transport of charge. To the extent that the membrane is permeable to only a single ion at a time (a reasonable approximation for the resting state of an axon), the equilibrium voltage value V_K can be calculated from the so-called Nernst equation

$$V_K = \frac{k_B T}{q} \ln \frac{[K^+]_{\text{outside}}}{[K^+]_{\text{inside}}}, \quad (11.1)$$

which is a statement of thermodynamic equilibrium. Here k_B is the Boltzmann constant, T is the absolute temperature in kelvins, and q is the magnitude of the ion's charge in coulombs ($q = e$ for a K$^+$ ion). For a squid giant axon at room temperature ($T = 293$ K) and for the experimentally measured intracellular and extracellular K$^+$ concentrations (400 mM and 20 mM respectively), Eq. (11.1) gives a value $V_K \approx -77$ mV, while the squid giant axon resting voltage has a value of about -68 mV, so the experimental resting voltage is shifted somewhat from the potassium equilibrium value Eq. (11.1) by the presence of the other ions.

When a propagating nerve pulse starts to arrive at a particular region of the axon, the membrane voltage of that region starts to increase from its rest value of -68 mV. (An increase of the membrane voltage from its resting value is called depolarization.) When the local voltage exceeds a threshold value of about -50 mV, an excitable response is generated such that the voltage rapidly increases by about 100 mV and then decays back close to the original rest value after about two milliseconds. (See Fig. 11.2.) This short-lived large change in voltage is called the action potential and constitutes the basic unit of information that neurons transmit to one another. For example, many sensory neurons like the touch-sensitive neurons (Meissner's corpuscles) in your fingertips transmit action potentials toward the brain at a rate that increases monotonically with the amount of pressure on the fingertip. (And other neurons somehow read out the rate of arriving action potentials and send out their own action potentials, and somewhere the brain eventually translates the pattern of action potentials into the sensation of pressure on the fingertip.) Action potentials propagate at constant speeds that vary from about 0.5 to 50 m/s depending on the type of neuron.

Mechanistically, the action potential arises from the opening and closing of potassium and sodium channels with different time scales (fast for sodium, slow for potassium) and at different times. Thus as the leading edge of an action potential

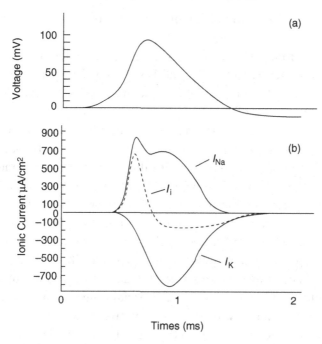

Fig. 11.2 (a) Action potential of the squid giant neuronal axon as calculated numerically by Hodgkin and Huxley for an axon temperature of 18.5 °C. (The simulation is sufficiently accurate that there is no practical difference between this figure and the experimental action potential.) This excitable pulse occurs whenever the membrane voltage V of any particular region of the axon rises above a threshold value of about -50 mV (about 17 mV in this panel, for which the resting state voltage has been set to zero). The calculation predicts a propagation speed of 18.8 m/s which is close to the experimentally observed speed of 21.2 m/s. (b) Calculated time course of ionic currents across the axon's membrane. The sodium current I_{Na} of Na^+ ions into the axon increases rapidly and causes the the membrane voltage to increase (depolarize). On a slower time scale, the potassium current of K^+ ions out of the axons (hence the negative sign of the I_K current) returns the membrane voltage to its original resting value. The dotted line labeled I_i is the sum of the sodium and potassium currents. (After Hodgkin and Huxley [46].)

arrives at some point on the axon and depolarizes the local voltage until it just crosses threshold, the previously closed sodium channels open rapidly and Na^+ ions can now diffuse down their large gradient into the axon. The resulting flow of Na^+ ions depolarizes the membrane voltage further which causes more Na^+ channels to open which further increases the Na^+ current. This positive feedback mechanism is the amplification mechanism that produces the excitable response of the axon. As the membrane voltage increases, even more potassium channels open but after a delay of about 1 ms and on a slower time scale. When the membrane voltage reaches about $+60$ mV, about 1 ms after the Na^+ channels first open, the sodium channels rapidly close and the outward potassium current restores the membrane voltage to it original rest value of −68 mV.

The dynamics of the sodium channels are such that, after closing, they cannot open again until more than a millisecond has passed. This period of inactivity is called the refractory phase of the action potential and has several implications. First, the refractory period of about 1 ms establishes an upper bound of about 1000 action potentials per second on the rate at which information can be transmitted between neurons. One kilohertz is slow compared to the gigahertz speeds of current computer chips, and brains compensate for this slow transmission rate by having a massively parallel network.[4] Another consequence of the refractory period is that if two action potentials approach each other on the same axon, they annihilate since neither one can cross the refractory region behind the leading edge of the approaching pulse, where the Na^+ channels recently closed and cannot yet open.[5] We see that biological excitable pulses in neurons and in heart tissue differ greatly from pulses of linear wave systems since the latter can pass through each other without difficulty. Refractory phases are especially important in heart dynamics since they can cause another wave to break into two pieces (see Fig. 11.13 later in this chapter), which is one mechanism that can trigger a cardiac arrhythmia.

The basic equation that describes pulse propagation is a statement about net charge conservation as various ions flow into, out of, and along the interior of a small piece of axon. It can be written as an evolution equation for the membrane voltage $V(x, t)$

$$c_m \, \partial_t V = i_{\text{ext}} + i_{\text{ions}}(V, \mathbf{u}) + \frac{a}{2r_L} \partial_x^2 V, \tag{11.2}$$

where the coordinate x indicates location along the axon and where we have assumed the simplest case of a spatially uniform axon. The term on the left side of Eq. (11.2) gives the charging of the lipid bilayer, which acts like a dielectric-filled capacitor of specific capacitance c_m (this is total capacitance divided by membrane area), whose value is about $10 \, \text{nF/mm}^2$ for all neurons. The charging arises from the currents that appear on the right side of Eq. (11.2). The specific current i_{ext} (current per membrane area) is an external source that the experimentalist might apply via an electrode. The specific current $i_{\text{ions}}(V, \mathbf{u})$ is associated with ions that diffuse through channels. Its value depends on the voltage V and also on a vector $\mathbf{u}(t)$ of

---

[4] In the mammalian cortex, each neuron connects to about 5000 other neurons on average, while the Purkinje cells in the cerebellum receive inputs from over 100 000 other neurons. This is three orders of magnitude or more higher connectivity than what can be currently achieved in integrated circuits.

[5] Action potentials usually travel in just one direction along an axon since only the neuron to which the axon belongs can initiate an action potential along the axon. However, neurobiologists often use two stimulating electrodes in different brain regions to intentionally create action potentials traveling in opposite directions on the same axon. This technique, called antidromic stimulation, is widely used to determine whether two neurons are directly connected in a living brain, by observing the failure of an action potential to reach the other electrode because of a collision with an action potential moving in the opposite direction on the same axon.

so-called gating variables $u_i(t)$, each with value between 0 and 1, that characterize the probability[6] for a channel of a certain type to be open at any give time t when the voltage V has a particular value. Hodgkin and Huxley used three variables $\mathbf{u} = (n, m, h)$ to describe the ionic currents. The gating variable $n(t)$ describes the dynamics of the K^+ channel while two gating variables m and h are needed to describe the more complex dynamics of the Na^+ channel. Finally, the last term in Eq. (11.2), $(a/(2r_L))\partial_x^2 V$, represents the longitudinal flow of current along the interior of the axon that is driven by spatial variations of the voltage along the axon. The parameter a is the radius of the axon (here assumed constant) and r_L is the intracellular resistivity (so that the longitudinal resistance R of a small length l of axon is given by $R = r_L l/(\pi a^2)$).

In the Hodgkin–Huxley model, the specific current for potassium ions is given by

$$i_K = \bar{g}_K n^4 (V - V_K), \tag{11.3}$$

where $\bar{g}_K$ is the maximum possible specific conductance (conductance per unit area) for K^+ ions and where $V_K \approx -77$ mV is the equilibrium K^+ voltage (given by Eq. (11.1)) such that there is no net flow of K^+ ions across the membrane for the experimentally observed interior and exterior K^+ concentrations. The parameters $\bar{g}_K$ and V_K are assumed to be independent of the membrane voltage although they can vary with temperature. The proportionality of the potassium current to the fourth power of n was deduced from fits to experimental data. Chemical kinetics then suggest that the potassium channel might consist of four subunits, each of which have to change their shape for the channel to open, and this turns out to be the case.

The variable n in Eq. (11.3) evolves according to a simple relaxational dynamics

$$d_t n(t) = \frac{n_\infty(V) - n}{\tau_n(V)}. \tag{11.4}$$

where the function $n_\infty(V)$ (to which n relaxes at long times) and the relaxation time constant $\tau_n(V)$ depend on the voltage and have functional forms that are deduced by fits to experimental data. The parameter $n_\infty(V)$ varies from zero at low voltages to unity at high voltages, with a dependence given by fitting to experiment, but roughly of the form of $\tanh(V - V_1)$ with the "turn on" voltage V_1 some empirical constant. It turns out that ion currents large enough to change significantly the membrane voltage V are too small to change the ion concentrations substantially. In particular, the propagation of a single nerve pulse does not significantly change the nonequilibrium ion concentrations so that the zero-current equilibrium voltages V_{Na}

---

[6] A picture that you should have in mind is that a channel is a flexible highly folded polymer that is constantly knocked about by thermal collisions with nearby molecules. These molecular collisions make the opening and closing of a channel stochastic and so a probabilistic description is needed.

and V_K given by Eq. (11.1) may be taken as constants in the dynamics. Over many pulses, of course the ion concentrations would be depleted as the passive ion channel currents move the system toward equilibrium but the ion pumps remain active and maintain the nonequilibrium ion concentrations.

The specific current for Na$^+$ ions is given by

$$i_{Na} = \bar{g}_{Na} m^3 h (V - V_{Na}), \tag{11.5}$$

where the constant $\bar{g}_{Na}$ is the maximum specific conductance for Na$^+$ ions. Here the gating variable m describes an activating process similar to the gate variable n for the potassium current. The gating variable h on the other hand describes an inactivating process, so that it is perhaps clearer to write $h = 1 - \bar{h}$. The variables m and $\bar{h}$ then obey dynamical equations analogous to Eq. (11.4), with additional functions $(m_\infty(V), \bar{h}_\infty(V))$ and $(\tau_m(V), \tau_h(V))$ fit to experiment, and that turn out to have a roughly similar form to the functions $n_\infty(V)$ and $\tau_n(V)$. Experimentally, the sodium channel responds much more rapidly[7] than the potassium channel so that

$$\max \tau_m \ll \max \tau_n, \max \tau_h. \tag{11.6}$$

There are other ion channels that move across the membrane but these ions play a relatively minor role in axonic action potentials and their influence is fit to a simple linear dependence

$$i_l = \bar{g}_l (V - V_l), \tag{11.7}$$

where the specific conductance $\bar{g}_l$ is a constant. Equation (11.7) is often called the leakage current since it corresponds to a small steady leak of charge (mainly due to Cl$^-$ ions) across the membrane whenever the membrane voltage is not equal V_l, which is the case for the unexcited rest state. The final expression for the total transmembrane ionic current i_{ions} in Eq. (11.2) is therefore

$$i_{ions} = \bar{g}_K n^4 (V - V_K) + \bar{g}_{Na} m^3 h (V - V_{Na}) + \bar{g}_l (V - V_l), \tag{11.8}$$

with approximate values[8]: $\bar{g}_K = 0.036\,\text{mS/mm}^2$, $V_K = -77\,\text{mV}$; $\bar{g}_{Na} = 1.2\,\text{mS/mm}^2$, $V_{Na} = 50\,\text{mV}$; and $\bar{g}_l = 0.003\,\text{mS/mm}^2$, $V_l = -54\,\text{mV}$. Hodgkin and Huxley deduced the detailed description represented by Eq. (11.2) and Eq. (11.8) by many experiments. In one kind of experiment, they would vary the internal and external concentrations of various ions such as Na$^+$ or K$^+$ (only possible with a big axon like the squid giant axon) and measure the change in

---

[7] Two time constants may be associated with a particular ionic current: the time constant τ of the gating variable and a resistor-capacitance RC time constant $C/\bar{g}$ of the current to reach its maximum value. For the sodium current, both time scales are short.

[8] Conductance is measured in SI units of siemens S, which is the reciprocal of the ohm unit for resistance. $1\,\text{mS} = (10^3\,\Omega)^{-1}$.

membrane voltage. In other experiments, they used an electronic feedback control method called voltage-clamp to measure the current flow whenever the membrane voltage V was changed to a new constant value. We encourage you to read the beautifully written classic papers of Hodgkin and Huxley for the full story in their own words.

Axonal membranes are not the only parts of the nervous system that have channels that respond to changes in the membrane voltage, there are also the membranes of dendrites (regions of neurons specialized to receive information from other neurons via synapses), membranes of synapses (micrometer-size objects that connect an axon to a receiving neuron and that amplify incoming action potentials by converting the electrical pulse into a chemical signal which diffuses to the receiving neuron and is then converted back into an electrical signal), and the membranes of somas, which are the neuronal cell body that contains the nucleus. Studies by many scientists after Hodgkin and Huxley, especially investigations of various genomes for the genes that code for channels, have revealed a bewildering diversity of channels. For example, neurons in mouse brains use over 75 kinds of potassium channels, over 9 kinds of sodium channels, 10 kinds of calcium channels, and other channels that are not easily characterized. Each neuron seems to use its own particular brew of channels. The number, kind, and spatial distribution of channels associated with a given neuron can evolve over time, in response to the pattern of action potentials received and to genetic instructions from the nucleus. While Hodgkin and Huxley explained quantitatively the excitable property of the squid giant axons and gave us the mathematical tools to analyze other biological excitable media, many mysteries remain about why so many different kinds of channels exist and how the variations in excitability play a role in brain function and brain disease.

11.1.2 Models of electrical signaling in the heart

Versions of the Hodgkin–Huxley model have been proposed to understand propagating electrical waves in heart muscle.[9] As in the squid axon, the excitability arises from the large self-stimulating sodium ion current. The degrees of freedom are again the voltage V across the membrane, and ionic currents through the membrane, with sodium, potassium, and calcium ions constituting the most important currents. The interconnected mesh of muscle cells that comprises heart muscle can be modeled

---

[9] Mammalian hearts also contain a network of specialized muscle cells called Purkinje fibers that propagate waves more rapidly than muscle cells and that are used by the heart to initiate a spatially coherent wave of contraction. Because an isolated Purkinje fiber can be used to study one-dimensional dynamics much like an axon, Hodgkin–Huxley models similar to those for heart muscle have been proposed and tested for Purkinje fibers.

on a macroscopic scale as an anisotropic three-dimensional continuous[10] material. Thus the electrical dynamics of heart muscle may be modeled by an equation of the form

$$c_m \partial_t V(\mathbf{x}, t) = i_{\text{ext}} + i_{\text{ions}}(V, \mathbf{u}) + \sum_{jk} \frac{\partial}{\partial x_j}\left(g_{jk} \frac{\partial V}{\partial x_k}\right), \quad (11.9)$$

where the total ionic specific current $i_{\text{ions}}(V, \mathbf{u})$ and the voltage-dependent dynamical equations for the internal gating variables $\mathbf{u}(t)$ contain the complexity of the ionic transport processes. The summation indices j and k go over spatial coordinates, from 1 to 2 for a two-dimensional sheet and from 1 to 3 for a volume. The last term with the summation is similar to the diffusive $\partial_x^2 V$ term in Eq. (11.2) and represents the coarse grained description of the current through the anisotropic electrically resistive muscle fiber network. The entity g_{jk} is a symmetric tensor that is proportional to the conductivity tensor. As a first approximation, this tensor has different components along the predominant fiber direction (which we denote by the unit vector $\hat{\mathbf{n}}$) and perpendicular to this direction

$$g_{jk} = g_L \hat{n}_j \hat{n}_k + g_T(\delta_{jk} - \hat{n}_j \hat{n}_k). \quad (11.10)$$

Real heart muscle is inhomogeneous and so the constants g_L and g_T can vary spatially which explains why the tensor g_{jk} is acted upon by the derivative $\partial/\partial x_j$ in Eq. (11.9).

Equation (11.9) is an example of what is called a monodomain heart model, which describes cardiac dynamics in terms of a single potential, namely the transmembrane voltage V. More general and accurate, but more complicated to work with, are bidomain models that use two separate equations to describe potentials and currents inside and outside the cardiac cells. Bidomain models are especially needed when investigating the influence of extracellular currents such as those generated by a defibrillator applied to the surface of a person's chest. Monodomain models cannot treat the effects of such currents on heart muscle quantitatively.

Cardiac action potentials differ greatly from neuronal action potentials in that the high-amplitude excited phase lasts several hundred milliseconds instead of about one millisecond. To model this feature, researchers have postulated additional slow ionic transport mechanisms beyond those discovered by Hodgkin and Huxley in the squid axon. One modification (introduced by Noble in 1962 in his study of Purkinje fibers) is to suppose that the potassium current is composed of two parts: a first part that rapidly decreases on depolarization, and that is treated as

---

[10] The extent to which it is appropriate to treat the heart as a continuous medium is still being determined. One concern is that the action potentials in hearts have narrow fronts (regions where the sodium currents initiate the action potential) that span just a few cardiac cells so it is possible that details of how the muscle cells are connected to one another or small inhomogeneities in muscle cells could influence the properties and stability of the front.

instantaneously following the membrane voltage; and a second part that slowly rises with the depolarization. In the Noble model, the potassium and sodium currents are then given by

$$I_{\text{ions}} = \left[g_{K_1}(V) + g_{K_2}n^4\right](V - V_K) + \left[g_{Na}m^3h + \tilde{g}_{Na}\right](V - V_{Na}). \quad (11.11)$$

The gating variables n, m, and h obey the same first-order kinetics, Eq. (11.4), as in the Hodgkin–Huxley model although with different parameter values that are obtained by fitting functional forms to cardiac data. A small constant sodium conductance $\tilde{g}_{Na}$ is also included.

As the amount of experimental information has increased, the complexity of the models has correspondingly grown. Although ionic currents are still described using expressions of the form $g_{\max} r^m s^n (V - V_{\text{eq}})$ – where $g_{\max}$ is the maximum channel conductance, r and s are gating variables raised to suitable powers, and V_{eq} is the equilibrium voltage given by Eq. (11.1) – several new features must be added to the basic Hodgkin–Huxley type description. For example, cardiac models now include a slow inward calcium ion current. The rest concentrations are such that this current is large enough to affect the Ca^{2+} ion concentration inside the muscle fibers, so that the calcium equilibrium voltage V_{Ca} also changes dynamically and one further has to include the influence of the cellular ion pumps, which we could ignore for neurons. Another complication is that for some of the ion currents such as the instantaneous potassium current $g_{K_1}(V)$ introduced by Noble, the current is no longer linear in the membrane voltage but has nonlinear rectifying properties that must be determined empirically.

Because of the spatiotemporal complexity of cardiac wave states, especially transitions of waves to novel states, it remains difficult to determine which of the many details of these many-variable empirically fit physiologically detailed cardiac models are essential for understanding the onset of arrhythmias and the resulting dynamics. For this reason, reduced models with two to four variables, such as those discussed in the next section, continue to play an important role in understanding cardiac dynamics.

11.1.3 FitzHugh–Nagumo model

A few years after the work of Hodgkin and Huxley, FitzHugh and independently Nagumo introduced a reduced version of the Hodgkin–Huxley model that retains the basic properties of fast excitability and slow recovery and that was simple enough that most of the key dynamical properties could be understood conceptually. The FitzHugh–Nagumo model has become a canonical set of equations for understanding the spatiotemporal dynamics of an excitable medium.

The starting point for FitzHugh's work was a famous nonlinear oscillator equation known as the van der Pohl equation[11]

$$d_t^2 x + \gamma\left(x^2 - 1\right) d_t x + x = 0. \tag{11.12}$$

where the real parameter γ is positive, $\gamma > 0$. For small values of x such that $x^2 - 1 \approx -1$, Eq. (11.12) becomes the familiar harmonic oscillator equation but with negative damping in which case the solution $x = 0$ is unstable to oscillations. These oscillations grow in magnitude until the positive nonlinear damping balances the negative linear damping on average. The van der Pohl equation yields approximately sinusoidal oscillations for small γ but highly anharmonic relaxation oscillations for large γ.

It is these relaxation oscillations that provide the basis for FitzHugh's model of nerve pulse propagation. To understand the relaxation oscillations mathematically, it is useful to introduce a variable y

$$y = \gamma^{-1} d_t x + x^3/3 - x, \tag{11.13}$$

to transform Eq. (11.12) into two first-order equations

$$d_t x = \gamma\left(y + x - x^3/3\right), \tag{11.14}$$

$$d_t y = -x/\gamma. \tag{11.15}$$

FitzHugh studied a version of these equations that he called the BoenHoeffer–van der Pohl (BVDP) model

$$d_t x = \gamma\left(y + x - x^3/3 + z\right), \tag{11.16a}$$

$$d_t y = -(x - a + by)/\gamma, \tag{11.16b}$$

where the parameters a and b are constants, and the variable z represents the external driving current I_{ext} in the Hodgkin–Huxley equations Eq. (11.2).

FitzHugh noticed that the four variables (V, m, h, n) in the Hodgkin–Huxley equations had different dynamical behaviors, with the variables V and m varying quickly during an action potential and the variables h and n more slowly. He further noticed that when he plotted two-dimensional phase portraits of a fast variable (V or m) combined with a slower variable (h or n) – for example the path traced out by the vector $(m(t), h(t))$ in the mh plane – he saw trajectories that were qualitatively reminiscent of trajectories traced out by the variables x and y of the BVDP model of

---

[11] It is interesting to note that, in 1928, van der Pohl and van der Mark built an electronic analog of a heart using a nonlinear vacuum-tube-based circuit to mimic the van der Pohl equation. This makes Eq. (11.12) one of the earliest published cardiac models.

Eqs. (11.16). FitzHugh therefore proposed to model either of the two fast Hodgkin–Huxley variables with the x variable of the BVDP, and to model either of the two slow Hodgkin–Huxley variables with the y variable. For $z = 0$ in Eq. (11.16a), a simple rescaling of variables and the addition of the diffusion term to the u equation changes the BVDP model into the FitzHugh–Nagumo model

$$\partial_t u = f(u, v) + \nabla^2 u, \qquad (11.17a)$$

$$\partial_t v = g(u, v), \qquad (11.17b)$$

where what look like reaction rates are given in terms of two functions $f(u, v)$ and $g(u, v)$

$$f(u, v) = \eta^{-1}\left(3u - u^3 - v\right), \qquad (11.18a)$$

$$g(u, v) = u - a - bv. \qquad (11.18b)$$

A small value of the parameter η in Eq. (11.18a) corresponds to fast dynamics of the variable u compared to the variable v. The reader should be aware that many different forms of the FitzHugh–Nagumo model have been published, although they are mathematically equivalent (see Exercise 11.7).

We can understand qualitatively the dynamics of two-variable equations like Eqs. (11.17) by looking at its nullclines. As we mentioned briefly in Section 3.1.2, nullclines are curves in the uv phase space that indicate where one of the right sides of Eqs. (11.17) vanishes in the absence of diffusion. Thus the nullcline for the u-evolution equation, Eq. (11.17a), is obtained by dropping the diffusion term (which corresponds to the assumption of a spatially uniform solution) and by setting the right side to zero: $f(u, v) = \eta^{-1}(3u - u^3 - v) = 0$. The set of points (u, v) that satisfy this equation is called the nullcline of Eq. (11.17a) or more simply the nullcline for the variable u or the u-nullcline. Similarly, the set of points (u, v) for which $g(u, v) = 0$ gives the v-nullcline.

A condition of the form $f(u, v) = 0$ allows the variable v to be defined as an implicit function of the variable u and sometimes the mathematical form of f is simple enough that one can solve for v explicitly in terms of u in which case it is easy to determine analytically the shape of the nullcline. You can easily verify that the u-nullcline is a cubic polynomial in u

$$v = 3u - u^3, \qquad (11.19)$$

while the v-nullcline is a line:

$$v = \frac{u - a}{b}. \qquad (11.20)$$

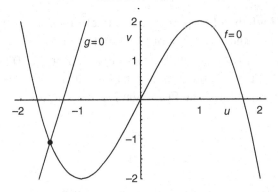

Fig. 11.3 Nullclines Eq. (11.19) and Eq. (11.20) of the FitzHugh–Nagumo model Eqs. (11.17). The intersection of these two nullclines at the black dot corresponds to a spatially uniform fixed point of the model. The parameter values $a = -1.3$ and $b = 0.2$ were used to plot the v nullcline. For small values of the parameter η, the fixed point is linearly stable, which is a necessary condition for excitable behavior.

For dynamical systems such that the nullclines cannot be obtained explicitly, numerical approximations can sometimes be found as described in Exercise 12.16 of Chapter 12.

The nullclines Eq. (11.19) and Eq. (11.20) of the FitzHugh–Nagumo model are shown in Fig. 11.3. The cubic u-nullcline is independent of all parameters and the linear v-nullcline moves about and changes slope as the parameters a and b are varied but is independent of the parameter η. We immediately see from the geometry of these curves that the FitzHugh–Nagumo equations always have one spatially uniform fixed point but not more than three. Although the nullclines and their points of intersection do not vary with the parameter η, this parameter does influence the stability of the uniform fixed points. For example, you should be able to show that, for the parameter range $0 \le b < 1/3$, there is exactly one uniform fixed point; that, for the constraint $-\sqrt{3} < a < -1 + 2b$, the fixed point lies to the left of the minimum of the u-nullcline; and that, for sufficiently small values of η, the fixed point is linearly stable. This stability is a necessary although not sufficient condition for the system to be excitable. To demonstrate excitability, we need to show that a sufficiently large perturbation about the fixed point will grow rapidly in magnitude and then decay back to the fixed point. We will do this using general qualitative arguments in the next section.

11.2 Oscillatory or excitable

Motivated by our discussion of the FitzHugh–Nagumo model in the previous section and by the simple two-variable chemical models like the Brusselator that were

11.2 Oscillatory or excitable

introduced in Chapter 3, we are led to investigate the following class of reaction–diffusion systems

$$\partial_t u = \eta^{-1} f(u, v) + \nabla^2 u, \tag{11.21a}$$

$$\partial_t v = g(u, v) + D \nabla^2 v, \tag{11.21b}$$

for two concentrations fields $u(\mathbf{x}, t)$ and $v(\mathbf{x}, t)$.[12] Of the two different possible diffusion constants for u and v that could appear, D_u and D_v, we have used D_u to set the length scale for both equations in which case the remaining diffusion constant D in the v equation gives the ratio D_v/D_u. The parameter D may be taken to be $O(1)$ or set to zero as in the FitzHugh–Nagumo model. The parameter η gives the ratio of reaction time scales, and is taken to be small since this will allow us to develop approximation schemes using perturbation theory. The reaction rates are given by the nonlinear functions $f(u, v)$ and $g(u, v)$ in Eqs. (11.21). The u- and v-nullclines are obtained respectively from the equations $f(u, v) = 0$ and $g(u, v) = 0$.

Equations (11.21) provide a tractable approach to study propagation phenomena and are directly motivated by the microscopic description of some chemical and biological systems. Many analogous phenomena can also be found in the amplitude equations developed around the instability to spatially uniform oscillations (type III-o) discussed in Chapter 10. However, these two approaches involve quite different approximations. As we will see, Eqs. (11.21) describe highly nonlinear phenomena, and may give propagating structures even when the uniform state is linearly stable. On the other hand, the amplitude equation approach describes weakly nonlinear phenomena near a linear instability to an oscillatory state and the phenomena are universal (not dependent on details of the particular system) within the validity of the amplitude equations. Nevertheless, we will often find similar qualitative results resulting from the two methods, and combining intuition gained from both descriptions is often profitable. We need to keep in mind that both descriptions are approximations and more elaborate equations are often needed to develop a quantitative understanding of nature. But since the more elaborate equations are usually not analytically tractable, understanding gained from the analytic approximations to the simpler equations remains valuable.

We investigate Eqs. (11.21) for the nullclines schematically sketched in Fig. 11.4, which qualitatively resemble those of the FitzHugh–Nagumo model, Fig. 11.3, in that there is a cubic-like nullcline and a linear-like nullcline. The reaction rates f and g change sign across the nullclines, leading to the directions of evolution of u and v as depicted by the arrows in the figure. For small values of the parameter η in Eq. (11.21a), the dynamics can be constructed graphically from these nullcline plots.

---

[12] These equation are of course the same general form as in Turing's analysis discussed in Chapter 3. However, here our focus is on dynamic phenomenon rather than on the linear stability of the uniform states.

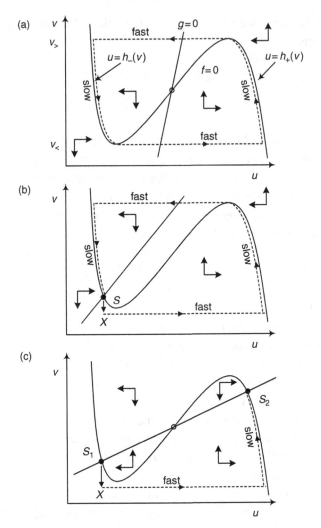

Fig. 11.4 Nullclines $f(u, v) = 0$ and $g(u, v) = 0$ for typical reaction–diffusion systems: (a) oscillatory, (b) monostable excitable, and (c) bistable excitable. Dashed lines show the trajectories of the dynamics. The arrows indicate the signs of the functions f and g that give the dynamics of u and v respectively. Solid and empty circles denote respectively stable and unstable steady states. For each v in the range $v_< < v < v_>$ indicated on the upper panel, there are three solutions for u that satisfy $f(u, v) = 0$. The largest and smallest solutions (these are the stable ones) are labeled $u = h_\pm(v)$.

The spatially homogeneous dynamics are given by the equations

$$d_t u = \eta^{-1} f(u, v), \qquad (11.22a)$$

$$d_t v = g(u, v). \qquad (11.22b)$$

11.2 Oscillatory or excitable

For small η, the dynamics can be split into fast and slow portions by the following argument. If $f(u, v)$ is nonzero, then because of the large η^{-1} factor in Eq. (11.22a), u will have a big time derivative and so vary rapidly in time. Thus if u is not located on its nullcline, it will rapidly evolve toward its nullcline in the direction determined by the sign of f, in a time of order η. (This direction is indicated by the horizontal dashed lines in Fig. 11.4 labeled "fast.") During this short time, v is effectively constant. This is the fast dynamics. Once u has reached the nullcline, evolution of v will occur at the slower $O(1)$ rate determined by Eq. (11.22b). (This is the evolution along the u-nullcline labeled "slow" in Fig. 11.4.) Equation (11.22a) is only consistent with slow dynamics if

$$f(u, v) \simeq 0, \tag{11.23}$$

and this relationship fixes $u(v)$ as v evolves. Since Eq. (11.23) has multiple solutions for the typical nullcline curve as shown in Fig. 11.4, it is useful to introduce an explicit notation for the two relevant solutions, $u = h_+(v)$ corresponding to the right branch of the nullcline, and $u = h_-(v)$ on the left. The middle portion of the u-nullcline leads to unstable solutions and is not relevant to the dynamics. Thus in the slow portion of the dynamics, the trajectory $(u(t), v(t))$ evolves along the u-nullcline $f = 0$ with $u = h_\pm(v)$.

We can now consider the dynamics for various nullcline configurations, with the important assumption that η is small.

11.2.1 Relaxation oscillations

Consider case (a) in Fig. 11.4. The single intersection point of the nullclines is a stationary point, that is easily seen to be unstable for small η. For example, a perturbation slightly increasing v from the fixed point will lead to dynamics that rapidly snaps to the branch $u = h_-(v)$, with u decreasing at essentially constant v. The dynamics will be attracted to a highly nonlinear "relaxation oscillation" corresponding to the dashed line trajectory in Fig. 11.4(a). We conclude that for these parameters, Eqs. (11.21) have oscillatory dynamics.

For small η, the orbit can be quantitatively constructed by piecing together successive slow and fast portions. For the slow portions, $u \simeq h_\pm(v)$ and then the dynamics of v is completely determined by Eq. (11.22b)

$$d_t v = g(h_\pm(v), v). \tag{11.24}$$

For example, on the left branch of the u-nullcline, the concentrations (u, v) will follow the u-nullcline downwards as v decreases according to

$$d_t v = g(h_-(v), v). \tag{11.25}$$

This continues until the minimum $v = v_<$ is reached, after which the slow motion ($d_t v < 0$) is no longer consistent with $f = 0$, and the dynamics must switch to the fast time scale. The fast dynamics consists of the rapid change of u according to Eq. (11.21a) over a time scale of order η as u evolves from the h_- to the h_+ branch. On this time scale, v does not significantly change and so the fast portion of the oscillation is given by the equation

$$d_t u = \eta^{-1} f(u, v_<). \tag{11.26}$$

This is followed by the slow evolution of v from $v_<$ to $v_>$ on the h_+ branch of the u-nullcline according to

$$d_t v = g(h_+(v), v), \tag{11.27}$$

and then the fast evolution back to the h_- branch at essentially fixed $v = v_>$

$$d_t u = \eta^{-1} f(u, v_>). \tag{11.28}$$

The orbit has the characteristic form of a relaxation oscillation with rapid switching followed by slow evolution, which qualitatively resembles a cardiac action potential. The period T is dominated by the slow portions so integrating Eqs. (11.25) and (11.27) gives us an estimate of the period

$$T \approx \int_{v_>}^{v_<} \frac{dv}{g(h_-(v), v)} + \int_{v_<}^{v_>} \frac{dv}{g(h_+(v), v)}. \tag{11.29}$$

11.2.2 Excitable dynamics

For other intersections of the two nullclines, the system Eqs. (11.21) may not have persistent oscillations but instead may undergo a large but temporary fluctuation driven by a relatively small perturbation. Panels (b) and (c) in Fig. 11.4 show two representative situations.

In Fig. 11.4(b), there is a single stationary point S, which from the evolution given by the signs of f and g can be seen to be stable to small perturbations. On the other hand, a larger perturbation, say starting at the point X, will lead to the orbit depicted by the dashed line that yields a large disturbance before the concentrations relax back to the fixed point. Again the dynamics can be constructed from portions of rapid motion of u at constant v and of slow motion along the $f = 0$ nullcline, as traced out by the dashed curve in the figure. Since an infinitesimal perturbation leads to a relaxation back to the steady state, whereas a small finite perturbation leads to a large response, this type of system is called excitable. The steady state is linearly stable, and so an excitable system is conceptually quite distant from one showing a type-III-o instability.

In Fig. 11.4(c), there are two fixed points S_1 and S_2 that are stable to small perturbations, as well as one unstable fixed point in the middle. Again a finite perturbation from one of the stable fixed points, e.g. from S_1 to X, will lead to a large disturbance, but now the disturbance relaxes to the second fixed point S_2. This situation corresponds to a bistable excitable medium.

11.3 Front propagation

We now return to Eqs. (11.21) and restore the diffusion terms. We will show that the full equations support various types of propagating disturbances that include fronts, pulses, and waves. These are all one-dimensional disturbances that propagate at some speed c (to be determined) without change of shape. They are therefore solutions of the form

$$u = u(\xi), \quad v = v(\xi), \quad \text{with} \quad \xi = x - ct, \tag{11.30}$$

where we assume propagation in the x-direction and that there is no dependence of the structures on the transverse coordinates. These solutions satisfy coupled *ordinary* differential equations given by substitution into Eqs. (11.21)

$$d_\xi^2 u + c\, d_\xi u + \eta^{-1} f(u, v) = 0, \tag{11.31a}$$

$$D d_\xi^2 v + c\, d_\xi v + g(u, v) = 0. \tag{11.31b}$$

Again, solutions may be constructed from rapidly varying portions for which u varies at essentially fixed v according to Eq. (11.31a), and from slowly varying portions for which v evolves with $u = h_\pm(v)$ according to Eq. (11.31b). The solutions will be developed from the plot Fig. 11.5 which is analogous to Fig. 11.4. However, "fast" and "slow" are now with respect to the ξ variable.

We first look at the propagation of fronts of u which are interfaces between regions where u takes on the values $u = u_+ = h_+(v_\text{f})$ and $u = u_- = h_-(v_\text{f})$ for some fixed $v = v_\text{f}$. The variation of u occurs on the fast scale so that it is consistent to take v fixed. The concentrations u and v trace out a path such as ABC in Fig. 11.5. The equation for the uniformly moving front (at speed c) is

$$d_\xi^2 u + c\, d_\xi u + \eta^{-1} F(u) = 0, \tag{11.32}$$

where $F(u) = f(u, v_\text{f})$. It is useful to rescale variables to eliminate the small parameter η. Thus we define $Z = \eta^{-1/2}\xi$ and $C = \eta^{1/2}c$ to obtain

$$d_Z^2 u + C\, d_Z u + F(u) = 0. \tag{11.33}$$

For $Z \to \pm\infty$, we have $u \to h_\pm(v_\text{f})$, where $F(u) = 0$. The form of $F(u)$ implied by the nullcline graph and by the signs of $f(u, v)$ is sketched in Fig. 11.6(a). This

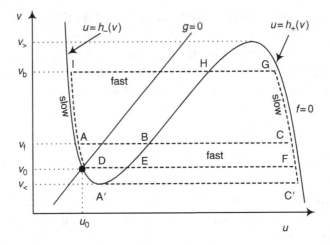

Fig. 11.5 Front, pulse, and wave train variations plotted on the nullcline graph for an excitable medium. The variation ABC gives a front of state $u = h_+(v_f)$ propagating into the state at $u = h_-(v_f)$ at some fixed $v = v_f$. The path DEFGHID gives a pulse propagating through the state at the fixed point (u_0, v_0). The path ABCGHIA gives a pulse of a wave train. Since the fixed point D is not visited in this orbit, a similar construction gives a pulse in a wave train in an oscillatory medium.

is exactly the same equation as studied in Section 8.3.1 where we were looking at fronts in the real amplitude equation, only the form of $F(u)$ is different. As there, we can use a "rolling ball" analogy to understand the solutions.

We use the replacements $Z \to T, u \to X$ and $F(u) \to d\Phi/du$ to map Eq. (11.32) into the equation for the "position" X of a unit mass fictional particle as a function of "time" T moving in a "potential" Φ and acted on by a "frictional damping" with strength C, with the form of Φ sketched in Fig. 11.6(b). For the general case $v_< < v < v_>$, the solution we are seeking starts at one of the maxima (e.g. u_+) at $T = -\infty$ and must run "down" the potential hill and up to the second maxima (at $u = u_-$), which it must just approach at $T = \infty$. It is clear from our intuition of frictional motion that there is a single value of the "damping" C for which this is possible, and we could calculate this value by integrating the equation of "motion." For $\Phi(u_+) > \Phi(u_-)$, the "damping constant" C for this type of solution to exist is positive.

On the other hand, if $\Phi(u_+) < \Phi(u_-)$, the value of C must be negative. Alternatively, in this case we could construct a solution with positive C running from u_- to u_+. Either description corresponds to the front moving in the reverse direction, with u_- invading u_+ rather than u_+ invading u_-. Furthermore it is clear that the "time" scale of the motion and the value of C will be $O(1)$ in general. There is an exception to this latter statement for the particular value $v = v^*$ known as the

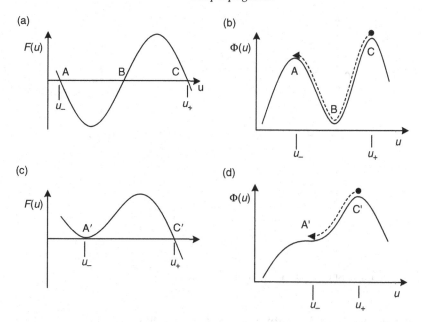

Fig. 11.6 Sliding ball analogy for constructing front solutions: (a) and (b) typical situation ("trigger" front) corresponding to path ABC in Fig. 11.5; (c) and (d) special situation ("phase" front) (corresponding to path A'C' in Fig. 11.5. The letters AB... label the same points as in that figure. The dashed line in (b) and (d) depicts the trajectory of the fictitious ball.

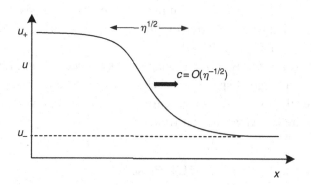

Fig. 11.7 Qualitative front solution of Eqs. (11.21) for small η.

stall speed at which $\Phi(u_+) = \Phi(u_-)$ so that the potential maxima are at the same height. In this case, $C = 0$ although the "time" scale of the motion remains $O(1)$.

Translating back to the unscaled units, we find a narrow front of width $O(\eta^{1/2})$ propagating at speed c which is of magnitude $O(\eta^{-1/2})$, a large value. The variation $u(\xi)$ for a front is sketched in Fig. 11.7. The speed is a function of the

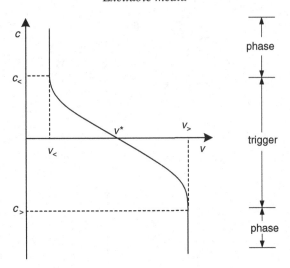

Fig. 11.8 Front speed c as a function of v concentration.

local concentration v of the slow species. The motion reverses as v passes through a particular value $v = v^*$ known as the stall point, for which $c = 0$.

As v passes through $v_<$ or $v_>$, one of the maxima of the "potential" $\Phi(u)$ disappears. Let us look at the case $v = v_<$. The potential well analogy for this case is sketched in Fig. 11.6(c) and (d). Our intuition for this situation tells us that any sufficiently large value of the damping C is sufficient to give a solution connecting u_+ at $T = -\infty$ to u_- at $T = \infty$. In fact, with further analysis it can be shown that there is a solution for any C greater than the limiting value of the unique $C(v)$ as v approaches $v_<$, $\lim_{v \to v_<} C(v)$.

The form of $C(v)$ resulting from the analysis is sketched in Fig. 11.8. For $v = v_>$ or $v = v_<$, where a range of speeds for each v is possible, the fronts are known as phase fronts. For the case $v_< < v < v_>$, for which there is a unique front velocity for each v, they are known as trigger fronts.

11.4 Pulses

Excitable media can have a localized region of excitation called a pulse that propagates at constant speed without change of shape. A physical example is the action potential that we discussed in Section 11.1.1. As we now describe, the structure of a pulse can be constructed from two front solutions.

Consider an excitable medium at the stable fixed point (u_0, v_0) for the nullclines shown in Fig. 11.5. A propagating pulse, Fig. 11.9(a), can be constructed from the four portions of the orbit DF, FG, GI, and ID that are sketched in the figure. For $\xi \to +\infty$ in front of the pulse, we have $u = u_0$ and $v = v_0$. The leading edge of the

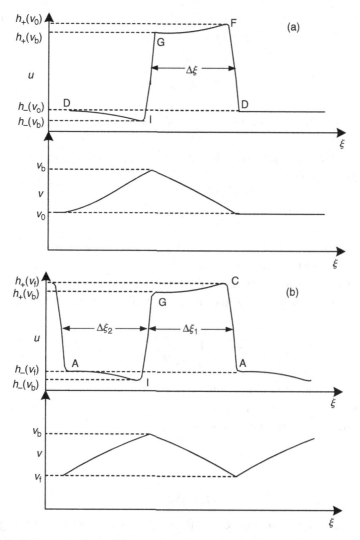

Fig. 11.9 Schematic plots of the u, v concentrations for (a) a pulse and (b) a wave train. Letters AB...I show the correspondence to points on the nullcline plot, Fig. 11.4.

pulse is a front on the short scale $\eta^{1/2}$ that connects $u_0 = h_-(v_0)$ and $u = h_+(v_0)$ at essentially fixed $v = v_0$ (DF in Fig. 11.5). The front velocity $c = \eta^{-1/2}C(v_0)$ can be read off from Fig. 11.8. Behind the front, v evolves slowly, with u following adiabatically $u = h_+(v)$, maintaining $f(u, v) \simeq 0$ (FG in Fig. 11.5), as governed by Eq. (11.31b). Note that since $c \gg 1$, diffusion plays no role and this equation reduces to

$$c\, d_\xi v + g = 0. \tag{11.34}$$

Equation (11.34) can be directly integrated far away from the front position ξ_f to give an implicit expression for $v(\xi)$

$$\xi_f - \xi = c \int_{v_0}^{v} \frac{dv}{g(h_+(v), v)}. \tag{11.35}$$

The back of the pulse must be a connection between $u = h_+(v)$ and $u = h_-(v)$ that moves at the *same* speed $c(v_0)$ as the front of the pulse. Since this is a front in the reverse sense, we must find the value of $v = v_b$ for which the speed given by the velocity curve Fig. 11.8 satisfies

$$c(v_b) = -c(v_0). \tag{11.36}$$

This is the portion GI in Fig. 11.5. The back of the pulse may be either trigger ($v_b < v_>$), as shown in the figure, or phase ($v_b = v_>$, $c(v_b) < c(v_>)$). Finally, v will evolve slowly back to v_0 along the nullcline $f = 0$ (portion ID in the figure). The distance between the front and the back, which is the width of the pulse, is given by

$$\Delta\xi = c \int_{v_0}^{v_b} \frac{dv}{g(h_+(v), v)}, \tag{11.37}$$

and is $O(\eta^{-1/2})$ and so is large. Similarly, the relaxation back to u_0 and v_0 behind the pulse occurs on this long length scale. In contrast, the widths of the fronts that form the leading and trailing edges of the excited region are $O(\eta^{1/2})$ and so are small.

11.5 Waves

Waves are made up of a sequence of pulses, and constructed in an analogous way, Fig. 11.9(b), except that the values of v_b and v_f at the back and front of the pulse are determined self consistently from the relation

$$c(v_b) = -c(v_f) = c, \tag{11.38}$$

with the wave speed c a free parameter. This gives a trajectory on the nullcline plot Fig. 11.5 such as ABCGHIA... Thus u varies on the length scale $\eta^{1/2}$ at constant $v = v_b$ or $v = v_f$ at the leading or trailing edge of each pulse, and v relaxes between these two values along the nullclines $f = 0$ over the long scales $\eta^{-1/2}$. The wavelength λ of the waves (the distance between two successive leading edges for example) is dominated by the slow portions, and is given by an expression analogous to Eq. (11.37). From this, we may immediately construct the temporal

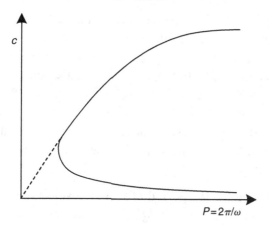

Fig. 11.10 Dispersion relation for wave propagation plotted in form of speed c as a function of the period $P = 2\pi/\omega$. For small speeds, the calculation ignoring the diffusion of v, Eq. (11.39) gives the dashed line, whereas more complete calculations suggest the slow branch extending to large periods for small speeds given by the solid line.

period $P = 2\pi\omega^{-1} = \lambda/c$

$$P = \frac{1}{c}(\Delta\xi_1 + \Delta\xi_2) = \int_{v_f}^{v_b} \frac{dv}{g(h_+(v),v)} + \int_{v_b}^{v_f} \frac{dv}{g(h_-(v),v)}, \tag{11.39}$$

which, with Eq. (11.38), is an implicit relation for $c(P)$, Fig. 11.10. The more usual form for a dispersion relation $\omega(k)$ can be derived from this plot. Note that for typical values of v_f, v_b the propagation speed c and the wavelength λ are both $O(\eta^{-1/2})$, whereas the period P, and so the frequency ω are $O(1)$.

We have focused on waves in an excitable medium. However, since the u, v trajectory ACGIA in Fig. 11.5 does not approach the fixed point D, the existence of this fixed point is irrelevant to the argument. This means that the same construction yields propagating waves in both excitable and oscillatory media. In the former case, there is a maximum wave velocity set by the pulse velocity, which gives the velocity at long wavelengths (the wave in this limit is formed of well-separated pulses). In the oscillatory case, the velocity is unlimited within the present approximations.

The small speed limit of wave propagation corresponds to the loop ACGIA in Fig. 11.5, shrinking to a small height so that the concentration of v at the leading and trailing edges v_f, v_b approach the stall value v^* (the trailing and leading edges must have opposite velocities and so the loop must straddle the zero velocity value). The analysis leading to Eq. (11.39) would then give a period and speed both proportional to the height of the loop, so that $P(c)$ goes to zero linearly (the dashed line in Fig. 11.10). However, the approximations we have used break down in this limit because the wavelength becomes small and the diffusion of v cannot be ignored, as was done in Eq. (11.34). The true behavior in this region is complicated, and a branch of slower moving waves is often seen in numerical simulations, as shown by the solid line in the figure. For very small η, a new scaling with η of the speed, period, and wavelength can be found, known as Fife scaling. This limit will turn out to be important in our discussion of spiral wave sources in Section 11.6, and so we briefly describe it in the following Etude. Numerical tests on various reaction–diffusion models do show however that the range of η over which the approximation is reliable is restricted to exceedingly small values that may not be appropriate for experimental systems.

Etude 11.1 Fife scaling for the period of waves at small speeds

Fife scaling is derived as follows. If diffusion is important in the v equation, we suppose that all terms in Eq. (11.31b) are of the same order of magnitude, which leads to the estimate

$$D\,\delta v/\lambda^2 \sim c\,\delta v/\lambda \sim 1, \tag{11.40}$$

where δv is the size of the variation from the stall value v^* for the concentration v between the fronts forming the pulses in the wave, and the last term is $g(u, v^*)$, which is $O(1)$. In the calculation of the front solutions, where we used the ball-in-potential analogy that was developed from the scaled form of the u equation Eq. (11.33), the difference in the "potentials" $\Phi(u_+)$ and $\Phi(u_-)$ will be zero for $v = v^*$ and proportional to δv for small δv. The "damping" C, which leads to the "dynamics" that just connects the two maxima, will scale as $C \sim \delta v$, giving a speed

$$c \sim \eta^{-1/2}\delta v. \tag{11.41}$$

However the "time" T for the "motion" does not depend sensitively on δv and remains of order unity, so that the front widths remain small, $O(\eta^{1/2})$. Putting together Eqs. (11.40) and (11.41) leads to

$$c \sim \lambda^{-1} \sim \eta^{-1/6}, \tag{11.42a}$$

$$P = \lambda c^{-1} \sim c^{-2} \sim \eta^{1/3}, \tag{11.42b}$$

$$\omega \sim \eta^{-1/3}. \tag{11.42c}$$

Notice that now $P \propto c^{-2}$, so that the period becomes large *for small speeds, consistent with the solid curve in Fig. 11.10.*

We have now shown that wave solutions exist in an excitable medium. Does this mean that we should expect to see sustained waves in these systems? The first thing to remember is that the spatially uniform state is stable in an excitable system. Furthermore, a wave or pulse excitation will tend to disappear from the system since, unlike linear waves, they do not propagate through one another and on collision tend to annihilate each other. At boundaries, reflection effects are usually small, and the reflected disturbance is typically not large enough to re-excite the medium, or is quenched by nonlinear interaction with the incoming disturbance. Sources of the propagating disturbances are therefore crucial to the dynamic state in excitable media. Of course, this is familiar in the biological examples of nerve fibers and heart tissue. The propagating action potential pulse in a nerve fiber is produced by the firing of the neurons, and, in a properly functioning heart, the pulse of electrical activity that stimulates the muscle contraction is produced by a specialized pacemaker region known as the sinus node, where the dynamics is oscillatory rather than excitatory. In these systems, the presence of self-sustained waves from sources localized within the tissue itself correspond to pathology, for example tachycardia arrhythmias in the heart.

In one-dimensional versions of the two-variable reaction–diffusion equations that we are using to understand excitable systems, Eqs. (11.21), there do not appear to be any stable intrinsic sources of waves. Waves must be produced by higher-order systems of equations or extrinsically, for example through inhomogeneities in the parameters. In two-dimensional systems, the evidence suggests that there are also no intrinsic point sources of azimuthally symmetric waves. Experiments in dishes of chemical reagents often show such target patterns of point sources radiating azimuthally symmetric waves. However, these are thought to be due to inhomogeneities producing local oscillations, such as specks of dust, and numerical and analytic investigation lead to the conclusion that there are no stable axisymmetric sources within the spatially homogeneous two-component equations Eqs. (11.21).

On the other hand, a rotating nonaxisymmetric source, which takes the geometric form of a spiral, may produce a train of waves at large distances, and have a persistence that we can understand by topological arguments, as in Section 10.2.5. Thus spiral defects play a vital role in wave propagation in two-dimensional excitable reaction–diffusion systems. We have already seen examples in Chapter 1. Figure 1.18(a) shows a spontaneous pattern of spirals in a thin layer of chemicals while Fig. 1.10 shows what may be spiral waves of electrical activity in a fibrillating heart.

The extension of the spiral source to three dimensions is known as a scroll wave. One way to think of a scroll wave is to draw a continuous one-dimensional curve in space, for example a straight line, a circle, a spiral or even a trefoil knot. Then imagine that each plane that intersects this line roughly perpendicular to the line contains a rotating spiral whose center lies on the line; you end up with a continuous stack of spirals that twist about in space with the line. The line that links the spiral cores is a line defect called a filament. Analytical, numerical, and experimental studies of scroll waves show that the filament is often dynamic, e.g. a circular filament might contract or expand depending on the model and choice of parameters, or a filament might give birth to multiple filaments via some instability (this has been proposed as one mechanism for the onset of ventricular fibrillation in a heart). Whether the waves seen on the surface of the heart such as Fig. 1.10 can be approximately understood in terms of two-dimensional structures that are roughly uniform over the thickness of the muscle, or whether there is a complex three-dimensional scroll wave inside the ventricular wall remains an open question of some importance since the answer would lead to different clinical strategies. These questions motivate further research regarding the properties of filaments and the waves that they emit.

Thus, in two- and three-dimensional excitable systems, we can expect to see self-sustained wave patterns, but only in the presence of spiral sources. An important question then arises, as we discussed in Section 10.2.5: for a given set of control parameters can spirals rotate at any frequency? Or is there a unique frequency or perhaps a discrete spectrum of possible frequencies? If there is a unique frequency, what is its value, and how does it depend on the parameters of the equations describing the system? In particular we might ask how the frequency scales with the small parameter η in Eqs. (11.21)?

We will focus our investigation of these issues on the simpler case of a two-dimensional excitable system. In the next section, we will learn how to construct spiral solutions for the two-variable reaction–diffusion model Eqs. (11.21). In the following section, we then discuss the stability of spirals.

11.6 Spirals

11.6.1 Structure

As was the case for other propagating solutions that we have discussed, spiral wave solutions to Eqs. (11.21) for small η can be constructed from the motion of fronts, as sketched in Fig. 11.11. The propagating region of the excited state (shaded in the figure) lies between two fronts, a leading and trailing edge, that both have the form of a spiral.

The new feature of the fronts forming a spiral from the ones considered so far is that they are curved rather than straight. Using the fact that the widths of the fronts are small compared to other lengths in the problem, the effect of the curvature on the dynamics is given simply by the so-called eikonal approximation[13]

$$c_n(v) = c(v) - K, \qquad (11.43)$$

---

[13] Note that curvature *away* from the direction of propagation *decreases* the propagation speed. You may find the eikonal equation written in the literature with either sign of the curvature correction term, depending on different conventions for how the sign of the curvature is defined.

11.6 Spirals

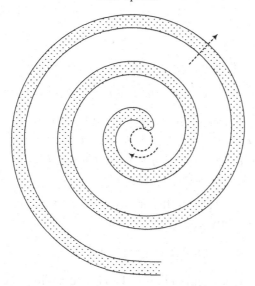

Fig. 11.11 Sketch of a spiral. The full line denotes the narrow front connecting the excited region ($u = h_+(v)$) (shaded) with the unexcited ($u = h_-(v)$) region (i.e. the "front" and the "back" of the pulse of excitation). The spiral tip rotates around a circular core indicated by the dashed line, leading to a propagating pulse train at large distances from the core.

which we derive in the Etude below. This equation expresses the normal velocity of a curved front c_n in terms of the propagation speed $c(v)$ of a plane front and in terms of the curvature K of the front (here the inverse of the radius of the curved front). The plane-front speed $c(v)$ is evaluated at the value v of the slow variable that coincides with the location of the front for the fast species u.

Etude 11.2 Derivation of the eikonal approximation
We are solving Eq. (11.21a) for a curved front in $u(\mathbf{x}, t)$ for a fixed v concentration $v = v_f$

$$\partial_t u = \eta^{-1} f(u, v_f) + \nabla^2 u. \tag{11.44}$$

Consider a circular front at a radius R. Then the Laplacian acting on u becomes

$$\nabla^2 u = \partial_r^2 u + r^{-1} \partial_r u \simeq \partial_r^2 u + R^{-1} \partial_r u, \tag{11.45}$$

where the second approximate equality is valid if the radius is large compared to the narrow width of the front so that we can neglect the variation of r over the extent of the front. In this limit, we can also approximate the motion as steady, and write in the vicinity of the front where u is changing

$$u(r, t) \simeq u(\xi) \text{ with } \xi = r - ct. \tag{11.46}$$

so that

$$\partial_t \to -c\, d_\xi. \tag{11.47}$$

Equation (11.44) becomes

$$d_\xi^2 u + (c + R^{-1})d_\xi u + \eta^{-1} f(u, v_f) = 0, \tag{11.48}$$

which is the same equation as for the propagation of a straight front Eq. (11.32) except for the replacement $c \to c + R^{-1}$. This then gives Eq. (11.43) with $K = R^{-1}$ the front curvature. The coefficient of unity in front of the curvature derives from the diffusion constant of unity in Eq. (11.21a).

To determine the value of v at the fronts we need to solve the slow evolution equation in the region between them

$$\partial_t v = g(h_\pm(v), v) + D\nabla^2 v, \tag{11.49}$$

with h_+ used in the shaded region in Fig. 11.11, and h_- in the unshaded region.

At large distances from the central core, a spiral rotating at frequency ω_s will produce waves that are approximately planar, with the wave number q_s fixed by the dispersion relation $\omega(q_s) = \omega_s$. The values of v at the front and back of the excited region are then the values v_f and v_b that are consistent with a wave of period $2\pi/\omega_s$, using the dispersion relation produced by the calculation of Section 11.5. For example, the concentration profiles along the arrow in Fig. 11.11 will be as in Fig. 11.9.

If we now follow the line in Fig. 11.11 that corresponds to the front or back of the excitation pulse, from the distant region where the straight front approximation is good, inwards toward the core region, the value of v will begin to change away from v_f or v_b as the two-dimensionality of the structure becomes important. Through the core region, the value of v must vary continuously, interpolating between v_f and v_b. This means that the concentration must pass through the stall value v^* somewhere. If we could ignore the curvature correction in Eq. (11.43), there would be no forward motion of the front at this point, and this would identify a point rotating around the core circle in Fig. 11.11. In practice, the curvature correction is likely to be significant and so a different value $v = v_c$ identifies this circumnavigating point.

What about the frequency ω_s of the spiral, and the question of whether there is a particular value determined by the construction of the solution, or whether all frequencies or a continuous range of frequencies are possible? We can argue phenomenologically that, if a particular value of the frequency is determined, then for small enough η it is likely to be determined by the Fife scaling limit, in which the frequency scales as $\omega_s \sim \eta^{1/3}$.

The argument goes something like this. In the core of the spiral, the curvature of the wave fronts becomes important. The simplest assumption is that the curvature in the core, which has the dimensions of inverse length, is of the same order of magnitude as the wave number q_s of the waves far away, i.e. that a single length scale defines both the core structure and the asymptotic wavelength. On the other hand, analyzing the effect of wave front curvature K on the front propagation velocity gave the correction Eq. (11.43). We expect the core radius to be set by the length scale at which the curvature correction becomes comparable to the planar velocity, $K \sim c$. With the single length scale assumption this means that $q_s \sim c$. For a general value of v away from the stall value, we found that $c \sim \eta^{-1/2}$ and $q \sim \eta^{1/2}$ which is incompatible with this relation. With v near the stall value v^* that gives Fife scaling, we have $c \sim q \sim \eta^{-1/6}$, which is now consistent with this relationship.

Unfortunately, in the Fife limit the diffusion of v cannot be ignored as it was in the simple discussion of pulses and waves in Section 11.4 and Section 11.5; this makes the analysis of the structure of the spiral quite hard in general. However, note that, even in this limit, the width of the front is still small compared with the radius of curvature, and so approximations based on the narrowness of the front are still good. With significant effort, it has been shown that a unique spiral frequency exists in various simplifying limits, for example for small D in Eqs. (11.21). The case of a "singly diffusive medium" given by $D = 0$ is discussed in the following Etude.[14] It is worth remarking that the same conclusion was found for spiral structures in the CGLE which applies to oscillatory media in the limit of weakly nonlinear behavior, as described in Section 10.2.5. Thus it is likely that this is the general result.

Etude 11.3 Spirals in the singly diffusive limit
In the singly diffusive limit, given by Eqs. (11.31) with $D = 0$, a number of simplifications make the full solution for the spiral more straightforward. Rather than carefully justify the assumptions involved, we will simply state them, and let the reader check their consistency with the final results.

An important simplification for $D = 0$ is that the concentration of the slow species v is constant along the front (at some value v_f) and back (at some value v_b) of the propagating pulse that forms the spiral. Thus Eq. (11.43) is to be solved with a constant speed $c(v_f)$ or $c(v_b)$. Furthermore v_f and v_b are close to the stall value v^* (i.e. the orbit ABCGHI in Fig. 11.5 forming the propagating pulses is thin straddling $v = v^*$) and the properties of $f(u, v)$ and $g(u, v)$ are needed only near this value.

Equation (11.49) with $D = 0$ is used to connect the values v_f and v_b. The spiral rotating at frequency ω_s can be parameterized through the coordinates of the front

---

[14] This Etude is derived from the paper "Scaling regime of spiral wave-propagation in single-diffusive media" by Karma [53]. You can find more details of the calculation there.

at distance r from an origin

$$x = r\cos(\theta_f(r) - \omega_s t), \quad y = r\sin(\theta_f(r) - \omega_s t), \tag{11.50}$$

where we have introduced the polar angles of the front and back as a function of radius $\theta_f(r)$ and $\theta_b(r)$. Equation (11.49) now reduces to

$$\omega_s \, \partial_\theta v = g(h_\pm(v), v), \tag{11.51}$$

independently of the radius (where we choose h_+ or h_- depending on the portion of the u-nullcline that is relevant to the front or the back propagation). Integrating between the front and the back, we get

$$\omega_s(v_b - v_f) = g_+^*(\theta_b - \theta_f), \tag{11.52}$$

where, since v_f and v_b are both near v^* we can simply evaluate the right-hand side of Eq. (11.51) using $v = v^*$ to give $g(h_+(v^*), v^*)$, which we write as g_+^*. Similarly integrating between back and front gives

$$\omega_s(v_f - v_b) = -g_-^*[2\pi - (\theta_b - \theta_f)], \tag{11.53}$$

where $g_-^* = -g(h_-(v^*), v^*)$ (the minus sign is inserted so that g_-^* is positive). These equations are easily solved to give

$$v_b - v_f = \pi \omega_s^{-1} \frac{g_+^* g_-^*}{g_+^* + g_-^*}, \tag{11.54a}$$

$$\theta_b - \theta_f = \pi \frac{g_-^*}{g_+^* + g_-^*}. \tag{11.54b}$$

The angle between the front and the back is independent of radius, so that the spiral has the form shown in Fig. 11.12. To leading order in $|v_{b,f} - v^*|$ we can approximate the propagation speed as

$$c(v_f) = -c(v_b) = c'\delta, \tag{11.55}$$

where

$$c' = -\left.\frac{dc}{dv}\right|_{v=v^*}, \tag{11.56}$$

and

$$v_f = v^* - \delta, \tag{11.57a}$$

$$v_b = v^* + \delta. \tag{11.57b}$$

Then according to Eq. (11.54a)

$$\delta = \frac{\pi \omega_s^{-1}}{2} \frac{g_+^* g_-^*}{g_+^* + g_-^*}. \tag{11.58}$$

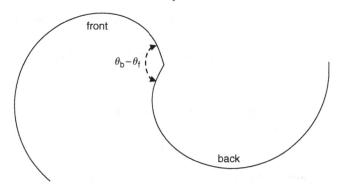

Fig. 11.12 Spiral in the singly diffusive limit. The heavy lines are the front and back of the pulse of enhanced "fast" concentration u. They are spaced by a constant angle $\theta_b - \theta_f$ in this case. Note that unlike the spiral in Fig. 11.11 the front and the back meet at a sharp junction, rather than the smooth curve there. This singularity is removed at smaller scales than analyzed here.

As yet, the frequency ω_s is unknown. This is determined by constructing the full shape $\theta_f(r)$ (which, since according to Eq. (11.54b) the angle between front and back is independent of the radius, immediately gives $\theta_b(r)$ as well) using the eikonal equation Eq. (11.43). It is convenient to introduce the variable

$$\psi = r \partial_r \theta_f. \tag{11.59}$$

In terms of this function, it can be shown after some algebra that the quantities appearing in the eikonal equation are

$$c_f n = \frac{\omega r}{(1+\psi^2)^{1/2}}, \tag{11.60}$$

$$K = \frac{\partial_r \psi}{(1+\psi^2)^{3/2}} + \frac{\psi}{r(1+\psi^2)^{1/2}}. \tag{11.61}$$

From Eq. (11.43), we can then write down a differential equation for $\psi(r)$. The variables can be rescaled

$$\bar{r} = cr, \quad \bar{\omega} = \omega_s/c^2, \tag{11.62}$$

with $c = c(v_f)$ the speed the front would have in the absence of curvature, to give

$$\partial_{\bar{r}} \psi = \bar{\omega} \bar{r}(1+\psi^2) - (1+\psi^2)^{3/2} - \frac{\psi(1+\psi^2)}{\bar{r}}. \tag{11.63}$$

The boundary condition at large distances is defined by the approach to plane waves

$$\partial_r \theta_f = q_s = \omega_s/c, \quad r \to \infty \tag{11.64}$$

so that in terms of ψ and the scaled variables the boundary conditions are

$$\psi \to \bar{\omega}\bar{r}, \quad \bar{r} \to \infty, \tag{11.65a}$$

$$\psi = 0, \quad \bar{r} = 0. \tag{11.65b}$$

Equation (11.63), with (11.65), leads to an eigenvalue condition for $\bar{\omega}$. This can be evaluated numerically to give

$$\bar{\omega} = 0.331. \tag{11.66}$$

In unscaled units this gives

$$\omega_s = 0.331 c^2. \tag{11.67}$$

Equations (11.55), (11.58), and (11.67) lead to the final result

$$\omega_s = 0.93 \left(\frac{g_+^* g_-^*}{g_+^* + g_-^*} \right)^{2/3} (c')^{2/3}, \tag{11.68}$$

which determines the spiral frequency in terms of parameters of the reaction–diffusion equations. Thus this analysis predicts a unique frequency of the spiral, determined by the reaction terms. Note that according to Eq. (11.41) $c' = O(\eta^{-1/2})$ and so $\omega_s = O(\eta^{-1/3})$ and $|v_{f,b} - v^*| = O(\eta^{1/3})$. These values are as predicted by the Fife scaling. They can also be used to justify the assumptions made at the beginning of the analysis.

The form of the solution shown in Fig. 11.12 is disturbing since the front and back form a sharp junction and v is discontinuous at the center, whereas we would expect a smooth solution to the equations. However, it must be remembered that the solution has been constructed on a length scale large compared with the $O(\eta^{1/2})$ width of the front and back. Further analysis shows that the apparent discontinuities are indeed resolved on this finer length scale.

The existence of a unique spiral frequency has important ramifications for predicting the behavior of excitable or oscillatory media. In many situations, instead of basing our understanding on the spatially uniform state and its instabilities, or the whole range of propagating wave states, we can instead focus on the properties of the waves at the single frequency ω_s and corresponding wave number q_s, and their spiral sources, since these will dominate the persistent state. This approach has already been described for waves and spiral sources in the CGLE in Section 10.2.5.

11.6.2 Formation

We have seen that spirals are persistent structures, but how do they form initially? A common mechanism is from breaks in wave fronts, induced by inhomogeneities

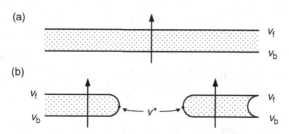

Fig. 11.13 (a) Single propagating pulse. Region shaded is the excited medium between the front and back, with $u = h_+(v)$ and v varying between v_f and v_b. (b) Broken pulse. The curved parts of the wave front may either retract, advance so that the pulse rejoins, or curl up to form a spiral pair.

in the medium, external perturbation (e.g. a hot wire in a chemical reaction or a damaged region in heart tissue), or by the collision of two wave fronts.

The simplest case to consider is a single straight pulse in an excitable medium. Suppose the wave front is cut into two separated pieces by some mechanism, as shown in Fig. 11.13. The tips may either retract, when the pulse will eventually dissipate, or may tend to grow. Which happens will depend on parameters such as the excitability of the resting medium. For the case of tip growth, we can argue as before that one point on the tip interface will instantaneously have zero normal velocity, so that the pulse will tend to pivot about this point. (A too large curvature leads to the tip retraction.) As time advances this can lead to a pair of counter-rotating spirals. If the break in the pulse is not large enough, an alternative outcome is that the two halves of the pulse reconnect.

11.6.3 Instabilities

The existence of a unique spiral frequency has important ramifications for predicting the behavior of excitable or oscillatory media. In many situations, instead of basing our understanding on the spatially uniform state and its instabilities, or the whole range of propagating wave states, we can instead focus on the properties of the waves at the single frequency ω_s and corresponding wave number q_s, and their spiral sources, since these will dominate the persistent state. Various instabilities have been identified in spiral wave systems. These instabilities may be important in biological applications, for example signaling the onset of fibrillation in the heart muscle. There are two different classes: instabilities of the waves far away from the core where the curvature has a small effect, and instabilities of the core itself.

The instabilities of the distant wave trains can be understood in terms of the secondary instabilities of the spatially periodic plane wave states. This is analogous to the stability balloon for type I-s systems discussed in Section 4.2.1: instability occurs when the wave number q_s produced by the spiral source passes outside of the stability balloon of plane wave states. Thus the usual types of instabilities must be considered, remembering that these are secondary instabilities about a spatially periodic state so that a Bloch analysis is needed. We might expect long wavelength longitudinal (Benjamin–Feir) and transverse instabilities as in the complex Ginzburg–Landau amplitude equation for oscillatory instabilities discussed in Chapter 10. In addition, there might be finite (Bloch) wave number instabilities, and these are likely to depend on the details of the dynamics of the excitable medium. One interesting case is when the wave vector of the instability is twice that of the wave itself – a "zone boundary" instability. This leads to a period doubling of the waves, known as alternans in the heart literature. As in the case of instabilities of spirals in the CGLE, the instability may be benign if it is convective, and then we must seek the point of absolute instability.

An example of a core instability is the meandering instability for which the spiral tip begins to meander periodically at some new independent frequency in a flower-like pattern, as in Fig. 11.14, rather than rotating uniformly around a circle. One can find a rotating frame of reference for which the path of the tip forms a closed cycle but in the laboratory frame, the path will typically not be closed. The meandering instability is often encountered in FitzHugh–Nagumo type models as a continuous Hopf bifurcation that introduces a new frequency ω_m, the meander frequency. The

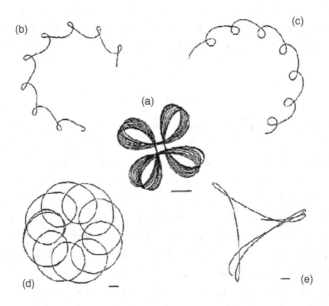

Fig. 11.14 Paths traced out by the spiral tip after the meander instability for various parameters in the FitzHugh–Nagumo reaction diffusion model. (From Winfree [113].)

pattern of the flower traced out by the tip is determined by the ratio of ω_m to the spiral frequency ω_s. For the degenerate case, $\omega_m = \omega_s$, the tip advances progressively in a cycloid motion in some particular direction. For ω_m/ω_s irrational, the motion is quasiperiodic. The tip may even develop a chaotic motion away from the onset of the instability, when the meander becomes large.

11.6.4 Three dimensions

In a three-dimensional reaction–diffusion system, the spiral core defines a line in the medium. To define the spiral defect, we need to specify the location of this line, and also a twist variable that defines how the phase of the spiral varies along the core line. Clearly, complicated structures can be imagined where the core forms various closed loops and knots. Often only the surface of the three-dimensional medium is readily accessible to experimental probes. The experimenter is then left with the difficult task of guessing the three-dimensional structure underlying observations on the two-dimensional surface. For example, a counter-rotating spiral pair on the surface may represent opposite ends of the same spiral core line intersecting the surface, or instead may be ends of two different three-dimensional spirals. The role of the three-dimensional structure on instabilities of the spirals on the two-dimensional section is also an important question.

11.6.5 Application to heart arrhythmias

An issue of considerable interest to medical doctors and biomedical engineers is whether the general behavior of spirals in excitable media is associated with heart arrhythmias in the intact undamaged heart, including perhaps the fatal ventricular fibrillation. An appealing scenario that is proposed in the cardiology literature assigns various medical symptoms diagnosed via electrocardiogram measurements to particular dynamical states of the heart muscle excitable media, see Fig. 11.15. Thus various types of tachycardia (an unusually rapid heart beat) may correspond to the development of a self-sustaining spiral structure, or to a counter-rotating pair of spirals. Fibrillation is then the breakdown of the simple spiral structure to a spatially disordered state, with perhaps many spirals forming and annihilating, and moving in a complex way.[15]

We would then like to know what are the conditions that render the heart susceptible to spiral formation? And what are the conditions that lead to the breakdown of this spiral to a disordered structure? From the general perspective of the dynamics of excitable media, neither of these phenomena are unexpected and indeed they are

---

[15] In small hearts such as the frog heart, there is evidence that an electrocardiogram showing fibrillation may correspond to a single, rapidly moving spiral.

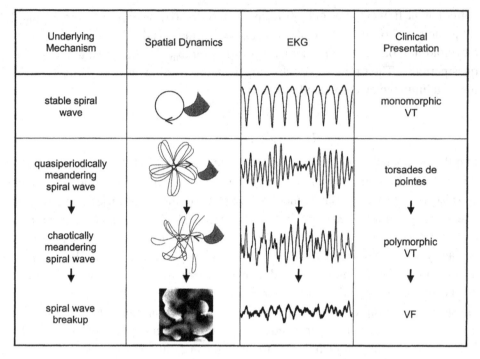

Fig. 11.15 Possible relationship between dynamical states and arrhythmias in the heart. (From Qu and Garfinkle [90].)

readily found in numerical simulations of reduced models and of ionic models in simple or realistic geometries, as parameters and external stimuli are varied. For example, one could easily imagine that the initial spiral might form by a wave breaking at a gross inhomogeneity produced by muscle damage due to a lack of oxygen in a heart attack. Or a spiral might form from an increased sensitivity to some change in the dynamical properties of the excitable medium, at smaller but always present inhomogeneities. Similarly, the well-documented sensitivity of the heart to tachycardia if an electrical stimulus is applied in the "vulnerable period" after an action potential pulse has propagated through the heart can be understood in terms of the resulting secondary wave front propagating into the refractory region behind the primary wave front and consequently breaking as in Fig. 11.13. The breakdown of the single spiral to a disordered state might result from any of the various mechanisms of spiral instability discussed in Section 11.6.3, either at the core or further out.

It is much more difficult to make definite statements regarding whether spiral formation and breakdown actually do occur in apparently healthy hearts, and to predict how the occurrence can be prevented through the effect of drugs on the properties of the heart tissue, or the consequences counteracted by judicially applied, small electrical perturbations. There is some experimental evidence and related numerical simulations of reduced equations that the breakdown only occurs in sufficiently thick heart muscle. This suggests that ventricular fibrillation is a three-dimensional phenomenon. Another suggestion is that the strongly anisotropic electrical conductivity properties of the heart muscle due to the fibrous nature, and the rotation of the anisotropy direction through the depth of the heart muscle wall, are also likely to be important. It is currently a highly active field of theory, numerics, and experiment to develop and test these ideas. Issues that immediately arise are the applicability of ideas developed for a spatially uniform excitable medium to the heart, which has a complicated topology and geometry, and is by no means homogeneous at the scale of the wave fronts, which may be only a few cell lengths in width. Theory and numerics to address these questions, and systematic experiments both on hearts in situ in the body and portions of the muscle sustained in a living state outside of the body, or cultured layers of muscle cells, show the promise of allowing researchers to answer these questions in the near future.

11.7 Further reading

(i) The ground breaking Hodgkin–Huxley work is described in their paper "A quantitative description of membrane current and its application to conduction and excitation in nerve" [46].
(ii) Chapters 5 and 6 of the book *Theoretical Neuroscience: Computational and Mathematical Modeling of Neural Systems* by Dayan and Abbott [29] give a modern discussion of neuron excitability and of the Hodgkin–Huxley equations.
(iii) Classic papers on the mathematical treatment of propagating fronts, pulses and waves, and spirals in excitable media from which our discussion is derived are "Waves in excitable media" [50] and "A geometrical-theory for spiral waves in excitable media" [54] by Keener.
(iv) A discussion of the application of the ideas of excitable media to arrhythmias in the heart is given in the article by Qu and Garfinkle [90].

Exercises

11.1 **Thought questions about excitable media:**

(a) Do all excitable media have the same property as neurons and heart tissue, that two pulses that approach each other will annihilate and disappear? If not, find a counter-example.
(b) Linear waves like light waves and sound waves can be refracted (their direction of propagation altered) as they pass from a medium of one index

442   Excitable media

of refraction to a medium with a different index of refraction. Is it possible for a nonlinear wave in an excitable medium to be refracted? Does Snel's law of refraction hold for waves in an inhomogeneous excitable medium?

11.2 **Absence of reflected action potentials:** Explain why an action potential cannot reflect from the end of an axon to produce an action potential traveling in the reverse direction. Similarly, explain why a propagating pulse in an experimentally isolated piece of heart tissue cannot produce a reflected pulse when the pulse reaches a physical boundary of the muscle.

11.3 **Nullclines for the Oregonator:** Sketch the nullclines $f(u, v) = 0$ and $g(u, v) = 0$ for the two-variable reduction of the Oregonator model of Exercise 3.9 for representative values of the parameters b and q. (Remember that we are interested only in positive u and v.) Find a set of values of b and q for which you expect relaxation oscillations for small values of η. (Hint: look at small q.)

11.4 **Relaxation oscillations in the Oregonator:** For $\eta = 0.01$ and for the values of b and q that you chose in the previous exercise to lead to relaxation oscillations, solve for the time evolution of the two-variable reduction of the Oregonator model. Make a plot of $u(t)$ and $v(t)$ and describe the oscillations. (If you do not see oscillations, reconsider your choice for the values of b and q!) Also, plot the $u(t) - v(t)$ trajectory on the nullcline plot and compare what you see with the discussion in Section 11.2.1.

11.5 **Van der Pohl model:** Investigate the oscillatory dynamics of the van der Pohl equation (11.12) for small and large values of γ. Plot $x(t)$ and also "phase space plots" of $\dot{x} = d_t x$ against x as t varies. The relaxational oscillations for large γ as a function of time are reminiscent of the dynamics at a fixed point in space during pulse-train propagation in the FitzHugh–Nagumo model.

11.6 **FitzHugh–Nagumo model:** The FitzHugh–Nagumo model is described by the equations

$$\partial_t u = \eta^{-1}\left(3u - u^3 - v\right) + \nabla^2 u, \tag{E11.1}$$

$$\partial_t v = u - a - bv. \tag{E11.2}$$

In the small η limit, for what positive values of a and b is the system (i) oscillatory? (ii) excitable monostable? (iii) excitable bistable?

11.7 **Different forms of the FitzHugh–Nagumo model:** There are many different forms, other than Eqs. (11.17), in which you will find the FitzHugh–Nagumo equations written in the literature. Another common form (used, in the source of Fig. 11.14 for example) is

$$\partial_t u = \varepsilon^{-1}(u - u^3/3 - v) + D\nabla^2 u, \quad \text{(E11.3a)}$$

$$\partial_t v = \varepsilon(u + \beta - \gamma v). \quad \text{(E11.3b)}$$

By appropriately rescaling x, t, u, v variables, transform these equations to the form of Eqs. (11.17) and relate the parameters $\varepsilon, \beta, \gamma$ to the parameters η, a, b there.

11.8 **Relaxation oscillations in the FitzHugh–Nagumo model:** Sketch $u(t)$ and $v(t)$ for the relaxation oscillations described by Eqs. (11.22) for the FitzHugh–Nagumo model with

$$f(u, v) = 3u - u^3 - v, \quad \text{(E11.4a)}$$

$$g(u, v) = u - a - bv, \quad \text{(E11.4b)}$$

with $a = 0$ and $b = 1/5$.

11.9 **Nullclines for the Barkley model:** In 1991, Dwight Barkley proposed a simplified version of the FitzHugh–Nagumo model that could be simulated with great efficiency in large two- and three-dimensional domains. His model has the form

$$\partial_t u = \eta^{-1} u(1 - u)(u - u_{\text{th}}(v)) + \nabla^2 u, \quad \text{(E11.5a)}$$

$$\partial_t v = u - v, \quad \text{(E11.5b)}$$

with

$$u_{\text{th}}(v) = \frac{v + b}{a}. \quad \text{(E11.6)}$$

For this model

(a) Produce a plot similar to Fig. 11.4 by plotting the nullclines and by indicating with arrows the signs of $\partial_t u$ and $\partial_t v$ in the various areas bordered by the nullclines.
(b) Assuming a small value of η, determine for what parameter values a and b this model has excitable dynamics and oscillatory dynamics.
(c) Use the potential construction of Section 11.3 to determine when the Barkley model supports the propagation of pulse trains (waves).

We encourage you to read the Barkley paper [10] as a nice example of how one can use insights from nonlinear dynamics and numerical analysis to invent a mathematical model for efficient simulation. From his website at the University of Warwick in England, you can also download a computer code called EZ-Scroll that integrates his equations in three-dimensional domains and that visualizes the output.

11.10 **Piecewise linear model:** A "piecewise linear" form of the reaction kinetics Eqs. (11.21) allows many calculations to be done more easily. The kinetics are defined by the functions

$$f(u, v) = \begin{cases} -u - v & \text{for } u < a, \\ 1 - u - v & \text{for } u > a, \end{cases} \quad \text{(E11.7a)}$$

$$g(u, v) = u - bv. \quad \text{(E11.7b)}$$

Sketch the u- and v-nullclines for $a = 0.25$ and for (i) $b = 0.2$ and (ii) $b = 1$. For what range of b is the reaction kinetics (i) monostable and (ii) bistable for this value of a and for small η?

11.11 **Effective potential for fronts in the piecewise linear model:** For the reaction kinetics of Exercise 11.10 with $a = 0.3$ and η small, consider the front connecting the rest state $u = 0, v = 0$ with the excited state at this value of v, i.e. $u = 1, v = 0$. Plot the effective potential $\Phi(u)$ for the "rolling ball" analogy (cf. Fig. 11.6).

11.12 **Stationary front in the piecewise linear model:** For the reaction kinetics of Exercise 11.10 with $a = 0.25$ and η small, what is the value of v for which the front connecting the small-u and large-u portions of the u-nullcline is stationary (i.e. $v = v^*$ giving the stall solution)?

11.13 **Moving front in the piecewise linear model:** Using the insight given by Exercise 11.11 calculate as a function of the parameter a the speed of the front connecting the rest state $u = 0, v = 0$ with the excited state at this value of v, i.e. $u = 1, v = 0$ for the reaction kinetics of Exercise 11.10 for η small. Verify that in the approximation of Section 11.3 this front speed is zero for $a = 0.5$.

11.14 **Pulse in the piecewise linear model:** Find the propagation speed c and calculate and plot the pulse shapes $u(x - ct)$ and $v(x - ct)$ as a function of $\xi = x - ct$ for the excitation pulse that propagates in the rest state of the reaction–diffusion system with the kinetics of Exercise 11.10 with $a = 0.25$ and $b = 0.2$ in the small-η approximation.

11.15 **Dispersion relation of waves in the piecewise linear model:** Calculate and plot expressions for the dispersion relationship $C(P)$ for the scaled speed $C = \eta^{1/2} c$ as a function of the temporal period P for waves propagating in the reaction–diffusion system with the kinetics of Exercise 11.10 with $a = 0.25, b = 0.2$, and η small. Use the scalings of Section 11.3 and do not worry about the breakdown of this scaling that occurs for small C.

12

Numerical methods

12.1 Introduction

Three kinds of mathematical problems have appeared frequently earlier in the book: the time evolution of a pattern-forming system, the identification of stationary states (e.g. a uniform state or a periodic hexagonal lattice), and the calculation of growth rates σ_q (eigenvalues) for small-amplitude perturbations of a stationary state. Except for simplified mathematical models that often cannot be compared quantitatively with experiment, and except for rather narrow parameter regimes such as just beyond the onset of a supercritical bifurcation, these three classes of problems cannot be solved analytically. It can then be helpful to use numerical methods on a digital computer.

In this chapter, we discuss some numerical ideas and algorithms to solve the first two of these three kinds of problems.[1] The discussion will be useful in several ways. First, many difficult concepts associated with pattern formation such as spatiotemporal chaos can often first be conveniently studied using a numerical method since the alternatives of experiments or analytics can be more time consuming, expensive, or difficult. Second, the great power of current computers and of modern numerical algorithms increasingly allow the investigation of evolution equations that describe a nonequilibrium system *quantitatively* and sometimes provide the only way to obtain information about a system. Simulations thus complement theory and experiment as an important third way of exploring and understanding nonequilibrium phenomena. Third, the following discussion should help you to understand the assumptions that underlie some of the numerical methods used to study pattern-forming systems and so give you a sense of when you can trust the simulations.

---

[1] For solving scientific problems, there is relatively little to learn about the numerical calculation of eigenvalues of a $N \times N$ matrix since the software is mature and can be invoked knowing just the matrix elements and the type of the matrix (e.g. whether it is banded or symmetric). Further information can be found in Refs. [89] and [28].

The diversity of pattern-forming systems and of their evolution equations is so great that it might seem impossible to describe in one chapter ideas that could be broadly useful. And it is the case that for the Boussinesq equations that describe a three-dimensional convecting flow or for a detailed model of electrical wave propagation in three-dimensional heart tissue, many years of computational experience are needed to develop a suitable algorithm, to implement the algorithm as a computer code, to debug and to validate the code, to optimize the code for a particular computer architecture, and finally to obtain scientific insights from results produced by the code. If the need arises for you to carry out a simulation, you should be prepared to talk to or to collaborate with experts in computational science and in numerical analysis.

However, a strength and elegance of numerical methods is that a few simple ideas provide a foundation to understand many algorithms, and these ideas can be usefully explained within the space of a chapter. For example, two ideas used in many numerical algorithms are iteration and Taylor series. Iteration refers to the idea of generating a sequence of vectors that converge to some desired but unknown answer, and the Taylor series provides a way to approximate a function with a simpler form over small space-time regions. (In Section 12.4, we will combine these two ideas to derive Newton's method, an algorithm that can efficiently find approximate solutions of nonlinear equations.) A further strength of many numerical algorithms is that their mathematical structure depends only weakly on the details of the equations of interest. For example, Newton's method has the same form when used to find a root of a polynomial or of some arbitrary transcendental equation such as $x - \cos(x) = 0$.

Numerical investigations are similar to laboratory experiments in that they provide only approximate answers to specific questions. For a particular choice of parameter values, for a particular choice of boundary conditions, for a particular initial state, and for some finite observation time, we can learn something about the behavior of a specific solution to the equations of interest. To discover trends in behavior – the more prized form of scientific insight – a numerical study has to be repeated, often many times, for different values of parameters spanning some range of interest and even then only an approximate form of the behavior can be deduced. In contrast, an analytical method can often predict explicitly how some phenomenon varies with a parameter. For example, the amplitude equation technique discussed in Chapter 6 predicts that the intensity of a pattern-forming field should increase according to the functional form $\sqrt{p - p_c}$ in the limit that some bifurcation parameter p approaches its critical value p_c. However, explicit mathematical predictions do not make analytical methods automatically superior to numerical or experimental approaches since analytical methods often depend on mathematical assumptions (e.g. that some quantity like $p - p_c$ is sufficiently small) or on physical assumptions (e.g. that some poorly understood physical effect can be ignored in the evolution equations) and the range of validity of these assumptions may not be known beforehand. Necessarily, progress in understanding nonequilibrium pattern formation requires repeated comparisons of analytics, simulation, and experiment.

The rest of this chapter is divided into the following sections. In Section 12.2, we discuss the central issues of representation and discretization, how to reduce the infinitely many degrees of freedom associated with a continuous field to the finitely many numbers that a computer can work with. In this section, we also discuss finite-difference approximations to derivatives, the convergence rates of finite-difference approximations, and the implications of floating-point arithmetic on the convergence of these approximations. In Section 12.3, we discuss explicit and implicit time integration algorithms and how the largest discrete time step is determined by a balance of numerical stability and numerical accuracy. We then show how a technique called operator splitting combines the explicit and implicit methods into a practical and broadly useful numerical method. We finish this section with some comments about how to estimate the space-time resolution needed to simulate some pattern-forming system. In Section 12.4, we discuss how the time-independent (stationary) states of known evolution equations can be found using iterative methods. A particularly important and widely used iterative algorithm is Newton's method, which converges rapidly provided that a good initial approximation for the solution is known.

12.2 Discretization of fields and equations

12.2.1 Finitely many operations on a finite amount of data

To solve any given problem, a digital computer can only carry out a finite number of operations on a finite amount of data in a finite amount of time.[2] This elementary observation has profound implications regarding how to represent and to solve equations on a computer. Many concepts that you take for granted from your experience with real numbers, continuity, calculus, and linear algebra cannot be used directly to solve a problem on a computer.

Let us mention a few examples. A number in a digital computer is stored as a fixed group of bits (zeros and ones) called a computer word. The fixed size of a computer

---

[2] All numerical calculations on a digital computer reduce to a finite (although possibly long) sequence of elementary operations that act on one or two computer words at a time to produce some new computer word. For example, the binary operation of adding two 64-bit floating-point numbers corresponds to carrying out finitely many manipulations of the 128 bits of the two numbers to produce a final 64-bit number that approximates their sum. The details of carrying out the sequence of elementary operations are hidden from the user by the compiler and by the computer's hardware.

word implies that a computer has only a finite set of numbers available – called its floating point numbers – to represent all real numbers. For example, many personal computers currently use 64-bit words and so can store only about $2^{64} \approx 10^{19}$ floating point numbers. Since there are only finitely many numbers that a computer can work with, a small error is typically incurred each time the value of a mathematical expression is stored by the process of rounding, which replaces the value by the closest floating point number. Rounding errors can perturb mathematical formulas and computer algorithms in unexpected ways, and are one of the many details to consider when trying to solve a mathematical problem with a digital computer. As one example, we will see in the next section how rounding prevents finite-difference approximations from converging toward the derivatives that they approximate (see Fig. 12.1 and the related discussion).

As a second example, consider a set of N linear equations in N unknowns written in the form $\mathbf{M}\mathbf{x} = \mathbf{b}$ where $\mathbf{M}$ is an $N \times N$ matrix, $\mathbf{b}$ is a given N-dimensional vector, and $\mathbf{x}$ is the N-dimensional solution that we would like to find. In your linear algebra course, you learned that a unique solution $\mathbf{x}$ exists for each right side $\mathbf{b}$ provided that the determinant of M is nonzero:[3] $\det(\mathbf{M}) \neq 0$. However, this analytical criterion is unreliable when used in a computer program. The determinant of a nonsingular matrix can easily turn out to be smaller in magnitude than half the smallest positive floating point number (about 10^{-323} for a 64-bit computer number) in which case the numerical value of the determinant rounds to zero (the nearest floating point number), giving the wrong conclusion that the matrix is singular. For example, the 400×400 diagonal matrix $\mathbf{D} = 0.1\mathbf{I}$ (0.1 times the 400×400 identity matrix $\mathbf{I}$) analytically has a nonzero determinant of $(0.1)^{400} = 10^{-400}$ which rounds to zero. Yet $\mathbf{D}$ is obviously nonsingular since it has an explicit inverse matrix given by $\mathbf{D}^{-1} = 10\mathbf{I}$. Computational scientists instead use a criterion based on the condition number of a matrix to determine numerically whether the corresponding set of linear equations has a solution. (See Exercises 12.7 and 12.8.) The condition number is insensitive to the overall magnitude of matrix elements and predicts correctly that the diagonal matrix $0.1\mathbf{I}$ is nonsingular for any matrix size N.

As a final example, we note that the familiar Taylor series for the exponential function

$$e^x = \sum_{n=0}^{\infty} \frac{x^n}{n!}, \qquad (12.1)$$

---

[3] A matrix with a nonzero determinant has a matrix inverse $\mathbf{M}^{-1}$ in which case the solution can be written formally as $\mathbf{x} = \mathbf{M}^{-1}\mathbf{b}$. However, this is not the recommended way to solve linear equations on a computer since there are numerical algorithms (e.g. Gaussian elimination) that determine $\mathbf{x}$ more efficiently and more accurately by avoiding the explicit construction of the matrix inverse [55].

requires the summation of infinitely many terms and so cannot be evaluated with finitely many operations. To evaluate this series for some argument x, a computer program must give up the mathematical goal of obtaining an exact answer and instead employ some strategy that yields an *approximate* answer after finitely many operations. A similar approach is needed for many mathematical problems solved on a computer.

12.2.2 The discretization of continuous fields

Because a computer can work with only a finite amount of data, a first step in preparing a mathematical problem for computer solution is to discretize the problem by reducing it to finitely many pieces. For pattern-forming systems, this involves choosing some representation for each field $u(\mathbf{x}, t)$ and then truncating the representation to some finite set of numbers. There are two widely used strategies for discretizing continuous fields: the Galerkin method, in which a field is expanded in terms of a specified set of basis functions and the finite-difference method, in which the values of the function are assumed to be known at a finite set of space-time points. We discuss these in turn and then compare them. For simplicity, we restrict our discussion to the discretization of a scalar field $u(x)$ of a single variable x. The generalization to vector fields that depend on several coordinates is straightforward.

In the Galerkin method, a basis[4] of functions $\phi_n(x)$ is identified and the field $u(x)$ is expanded in that basis as follows:

$$u(x) = \sum_{n=0}^{\infty} u_n \phi_n(x). \tag{12.2}$$

The coefficients u_n define a representation of the field u and this representation has the nice property of varying linearly with u, since the representation of the linear combination $cu(x) + dv(x)$ consists of the coefficients $cu_n + dv_n$. (But the representation of the product of two fields is not the product of the corresponding coefficients, an important point that we return to in a moment.) Two familiar examples of such basis expansions (say for fields defined on an interval $[0, l]$ of length l) are Fourier analysis with basis functions $\phi_n = \exp(2\pi i n x / l)$ and Taylor series with basis functions consisting of the monomials $\phi_n = (x - a)^n$ centered on some point a.

For a given basis $\phi_n(x)$, an arbitrary field $u(x)$ will generally have infinitely many nonzero coefficients u_n. A finite representation suitable for a computer algo-

---

[4] Recall that a basis for some linear vector space is a set of vectors that are linearly independent and that are complete in that every vector in the space can be written as a linear combination of the basis vectors.

rithm can be obtained by retaining only the first $N+1$ coefficients so we can write

$$u(x) \approx \sum_{n=0}^{N} u_n \phi_n(x). \qquad (12.3)$$

(An example was given in Section 4.1.3, in which we showed that, sufficiently close to onset, only the first of an infinite number of Fourier terms was needed to approximate the stationary solution $u(x)$ of the Swift–Hohenberg equation.) If the sum Eq. (12.2) converges, we can approximate the field u arbitrarily well with Eq. (12.3) by taking the integer N sufficiently large.

For any vector space of functions, there are infinitely many different bases that span the space, and in the literature you will see bases using Fourier modes, finite elements, Chebyshev polynomials, wavelets, and combinations of such bases (e.g. a code might use Fourier modes for the x-dependence of a field $u(x, z)$ and Chebyshev polynomials for the z-dependence). Which basis to use is a complex question that involves mathematical, computational, and scientific considerations that lie beyond what we can discuss in this book. The appropriate choice of basis can differ from problem to problem and even from parameter value to parameter value for a given problem. We can say briefly that an overall goal is to solve a problem to a specified level of accuracy with the least amount of work. A first step in this direction is to identify a basis that is "close to the physics" in the sense that Eq. (12.3) converges rapidly, allowing the integer N and so the overall amount of computational work to be small. Symmetries of the evolution equations, solutions, and boundary conditions also often play a role in the choice of the basis.

A second way to discretize the field $u(x)$ is the finite-difference method. Here the discretization is achieved by truncating the spatial domain to a finite set of N spatial points x_n and by assuming that the values of the field are known only at these finitely many points:

$$u_n = u(x_n), \quad n = 1, \ldots, N. \qquad (12.4)$$

Just like the Galerkin method,[5] each field u is represented by a finite-dimensional vector of coefficients $(u_1, \ldots, u_N)$ and this representation varies linearly with u. However the physical meaning of the numbers Eq. (12.4) differs from the meaning of the Galerkin coefficients.

Just as there is flexibility regarding how to choose the basis functions ϕ_n in Eq. (12.3), there is flexibility regarding where to place the points x_n. In the absence

---

[5] A finite-difference method can be considered formally as a Galerkin method with basis functions $\phi_n(x) = \delta(x - x_n)$ where $\delta(x)$ is a Dirac delta function. In practice, the two methods are treated as distinct since they lead to different kinds of algorithms.

of any prior knowledge of u, a convenient choice is to spread the points x_n uniformly through space so we can write

$$x_n = x_1 + (n-1)\Delta x, \quad 1 \le n \le N. \tag{12.5}$$

The mesh points x_n are then characterized by their constant spatial separation $\Delta x = x_n - x_{n-1}$, which also defines the spatial resolution or mesh size. For some problems, the field u might have a special structure, e.g. it may have many oscillations near a wall and be almost constant away from the wall. In this case, it would make sense to distribute the points x_n nonuniformly, with a higher density near the wall and a smaller density elsewhere. There are adaptive algorithms that can automatically vary the number and positions of points to achieve a given accuracy with the least amount of work, but these are too complicated to discuss in this chapter and also are quite complicated to code.

12.2.3 The discretization of equations

Our discussion so far has concerned how to discretize a field by a finite truncation of a Galerkin expansion or of the spatial domain. We next consider how to discretize the equations that express how the field and its various derivatives are related at each space-time point. Restricting our discussion now just to finite-difference methods, we can discretize equations in the same way as the fields, by assuming that the equations are valid only at some finite number of spatial points x'_n. The discretization points x'_n for the equations do not have to coincide with the discretization points x_n for the fields that satisfy the equations (although this is often the case), but their numbers must be about the same so that there are as many discretized equations and discretized boundary conditions as there are unknown field values to solve for.

For example, we may be interested in solving the one-dimensional nonlinear equation

$$u - u^3 + \frac{d^2u}{dx^2} = 0, \tag{12.6}$$

on the interval $[0, l]$ for the unknown solution $u(x)$ with boundary conditions $u(0) = 0$ and $u(l) = 0$. We can discretize this equation by requiring that it hold at the interior[6] discrete points $x_n = n\Delta x$ (with $0 < n < N$) at which the field u has been previously discretized:

$$\left(u - u^3 + \frac{d^2u}{dx^2} \right)\bigg|_{x=x_n} = u_n - u_n^3 + \frac{d^2u}{dx^2}\bigg|_{x=x_n} = 0. \tag{12.7}$$

---

[6] Eq. (12.6) does not have to be discretized at the points $x = 0$ and $x = l$ since the boundary conditions $u(0) = 0$ and $u(l) = 0$ provide the field values at those points.

The discrete formulation of Eq. (12.6) with its boundary conditions can be completed to yield $N-1$ equations for the $N-1$ unknown values u_n provided that we can find a way to express the derivatives $d^2u/dx^2|_{x_n}$ in terms of the field values u_n.

In this section, we show by a worked example how to derive finite-difference approximations to derivatives of a field that has been discretized on some mesh of points x_n. A finite-difference approximation is a special linear combination of finitely many field values u_n that converges to some mathematical expression involving the field (not necessarily a derivative but this is the most common case) in the limit that the spatial resolution Δx becomes sufficiently small. The idea is simple although the algebra can be tedious: to estimate a kth-order derivative $u^{(k)}(x)$ of a field u at a point x, we find a polynomial $p(x)$ of order $m \geq k$ that interpolates (passes exactly through) the $m+1$ pairs of points (x_n, u_n) closest to the point x, and then use the kth derivative of the polynomial at x, $p^{(k)}(x)$, to estimate the value of $u^{(k)}(x)$.[7] The resulting estimate will yield a finite-difference approximation to $u^{(k)}(x)$ in the form of a linear combination of the $m+1$ field values u_n closest to the point x. Generally, the higher the order m of the polynomial, the more accurate the corresponding finite-difference expression and the more rapidly the numerical solution u_n converges to the unknown mathematical solution.

Etude 12.1 Derivation of some finite-difference approximations

Let us assume that we know the values u_n of a function $u(x)$ only on the uniform mesh Eq. (12.5) with spatial resolution Δx. An estimate for the first derivative $u'(x_n)$ at the point x_n in terms of these grid values can be obtained by finding the linear polynomial $p(x) = a + bx$ that interpolates the pairs of points (x_n, u_n) and (x_{n+1}, u_{n+1}) so that

$$p(x_n) = u_n \quad \text{and} \quad p(x_{n+1}) = u_{n+1}. \tag{12.8}$$

These constitute two linear equations for the two unknown coefficients a and b and you can verify that

$$p'(x_n) = b = \frac{u_{n+1} - u_n}{x_{n+1} - x_n}, \tag{12.9}$$

provides an estimate for the first-order derivative $u'(x_n)$ at the point x_n. (Alternatively, we could have chosen as the interpolation points the values at x_{n-1} and x_n or at x_{n-1} and x_{n+1} to derive other finite-difference approximations.) It is traditional to write this finite-difference approximation in the form

$$\frac{u_{n+1} - u_n}{\Delta x}. \tag{12.10}$$

---

[7] The interpolating polynomial $p(x)$ is useful for solving other numerical problems associated with the field $u(x)$. For example, $p(x)$ can be integrated analytically to estimate the integral of $u(x)$ over some small interval about x. It can also be used to estimate the position and value of an extremum of u.

Some finite-difference approximations provide better estimates than others. The accuracy of the finite-difference approximation Eq. (12.10) can be obtained by substituting the Taylor series for u_{n+1} about the point of evaluation $x = x_n$ up to second order,[8]

$$u_{n+1} = u(x_n + \Delta x) = u_n + \left[u'(x_n)\right] \Delta x + \left[\frac{u''(x_n)}{2!}\right] \Delta x^2 + \cdots, \qquad (12.11)$$

to find

$$\frac{u_{n+1} - u_n}{\Delta x} = u'(x_n) + \left[\frac{u''(x_n)}{2}\right] \Delta x + \cdots, \qquad (12.12)$$

where the dots $\cdots$ denote terms of higher order in the mesh size Δx. Equation (12.12) says that the finite-difference expression on the left differs from the derivative $u'(x_n)$ by an error term $\left[u''(x_n)/2\right] \Delta x + \cdots$ that goes to zero in the limit $\Delta x \to 0$. Because the lowest-order term in the error is first-order in Δx, the finite-difference expression $(u_{n+1} - u_n)/\Delta x$ is said to be a "first-order-accurate approximation" to $u'(x_n)$. It is also called a "two-point" finite-difference expression since the linear combination spans at most two mesh points.

A more rapidly converging finite-difference approximation to $u'(x_n)$ can be obtained by using a higher-order interpolation polynomial $p(x)$ that passes through more points close to x_n. For example, we can solve three linear equations in three unknowns to find the quadratic polynomial $p(x) = a + bx + cx^2$ that interpolates the three pairs of points (x_{n-1}, u_{n-1}), (x_n, u_n), and (x_{n+1}, u_{n+1}) closest to the point of interest x_n. You can verify that the derivative $p'(x_n)$ leads to the following "second-order-accurate three-point finite-difference approximation" for the first derivative $u'(x_n)$:

$$\frac{u_{n+1} - u_{n-1}}{2\Delta x} = u'(x_n) + \left[\frac{u^{(3)}(x_n)}{6}\right] \Delta x^2 + \cdots. \qquad (12.13)$$

The leading part of the error term $\left[u^{(3)}/6\right] \Delta x^2 + \cdots$ was found by substituting Taylor series about x_n up to third order for the expressions $u_{n\pm1} = u(x_n \pm \Delta x)$ on the left side. The second derivative $p''(x_n)$ of the same quadratic interpolating polynomial yields a second-order-accurate three-point finite-difference approximation for the second derivative $u''(x_n)$:

$$\frac{u_{n+1} - 2u_n + u_{n-1}}{\Delta x^2} = u''(x_n) + \left[\frac{u^{(4)}(x_n)}{12}\right] \Delta x^2 + \cdots. \qquad (12.14)$$

---

[8] The notation Δx^k here means $(\Delta x)^k$, the kth power of the mesh spacing Δx. Some computational science books and articles use the symbol h instead of Δx to simplify the notation.

The left-hand side of this equation solves the problem of how to complete the discretization of Eq. (12.7): at each point x_n, we replace the second derivative $u''(x_n)$ with the three-point difference on the left side of this equation.

For use further below in this section, we list the following two five-point finite-difference approximations:

$$\frac{u_{n-2} - 8u_{n-1} + 8u_{n+1} - u_{n+2}}{12 \Delta x} = u'(x_n) - \left[\frac{u^{(5)}(x_n)}{30}\right] \Delta x^4 + \cdots,$$

(12.15a)

$$\frac{u_{n-2} - 4u_{n-1} + 6u_n - 4u_{n+1} + u_{n+2}}{\Delta x^4} = u^{(4)}(x_n) + \left[\frac{u^{(6)}(x_n)}{6}\right] \Delta x^2 + \cdots,$$

(12.15b)

which you should try to derive for yourself. Note that the first expression is a fourth-*order*-accurate approximation for the first-order derivative $u'(x)$.

In the limit $\Delta x \to 0$, the left side of Eq. (12.13) asymptotically provides a more accurate estimate of $u'(x_n)$ than the two-point difference Eq. (12.12) since the leading term of its error is smaller by a factor Δx. You might deduce from this that it would be best to use the highest-order finite-difference expression possible to approximate some derivative. However, because interpolation polynomials of higher and higher order become more and more oscillatory, using finite-difference expressions whose order of accuracy is substantially greater than the order of the derivative can render a numerical algorithm unstable and is not recommended. In practice, second-order- or fourth-order-accurate expressions are used in most finite-difference codes. If needed, higher accuracy can be achieved by using more mesh points (increasing the integer N) or by switching to a Galerkin method.

To give you a sense of the accuracy of these finite-difference expressions as a function of the spatial resolution Δx, we have plotted in Figure 12.1 the relative errors of Eqs. (12.13), (12.15a), and (12.15b) for first- and fourth-order derivatives of the field $u(x) = \sin(x)$ at the point $x = \pi/4$ (for which $u^{(1)}(\pi/4) = u^{(4)}(\pi/4) = 1/\sqrt{2} \approx 0.707$). The finite-difference expressions were evaluated by the computer mathematics program Mathematica using 64-bit floating point numbers, for values $\Delta x = 10^{i/4}$ with i an integer satisfying $-80 \le i \le 2$. Since the precise way that floating point expressions are evaluated depends on the computer language, on the choice of compiler options, and on the CPU hardware, you will get similar but not identical curves if you try to reproduce these curves yourself.

Before discussing this figure, it is helpful to introduce some notation. Let $\delta_{m,k} u(x_n)$ denote an mth-order-accurate finite-difference approximation of the kth-order derivative of the field u at the point x_n and let $\epsilon_{m,k}$ denote the magnitude of the corresponding dimensionless relative error so that

$$\epsilon_{m,k} = \left| \frac{\delta_{m,k} u(x_n) - u^{(k)}(x_n)}{u^{(k)}(x_n)} \right|. \tag{12.16}$$

By definition, the numerator $\delta_{m,k} u - u^{(k)} \propto \Delta x^m$ as $\Delta x \to 0$ and so we expect that:

$$\log_{10} \epsilon_{m,k} = c + m \log_{10} \Delta x, \tag{12.17}$$

where c is some constant, i.e. a plot of $\log_{10} \epsilon_{m,k}$ versus $\log_{10} \Delta x$ for sufficiently small Δx should be a straight line with slope m. Further, the quantity $-\log_{10} \epsilon_{m,k}$ tells us directly the number of significant digits in the finite-difference approximation $\delta_{m,k}$.

Starting with values of Δx of order one (the right-hand side of the graph) and then following Δx toward zero (the left-hand side of the graph), we see from Fig. 12.1 that all three relative errors $\epsilon_{2,1}$, $\epsilon_{4,1}$, and $\epsilon_{2,4}$ start to decrease toward zero as expected and satisfy the linear behavior Eq. (12.17) with slope m (slope 2 for the second-order-accurate differences $\delta_{2,k}$, slope 4 for the fourth-order difference $\delta_{4,1}$). Over this initial range of decreasing linear behavior, $\epsilon_{2,1} > \epsilon_{4,1}$, which implies that the fourth-order-accurate approximation $\delta_{4,1}$ for the first-derivative is indeed more accurate than $\delta_{2,1}$. However, a higher-order finite-difference expression for some derivative is not automatically more accurate than a lower-order expression for all spatial resolutions Δx. It can be the case that $\epsilon_{m,k} > \epsilon_{n,k}$ for $m > n$ over some range of Δx although we expect the inequality to become reversed for sufficiently small Δx.

From Fig. 12.1, we can read off the spatial resolution needed for a finite-difference approximation to attain at least three significant digits, corresponding to a relative error smaller than 10^{-3}. We would need to choose a spatial resolution $\Delta x < 0.08$ for $\delta_{2,1}$, a five times coarser resolution of $\Delta x < 0.4$ for $\delta_{4,1}$, and a resolution $\Delta x < 0.1$ for $\delta_{2,4}$. For the differences $\delta_{2,1}$ and $\delta_{4,1}$, these resolutions are equivalent respectively to using about 80 and 16 uniformly spaced mesh points respectively to span a period 2π of the sine curve $u(x) = \sin(x)$. The fact that fewer spatial points are needed with a higher-order finite difference to achieve a given accuracy implies generally (but not always) that higher-order differences can lead to more efficient computer algorithms.

Figure 12.1 shows a surprising result as Δx decreases to zero: the relative errors start to decrease according to Eq. (12.17), but then the errors reach a minimum and

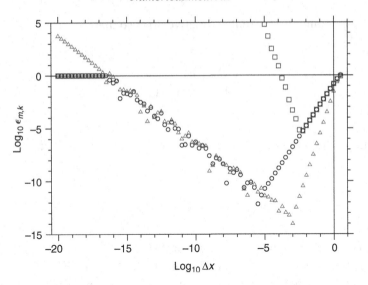

Fig. 12.1 A log-log plot of the relative error $\epsilon_{m,k}$, Eq. (12.16), for $\epsilon_{2,1}$ (circles), $\epsilon_{4,1}$ (triangles), and $\epsilon_{2,4}$ (squares) versus the spatial resolution Δx for the field $u = \sin(x)$ at the point $x = \pi/4$. The negative of the vertical coordinate gives the number of significant digits attained by the corresponding finite-difference $\delta_{m,k}$. As $\Delta x \to 0$, all three finite-difference expressions converge toward and then diverge away from their respective limit $u^{(k)}(x)$, in contrast to the expected behavior Eq. (12.17). These curves suggest two tendencies that hold generally: higher-order derivatives $u^{(k)}$ are more difficult to approximate to a given order of accuracy (compare the circles, $\epsilon_{2,1}$ with the squares, $\epsilon_{2,4}$), and higher-order finite-difference approximations for a given derivative start to diverge at larger spatial resolutions (compare the circles, $\epsilon_{2,1}$ with the triangles, $\epsilon_{4,1}$).

increase in magnitude. We conclude that the mathematical limit

$$\lim_{\Delta x \to 0} \delta_{m,k} u(x) - u^{(k)}(x), \qquad (12.18)$$

does not exist when evaluated using floating point arithmetic, despite the "obvious" fact that the difference $\delta_{m,k} u - u^{(k)}$ mathematically vanishes as Δx^m in this limit. In fact, the approximately linear increase of the relative errors on this log-log plot suggests instead that $\delta_{m,k} u - u^{(k)}$ diverges as Δx^{-k} in the limit $\Delta x \to 0$. Because of the minima in Fig. 12.1, the differences $\delta_{2,1}$, $\delta_{4,1}$, and $\delta_{2,4}$ cannot achieve better than 12, 14, and 5 significant digits respectively when evaluated with 64-bit numbers, no matter how small the spatial resolution Δx.

A brief explanation for the strange divergence of the finite-difference approximations as $\Delta x \to 0$ is the following. The floating point value of the function $\sin(x)$ can differ from the true mathematical value by a round-off error that can be expressed as a multiplicative factor $1 + \epsilon_x$, where ϵ_x is a tiny quantity that varies with x but that is never larger than the machine precision ϵ_{mach} of the floating point system. Thus the floating point value of the numerator in $\delta_{2,1}$ (Eq. (12.13)) can be written in the form

$$u(x+\Delta x)(1+\epsilon_1) - u(x-\Delta x)(1+\epsilon_2)$$
$$\approx [u(x+\Delta x) - u(x-\Delta x)] + (\epsilon_1 - \epsilon_2)u(x), \qquad (12.19)$$

provided that Δx is so small that $\epsilon_1 u(x+\Delta x) - \epsilon_2 u(x-\Delta x) \approx (\epsilon_1 - \epsilon_2)u(x)$. The term in brackets goes to zero as expected, but the second term involving the epsilons is generally a nonzero term (of magnitude $\epsilon_{mach}|u(x)|$) that causes a divergence in the limit $\Delta x \to 0$, since the numerator Eq. (12.19) is divided by Δx.

Although you might find the divergence of the finite-difference expressions $\delta_{m,k}$ from their mathematical limits $u^{(k)}$ alarming, this divergence is not a problem for most finite-difference codes. First, scientists rarely need so much accuracy that they would choose a spatial resolution so small that it would approach minima like those shown in Fig. 12.1. Second, for two and three space dimensions, most researchers would not have adequate computer resources to solve a discretized pattern-forming problem with such a fine spatial resolution since the associated computational effort grows at least as rapidly as $(l/\Delta x)^d$, where l is the size of the system and d is the number of spatial dimensions. Still, it is important for you to appreciate that many mathematical expressions do not behave as expected on a digital computer for subtle reasons associated with floating point arithmetic. You may well face a situation in which you or a collaborator will have written a computer code, verified painstakingly that the algorithm for the equations was programmed correctly, found that the code compiles without error, and yet your code does not produce correct results because of "invisible" floating-point errors.

12.3 Time integration methods for pattern-forming systems

12.3.1 Overview

Now that you have some familiarity with how the finite-difference method can be used to discretize fields and equations, we turn to the first of the two mathematical problems that are central to this chapter, the numerical solution of evolution equations for pattern-forming systems. Nearly all the evolution equations discussed in this book are *initial-boundary-value problems* such that the future state of the system – the values of all the fields within the domain of the continuous medium – can be determined from the equations provided that we know the boundary conditions for the fields and the initial values of the fields at some starting time t_0. Without loss of generality, we can assume that the initial time $t_0 = 0$ since most evolution equations that we consider are time-translation invariant (the coefficients in the equations are independent of time) and so the dynamics does not depend on when the calculation is started.

The first step in solving an evolution equation numerically on a computer is to discretize the fields $\mathbf{u}(\mathbf{x}, t)$, the equations, and the boundary conditions so that the mathematical problem is reduced to finitely many numbers. Considering for now the simpler case of a scalar field $u(x, t)$ of a single spatial variable x, we can use the ideas discussed in Section 12.2.2, to discretize in time just as we did in space. We therefore assume that the values of u are known only on a finite uniform set of space-time points given by

$$(x_n, t_i) = (n\, \Delta x, i\, \Delta t), \quad i \geq 0, \tag{12.20}$$

with constant temporal resolution Δt (also called the time step of the algorithm) and uniform spatial resolution Δx. The space-time resolutions Δx and Δt must be specified as input to the numerical algorithm and the dependence of the algorithm's output on these parameters should be explored in a manner similar to Fig. 12.1. We will use the notation

$$u_{n,i} = u(x_n, t_i) = u(n\, \Delta x, i\, \Delta t), \tag{12.21}$$

to denote the exact field values at these discrete points and a tilde notation $\tilde{u}_{n,i}$ to denote an approximate numerical solution for the field values at the same space-time point.

A time integration algorithm uses the initial data $u_{n,0}$ at time $t = 0$ (we also say "time level 0") together with the discretized evolution and boundary equations to deduce approximate field values $\tilde{u}_{n,1}$ at time $t = \Delta t$ (time level 1) in the future. The new values $\tilde{u}_{n,1}$ are then used as initial data for an "initial" time $t = \Delta t$ and the same algorithm is used to deduce approximate field values $\tilde{u}_{n,2}$ at time $t = 2\,\Delta t$. In this way, the discrete spatial structure of the field is calculated at successive time steps from the initial data at time $t = 0$ to some future time $T = N_t\, \Delta t$, where the integer N_t is the total number of time steps.

Provided that the discretization and the algorithm are well chosen and programmed correctly, the approximate discrete values $\tilde{u}_{n,i}$ at time level i will converge to the unknown exact mathematical values $u_{n,i}$ in the limits $\Delta t \to 0$ and $\Delta x \to 0$ (keeping $n\, \Delta x$ and $i\, \Delta t$ constant so that the mesh point corresponds to a fixed physical time and location). We will discuss in Section 12.3.5 how to identify an adequate space-time resolution for an actual calculation. The rate of convergence is determined by the orders of the finite-difference expressions used to discretize the time and space derivatives in the equations and boundary conditions. For simplicity of the discussion, in this chapter we will discuss mainly discretizations that involved second-order-accurate spatial derivatives and first- or second-order-accurate time derivatives.

12.3 Time integration methods for pattern-forming systems

As we discuss in the next two subsections, there are two basic kinds of numerical algorithms for integrating evolution equations, explicit and implicit. Explicit time integration methods are the simplest to program but have the property that their output becomes unstable over time[9] (tending to grow exponentially in magnitude) unless the time step Δt is smaller than some power of the spatial resolution,

$$\Delta t < C \, \Delta x^\alpha, \qquad (12.22)$$

where C is some positive constant and where the exponent α is usually equal to the order of the highest-order spatial derivative present in the evolution equations. Since dissipative pattern-forming problems typically involve a diffusive process with $\alpha \geq 2$, explicit time integration methods have the drawback that they can achieve higher accuracy in space (smaller Δx) only by taking many more time steps of smaller size to span a given observation time T. The bound Eq. (12.22) on Δt is especially severe for equations like the Swift–Hohenberg model Eq. (2.3) for which $\alpha = 4$.

The stability condition Eq. (12.22) is an unphysical constraint since a mathematical solution of the evolution equations has its own time scales (e.g. the period of oscillation for a limit cycle or the decay time for a transient) that have nothing to do with the choice of the spatial mesh used to discretize the equations. For example, even if the dynamics is slowly varying or stationary so that the fields change only a little over a long time, Eq. (12.22) still forces the time step to be small. Implicit time integration methods avoid the restriction Eq. (12.22) so that the time step is not constrained by the fineness of the spatial mesh. The largest possible time step can be comparable with the physical time scales and is determined mainly by the desired accuracy of the solution since the larger the time step Δt, the larger the error.

For nonlinear evolution equations, implicit integration methods have the drawback that a set of N nonlinear equations has to be solved each time step, where N is the total number of mesh points. As we will see in Section 12.4, finding the numerical solution to a set of nonlinear equations can be difficult to code and computationally expensive. Because explicit codes are much easier to write and to debug, and because analytical considerations rarely suffice to determine whether an implicit algorithm (which takes relatively few but expensive time steps) will be more efficient and accurate than an explicit code (which takes relatively more but inexpensive time steps), most researchers first try an explicit algorithm.

---

[9] An adequate discussion of what is meant by a "stable" numerical algorithm for an evolution equation would exceed the space that we want to spend on this topic. An accessible reference for further information is the book by John Strikwerda [98]. We will use an informal sense of stability, that if all analytical solutions decay asymptotically to zero, then so should any numerical solution. This weaker definition is sufficient to illustrate the main ideas.

Fortunately, for many pattern-forming evolution equations, the highest-order spatial derivatives occur as linear operators, e.g. as a Laplacian ∇^2 or biharmonic ∇^4. We can then use the technique of operator splitting (discussed in Section 12.3.4) to integrate the nonlinear terms by an explicit method and the linear terms by an implicit method and then combine the output of the two methods to advance the fields in time by one time step. The overall algorithm acts like an explicit time integration method but with an exponent α in Eq. (12.22) that is smaller than the order of the highest-order spatial derivative.

12.3.2 Explicit methods

Explicit time integration algorithms can be derived by requiring that the evolution equations hold at the present time level i (time $t_i = i \, \Delta t$) and then by using finite-difference approximations for the time derivatives that couple the unknown future field values at time level $i + 1$ to known values at level i and possibly to several previous levels[10] to predict the field values at time $i+1$. Because the time derivatives enter linearly in most evolution equations, simple algebra suffices to calculate the new field values in terms of the "explicitly" known current and recent field values. We illustrate the main idea through two worked examples.

Etude 12.2 Forward-Euler method for a set of linear constant-coefficient ordinary differential equations

As a first example, let us consider an evolution equation for which there are no spatial derivatives and no boundary conditions, a set of K constant-coefficient linear odes in the form

$$\frac{d\mathbf{u}}{dt} = \mathbf{M}\mathbf{u}, \tag{12.23}$$

where M is a $K \times K$ matrix of real numbers. We want to find a K-dimensional vector solution $\mathbf{u}(t)$ for $t \geq 0$ that passes through a specified K-dimensional initial vector $\mathbf{u}_0$ at $t = 0$ so that

$$\mathbf{u}(0) = \mathbf{u}_0. \tag{12.24}$$

We obtain an explicit algorithm in two steps. First, we assume that Eq. (12.23) holds at time level i. Second, we use an equation analogous to Eq. (12.10) to approximate the time derivative at time t_i in terms of the current numerical values $\tilde{\mathbf{u}}_i$ and future unknown values $\tilde{\mathbf{u}}_{i+1}$ at time $t_{i+1} = t_i + \Delta t$. We have

$$\left(\frac{d\mathbf{u}}{dt}\right)_i \approx \frac{\tilde{\mathbf{u}}_{i+1} - \tilde{\mathbf{u}}_i}{\Delta t} = (\mathbf{M}\mathbf{u})_i = \mathbf{M}\tilde{\mathbf{u}}_i, \tag{12.25}$$

---

[10] So-called multistep explicit algorithms use field values at time levels $i, i - 1, \ldots, i - k$ over k steps into the past to obtain an even more accurate estimate of the future field at time level $i + 1$. Multistep algorithms use extra memory (the storage of the field values at past times) to reduce the amount of work per time step.

12.3 Time integration methods for pattern-forming systems

or equivalently

$$\tilde{\mathbf{u}}_{i+1} = \tilde{\mathbf{u}}_i + \Delta t\, \mathbf{M} \tilde{\mathbf{u}}_i, \qquad (12.26)$$
$$= (\mathbf{I} + \Delta t\, \mathbf{M}) \tilde{\mathbf{u}}_i, \qquad (12.27)$$

where $\mathbf{I}$ is the $K \times K$ identity matrix. This is an "explicit" algorithm since the future numerical vector $\tilde{\mathbf{u}}_{i+1}$ can be calculated simply in terms of the explicitly known present vector $\tilde{\mathbf{u}}_i$. An explicit algorithm that is derived by using Eq. (12.10) to approximate the time derivative at time level i is known as a forward-Euler method, named after the Swiss mathematician Leonhard Euler who first proposed its use in the eighteenth century.

There is a useful insight that we can obtain with modest effort regarding when the forward-Euler method will be stable (not diverge from the unknown exact mathematical solution). Let us assume that the matrix $\mathbf{M}$ can be diagonalized and that it has K real eigenvalues λ_n that are all negative. Then any solution of Eq. (12.23) will decay to zero since it can be written as a linear superposition of decaying exponentials $\exp(\lambda_n t)$.

Does the numerical solution $\tilde{\mathbf{u}}_i$ generated by Eq. (12.26) have this same property of decaying asymptotically for all initial conditions? By iterating Eq. (12.27) successively starting with $i = 0$ then with $i = 1$ and so on, we see that

$$\tilde{\mathbf{u}}_i = (I + \Delta t\, \mathbf{M})^i \mathbf{u}_0, \qquad (12.28)$$

A theorem of linear algebra says that the ith power of a matrix $I + \Delta t\, \mathbf{M}$ will converge to zero as $i \to \infty$ if and only if all the eigenvalues of that matrix (here $1 + \Delta t \lambda_n$) have magnitude less than one. (You have perhaps seen this theorem used to derive the linear stability condition for the fixed point of a vector map, leading to the condition that all eigenvalues of the Jacobian matrix have magnitude less than one.) A necessary condition for all numerical solutions $\tilde{\mathbf{u}}_i$ to decay to zero is therefore that

$$|1 + \Delta t\, \lambda_n| < 1. \qquad (12.29)$$

Since we have assumed that the eigenvalues λ_n are real and negative while the time step Δt is real and positive, the K inequalities Eq. (12.29) lead to the condition

$$\Delta t < \frac{2}{|\lambda_{\max}|}, \qquad (12.30)$$

for all solutions to decay asymptotically, where $\lambda_{\max}$ is the eigenvalue of largest magnitude, corresponding to the eigenmode that decays most rapidly. Equation (12.30) makes sense since the algorithm should take time steps that are smaller than the fastest time scale in the mathematical problem, which is of order $1/|\lambda_{\max}|$

for this constant-coefficient linear ode problem. If Δt exceeds this bound, some initial conditions will grow exponentially in magnitude, contrary to the correct mathematical behavior.

For the more general case of K nonlinear odes,

$$\frac{d\mathbf{u}}{dt} = \mathbf{f}(\mathbf{u}), \quad (12.31)$$

the forward-Euler algorithm takes the form

$$\tilde{\mathbf{u}}_{i+1} = \tilde{\mathbf{u}}_i + \Delta t\, \mathbf{f}(\tilde{\mathbf{u}}_i). \quad (12.32)$$

This can be programmed using just two arrays, say u_old *to store the current field values and* u_new *for the future field values. Fig. 12.2 shows a schematic example of C++ code. The first* for *loop over* n *uses Eq. (12.32) to calculate the new field values. These can then be processed in various ways, e.g. analyzed statistically, written to some external file, or plotted on the screen. Afterwards, the second* for *loop over* n *copies the future values to the old values, creating the initial condition for the next time step.*

The largest time step that the forward-Euler algorithm can take before the numerical solution Eq. (12.32) diverges from the mathematical solution to Eq. (12.31) (usually by growing exponentially in magnitude) is typically of order $1/|\lambda_{max}|$, where λ_{max} is the eigenvalue of largest magnitude belonging to the Jacobian matrix $\partial \mathbf{f}/\partial \mathbf{u}$ evaluated at $\mathbf{u}(t)$. Since $\mathbf{u}(t)$ changes with time, so do the eigenvalues of the Jacobian and hence so does the largest possible time step allowed by stability. So-called time-adaptive algorithms can take advantage of the variation of λ_{max} with time to adjust automatically each time step to the largest possible value consistent with stability and accuracy. The evolution equations can then be integrated over some specified interval of time to a desired accuracy with a decreased amount of work.

```
for ( i = 1 ; i <= Number_of_times_steps; i++ ) {
   for ( n = 0 ; n < K ; n++ )
      u_new[n] = u_old[n]  +  Delta_t * f(n, u_old) ;

   // ... analyze, store, plot values of u_new

   for ( n = 0 ; n < K ; n++ )    // new initial condition
      u_old[n] = u_new[n] ;
}
```

Fig. 12.2 Schematic C++ code for integrating the evolution equation $d\mathbf{u}/dt = \mathbf{f}(\mathbf{u})$ using the forward-Euler explicit method. The quantity f(n,u) is a function of two arguments that evaluates the nth-component $f_n(\tilde{\mathbf{u}})$ of the vector field $\mathbf{f}(\tilde{\mathbf{u}})$.

12.3 Time integration methods for pattern-forming systems

Etude 12.3 Forward-Euler method for a reaction–diffusion equation

Let us consider now the numerical integration of a representative evolution equation for a pattern-forming system, a reaction–diffusion evolution equation

$$\partial_t u(x,t) = r(u) + D\,\partial_x^2 u, \tag{12.33}$$

for a scalar concentration field $u(x,t)$, where $r(u)$ is some nonlinear reaction rate and D is a positive diffusion constant. (You may wish to review Chapter 3, where reaction–diffusion equations were first discussed.) On a spatial interval $[0,l]$ of length l, we want to find a solution $u(x,t)$ for times $t \geq 0$ that satisfies the boundary conditions

$$u(0,t) = c_0, \quad \text{and} \quad \partial_x u(l,t) = f_0, \tag{12.34}$$

and that passes through a given initial condition $u_0(x)$ at time $t = 0$ so that

$$u(x,0) = u_0(x). \tag{12.35}$$

The boundary condition Eq. (12.34) at $x = 0$ corresponds to an imposed constant concentration with value c_0, while the boundary condition at $x = l$ corresponds to an imposed constant flux with value Df_0. The initial state $u_0(x)$ is some specified function that satisfies the boundary conditions Eq. (12.34).

We discretize Eqs. (12.33) and (12.34) by assuming that the field values $u(x,t)$ are known only on the discrete uniform space-time mesh Eq. (12.20) with time step Δt and spatial resolution Δx and with indices $i \geq 0$ and $0 \leq n \leq N$. To obtain a forward-Euler algorithm, we first require that Eq. (12.33) hold at time t_i at each spatial mesh point x_n. Second, we approximate the time derivative with the first-order finite difference expression analogous to Eq. (12.10) in the form

$$\partial_t u|_{n,i} \approx \frac{u_{n,i+1} - u_{n,i}}{\Delta t}, \tag{12.36}$$

and approximate the second-order spatial derivative in Eq. (12.33) with Eq. (12.14) in the form

$$\partial_x^2 u\bigg|_{n,i} \approx \frac{u_{n+1,i} - 2u_{n,i} + u_{n-1,i}}{\Delta x^2}. \tag{12.37}$$

On substituting these finite-difference approximations into Eq. (12.33) and rearranging, we obtain the following forward-Euler algorithm for the future field $\tilde{u}_{i+1,n}$:

$$\tilde{u}_{n,i+1} = \tilde{u}_{n,i} + \Delta t\, r(\tilde{u}_{n,i}) + \frac{\Delta t D}{\Delta x^2}\left(\tilde{u}_{n+1,i} - 2\tilde{u}_{n,i} + \tilde{u}_{n-1,i}\right). \tag{12.38}$$

This can be expressed in C++ in a way similar to Fig. 12.2:

```
c1 = Delta_t * D / (Delta_x * Delta_x) ;
for ( n = 2 ; n < N ; n++ )
   u_new[n] = u_old[n]  +  Delta_t * r(u_old[n])
       + c1 * ( u_old[n+1] - 2. * u_old[n] + u_old[n-1] ) ;
```

The loop goes over all spatial points except the boundary points at $n = 0, 1$, and N since these need to be treated specially as explained in the next two paragraphs. The expression r(u) *is a function of a variable* u *that evaluates the reaction rate. After analyzing, storing, or plotting the array* u_new*, its contents are copied to* u_old *to prepare for the next time step (see Fig. 12.2).*

The boundary conditions Eq. (12.34) are taken into account as follows. The field value $u(0, t) = c_0$ at $x = 0$ is known for all time so there is no need to impose Eq. (12.38) at $x = 0$. Instead, we simply include the boundary value when evaluating Eq. (12.38) at $n = 1$ ($x = \Delta x$) as a special case:

$$\tilde{u}_{1,i+1} = \tilde{u}_{1,i} + \Delta t\, r(\tilde{u}_{1,i}) + \frac{\Delta t\, D}{\Delta x^2} \left(\tilde{u}_{2,i} - 2\tilde{u}_{1,i} + c_0 \right). \tag{12.39}$$

If we impose Eq. (12.38) at $n = N$ ($x = l$), we run into the difficulty that we need to know the value $u_{N+1,i}$ at a coordinate $x_{N+1} = l + \Delta x$ that lies outside the interval $[0, l]$ over which the field u is defined. Rather than reject this situation, it is actually advantageous to extend the spatial mesh by one point beyond the boundary because the resulting discretization and numerical solution $\tilde{u}_{n,i}$ will be more accurate.[11] These extra spatial points are called ghost points, and boundary conditions can be used to eliminate the field values at these ghost points in terms of boundary data and interior field values. Thus if we discretize the flux boundary condition Eq. (12.34) at $x = l$ using the second-order-accurate approximation Eq. (12.13), we have

$$f_0 = \partial_x u(x,t)|_{N,i} \approx \frac{u_{N+1,i} - u_{N-1,i}}{2\,\Delta x}. \tag{12.40}$$

Eliminating $u_{N+1,i}$ in terms of $u_{N-1,i}$ and f_0, we obtain the following special case of Eq. (12.38) for $n = N$:

$$\tilde{u}_{N,i+1} = \tilde{u}_{N,i} + \Delta t\, r(\tilde{u}_{N,i}) + \frac{\Delta t\, D}{\Delta x^2}(2\,\Delta x f_0 - 2\tilde{u}_{N,i} + 2\tilde{u}_{N-1,i}). \tag{12.41}$$

It is important that the initial condition Eq. (12.35) satisfy the boundary condition $\partial_x u_0 = f_0$ for the forward-Euler algorithm (and other explicit algorithms) to

---

[11] For the implicit algorithms discussed in the next section, using field values just outside the domain is also helpful, e.g. it can yield symmetric positive-definite matrices that lead to more rapidly convergent numerical algorithms.

work. *If this condition is not satisfied at t = 0, the algorithm will not force this condition at later times and the field will satisfy the wrong boundary condition at x = l arbitrarily far into the future. In contrast, the implicit methods we discuss in the next section will force an initial field to satisfy the boundary condition $\partial_x u = f_0$.*

If we assume that the reaction rate r(u) has the linear form $r_0 u$ for r_0 a negative constant, that the diffusion constants D are positive, and that we solve the equation on an infinite domain, then all solutions of Eq. (12.33) will decay asymptotically to zero. Using a Fourier method as explained in Exercise 12.9, you can show that the numerical solution $\tilde{u}_{n,i}$ generated by the forward-Euler algorithm Eq. (12.38) on an infinite interval will also decay for all initial conditions if and only if the time step is bounded by an expression that depends on the strength of diffusion and on the rate of decay r_0:

$$\Delta t < \frac{2}{4D/\Delta x^2 + |r_0|}. \tag{12.42}$$

If the concentration u decays rapidly ($|r_0| \gg 1$), the term $4D/\Delta x^2$ in the denominator can be ignored over a substantial range of Δx and so the largest possible time step is not influenced by the spatial mesh size. In this case, a forward-Euler algorithm is practical and, indeed, some sophisticated models of cardiac dynamics are integrated satisfactorily with the unsophisticated forward-Euler method since the reaction dynamics is fast compared to diffusion.

In the limit $\Delta x \to 0$, the $|r_0|$ term can be neglected in the denominator and the time step is restricted purely by diffusion:

$$\Delta t \leq \frac{1}{2D} \Delta x^2, \tag{12.43}$$

which is indeed of the form Eq. (12.22) with a constant $C = (2D)^{-1}$. Numerical experiments with many nonlinear evolution equations show that a criterion similar to Eq. (12.43) holds generally, even on finite domains and for different kinds of boundary conditions, so that our analysis of the linear problem illustrates the essential mechanism of how stability limits the largest possible time step.

12.3.3 Implicit methods

For a small spatial resolution Δx, explicit evolution algorithms for pattern-forming systems are unstable unless the time step Δt satisfies an inequality of the form Eq. (12.22) and is therefore tiny. For many evolution problems, this restriction can be avoided by an ingenious insight, which is to require that the evolution equations hold not at the present time level i but at some time a little bit into the future. This eliminates a bound on the time step associated with diffusion so that the largest possible time step is now restricted only by accuracy considerations. Such integration algorithms are called implicit because the future field values $\tilde{u}_{n,i+1}$ now satisfy some set of nonlinear equations and so cannot be determined directly by simple algebra.

The most widely used implicit algorithms for pattern-forming systems are the backward-Euler or Crank–Nicolson methods. The backward-Euler method is derived by imposing the evolution equation at time t_{i+1} in the future and by using the same two-point finite-difference expression as in Eq. (12.25) to discretize the time derivative. If the evolution equation has the form:

$$\partial_t \mathbf{u}(x, t) = \mathbf{f}(\mathbf{u}) \qquad (12.44)$$

(where $\mathbf{f}$ is now some nonlinear vector operator that can include derivatives acting on $\mathbf{u}$), then the backward-Euler method gives the following relation:

$$\frac{\tilde{\mathbf{u}}_{i+1} - \tilde{\mathbf{u}}_i}{\Delta t} = \mathbf{f}(\tilde{\mathbf{u}}_{i+1}). \qquad (12.45)$$

Here we have discretized only with respect to time so that $\tilde{\mathbf{u}}_i(x) = \tilde{\mathbf{u}}(x, t_i)$. Equation (12.45) is consistent with the mathematical problem Eq. (12.44) since it converges to that equation in the limit $\Delta t \to 0$. After discretization in space, Eq. (12.45) becomes a set of MN nonlinear equations for the N mesh values of each of the M components of $\tilde{\mathbf{u}}_{i+1}$. In practice, a solution can be found only by numerical means as discussed in Section 12.4.

The Crank–Nicolson algorithm is derived by imposing the evolution equation at the future time $t_{i+1/2} = t_i + \Delta t/2$ that is midway between time levels i and $i + 1$:

$$(\partial_t \mathbf{u})_{i+1/2} = \mathbf{f}(\mathbf{u})|_{i+1/2} \quad \text{or} \quad \frac{\tilde{\mathbf{u}}_{i+1} - \tilde{\mathbf{u}}_i}{\Delta t} = \frac{1}{2}(\mathbf{f}(\tilde{\mathbf{u}}_{i+1}) + \mathbf{f}(\tilde{\mathbf{u}}_i)). \qquad (12.46)$$

Here we have used a finite-difference scheme in time analogous to Eq. (12.13) centered on $t_{i+1/2}$ with mesh size $\Delta t/2$ to approximate $(\partial_t \mathbf{u})_{i+1/2}$ to second-order accuracy, and we have approximated the quantity $\mathbf{f}(\mathbf{u})|_{i+1/2}$ to second-order accuracy in Δt by averaging the corresponding quantities at times t_i and t_{i+1}. The equation on the right in Eq. (12.46) again constitutes a set of nonlinear equations to solve for the field values $\tilde{\mathbf{u}}_{i+1}$ at time t_{i+1}.

The backward-Euler and Crank–Nicolson algorithms are first-order- and second-order-accurate respectively in time. Although the Crank–Nicolson method is more accurate in the limit $\Delta t \to 0$, for some evolution equations it can cause unphysical small-amplitude temporal oscillations since Crank–Nicolson does not damp dissipative modes as strongly as backward-Euler. The less accurate backward-Euler method would then be the better choice since it usually damps out these small oscillations. Further information about the stability properties of these algorithms can be found in the book by Strikwerda [98].

We present two worked examples to illustrate some of the implications and technical details of working with implicit algorithms.

Etude 12.4 Backward-Euler algorithm for a set of linear constant-coefficient odes

The backward-Euler algorithm Eq. (12.45) applied to the evolution equation Eq. (12.23) gives

$$\frac{\tilde{\mathbf{u}}_{i+1} - \tilde{\mathbf{u}}_i}{\Delta t} = \mathbf{M}\tilde{\mathbf{u}}_{i+1}, \tag{12.47}$$

which we can rewrite as

$$(\mathbf{I} - \Delta t\,\mathbf{M})\tilde{\mathbf{u}}_{i+1} = \tilde{\mathbf{u}}_i. \tag{12.48}$$

This is a set of K linear equations for the K components of the vector $\tilde{\mathbf{u}}_{i+1}$ in terms of the known vector $\tilde{\mathbf{u}}_i$.

Let us again assume that the matrix $\mathbf{M}$ is diagonalizable and has negative real eigenvalues, $\lambda_n < 0$. Then the matrix $\mathbf{I} - \Delta t\,\mathbf{M}$ is nonsingular since all of its eigenvalues $1 - \Delta t\,\lambda_n > 0$ are nonzero for $\Delta t > 0$. Eq. (12.48) can therefore be written in the form

$$\tilde{\mathbf{u}}_{i+1} = (\mathbf{I} - \Delta t\,\mathbf{M})^{-1}\tilde{\mathbf{u}}_i, \tag{12.49}$$

which has the same general form as Eq. (12.27). We conclude that the numerical solution $\tilde{\mathbf{u}}_i$ will decay to zero in the limit $i \to \infty$ if and only if each eigenvalue of the matrix $(\mathbf{I} - \Delta t\,\mathbf{M})^{-1}$ has magnitude less than one. Since the eigenvalues of an inverse nonsingular matrix are the inverses of the eigenvalues of the original matrix, we conclude that the numerical solution will decay as expected, provided that

$$\frac{1}{1 - \Delta t\,\lambda_n} < 1, \quad \text{for } 1 \le n \le K. \tag{12.50}$$

But these inequalities are always satisfied since $\lambda_n < 0$ and $\Delta t > 0$. Thus the backward-Euler algorithm eliminates the constraint Eq. (12.30) and the only restriction on the time step arises from accuracy.

Etude 12.5 Backward-Euler algorithm for a linear one-dimensional reaction–diffusion equation

Let us now apply the backward-Euler method to the linear reaction–diffusion system

$$\partial_t u = r_0 u + D\,\partial_x^2 u, \tag{12.51}$$

with the boundary conditions Eq. (12.34), where r_0 and D are constants. For many evolution problems, the operator splitting method described in the next section allows us to avoid applying implicit algorithms to nonlinear evolution equations so Eq. (12.51) is representative of the kind of linear problem that is solved implicitly when working with mathematical descriptions of pattern-forming systems. The results discussed here will also be useful in Section 12.4, when we discuss how to use Newton's method to calculate stationary states of Eq. (12.33).

The backward-Euler algorithm Eq. (12.45) applied to Eq. (12.51) becomes

$$\frac{\tilde{u}_{i+1} - \tilde{u}_i}{\Delta t} = r_0 \tilde{u}_{i+1} + D \partial_x^2 \tilde{u}_{i+1}. \tag{12.52}$$

Gathering the unknown future field $\tilde{u}_{i+1}(x)$ on the left-hand side gives

$$\left(1 - r_0 \Delta t - D \Delta t \partial_x^2\right) \tilde{u}_{i+1} = \tilde{u}_i. \tag{12.53}$$

Equation (12.53) together with Eq. (12.34) define a boundary-value differential problem for the future field $\tilde{u}_{i+1}(x)$. This is a constant-coefficient Helmholtz equation of the form $(c_1 + c_2 \partial_x^2) u = r$.

We discretize Eq. (12.53) in space by asking that it hold at spatial points $x_n = n \Delta x$ in the interval $[0, l]$ (with $1 \le n \le N$) and by using Eq. (12.37) to approximate the derivative $\partial_x^2 \tilde{u}_{i+1}$. Since we know the field value at $x_0 = 0$ by the first boundary condition in Eq. (12.34), the first discrete equation is obtained for $n = 1$ at $x_1 = \Delta x$ and takes the form

$$(1 - r_0 \Delta t) \tilde{u}_{1,i+1} - \frac{D \Delta t}{\Delta x^2} \left(\tilde{u}_{2,i+1} - 2 \tilde{u}_{1,i+1} + c_0\right) = \tilde{u}_{1,i}. \tag{12.54}$$

If we define the dimensionless parameter β by

$$\beta = \frac{D \Delta t}{\Delta x^2}, \tag{12.55}$$

Eq. (12.54) can be written in the form

$$m_{1,1} \tilde{u}_{1,i+1} + m_{1,2} \tilde{u}_{2,i+1} = b_1, \tag{12.56}$$

with

$$m_{1,1} = 1 - r_0 \Delta t + 2\beta, \quad m_{1,2} = -\beta, \quad b_1 = \tilde{u}_{i,1} + \beta c_0. \tag{12.57}$$

For $2 \le n \le N - 1$, you can show that imposing Eq. (12.53) at point x_n leads to the linear equations

$$m_{n,n-1} \tilde{u}_{n-1,i+1} + m_{n,n} \tilde{u}_{n,i+1} + m_{n,n+1} \tilde{u}_{n+1,i+1} = b_n, \tag{12.58}$$

with

$$m_{n,n-1} = -\beta, \quad m_{n,n} = 1 - r_0 \Delta t + 2\beta, \quad m_{n,n+1} = -\beta, \quad b_n = \tilde{u}_{n,i}. \tag{12.59}$$

Finally, we get a discrete equation for $n = N$ by imposing Eq. (12.51) at $x_N = l$ and by using the discretization Eq. (12.40) to eliminate the value $\tilde{u}_{N+1,i+1}$ at the ghost point $x_{N+1} = l + \Delta x$. This leads to the linear equation

$$m_{N,N-1} \tilde{u}_{N-1,i+1} + m_{N,N} \tilde{u}_{N,i+1} = b_N, \tag{12.60}$$

with

$$m_{N,N-1} = -2\beta, \quad m_{N,N} = 1 - r_0 \Delta t + 2\beta, \quad b_N = \tilde{u}_{n,i} + 2\beta f_0 \Delta x. \quad (12.61)$$

If we group the N unknown field values $u_{n,i+1}$ into a column vector $\tilde{\mathbf{u}}_{i+1}$ and if we group the N numbers b_n into a column vector $\mathbf{b}$,

$$\tilde{\mathbf{u}}_{i+1} = \begin{pmatrix} \tilde{u}_{1,i+1} \\ \tilde{u}_{2,i+1} \\ \vdots \\ \tilde{u}_{N-1,i+1} \\ \tilde{u}_{N,i+1} \end{pmatrix}, \quad \mathbf{b} = \begin{pmatrix} b_1 \\ b_2 \\ \vdots \\ b_{N-1} \\ b_N \end{pmatrix}, \quad (12.62)$$

then the linear equations Eqs. (12.56), (12.58), and (12.60) can be written in matrix form

$$\mathbf{M}\tilde{\mathbf{u}}_{i+1} = \mathbf{b}, \quad (12.63)$$

where the $N \times N$ matrix $\mathbf{M}$ is given by

$$\mathbf{M} = \begin{pmatrix} m_{1,1} & m_{1,2} & 0 & 0 & 0 & \cdots & 0 \\ m_{2,1} & m_{2,2} & m_{2,3} & 0 & 0 & \cdots & 0 \\ 0 & m_{3,2} & m_{3,3} & m_{3,4} & 0 & \cdots & 0 \\ \cdots & \cdots & \cdots & \cdots & \cdots & \cdots & \cdots \\ \cdots & 0 & m_{n,n-1} & m_{n,n} & m_{n,n+1} & 0 & \cdots \\ \cdots & \cdots & \cdots & \cdots & \cdots & \cdots & \cdots \\ 0 & \cdots & 0 & m_{N-2,N-3} & m_{N-2,N-2} & m_{N-2,N-1} & 0 \\ 0 & \cdots & 0 & 0 & m_{N-1,N-2} & m_{N-1,N-1} & m_{N-1,N} \\ 0 & \cdots & 0 & 0 & 0 & m_{N,N-1} & m_{N,N} \end{pmatrix}. \quad (12.64)$$

with matrix elements given by Eqs. (12.57), (12.59), and (12.61). Only the values b_1, b_N, and the matrix element $m_{N,N-1}$ are affected by the boundary conditions Eq. (12.34). Unlike an explicit algorithm such as forward-Euler, the solution $\tilde{\mathbf{u}}_{i+1}$ in Eq. (12.62) from an implicit algorithm will satisfy the boundary condition $\partial_x u(l,t) = f_0$ to second-order accuracy, even if the initial state Eq. (12.35) does not.

The matrix $\mathbf{M}$ in Eq. (12.64) is called a tridiagonal matrix since all elements are zero except for those lying on three diagonals. The fact that there are at most three nonzero matrix elements on each row reflects our choice of a three-point finite-difference expression $\delta_{2,2}$ to approximate the derivative $\partial_x^2 u$ in Eq. (12.33). If instead we had used a 5-point fourth-order-accurate finite difference $\delta_{4,2}$, the matrix $\mathbf{M}$ would be a pentadiagonal matrix with at most five nonzero elements per row. Pentadiagonal matrices also occur for one-dimensional models of pattern-forming systems that have fourth-order derivatives such as the Swift–Hohenberg and Kuramoto–Sivashinsky equations (see Chapter 5).

Tridiagonal and pentadiagonal matrices are examples of banded matrices, for which the nonzero matrix elements occur only in some finite band of diagonals containing the main diagonal. Finding the solution to a set of N linear equations Eq. (12.63) described by an $N \times N$ banded matrix Eq. (12.64) is a classic solved problem in numerical linear algebra [55]. A solution $\tilde{\mathbf{u}}_{i+1}$ can be computed with a relative error of order the machine precision ϵ_{mach} after $O(NB^2)$ operations where B is the matrix bandwidth (the number of diagonals spanning the right-most nonzero diagonal to the leftmost nonzero diagonal). For a fixed choice of finite-difference approximations (which fixes the bandwidth B), the amount of work to take one implicit time step for this one-dimensional problem grows linearly with N. A computational effort proportional to the number of unknowns is about as good as it gets for a numerical algorithm since a code has to touch each mesh value at least once in constructing an answer.

With a computer mathematics program like Maple, Mathematica, or Matlab, you can simply input the matrix elements of $\mathbf{M}$ and the components of the right side vector $\mathbf{b}$ into an intrinsic function to obtain the numerical solution $\tilde{\mathbf{u}}$. If instead a compiled language like C++ is used, you would write a code to define the nonzero matrix elements $M_{m,n}$ and components b_n, and then pass these data to some previously written library function for solving banded linear equations. A high-quality public-domain library for linear algebra is the LAPACK library [4] and its source code is available from the web site www.netlib.org.

12.3.4 Operator splitting

In this section, we combine the ideas of explicit and implicit time integration methods and discuss a widely used technique called operator splitting to integrate evolution equations. For many problems, operator splitting achieves a practical balance between competing goals of efficiency, accuracy, reduced memory storage, and the time and effort needed to develop a working code.

Operator splitting is useful when an evolution equation can be written in the additive form

$$\partial_t u(x,t) = \mathcal{L}[u] + \mathcal{N}[u], \qquad (12.65)$$

where $\mathcal{L}$ is a linear operator containing the highest-order spatial derivatives and $\mathcal{N}$ is a nonlinear operator containing the remaining terms. For example, for the nonlinear reaction–diffusion equation Eq. (12.33), we can identify $\mathcal{L}[u] = D \partial_x^2 u$ and $\mathcal{N}[u] = r(u)$ since most reaction rates are purely algebraic functions of the field and so have no spatial derivatives. Similarly, for the Swift–Hohenberg equation Eq. (2.3), we could choose $\mathcal{L}[u] = (r-1)u - 2\partial_x^2 u - \partial_x^4 u$ and $\mathcal{N}[u] = -u^3$.[12] The idea is then to integrate the nonlinear term with any convenient explicit algorithm, as if this were the only term in the evolution equation:

$$\partial_t u = \mathcal{N}[u]. \qquad (12.66)$$

Taking one time step of size Δt will produce some intermediate field that we will denote as $\tilde{u}^*$. We then use this intermediate value as the initial data for a second evolution problem in which only the linear operator appears:

$$\partial_t u = \mathcal{L}[u], \quad \text{with initial data } u^*. \qquad (12.67)$$

The output from this second step is then taken as the numerical approximation $\tilde{u}_{i+1}$ of the unknown analytical field u_{i+1} that evolved from the same initial condition $\tilde{u}_i$. The explicit step is usually done first since the implicit step that follows will enforce the boundary conditions on the solution $\tilde{u}_{i+1}$, up to the order of accuracy of the discretization for the boundary conditions.

Most evolution equations involve several coupled fields. In these cases, there is usually a separate evolution equation for each field and the evolution equations take the form:

$$\partial_t u_1(\mathbf{x}, t) = \mathcal{L}_1[\mathbf{u}] + \mathcal{N}_1[\mathbf{u}], \qquad (12.68a)$$

$$\ldots$$

$$\partial_t u_K(\mathbf{x}, t) = \mathcal{L}_K[\mathbf{u}] + \mathcal{N}_K[\mathbf{u}]. \qquad (12.68b)$$

Here K is the number of fields, the vector $\mathbf{u} = (u_1(\mathbf{x}, t), \ldots, u_K(\mathbf{x}, t))$, the operators $\mathcal{L}_k$ are the linear operators containing the highest-order spatial derivatives of the kth equation, and the $\mathcal{N}_k$ are the corresponding nonlinear operators containing the remaining terms of the kth equation.

The K fields $u_k(\mathbf{x}, t)$ can be advanced one time step Δt by applying operator splitting to each equation in turn. Starting with the first equation Eq. (12.68a), we integrate the nonlinear term $\mathcal{N}_1$ explicitly and then the linear term $\mathcal{L}_1$ implicitly to obtain the field values of $\tilde{u}_1(t_{i+1}, \mathbf{x})$ at the next time step. Then we turn to the second equation, advance $\mathcal{N}_2$ explicitly and then $\mathcal{L}_2$ implicitly which gives the field $\tilde{u}_2(t_{i+1}, \mathbf{x})$ at the next time step, and so on until all the fields have been updated by one time step. Depending on the nature of the linear and nonlinear

---

[12] This Swift–Hohenberg example points out an ambiguity in the choice of linear and nonlinear operators, e.g. we could have chosen instead $\mathcal{L} = -2\partial_x^2 - \partial_x^4$ and $\mathcal{N} = (r-1)u - u^3$. It is crucial that the highest-order spatial derivatives appear in the linear operator but not so for the lower-order linear operators. Adding a lower-order linear term to the nonlinear operator can sometimes improve the overall stability of an operator splitting method.

terms in each evolution equation, it may be appropriate to use different explicit and implicit algorithms for the different operators $\mathcal{N}_k$ and $\mathcal{L}_k$.

When advancing the evolution equations for each field in Eqs. (12.68), there is the option to always use the same initial data for each successive equation or to use the latest data from one equation as the initial data for a next equation. In the first case, the order of updating the equations does not matter. However, in the second case, the order of updating the equations can matter a lot and different orders can lead to different maximum time steps (before instability occurs), or even instability for any choice of time steps. For many nonlinear problems, the appropriate order cannot be determined analytically, and so numerical experiments are needed.

The operator-splitting method can require much less memory (RAM) than a fully implicit algorithm, although it often requires more memory than a fully explicit algorithm. Since each equation in Eqs. (12.68) is updated in turn, the implicit step for the kth equation requires solving a set of N linear equations where N is the total number of mesh points, i.e. an $N \times N$ matrix needs to be constructed and the nonzero elements stored at each time step. In contrast, a fully implicit time-stepping method would require gathering all the future field values, KN in all, into a single vector and constructing and storing the nonzero elements of a possibly huge $KN \times KN$ matrix. For three-dimensional pattern-forming problems, only a large parallel computer is capable of storing the many matrix elements needed to carry out a fully implicit time step.

We illustrate operator splitting with a worked example.

Etude 12.6 Operator-splitting method for the Brusselator

On page 105, we discussed the Brusselator reaction–diffusion evolution equations

$$\partial_t u_1 = a - (b+1)u_1 + u_1^2 u_2 + D_1 \partial_x^2 u_1, \tag{12.69a}$$

$$\partial_t u_2 = b u_1 - u_1^2 u_2 + D_2 \partial_x^2 u_2, \tag{12.69b}$$

where the parameters a, b, D_1, and D_2 are positive constants. Here we look at the numerical solution on the interval $[0, l]$ of length l. Typical boundary conditions at $x = 0$ and $x = l$ for the concentration fields $u_i(x, t)$ could be constant concentrations or constant fluxes.

Each of these evolution equations have the additive form Eq. (12.65) with the highest-order derivatives appearing as linear operators. One choice of operators for an operator-splitting algorithm could be

$$\mathcal{N}_1[\mathbf{u}] = a - (b+1)u_1 + u_1^2 u_2, \quad \mathcal{L}_1[\mathbf{u}] = D_1 \partial_x^2 u_1,$$
$$\mathcal{N}_2[\mathbf{u}] = b u_1 - u_1^2 u_2, \quad \mathcal{L}_2[\mathbf{u}] = D_2 \partial_x^2 u_2.$$

12.3 Time integration methods for pattern-forming systems

The nonlinear operator $\mathcal{N}_1$ in Eq. (12.69a) could be integrated with a second-order-accurate explicit algorithm such as an Adams–Bashforth or Runge–Kutta scheme to obtain an intermediate field $\tilde{u}_1^*$. This field is then used as initial data to integrate the linear operator $\mathcal{L}_1$ with, say, an implicit Crank–Nicolson scheme to produce the field $\tilde{u}_{i+1}$ at the future time step $t + \Delta t$. These explicit and implicit steps are then repeated for the operators $\mathcal{N}_2$ and $\mathcal{L}_2$ in Eq. (12.69b) at which point one full time step has been advanced.

For each equation, the implicit Crank–Nicolson step requires the solution of a set of linear equations corresponding to a tridiagonal matrix similar to Eq. (12.64). This solution constitutes the most time-consuming part of the operator-splitting algorithm.[13] For a one-dimensional reaction–diffusion problem in a large spatial domain, modern personal computers can easily handle hundreds of thousands of time steps with of order, say, 10 000 mesh points, and so the long-time behavior of the dynamics in rather large systems can be readily explored. Time integration becomes much more challenging for two- and three-dimensional pattern-forming systems since the corresponding linear equations of the implicit step may number in the tens of millions. Such large problems require sophisticated linear algebra techniques and often distributed data structures and algorithms on a parallel computer.

If the constants in Eqs. (12.69) are such that the time scales for the reaction rates are fast compared to diffusion (as can be quantitatively understood by generalizing the argument that led to Eq. (12.42)), then there is no advantage to using operator splitting, since a small time step is needed anyhow to resolve the fast dynamics. In this case, a forward-Euler or similar simple explicit algorithm will work well.

12.3.5 How to choose the spatial and temporal resolutions

We conclude this section on time integration methods by discussing how to choose the time step Δt and spatial resolution Δx when solving the evolution equations that describe a pattern-forming system. A well-designed and correctly implemented algorithm generates a numerical solution that converges to the unknown solution in the limits $\Delta t \to 0$ and $\Delta x \to 0$ (at least until Δt and Δx are so small that floating point errors become significant, as shown in Fig. 12.1). However, the smaller the values of Δt and of Δx, the more computational work needed to compute the solution and the more memory (RAM) and disk space needed to store data related to the simulation. A computational scientist's goal lies in the opposite direction of

---

[13] A technical note: since the linear operators are the same at each time step, a more efficient algorithm can be obtained by using the *PLU* factorization theorem [55] to factor the corresponding band matrices into lower- and upper-triangular matrices L and U as a preprocessing step, before any time steps are taken. This expensive factorization step is done only once over the entire time integration instead of at each time step.

these limits, namely to identify the largest (crudest) values of Δt and Δx that will lead to an acceptable answer.

There is no simple way to determine an adequate space-time resolution. You should use available knowledge from experiment and theory, use some common sense based on past experience, compare results of calculations based on several choices of space-time resolutions, and be skeptical that a code and its results are correct until a thorough effort has been made to validate them. The coarsest acceptable resolution also depends on the question being asked. If the question is "What basins of attraction exist?" then a lower resolution is likely acceptable compared to a question of the sort "For what specific parameter value does a bifurcation occur?" or "What is the exponent that governs how heat transport scales with Rayleigh number for large R?".

A first step toward identifying a suitable space-time resolution is to take advantage of existing experimental data. If time series are available for different fields of a pattern-forming system and all peaks are resolved, then the time step Δt should be smaller (say by a factor of 5) than the peak-to-peak distance in any of the time series. Equivalently, the power spectrum $P(\omega)$ can be calculated for a given time series and the high-frequency regime examined to determine if the spectrum is decreasing monotonically toward the base set by instrumental noise. A time step Δt could then be chosen such that the corresponding frequency $1/\Delta t$ is well into the regime of asymptotic decay in $P(\omega)$, say such that the magnitude of the power spectrum has dropped by 1000 compared to the magnitude of the largest peak. Similar considerations based on the spatial variation of the fields and on the asymptotic decay of their wave number spectra $P(k)$ can suggest a starting spatial resolution. If a simulation is not too expensive to carry out, all results should be repeated with finer resolutions (say by successive factors of 1/2) to verify that the qualitative and statistical properties have become invariant with respect to the space-time resolution. For chaotic solutions, changing the resolution will lead to a completely different solution in a time of order $1/\lambda_1$, where λ_1 is the largest Lyapunov exponent. You should then compare statistical properties of the solutions such as averages, standard deviations, probability distribution functions, and Lyapunov exponents to determine if the space-time resolution is adequate.

If analytical theory is available, particularly in the form of a stability analysis, then that theory can sometimes suggest starting choices for the space-time resolution. Thus the fastest growing (most unstable) linear mode about a stationary state is often the fastest dynamics even of the saturated nonlinear state (at least near onset). The reciprocal of the largest growth rate $1/\max_q \text{Re}\,\sigma_q$ is then a good bound for the time step. Similarly, the reciprocal $1/q_{max}$ of the wave number q_{max} that bounds the upper range of stable wave numbers in a Busse balloon can provide a useful first estimate of the magnitude of the spatial resolution Δx.

12.4 Stationary states of a pattern-forming system

We now turn to the second of the two central problems of this chapter, namely finding a solution to a set of nonlinear equations of the form

$$\mathbf{f}(\mathbf{u}) = \mathbf{0}, \qquad (12.70)$$

where $\mathbf{f}$ is a known N-dimensional vector function of an N-dimensional vector $\mathbf{u}$. Such a problem arises when calculating a stationary pattern $\mathbf{u}(\mathbf{x})$ in preparation for a linear stability analysis of that pattern. Upon setting the time derivatives of the evolution equations to zero and then discretizing the resulting equations in space by either a Galerkin or finite-difference method, a finite number N of nonlinear equations is obtained for the N values u_n that represent the field (see Eqs. (12.3) and (12.4)). Nonlinear equations of the form Eq. (12.70) also arise when implementing an implicit time integration method such as was discussed in Section 12.3.3. The variable $\mathbf{u}$ would then have the meaning of the future discretized field values $\mathbf{u}_{i+1}$.

Solving a set of nonlinear equations is among the most difficult mathematical and computational problems associated with pattern-forming systems. There are rarely theorems to indicate whether a solution of Eq. (12.70) exists and there are no systematic methods to find even a single solution. To complicate matters further, nonlinear equations that have a solution typically have many solutions (think of $\sin(u) = 0$) and finding a physically relevant solution can be challenging. The successful numerical solution of nonlinear equations often requires supplementary scientific knowledge, insight, and experience.

If a solution of Eq. (12.70) corresponds to a linearly stable fixed point of an evolution equation, the time integration methods of the previous section can be used to find an approximate answer by choosing an initial state within the basin of attraction of the fixed point (not always easily done), and then by integrating for a long time until the transient toward the stable fixed point has decayed sufficiently. However, only stable or fully unstable stationary states can be found by this approach and unstable states with some expanding and some contracting eigenmodes must be found by other methods such as Newton's method discussed in Section 12.4.2. Another difficulty of finding a stationary state by time integration is that many pattern-forming systems relax diffusively toward their fixed points. The time scale for a transient to decay scales as l^2, where l is the size of the system, which can lead to long relaxation times for large systems. A possibility might be to use a hybrid method, a time integration to obtain a rough approximation of a stationary state that is then used as the initial state of a more rapidly convergent Newton's method.

12.4.1 Iterative methods

Most numerical algorithms for finding solutions to Eq. (12.70) use iteration to generate a sequence of vectors $\mathbf{u}_k$ that converge to a solution in the limit $k \to \infty$. A strategy for discovering an iterative method is to rewrite Eq. (12.70) algebraically in the form

$$\mathbf{u}_{k+1} = \mathbf{g}(\mathbf{u}_k), \quad k \geq 1, \qquad (12.71)$$

such that a fixed point $\mathbf{u}^*$ of this map is a solution of Eq. (12.70). (Here $\mathbf{g}(\mathbf{u})$ is an N-dimensional vector function.) As a simple example, if we didn't know the formula for the roots of the quadratic equation

$$f(u) = u^2 + bu + c = 0, \qquad (12.72)$$

we could try to find a root numerically by one of the following three iterative methods whose fixed points you can easily verify to be roots of Eq. (12.72):

$$u_{k+1} = -\frac{c + u_k^2}{b}, \quad u_{k+1} = -\frac{c + bu_k}{u_k}, \quad u_{k+1} = \frac{u_k^2 - c}{2u_k + b}. \qquad (12.73)$$

Since Eq. (12.71) is a nonlinear map of the sort discussed in introductory nonlinear dynamics texts, a sufficient criterion that the sequence generated by Eq. (12.71) converges is that the fixed point $\mathbf{u}^*$ be linearly stable and that the initial vector $\mathbf{u}_1$ lie sufficiently close to $\mathbf{u}^*$. Linear stability of the fixed point in turn requires that the eigenvalues of the $N \times N$ Jacobian matrix $\partial \mathbf{g}/\partial \mathbf{u}|_{\mathbf{u}^*}$ evaluated at the fixed point all have magnitudes less than one. This criterion can be difficult to apply since an approximate initial guess for a fixed point may not be known and the components of $\mathbf{g}$ may be difficult to work with analytically.

An advantage of iterative algorithms of the form Eq. (12.71) is that they are often easy to program since each iteration requires only that each component of an explicitly known function be evaluated in turn. (Of course, for some problems the function $\mathbf{g}$ may be complicated and expensive to evaluate, e.g. some components may themselves be determined by running an iterative algorithm.) A potential drawback of the algorithm Eq. (12.71) is that the rate of convergence can be slow. The magnitude of the error vector $\mathbf{u}_k - \mathbf{u}^*$ decreases each iteration (asymptotically) by a constant factor corresponding to the largest eigenvalue magnitude of the Jacobian matrix $\partial \mathbf{g}/\partial \mathbf{u}$. In some cases, the constant factor can be close to one and many iterations are needed to obtain an accurate answer (see Etude 12.7 below).

12.4.2 Newton's method

Newton's method[14] is a clever way to choose the function **g** in the iterative method Eq. (12.71) such that:

(i) Convergence to a solution of Eq. (12.70) is guaranteed if an initial guess is sufficiently close to a fixed point. The problem of figuring out how to manipulate Eq. (12.70) into a convergent algorithm Eq. (12.71) is therefore avoided.
(ii) Convergence is rapid, with the number of significant digits roughly doubling per iteration.

The cost for achieving these capabilities is that a set of N linear equations must be solved at each step where N is the number of unknowns. For two- and three-dimensional pattern-forming problems, N can be a large number and then parallel computing and sophisticated numerical linear algebra algorithms may be needed to carry out each iteration. Still, because local convergence is guaranteed for all fixed points and because the convergence rate is rapid, Newton's method is often worthwhile to implement and it remains a method of choice for solving nonlinear equations.

The derivation of Newton's method is worth discussing since it is brief and illustrates how a powerful numerical algorithm can be invented with little more than the idea of iteration and the use of a Taylor series expansion. Let us assume that we somehow know a vector $\mathbf{u}_k$ that is close to an unknown solution $\mathbf{u}^*$ of Eq. (12.70). By close, we mean that there is a small vector $\delta \mathbf{u}_k$ that we can add to $\mathbf{u}_k$ such that $\mathbf{u}_k + \delta \mathbf{u}_k$ is an exact solution:

$$\mathbf{f}(\mathbf{u}_k + \delta \mathbf{u}_k) = \mathbf{0}. \qquad (12.74)$$

If the vector $\delta \mathbf{u}_k$ is sufficient small, we can Taylor expand the left side of Eq. (12.74) about $\mathbf{u}_k$ to linear order in $\delta \mathbf{u}_k$ to obtain

$$\mathbf{f}(\mathbf{u}_k) + \mathbf{J}_k \, \widetilde{\delta \mathbf{u}}_k = \mathbf{0}, \qquad (12.75)$$

where $\mathbf{J}_k = \partial \mathbf{f}/\partial \mathbf{u}|_{\mathbf{u}_k}$ is the $N \times N$ Jacobian matrix of $\mathbf{f}$ evaluated at the known point $\mathbf{u}_k$. We have also made a small change in notation, replacing the exact vector $\delta \mathbf{u}_k$ in Eq. (12.74) with an approximate vector $\widetilde{\delta \mathbf{u}}_k$ since we have dropped the higher-order terms in Eq. (12.75) that are needed for $\delta \mathbf{u}_k$ to satisfy Eq. (12.74) exactly. Rewriting Eq. (12.75) in the form

$$\mathbf{J}_k \widetilde{\delta \mathbf{u}}_k = -\mathbf{f}(\mathbf{u}_k), \qquad (12.76)$$

we can write down Newton's method as the following steps for generating a sequence of vectors $\mathbf{u}_k$ that converges toward a zero $\mathbf{u}^*$ of Eq. (12.70):

---

[14] The algorithm is named after Isaac Newton, who proposed it as a way to find the root of a certain polynomial. The algorithm goes by other names, such as Newton–Raphson in the context of finding a zero of a function of a single variable and Newton–Kantorovich in the context of solving systems of equations.

First, the right side of Eq. (12.76), $-\mathbf{f}(\mathbf{u}_k)$, is evaluated and stored. This vector is called the residual of the Newton method, and vanishes when convergence is achieved.

(ii) Second, the N linear equations of Eq. (12.76) are solved for the vector $\widetilde{\delta\mathbf{u}}_k$. (This vector is called the correction of a Newton method and is another quantity that vanishes when the algorithm has fully converged.) The success of this step presumes that the matrix $\mathbf{J}_k$ is nonsingular and so the condition number of the matrix $\mathbf{J}_k$ should be monitored during successive iterations. (See Exercise 12.8 for a brief introduction to the condition number of a matrix.)

(iii) A new and presumably better approximation $\mathbf{u}_{k+1}$ is obtained by adding the correction to the present vector:[15]

$$\mathbf{u}_{k+1} = \mathbf{u}_k + \widetilde{\delta\mathbf{u}}_k. \tag{12.77}$$

(iv) Steps (i) through (iii) are repeated until adequate convergence is attained. In practice, this means that the magnitudes of the residual and correction are acceptably small.

With rather general assumptions about the properties of the Jacobian matrix $\mathbf{J}^*$ evaluated at the fixed point, a theorem can be proved that Newton's method always converges (ignoring floating point effects) provided that the initial guess is sufficiently close to a solution [97]. Further, the convergence is quadratic, which has the consequence that the number of significant digits in each component of $\mathbf{u}_k$ approximately doubles after each iteration.

Newton's method is challenging to implement. During each iteration, the N components of the residual $-\mathbf{f}(\mathbf{u}_k)$ and the N^2 nonzero matrix elements $J_{ij} = \partial f_i / \partial u_j \big|_{\mathbf{u}_k}$ of the Jacobian matrix $\mathbf{J}_k$ need to be evaluated and then the N linear equations Eq. (12.76) must be solved for the correction $\widetilde{\delta\mathbf{u}}_k$. There is a large and sophisticated literature regarding how to implement Newton's method efficiently and how to stabilize the algorithm so that it doesn't diverge if a poor initial vector $\mathbf{u}_1$ is chosen. For most scientists, a good starting point would be a software library written by a computational expert, e.g. one of the nonlinear equation solvers available through www.netlib.org. Given such software, the most important first step is to find an initial vector $\mathbf{u}_1$ such that Newton's method converges and converges to a scientifically relevant answer. Finding an initial vector can be hard, but a possibly useful strategy is continuation, in which one solves a sequence of nonlinear problems by varying a parameter p in small increments from some initial value p_1 to some final value p_f such that the numerical problem for p_1 has a simple solution and p_f is the parameter value of interest. The converged nonlinear solution for parameter value p_{i-1} is used as the starting state for parameter value p_i. For example, a nonlinear stationary convecting state at large Rayleigh number might be

---

[15] Formally, Newton's algorithm can be written in a single line in the form $\mathbf{u}_{k+1} = \mathbf{g}(\mathbf{u}_k) = \mathbf{u}_k - \mathbf{J}_k^{-1}\mathbf{f}(\mathbf{u}_k)$ in terms of the matrix inverse of $\mathbf{J}_k$. But this is a poor way to think of the algorithm numerically since it is more efficient and accurate to solve the linear equations Eq. (12.76) for $\widetilde{\delta\mathbf{u}}_k$ without ever constructing the matrix inverse.

12.4 Stationary states of a pattern-forming system

found by solving the time-independent Boussinesq equations just above the onset of convection for which a solution is known analytically by some perturbation theory. One then solves a succession of nonlinear problems by increasing R in small increments until the desired Rayleigh number is attained.

Note that if Newton's method converges, so that $\mathbf{u}_k \to \mathbf{u}^*$, where $\mathbf{f}(\mathbf{u}^*) = \mathbf{0}$, then the residual $-\mathbf{f}(\mathbf{u}_k)$ converges to $\mathbf{0}$ and consequently the correction $\widetilde{\delta\mathbf{u}}_k = -\mathbf{J}_k^{-1}\mathbf{f}(\mathbf{u}_k)$ also converges to $\mathbf{0}$. Most implementations of Newton's method test for convergence of the algorithm by requiring that the magnitudes of the residual $-\mathbf{f}(\mathbf{u}_k)$ and of the correction $\widetilde{\delta\mathbf{u}}_k$ are simultaneously small:

$$\|\mathbf{f}(\mathbf{u}_k)\| \leq \epsilon_1 \quad \text{and} \quad \|\widetilde{\delta\mathbf{u}}_k\| \leq \epsilon_2, \tag{12.78}$$

and *you* need to figure out how to choose the convergence parameters ϵ_1 and ϵ_2. It is common for the magnitudes of the vectors $\mathbf{f}(\mathbf{u}_k)$ and $\widetilde{\delta\mathbf{u}}_k$ to differ by orders of magnitude as both converge to zero, and so some experience is needed in choosing the values of the ϵ_i. In particular, testing just the residual or the correction for smallness can lead to wrong conclusions about convergence being attained.

We illustrate Newton's method by two worked examples.

Etude 12.7 Finding a zero of the function $f(u) = u - \cos(u)$

Let us compare the convergence of a simple iteration scheme Eq. (12.71) with Newton's method Eq. (12.76) for the transcendental equation $f(u) = u - \cos(u)$ which has a unique zero $u^* \approx 0.739085$ (see Fig. 12.3). Easiest to try is the iterative method

$$u_{k+1} = g(u_k) = \cos(u_k), \tag{12.79}$$

which we expect to converge with a rate $|g'(u^*)| = |\sin(u^*)| \approx 0.67$ at each iteration, corresponding to about one new decimal digit every 6 iterations (since $0.67^6 \approx 0.1$). This algorithm can be executed on a calculator by typing some number (the initial value u_1) and hitting the cosine button over and over again, with the argument of the cosine function evaluated in radians. Alternatively, we can apply Newton's method, which you can verify takes the following form for a function of one variable:

$$u_{k+1} = u_k - \frac{f(u_k)}{f'(u_k)} \tag{12.80a}$$

$$= u_k - \frac{u_k - \cos(u_k)}{1 + \sin(u_k)}. \tag{12.80b}$$

(The third iteration formula in Eq. (12.73) can now be understood as Newton's method Eq. (12.80a) applied to the quadratic equation and is the most rapidly convergent of the three formulas.) Eq. (12.80a) has a simple geometric interpretation in that the new value u_{k+1} is the zero of the line tangent to $f(u)$ at u_k; this is

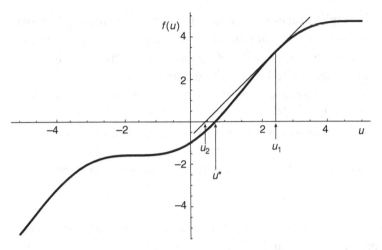

Fig. 12.3 Plot of the transcendental function $f(u) = u - \cos(u)$ with the line tangent to f at $u_1 = 2.5$, $y = f(u_1) + f'(u_1)(u - u_1)$. Newton's method uses the easily calculated zero of the tangent line $u_2 = u_1 - f(u_1)/f'(u_1)$ to obtain an improved estimate of a zero u^* of f.

illustrated in Fig. 12.3. Also from this figure, you can see that if an initial guess is chosen close to an extremum of f (e.g. $u_1 = 0.99(3\pi/2) \approx 4.7$), the tangent line is nearly horizontal and will intersect the u axis at a value u_2 that is far from u_1. For our function $f = u - \cos(u)$, this poor initial condition leads to a long-lived transient that eventually converges to u^*, but for most functions Newton's method will diverge to $\pm\infty$ when started near an extremum. Newton's method requires a good initial guess!

Table 12.1 shows how these two methods converge for the same starting value $u_1 = 2.5$ shown in Fig. 12.3. After ten iterations, Eq. (12.79) has achieved only two significant digits while after five iterations, Newton's method has attained seven significant digits, plenty for most scientific applications. The data also demonstrate how Newton's method approximately doubles the number of significant digits each iteration.

Etude 12.8 Newton's method for a stationary one-dimensional reaction–diffusion pattern

For our second example, we use Newton's method and finite differences to calculate a numerical stationary solution $u(x)$ to the time-independent reaction–diffusion equation

$$0 = r(u) + D \partial_x^2 u, \tag{12.81}$$

with reaction rate $r(u)$ and diffusion constant D given by

$$r(u) = u - u^3 \quad \text{and} \quad D = 1/2. \tag{12.82}$$

12.4 Stationary states of a pattern-forming system

Table 12.1. *Comparison of a simple iteration method Eq. (12.79) with Newton's method Eq. (12.80b) for the same starting value $u_1 = 2.5$. The quadratic convergence of Newton's method is much more rapid than the linear convergence of the simple iterative scheme.*

k	u_k from Eq. (12.79)	u_k from Eq. (12.80b)
1	2.5	2.5
2	−0.80114361554693371483	0.43481317286424721536
3	0.69588587435159201329	0.76701270536720584199
4	0.76748609945623035296	0.73925235862152956077
5	0.71965841403522752424	0.73908513938875372649
6	0.75203092182513546386	0.73908513321516065007
7	0.73030300581355770658	0.73908513321516064166
8	0.74497230276350045050	0.73908513321516064166
9	0.73510668004495655457	0.73908513321516064166
10	0.74175921091330173906	0.73908513321516064166

We solve this on the interval $[0, l]$ with boundary conditions[16]

$$u(0) = 0 \quad \text{and} \quad \partial_x u|_l = \text{sech}^2(l), \qquad (12.83)$$

for which there is an analytical solution

$$u(x) = \tanh(x) \qquad (12.84)$$

that we can compare with the numerical solution. As you learned in Chapter 6, Eqs. (12.81) and (12.82) are also the equations satisfied by a slowly varying envelope $u(x)$ that modulates a cellular state near the onset of a type-I-s instability.

Conceptually, it is somewhat cleaner to apply Newton's method directly to Eqs. (12.81) and (12.83) and then discretize rather than discretize and then use Newton's method. Let us assume that we somehow know a pattern $u_k(x)$ that is close to an unknown solution $u^*(x)$ of Eq. (12.81), and we wish to find a better solution. We then require that $u_k + \delta u_k$ be an exact solution where the correction $\delta u_k(x)$ is a small perturbation of u_k. Substituting $u_k + \delta u_k$ into Eq. (12.81) and linearizing to first order in the correction (in exact analogy to Eq. (12.75)), you can show that a slightly different correction $\widetilde{\delta u}_k$ satisfies the linear boundary-value differential equation:

$$\left(1 - 3u_k^2 + D\partial_x^2\right)\widetilde{\delta u}_k = -\left[r(u_k) + D\partial_x^2 u_k\right], \qquad (12.85)$$

---

[16] The function sech(l) decays to zero rapidly for $l \geq 6$ so we could also use the boundary condition $\partial_x u = 0$ with good accuracy in large domains.

with boundary conditions

$$\widetilde{\delta u}_k(0) = 0 \quad \text{and} \quad \partial_x \widetilde{\delta u}_k \big|_l = 0. \tag{12.86}$$

Equations (12.85) and (12.86) define an operator form of Newton's equation. Given the known residual on the right-hand side of Eq. (12.85), we can solve for the correction $\widetilde{\delta u}_k$ on the left-hand side and then obtain a more accurate solution by adding the correction to the currently known pattern u_k:

$$u_{k+1}(x) = u_k(x) + \widetilde{\delta u}_k(x). \tag{12.87}$$

In some rare cases, we might be able to solve the Newton step Eq. (12.85) analytically for the correction $\widetilde{\delta u}_k$ but here we will solve this equation numerically.

Equations (12.85) and (12.86) are nearly exactly the same as Eqs. (12.53) and (12.34) that we derived previously when we applied the backward-Euler method to the time-dependent reaction–diffusion equation Eq. (12.33). We can therefore use nearly all of Etude 12.5 to solve the Newton equation Eq. (12.85) numerically using finite differences. Upon discretization on a spatially uniform mesh, Eq. (12.85) becomes a tridiagonal set of linear equations Eq. (12.63) with the vectors $\mathbf{u}_{i+1}$ and $\mathbf{b}$ replaced respectively with the discretized versions of the correction $\widetilde{\delta u}_k$ and residual $-(r(u_k) + D \partial_x^2 u_k)$. The tridiagonal matrix Eq. (12.64) has the same matrix elements Eqs. (12.57), (12.59), and (12.61), provided we everywhere replace Δt with the value -1 and in the diagonal matrix elements $m_{n,n}$ we replace the expression $1 - r_0 \Delta t$ with the expression $1 - 3u_k^2(x_n)$. Thus we can carry out one iteration of Newton's method for Eq. (12.85) by solving a tridiagonal set of linear equations. This requires a computational effort that increases linearly with the number of mesh points N.

12.5 Conclusion

In this chapter, we have discussed the numerical solution of two of the more important mathematical problems associated with pattern-forming systems: how to integrate the evolution equations that describe how a system changes over time, and how to find stationary states of the evolution equations. Our discussion is too brief and incomplete to transform you into a practicing computational scientist, but you should now have a sense of how a computational scientist prepares a mathematical problem for solution on a computer and how the themes of floating point arithmetic, iteration, Taylor series, and linear algebra can be combined in different ways to derive algorithms that solve the resulting discretized problems.

The enormous improvements in computer hardware combined with the discovery and implementation of many efficient numerical algorithms helps to explain why pattern formation has blossomed as a branch of science over the last twenty years. With current techniques and computers, it is fair to say that almost any evolution equation that can be written down can be solved numerically, at least under restricted conditions of reduced dimensionality or high symmetry. Increasingly, the issue is not whether some pattern-forming system can be simulated but how to understand the flood of spatiotemporal data that is produced by a successful simulation.

As we discussed in Section 12.2, an essential first step in preparing a problem for computer solution is to discretize the fields, equations, and boundary conditions so that a finite number of relations involving a finite number of unknowns is obtained. As you should now appreciate, there are two kinds of errors that arise during this process. The smaller errors are the floating-point errors that arise when a mathematical expression is approximated by one of the finitely many floating-point numbers of a digital computer. This causes a relative error of order the machine precision or larger (about 10^{-16} for computers with 64-bit words). Floating-point errors can be reduced by using computer words with more bits (e.g. some computers support 128-bit words) and sometimes by reorganizing expressions algebraically, e.g. to avoid subtractions between expressions of comparable magnitude. The larger errors are usually the truncation error associated with approximating various derivatives by finite-difference approximations involving finite linear combinations of mesh values. Truncation errors can be reduced by using higher-order polynomial approximations, by using more mesh points within the region of interest, or perhaps by using a Galerkin-based method that converges more rapidly.

In discussing the numerical integration of evolution equations in Section 12.3, we discussed explicit and implicit time-stepping methods and how they can be combined by operator splitting to obtain an algorithm that can take time steps that are only weakly dependent on the spatial resolution. For calculating stationary states of evolution equations, we discussed in Section 12.4 the general strategy of iteration methods of the form Eq. (12.71) and a more specialized iteration algorithm known as Newton's method. The latter always converges if an initial state is known that is sufficiently close to an unknown stationary solution, and further convergence can be quite rapid. However, finding such an initial guess can be difficult for many scientific problems. Many initial guesses may lead to a sequence of vectors that diverge or that converge to some stationary state that is not of interest.

Newton's method reduces the solution of a set of nonlinear equations to the repeated solution of a set of linear equations (see Eqs. (12.76) and (12.77)). As a consequence, improving the simulation of pattern-forming systems largely reduces to advancing the frontiers of numerical linear algebra, especially on parallel computers. Two of the more important techniques for solving large sets of linear equations are called preconditioned conjugate gradient methods and multigrid. The latter is especially impressive since, for certain problems, it attains an optimal efficiency in that the amount of work needed to solve N equations is proportional to N.

We finish this chapter by making a few general observations regarding when and how numerical methods should be used for studying pattern-forming systems.

(i) *If possible, use numerical methods last, not first.* It is best to have a clear mathematical and experimental understanding of a given pattern-forming problem before using a computer to simulate the system. What are the mathematical theorems that identify the necessary and sufficient conditions to have a well-posed problem? What are the basic physical length and time scales? What kinds of instabilities and broken symmetries occur as parameters are varied? What qualitative trends of behavior can be expected as some parameter is varied?

Some prior knowledge of these kinds of questions can be invaluable since, inevitably, computer codes do not work correctly the first time they are written. It can be exceedingly difficult to determine why a code is producing incorrect results, especially since modern programs may be hundreds of thousands of lines long and they may call external libraries whose source code is not available or is incomprehensible. Possible errors might include the following: the evolution equations might be programmed incorrectly (e.g. a minus sign lost when typing in the equations), the equations might be programmed correctly but the algorithm used to solve them might be programmed incorrectly, the equations and algorithm might be programmed correctly but the algorithm is unstable or nonconvergent (e.g. a correct iterative method may not have been iterated sufficiently many times to attain a meaningful answer), and the code and algorithm might be correct and convergent but the visualization software is incorrect.

(ii) *Avoid writing your own software.* Many years of experience are needed to write, to debug, and to validate software. Further, writing and debugging software can be costly in terms of your own time. (Ten lines of correct code per day is considered an impressive achievement for professional programmers!) For these reasons, you should take advantage of previously written code as much as possible, e.g. functions available in computer mathematics environments like Mathematica, Matlab, or Maple, or code available in www.netlib.org. A corollary of not writing your own software is to use the ability of programs like Mathematica and Maple to translate complicated mathematical expressions directly into a computer language like C so that you do not have to type lengthy algebraic expressions in by hand.

(iii) *Gain experience with simple examples.* The algorithms for pattern-forming systems can be as difficult to understand as the evolution equations themselves. You should gain experience and insight by first solving the smallest or simplest versions of problems, especially ones for which analytical or perturbative solutions are available for comparison. Especially valuable is to print out raw data and to visualize intermediate results and fields.

(iv) *Be numerically defensive.* Because many scientists use previously written code, it is extremely important the data provided as input to those codes should be correct or valid. Functions that you write should actively test all known properties of input data and report an error (and possibly terminate) if a test fails. For example, the fluid temperature and density fields must be positive quantities everywhere but could become negative because of numerical errors. Similarly, the number of mesh points, space-time resolutions, and some parameters like the Prandtl number should always be positive parameters and these facts should be tested occasionally for confirmation.

(v) *Get several numerical answers for a given pattern-forming problem.* A single numerical answer is rarely useful since there is no way to judge its correctness. Any important numerical problem should be studied in several different ways and the answers from these different ways compared. For example, any result obtained from a time evolution code should be repeated with finer and coarser space-time resolutions so the results can be verified to have converged. A strong test of correctness is to compare the output of codes that solve the same problem but that use different algorithms (e.g. finite difference versus Galerkin representations) and that were written by different researchers. If possible, numerical output should be compared with available analytical results and experimental data.

(vi) *Talk to experts.* Computational science is a large active field of its own and it can be difficult to penetrate the literature and to identify the current best techniques for studying a given problem. Especially when solving some problem close to a frontier of science (e.g. the simulation of granular flows, cardiac dynamics, or neural tissue), it will be helpful to talk to people who have previously thought about the problem or who have experience from solving other problems.

12.6 Further reading

(i) The book *Numerical Recipes, Third Edition* by Press *et al.* [89] is a popular and readable survey of numerical methods that are widely used by scientists. The book *Numerical Analysis: Mathematics of Scientific Computing* by Kincaid and Cheney [55] gives a more mathematical and rigorous discussion of many of the same algorithms.

(ii) The *Journal of Computational Physics* is a good place to look for discussions, analyses, and applications of various algorithms that are used in pattern-forming systems

Exercises

12.1 Convergence of a finite-difference expression for the average of a field:

(a) Show that the two-point finite-difference expression

$$A_2[u(x)] = \frac{u(x + \Delta x) + u(x - \Delta x)}{2}, \quad \text{(E12.1)}$$

gives a second-order-accurate estimate of the function value $u(x)$ for a sufficiently small mesh size Δx.

(b) In analogy to our discussion of Fig. 12.1, predict whether the limit

$$\lim_{\Delta x \to 0} A_2 - u(x) \tag{E12.2}$$

exists and has the value zero when evaluated in floating point arithmetic. Verify your prediction with a suitable numerical calculation.

12.2 **Accuracy of a three-point approximation for a second derivative:** Show that the three-point finite-difference approximation on the left of Eq. (12.14) also provides an estimate of the second derivatives $u''(x_{n\pm 1})$ at coordinates $x_{n\pm 1}$. Is the error still second order in Δx at these points?

12.3 **Finite-difference approximation for a mixed partial derivative.** Find a second-order-accurate finite-difference approximation for the second-order *partial* derivative

$$\left. \frac{\partial^2 u}{\partial x \, \partial y} \right|_{(x_m, y_n)} \tag{E12.3}$$

of the field $u(x, y)$ at the mesh point (x_m, y_n). Your finite-difference expression will be some linear combination of field values $u_{m',n'} = u(x_{m'}, y_{n'})$ close to the point (x_m, y_n) where the two-dimensional uniform mesh is defined by $(x_m, y_n) = (m \, \Delta x, n \, \Delta y)$ for arbitrary integers m and n and spatial resolutions Δx and Δy.

(a) Derive and write down the leading term in the error.
(b) For some nontrivial function $u(x, y)$ whose partial derivatives are easily calculated, verify the correctness of your finite-difference expression by creating a plot analogous to Fig. 12.1, in which you show the relative error converges to zero in the limit that the spatial resolution $h = \Delta x = \Delta y$ of a square mesh goes to zero. Note: A "nontrivial function" means that you want to avoid the functions 1, x, y, and xy since the error will be zero for these choices. Can you explain why?

12.4 **The discretization error for the two-dimensional Laplacian on a square mesh is not rotationally invariant:** Continuing the previous example, consider a two-dimensional field $u(x, y)$ that is discretized on a *square* mesh with values $u_{m,n} = u(mh, nh)$, where $h = \Delta x = \Delta y$ is the mesh spacing. Using Taylor series, derive the lowest-order error of the following five-term finite-difference approximation

$$\frac{1}{h^2}(u_{m+1,n} + u_{m-1,n} + u_{m,n+1} + u_{m,n-1} - 4u_{m,n}), \tag{E12.4}$$

for the two-dimensional Laplacian $(\nabla^2 u)_{m,n}$ at point (x_m, y_n).

The Laplacian operator $\nabla^2 = \nabla \cdot \nabla$ is rotationally invariant as the dot-product of two vectors, and indeed ∇^2 retains the same form when the coordinate system is rotated by an arbitrary angle θ around the origin. Explain why the lowest-order error of the finite-difference approximation Eq. (E12.4) is *not* rotationally invariant. Thus if this five-term discretization is used to simulate an isotropic spatiotemporal problem with propagating waves (e.g. the wave equation $\partial_t^2 u = c^2 \nabla^2 u$ or some model of an isotropic excitable medium), the waves will propagate with slightly different speeds in different directions because of the lattice discretization.

Can you find a nine-term finite-difference approximation of ∇^2 whose discretization error is rotationally invariant to lowest order?

12.5 **Why finite-difference approximations to derivatives converge toward and then diverge from their limits:** Assume that the floating point value of a field $u(x)$ can be written in the form $u(x)(1 + \epsilon_x)$, where, ϵ_x is a small quantity that varies with x and that is bounded in magnitude by the machine precision ϵ_{mach} (which is of order 10^{-16} for 64-bit computer words). With this assumption, explain in detail why the finite-difference approximations $\delta_{m,k}$ in Fig. 12.1 converge toward and then diverge from their limits:

(a) Explain why $\delta_{m,k} u$ diverges as Δx^{-k} as $\Delta x \to 0$.
(b) Explain why the minimum in the relative error for $\delta_{m,k}$ occurs approximately for

$$\Delta x_{min} \approx \left(\epsilon_{mach} \frac{|u(x)|}{|u^{(m+k)}(x)|}\right)^{1/(m+k)} \approx \epsilon_{mach}^{1/(m+k)}, \quad \text{(E12.5)}$$

if we can assume that the function u and its first few derivatives are of order one in magnitude. Is Eq. (E12.5) consistent with Fig. 12.1?
(c) For sufficiently small Δx, explain why the relative error can become constant with magnitude 1. This is the case for the $\epsilon_{2,1}$ and $\epsilon_{4,1}$ curves in Fig. 12.1.

12.6 **Time-step stability condition for the Crank–Nicolson algorithm:** For the Crank–Nicolson method applied to Eq. (12.23) with **M** diagonalizable and having negative real eigenvalues $\lambda_n < 0$, show that the inequalities corresponding to Eq. (12.50) for the backward-Euler algorithm now take the form:

$$\left|\frac{1 + \frac{\Delta t}{2}\lambda_n}{1 - \frac{\Delta t}{2}\lambda_n}\right| < 1. \quad \text{(E12.6)}$$

Verify that this inequality is true for all Δt so that the time step is constrained only by accuracy.

12.7 **Condition number of a mathematical problem:** Consider a function $y = f(x)$ that produces an output number y for some input number x. If x is perturbed by a small amount δx, this will lead to a perturbation $\delta y = f(x + \delta x) - f(x) \approx f'(x)\delta x$ in the output. The condition number C of the function $f(x)$ at the value x is defined to be the magnitude of the ratio of the relative error of the output $\delta y/y$ to the relative error of the input $\delta x/x$:

$$C = \left|\frac{\delta y/y}{\delta x/x}\right| \approx \left|\frac{xf'(x)}{f(x)}\right|. \tag{E12.7}$$

A mathematical problem is said to be ill-conditioned if the condition number becomes large. In particular, if the condition number C becomes so large that $|\delta x/x|C = |\delta y/y| \approx 1$, then $\delta y/y \approx 1$ and there are effectively no significant digits in y, which therefore should not be used in further calculations. Since round-off errors produce relative errors of magnitude at least $\epsilon_{mach} \approx 10^{-16}$ for 64-bits words, *a mathematical calculation on a digital computer is in serious trouble if the magnitude of C becomes comparable to $1/\epsilon_{mach} \approx 10^{16}$, no matter which algorithm is used to evaluate the function f.*

As a matter of course, the order of magnitude of the condition number of any calculation carried out on a computer should be estimated – at least empirically – to determine whether the calculation is well-conditioned. This can be done by making two calculations, one with the original data and one with each value of the original data perturbed by a multiplicative factor $1+\epsilon$ where $\epsilon \ll 1$ is a specified relative error of order the known error of the input (e.g. this could be the relative error associated with experimental measurements or with truncation errors Eq. (12.12) and are typically orders of magnitude larger than round-off errors). The relative error of the output can then be obtained and the magnitude of the condition number estimated. If a calculation is ill-conditioned for all input values over the range of interest, then that algorithm should be replaced by an alternative one if possible.

(a) Using the formula Eq. (E12.7), determine for what values of x the following functions f are ill-conditioned: x^α, $\log(x)$, $\sin(x)$, and $\sin^{-1}(x)$.

(b) Consider the problem of finding a zero x^* of a mathematical function $f(x)$ so that $f(x^*) = 0$. The perturbations caused by evaluating $f(x)$ in floating point arithmetic can often be interpreted as adding some small new function $\epsilon g(x)$ to the unperturbed function f where ϵ is a small number and $g(x)$ is a known function of order 1 in magnitude. Thus finding numerically a root r of f is equivalent to finding an exact mathematical root of the related equation $f(x) + \epsilon g(x) = 0$.

1. By differentiating the expression $f(r) + \epsilon g(r) = 0$ implicitly with respect to the parameter ϵ, show that the condition number of the root $r = r(\epsilon)$ is

$$C \approx \left| \frac{\epsilon g(r)}{rf'(r)} \right|, \quad \text{(E12.8)}$$

to lowest order in the small quantity ϵ.

2. Consider the 20th-order polynomial $f(x) = \prod_{i=1}^{20}(x-i)$ whose roots are the integers $1, 2, \ldots, 20$ and consider making a small perturbation of size ϵ in the coefficient of x^p, i.e. choose $g(x) = x^p$. Show that the condition number for a root r has magnitude $C \approx \epsilon r^{p-1}/((r-1)!(20-r)!)$.

Note: This expression attains a largest value of $C \approx 10^{10}\epsilon$ for $r = 16$ and $p = 20$. Thus if there is an error in the 10th significant digit of the coefficient of x^{20} in the evaluation of $f(x)$, no significant digits can be expected in the root $r = 16$! In fact, such a tiny perturbation will cause many of the roots of this polynomial to become complex, with imaginary parts of order one! So this innocent-looking polynomial is a highly ill-conditioned function of its coefficients.

12.8 **Condition number of a symmetric matrix:** In Section 12.2.1, we saw that a nonzero determinant of a matrix **M** could round to zero, incorrectly suggesting that the matrix is singular. Numerical analysts have discovered that a superior numerical criterion to determine if a matrix is singular is whether its condition number cond(**M**) is large. Roughly speaking, cond(**M**) is the ratio of the order of magnitude of a relative error produced in a solution $\mathbf{x} = \mathbf{x}[\mathbf{M}, \mathbf{b}]$ (for a set of linear equations $\mathbf{Mx} = \mathbf{b}$) to the order of magnitude of a relative error in the matrix elements of **M** or in the components of the vector **b**. In the following, we simply introduce a plausible definition for the condition number of a symmetric matrix and let you explore its implications. Further information can be found in most books on numerical analysis.

A theorem of linear algebra says that an $N \times N$ symmetric real matrix **S** can be diagonalized to give an $N \times N$ diagonal matrix **E** of N real eigenvalues e_i of **S**. Let us relabel the eigenvalues to be in decreasing order of magnitude: $|e_1| \geq |e_2| \geq \cdots \geq |e_N|$. The condition number cond(**S**) of the symmetric matrix **S** is then defined to be the ratio of the largest eigenvalue magnitude to the smallest eigenvalue magnitude[17]

$$\text{cond}(\mathbf{S}) = \frac{|e_1|}{|e_N|}. \quad \text{(E12.9)}$$

[17] There are numerical algorithms that can estimate the condition number of a matrix quickly without having to calculate any eigenvalues. The condition number is therefore a practical as well as useful concept.

This definition implies that $1 \leq \text{cond}(\mathbf{S}) \leq \infty$.

(a) Explain why $\text{cond}(\mathbf{S}) = \infty$ for a singular matrix.
(b) Show that $\text{cond}(c\mathbf{S}) = \text{cond}(\mathbf{S})$, i.e. the condition number is scale invariant since multiplying a matrix by a constant c does not change its condition number.

In contrast, $\det(c\mathbf{S}) = c^N \det(\mathbf{S})$ so that the determinant is greatly changed after scaling a matrix by a constant. This scaling behavior explains why the determinant of a matrix can easily underflow (round to zero) or overflow (round to infinity) and so is a poor way to test for whether a matrix is singular.

(c) Using Mathematica or some similar program that allows the easy computation of the eigenvalues of a matrix, compute numerically the condition number $\text{cond}(\mathbf{T})$ of the symmetric tridiagonal matrix $\mathbf{T}$ that arises from solving the one-dimensional Poisson equation $\partial_x^2 u = r$ numerically on the interval $[0, 1]$ with boundary conditions $u(0) = u(1) = 0$, using a uniform spatial mesh and the three-point second-order-accurate finite-difference approximation Eq. (12.14). (Etude 12.5 explains how to derive the matrix elements of $\mathbf{T}$.) By plotting the condition number versus N^2 where N is the number of mesh points, verify that $\text{cond}(\mathbf{T}) \propto N^2$ for large enough N. This implies that the matrix becomes more ill-conditioned (more singular) as the mesh size $\Delta x = 1/N$ decreases.

12.9 **Fourier method for analyzing the stability of explicit and implicit time integrators (von Neumann stability method):** A Fourier analysis shows that all solutions $u(x, t)$ of the one-dimensional diffusion equation on the real line:

$$\partial_t u = D \partial_x^2 u, \quad D > 0, \tag{E12.10}$$

decay asymptotically to zero as $t \to \infty$. This is true since each Fourier mode $u_k(x, t) = c_k(t) \exp(ikx)$ of wave number k and coefficient $c_k(t) = c_k^0 \exp(-Dk^2 t)$ (where c_k^0 is some complex constant related to the initial state $u(x, 0)$) decays separately to zero.

Fourier analysis[18] can also be used to deduce conditions on the space-time resolutions Δx and Δt so that the same property of asymptotic decay (stability) holds for numerical solutions $\tilde{u}_{n,i} \approx u(x_n, t_i)$ generated by some

---

[18] The following technique is known in the numerical pde literature as a von Neumann stability analysis.

numerical integration scheme for Eq. (E12.10). Thus consider the forward-Euler algorithm for Eq. (E12.10)

$$u_{n,i+1} = u_{n,i} + \frac{\Delta t \, D}{\Delta x^2}(u_{n-1,i} - 2u_{n,i} + u_{n+1,i}), \quad \text{(E12.11)}$$

which we obtain by setting the reaction rate $r(u)$ to zero in Eq. (12.38). Because Eq. (E12.11) is linear in the mesh values $u_{n,i}$ and the coefficients are constants, any solution of Eq. (E12.11) is a superposition of discrete Fourier modes of the form

$$u_{n,i} = g_k^i e^{Ikn\Delta x}, \quad \text{(E12.12)}$$

where I is a square root of -1,[19] k is a real wave number, and the quantity g_k is called the growth factor since its magnitude determines whether the Fourier mode will grow ($|g_k| > 1$) or decay ($|g_k| < 1$) over time i. The asymptotic stability of Eq. (E12.11) can therefore be determined mode by mode.[20]

(a) By substituting Eq. (E12.12) into Eq. (E12.11), show that the growth factor g_k is given by

$$g_k = 1 - \frac{4 \Delta t \, D}{\Delta x^2} \sin^2\left(\frac{k \Delta x}{2}\right). \quad \text{(E12.13)}$$

(b) Show that $|g_k| < 1$ for all wave vectors k (stability) if and only if

$$\Delta t < \frac{1}{2D}\Delta x^2. \quad \text{(E12.14)}$$

This is a relation of the form Eq. (12.22) with $C = 1/(2D)$ and $\alpha = 2$.

(c) Using a similar stability analysis, show that a time integration method of Eq. (E12.10) based on the Crank–Nicolson method Eq. (12.46) is always stable.

(d) Using a similar stability analysis, show that the time integration algorithm

$$u_{n,i+1} = u_{n,i-1} + \frac{2 \Delta t \, D}{\Delta x^2}(u_{n-1,i} - 2u_{n,i} + u_{n+1,i}), \quad \text{(E12.15)}$$

---

[19] This is the only place in the book where we do not use i to represent the square root of -1, we want to continue to use the index i to label spatial mesh points.

[20] Eq. (E12.11) is a coupled map lattice of the sort discussed in Exercise 2.8 on page 91. Deriving a condition for asymptotic decay of an arbitrary solution is therefore the same as studying the linear stability of the fixed point $u_{n,i} = 0$.

for Eq. (E12.10), which is obtained by using the second-order-accurate discretization Eq. (12.13) to approximate $\partial_t u$, is *unstable* for any choice of time step Δt.

Thus one should not be too greedy for higher accuracy (second- instead of first-order) since the resulting algorithm is useless. A time integration method based on Eq. (12.13) is called a leap-frog method since the discretized time derivative "leaps" from time level $i-1$ over the present time level i to time level $i+1$. Explicit leap-frog methods tend to be unstable for pattern-forming evolution equations with a diffusion term. However, leap-frog is often useful for conservative dynamical systems such as the Newtonian evolution equations associated with a plasma of classical charged point particles, planets in the Solar System, or stars in a galaxy.

12.10 **Discretization and solution of a constant-coefficient one-dimensional biharmonic problem for the Swift–Hohenberg and Kuramoto–Sivashinsky models:** Several mathematical models of pattern-formation and spatiotemporal chaos involve biharmonic operators, e.g. the Swift–Hohenberg equation and the Kuramoto–Sivashinsky equation discussed in Chapter 5. When operator splitting is used to integrate the corresponding evolution equations in one-space dimension, an implicit algorithm (backward Euler or Crank–Nicolson) leads to a constant-coefficient generalized biharmonic equation of the form:

$$\left(\partial_x^4 + b_1 \partial_x^2 + b_2\right)\tilde{u}(x) = r(x), \qquad (E12.16)$$

for the field $\tilde{u}(x)$ at time t_{i+1}. Here b_1 and b_2 are constant coefficients and $r(x)$ is a known function. If Eq. (E12.16) is solved on an interval $[0, l]$ of length l, then typical boundary conditions on $\tilde{u}$ would be

$$\begin{aligned}&\text{for } x = 0: \quad \tilde{u} = c_1 \quad \text{and} \quad \partial_x \tilde{u} = d_1,\\ &\text{for } x = l: \quad \tilde{u} = c_2 \quad \text{and} \quad \partial_x \tilde{u} = d_2,\end{aligned} \qquad (E12.17)$$

where the constants c_i and d_i are assumed known.

(a) By using the finite differences Eq. (12.15b) ($\delta_{2,4}$), Eq. (12.14) ($\delta_{2,2}$), and Eq. (12.13) ($\delta_{2,1}$) to approximate the derivatives ∂_x^4, ∂_x^2, and ∂_x (the latter appears in the boundary conditions) on a uniform spatial mesh x_n of $N+1$ points spanning $[0, l]$, and by using appropriate ghost points as discussed in Etude 12.5, derive in full detail the matrix elements of an $N-1 \times N-1$ pentadiagonal matrix **M** and $(N-1)$-dimensional right-hand-side vector **b** such that the linear equations

$$\mathbf{M}\tilde{\mathbf{u}} = \mathbf{b}, \tag{E12.18}$$

describe the resulting discretized mathematical problem for the $N - 1$ unknown interior mesh values $\tilde{u}_n$ with $1 \leq n \leq N - 1$.

(b) Using a computer mathematics program like Mathematica, Maple, or Matlab, write a computer code to solve Eq. (E12.18). Use your code to study the correctness and convergence properties of your discretization. Is your algorithm second-order-accurate in space, i.e. does the relative error of your numerical solution $\tilde{u}$ go to zero as Δx^2?

(c) Write a code that uses operator splitting and the constant-coefficient biharmonic solver that you wrote to integrate the one-dimensional Swift–Hohenberg equation on an interval $[0, l]$ with boundary conditions $c_i = d_i = 0$ in Eq. (E12.17). Use Adams–Bashforth on the cubic nonlinear term and your biharmonic solver on the linear terms. Choose an initial condition of small amplitude positive and negative random numbers and study its evolution for $r = -0.1$ and for $r = 0.1$. Reading Section 12.3.5 will give you some ideas regarding how to identify space-time resolutions Δt and Δx.

12.11 **Thinking about iteration algorithms:**

(a) If $x_{k+1} = g(x_k)$ is a one-dimensional iteration formula such that the inverse function $g^{-1}(x)$ of g exists, explain why either $x_{k+1} = g(x_k)$ or $x_{k+1} = g^{-1}(x_k)$ will converge to a fixed point x^* for some initial condition x_1 sufficiently close to x^*. Is this also true for the vector case Eq. (12.71)?

(b) Determine analytically the conditions for the formulas in Eq. (12.73) to converge to a root of the quadratic equation Eq. (12.72) and test your conclusions numerically. Can both roots of Eq. (12.72) be found with one of these iteration formulas?

12.12 **Solving an implicit time integration step by simple iteration:** The backward-Euler method Eq. (12.45) and Crank–Nicolson method Eq. (12.46) can be written schematically in the form

$$\mathbf{u} = \mathbf{a} + \Delta t\, \mathbf{g}(\mathbf{u}), \tag{E12.19}$$

were **u** is the unknown vector at the next time step, **a** is some known constant vector, $\mathbf{g}(\mathbf{u})$ is some known nonlinear function of **u**, and Δt is the time step. Show that if we write Eq. (E12.19) in the form Eq. (12.71),

$$\mathbf{u}_{k+1} = \mathbf{a} + \Delta t\, \mathbf{g}(\mathbf{u}_k), \tag{E12.20}$$

and if the Jacobian matrix $\partial \mathbf{g}/\partial \mathbf{u}$ has bounded derivatives in the vicinity of the fixed point, then this iterative scheme will always converge provided that Δt is sufficiently small.

Thus one can always solve the nonlinear equations associated with an implicit time-stepping method by simple iteration on the unknown future field values, provided that the time step is sufficiently small and that a good starting value $\mathbf{u}_1$ can be identified. In practice, the time step needed for convergence may be too small and Newton's method will be preferred.

12.13 **Thinking about Newton's method:**

(a) Explain geometrically why Newton's method for the function $f(u) = u^2 + 1$ generates a diverging sequence u_k for any initial value u_1.
(b) With an appropriate numerical experiment, determine whether Newton's method converges rapidly (the number of significant digits roughly doubles per iteration) for a *multiple* root u^* of some function $f(u)$ of a single variable u. Recall that a multiple root has the property that $f'(u^*) = 0$ when $f(u^*) = 0$.

12.14 **Can one do better than Newton's method?** Newton's method is based on the geometric idea of using the line tangent to a function f at a point u_k to obtain an improved estimate u_{k+1} of a zero of f. An even better local approximation of the function f at u_k can be obtained by using more terms in a Taylor series of f about u_k, e.g. the quadratic polynomial that osculates f at u_k.

Derive an iterative algorithm Eq. (12.71) to solve the equation $f(u) = 0$ by using a local quadratic approximation of f. For some suitable test problem, investigate by numerical experiments how rapidly your algorithm converges compared to Newton's method Eq. (12.80a). Discuss whether your method is or is not more useful than Newton's method for finding roots of nonlinear equations.

12.15 **Basin of attraction for Newton's method:** A mathematical theorem assures us that Newton's method will converge for any initial condition u_1 that is sufficiently close to a root u^* of the equation $f(u) = 0$. But what about initial conditions that are not close to a root? Explore this situation by determining the set of all initial conditions u_1 that will converge via Newton's method Eq. (12.80a) to the root $u = 0$ of the function $f(u) = \sin(u)$.

12.16 **Calculating implicit functions and nullclines by Newton's method:** In many scientific problems, two variables u and v may be related implicitly to one another through some algebraic relation of the form $f(u, v) = 0$. It is often the case that one variable, say v, cannot be solved for explicitly in terms of the other variable u although one knows by the implicit function theorem that the function $v = v(u)$ exists over some appropriate domain. If the function $f(u, v)$ appears as the right side of some evolution equation $\partial_t u = f(u, v)$, then the relation $v = v(u)$ is called the nullcline for that equation. You saw in Section 11.2 how nullclines play an important conceptual role in the analysis of two-variable models of an excitable medium.

Newton's method can be used to generate a numerical approximation to the implicit relation $v(u)$ provided that at least one starting point (u_1, v_1) can be found such that $f(u_1, v_1) = 0$.

(a) Assume that a point (u_m, v_m) is known such that $f(u_m, v_m) = 0$ with $m \geq 1$. Derive a Newton method that will converge to a solution v_{m+1} such that $f(u_{m+1}, v_{m+1}) = f(u_m + \Delta u, v_{m+1}) = 0$, where Δu is some sufficiently small increment in u.

(b) Use your algorithm to calculate and then plot a numerical table of the function $v = v(u)$ over the interval $0 \leq u \leq 10$ for the relation

$$f(u, v) = 3u^7 + 2v^5 - u^3 + v^3 - 3 = 0, \qquad \text{(E12.21)}$$

starting with the point $(u_1, v_1) = (0, 1)$ and using an increment $\Delta u = 0.1$. Each value v_m should be calculated to at least five significant digits.

Appendix 1
Elementary bifurcation theory

This appendix provides some of the background for Chapter 4, especially for Eq. (4.14), by reviewing some of the elementary bifurcation theory that is often discussed in an introductory undergraduate course on nonlinear dynamics. Bifurcation theory is concerned with the *change* in the nature of solutions as parameters are varied. The changes can involve changes in the numbers or types of attractors, in the structure of the basins of attraction, or in even more subtle details of the phase space that are not easily detected by experiment. Sufficiently close to the onset of a bifurcation of a fixed point, a combination of a perturbation expansion and of nonlinear changes of variables can reduce the evolution equations to a much simpler dynamical system (usually a few odes) called a normal form. The normal form captures the essential behavior of the evolution equations sufficiently close to the bifurcation point and can be used to classify the possible bifurcations. For our purposes, the classification and associated language (e.g. pitchfork, Hopf, and other kinds) are the more important topics so we do not show how to reduce a set of equations describing a physical system to normal form, which can involve lengthy calculations, even with a computer mathematics program.

We begin our discussion by analyzing the bifurcations of some simple one-variable dynamical systems and then discuss how these systems are related to the normal forms of more complicated evolution equations. The simplest continuous-time dynamical system is a first-order ode in a single variable $u(t)$,

$$d_t u = f(u), \qquad (A1.1)$$

for some given function $f(u)$. The value of u describes the position of the system along some one-dimensional line. Because of the uniqueness theorem for initial value problems, the only possible attractors (bounded nontransient behaviors) in the one-dimensional phase space of Eq. (A1.1) are constants (fixed points) u_s that satisfy $f(u_s) = 0$.

Bifurcation theory classifies the changes in the nature of the solutions to Eq. (A1.1): fixed points may appear or disappear, new fixed points may grow out of

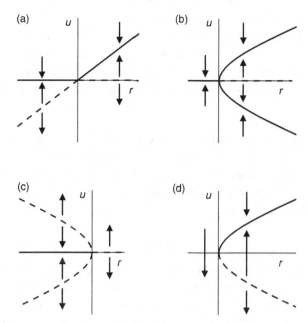

Fig. A1.1 Simple bifurcations of Eq. (A1.1) as the parameter r is varied: (a) transcritical; (b) forward pitchfork; (c) inverted pitchfork; (d) saddle node. The heavy full lines denote stable stationary solutions u, and the dashed lines denote unstable solutions. The arrows show directions of the flow du/dt at some particular values of r.

the old ones, fixed points may collide leading to a reduced number, or the stability of fixed points may change. In the vicinity of the bifurcations, the stability of the fixed points can be related to the type of bifurcation. The types of behavior are simply depicted in the bifurcation diagrams in Fig. A1.1, and by the corresponding normal forms that we list below. In these, we suppose that the bifurcation occurs near $u = 0$, which can always be arranged by a suitable change of variables. There then turn out to be four normal forms which are a Taylor expansion in small u together with rescaling t and u to eliminate unnecessary constants.[1]

The normal forms for bifurcations of fixed points in a one-dimensional phase space are then the following:

Transcritical bifurcation: A stable and unstable fixed point collide, with an exchange of stability (Fig. A1.1(a)). The corresponding normal form is

$$d_t u = ru - u^2. \tag{A1.2}$$

---

[1] In more complicated situations than our one-dimensional example, *nonlinear* changes of variables may also be needed to reduce the form to one of a few canonical ones, see for example the review article by Crawford [23].

Here r is the bifurcation parameter, and the bifurcation occurs at $r = 0$. The nature of the solutions is most easily seen from the graphical representation Fig. A1.1(a). For $r < 0$, there is a stable fixed point at $u = 0$ and an unstable one at $u = r$. At $r = 0$, the two fixed points collide and "exchange stability" so that the $u = 0$ fixed point becomes unstable and the $u = r$ fixed point is now stable. Note that the sign of the quadratic term in Eq. (A1.2) is unimportant since changing the sign does not change the qualitative behavior.

Forward pitchfork bifurcation: For a system that is unchanged under the substitution $u \to -u$ (inversion symmetry with respect to u), the quadratic nonlinearity in Eq. (A1.2) must be absent (the symmetry of the physical system translates into the invariance of the dynamical equation under this transformation) and the bifurcating solutions must be symmetric about $u = 0$ as shown in Fig. A1.1(b). The normal form is

$$d_t u = ru - u^3. \tag{A1.3}$$

For $r < 0$ there is a single stable fixed point at $u = 0$. For r positive the $u = 0$ fixed point remains, but is unstable, and two new, stable fixed points $u = \pm\sqrt{r}$ develop. As well as "forward," the adjectives normal, continuous, or second-order are used to describe this type of pitchfork bifurcation (the latter in analogy with phase transitions in equilibrium physics).

Backward pitchfork bifurcation: This is the same as the previous example, except that the cubic nonlinearity has the opposite sign

$$d_t u = ru + u^3. \tag{A1.4}$$

The $u = 0$ fixed point is again stable for $r < 0$ and unstable for $r > 0$, but now the additional fixed points exist for $r < 0$ and are unstable (Fig. A1.1(c)). The connection between the form of the bifurcation (new fixed points for $r > 0$ or $r < 0$) and the stability of these fixed points, shown in this and the previous example, demonstrates the power of bifurcation theory. This type of pitchfork bifurcation is also described as inverted. Other names are first-order, or discontinuous, coming from the jump in the physical solution that occurs at $r = 0$ from $u = 0$ to some finite value where nonlinearity eventually saturates the growth.

Saddle-node bifurcation: In the bifurcations considered so far, we have always had at least one fixed point present for all r. In the saddle-node bifurcation (Fig. A1.1(d)), two new fixed point solutions are created out of none. A saddle-node bifurcation can occur at any value of u, but again we shift the point of appearance to $u = 0$ by a change of variables. The normal form is

$$d_t u = r - u^2. \tag{A1.5}$$

There is no solution for $r < 0$, and two solutions $u = \pm\sqrt{r}$ develop for $r > 0$, one of which is stable and one unstable. Changing the sign of the u^2 term gives an analogous bifurcation such that two fixed points collide and annihilate with increasing r.

Notice that both the transcritical and backward pitchfork bifurcation normal forms lead to values of r for which the value of u diverges to large values without saturation. This unphysical behavior can be corrected by extending the Taylor expansion to higher order. For example, for the transcritical bifurcation we might then find

$$d_t u = ru + u^2 - gu^3 \tag{A1.6}$$

with g a positive constant. In this case the unstable solution $u = r$ for negative r "bends around" at a saddle-node bifurcation and becomes stable, as in Fig. A1.2(a). Now for $r > 0$, the growth of a negative perturbation in u saturates at a finite value, and there is no divergence to infinity. Note however that the saturation value of u is not in general small so that the truncation of $f(u)$ at a third-order Taylor expansion will not in general be reliable, and this procedure must be considered ad hoc rather than controlled. The saturation value is small for small r if the parameter g is also small, and in this case the expansion scheme is reliable.

Similarly for the backward pitchfork bifurcation, continuing the expansion to higher order (remembering only odd powers are allowed by the $u \to -u$ symmetry) might lead to

$$d_t u = ru + u^3 - gu^5. \tag{A1.7}$$

Again if g is positive, saddle-node bifurcations lead to stable branches of solutions away from $u = 0$ and no divergence to infinity as shown in Fig. A1.2(b). This saturation of the growth is an uncontrolled approximation unless g happens to be small. In either case, if the parameter g turns out to be negative, the additional nonlinear term further amplifies the growth, and no saturation to a fixed point occurs in the equation. To eliminate the unphysical divergence to infinity, we would have to rely on higher-order terms in the Taylor expansion, or to use the full functional form of $f(u)$.

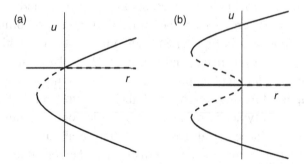

Fig. A1.2 Bifurcations with higher-order stabilizing terms added to the Taylor expansion: (a) transcritical; (b) inverted pitchfork.

For higher-dimensional dynamical systems, if the instability of a fixed point solution occurs through a single eigenvalue of the linear stability analysis that passes from negative to positive values, then the possible bifurcation types are again restricted to the same four: transcritical, forward and backward pitchfork, and saddle-node. This restriction occurs because, for control parameters near the bifurcation value and for values of the dynamical variables near the fixed point value, the dynamics can be reduced to the one-dimensional dynamics along the direction of slow dynamics corresponding to the eigenvector of the single eigenvalue that passes through zero. The dynamics in the other dimensions evolve rapidly enough to "adiabatically follow" the slow dynamics of the weakly stable or unstable direction, a result known as the center manifold theorem. Such a bifurcation is known as simple. The dynamics near the bifurcation point is the same as in the one-dimensional case, but in the higher-dimensional case, much richer dynamics is generally possible. As a consequence, if bifurcation theory does not predict the saturation onto a new stable fixed point near the original fixed point but instead predicts growth to values far away (e.g. the transcritical or backward pitchfork cases), then the dynamical state after the bifurcation may be stationary, periodic, quasiperiodic, or chaotic.

One additional type of bifurcation from a fixed point is typical in higher-dimensional dynamical systems. Instead of a single eigenvalue passing from (real) negative to positive values, a complex conjugate pair of eigenvalues may evolve from having negative real parts (exponential decay) to positive real parts (exponential growth). (For a real dynamical system, eigenvalues that are not real must come in complex conjugate pairs since the eigenvalues satisfy a characteristic polynomial with real coefficients.) This is the case of the oscillatory instability discussed in Chapter 2, and in the context of dynamical systems is known as a Hopf bifurcation. The normal form is now

$$d_t u = ru \mp |u|^2 u \tag{A1.8}$$

with $u(t)$ a complex variable that represents the magnitude and the phase of the oscillating solution that grows from the bifurcation point. The bifurcation is forward for the negative sign in Eq. (A1.8) leading to a stable oscillating solution of amplitude $|u| = \sqrt{r}$ for positive r, and is backward for the positive sign, with an unstable oscillating solution of amplitude $|u| = \sqrt{-r}$ for negative r.

The five simple bifurcation types listed above are the only bifurcations from a fixed point that will typically occur when a single control parameter is varied. Typical here means "in nearly all cases" and "in the absence of symmetry." A degenerate bifurcation occurs when more than one real eigenvalue or complex pair crosses the real axis together. Without symmetry, such a situation is a coincidence, and usually varying a second control parameter will eliminate this coincidence.

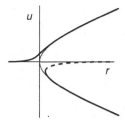

Fig. A1.3 Imperfect forward pitchfork bifurcation described by the normal form Eq. (A1.9) for a small value of h. The heavy curves show the stable (solid) and unstable (dashed) solutions for the imperfect bifurcation $h > 0$. The light curve is a reminder of the behavior of Fig. A1.1(b) for the perfect bifurcation $h = 0$.

Such bifurcations are called codimension-n bifurcations if we need to vary n control parameters to arrange the coincidence.

For pattern forming instabilities, the system almost always has symmetries from the very nature of the problem, so that we must nearly always look at the case of a degenerate bifurcation and go beyond the simplest classification described above. In simple examples, the additional complications are not too great and can be studied in an ad hoc way, as we discuss for example in Section 4.3. In more complicated examples, it is advantageous to use the combined mathematical methods of bifurcation theory and group theory know as equivariant bifurcation theory. This goes beyond the scope of the present book.

Bifurcations show a sharp transition between solutions, at a particular parameter value (the critical value). In practice, small experimental imperfections that violate the conditions of the theory may smooth out the sharpness. If there is still the appearance of a transition between different solutions at a coarse view, but on more delicate inspection smooth behavior is apparent, we call the bifurcation imperfect.

For example, in a Rayleigh–Bénard convection experiment, a small imperfection in the junction between the top or bottom plate and a side wall (for example, the glue holding the two together may have a different thermal conductivity than either wall) may lead to horizontal temperature gradients that tend to drive a small circulating fluid motion near the sidewall for any temperature difference across the layer of fluid, even far below the critical Rayleigh number for the onset of convection in an infinite geometry. The result is that a careful measurement of some order parameter such as the total heat transport (Nusselt number) shows that the transition from the motionless conducting fluid to convection does not occur abruptly for a single value of the temperature difference across the plates. Instead the order parameter has a small but nonzero value for parameter values below the theoretical critical value, and then smoothly increases across the critical value until, well above the critical value, it merges with the values expected in the idealized perfect case.

An imperfect bifurcation can often be modeled by adding a small constant term to the normal form. For example, an imperfect forward pitchfork bifurcation is described by the expression

$$d_t u = ru - u^3 + h, \tag{A1.9}$$

where the constant h is small. The solutions to this equation and their stability are shown in Fig. A1.3 for $h > 0$. The $u \approx 0$ solution for large negative r now connects continuously with the positive solution for $r > 0$. The negative u solution for $r > 0$ forms a separate branch with a change of stability at a saddle-node bifurcation. The crucial feature of an imperfect bifurcation is that small imperfections can have a relatively large effect near the critical parameter value, even though the effects may be negligible far away from this point.

Similar analyses can be done for imperfect transcritical and Hopf bifurcations. In bifurcations such as the pitchfork bifurcation where a symmetry is broken, the imperfection term h must correspond to an external physical effect that violates the symmetry. This makes it easier to predict whether a suspected non-ideal feature of an experiment will lead to a perfect or imperfect bifurcation.

A1.1 Further reading

(i) For an introduction to bifurcations in low-dimensional dynamical systems see *Nonlinear Dynamics and Chaos* by Strogatz [99].

(ii) A more advanced discussion of bifurcation theory is "Introduction to bifurcation theory" by Crawford [23].

(iii) A recent book emphasizing the unifying role that symmetry plays in bifurcation theory and dynamics and pattern formation in general is *The Symmetry Perspective: from Equilibrium to Chaos in Phase Space and Physical Space* by Golubitsky and Stewart [39].

(iv) A nice experimental study that illustrates many of the points in this appendix including subcritical and imperfect bifurcations is the paper "Tricritical phenomena in Rotating Couette-Taylor Flow" by Aitta *et al.* [2].

Appendix 2
Multiple scales perturbation theory

This appendix describes and gives some examples of multiple scales perturbation theory. This is a widely used technique in applied mathematics, physics, engineering, and other fields that systematically yields approximate solutions to ordinary and partial differential equations for which there is a small parameter ε such that the mathematical problem can be solved without too much effort when the small parameter is set to zero. In the context of pattern formation, the formalism provides a systematic way to analyze the spatiotemporal behavior of fields near the supercritical instability of a spatially uniform state. Near such a bifurcation, for reasons clarified by the multiple scales theory, the physical system can be accurately analyzed as a slowly varying spatiotemporal modulation of a fast oscillatory behavior in space or in time.

The perturbation theory is based on two key features. First is the idea of multiple scales, which is to introduce scaled space and time coordinates that capture the slow modulation of the pattern. These new scaled variables will be treated as mathematically independent of the original variables that are used to describe the pattern state itself. The second key feature is the use of what are known as solvability conditions. In the formalism, these conditions arise as mathematical statements that prevent a resonant driving of a higher-order term by a lower-order term that would cause the perturbation method to fail after a short time.[1] The lowest-order nontrivial solvability condition often ends up being an evolution equation for a slowly varying multiplicative factor of the unperturbed solution, what we have called an amplitude equation.

In the next section, we first discuss the theory in some generality and then we will demonstrate the theory for two concrete examples: for a set of three ordinary

---

[1] Roughly speaking, the lowest-order multiple scales equation has a form analogous to a harmonic oscillator being driven by a resonant force, $d_t^2 A_1 + \omega^2 A_1 = f(A_0)\cos(\omega t)$, where A_1 is a higher-order modulatory factor, A_0 is the zeroth-order modulatory factor that depends only on the slowly varying scaled variables, and $f(A_0)$ is some expression involving A_0 and its derivatives. To avoid a resonant response, we would set $f(A_0) = 0$, which then gives the lowest-order solvability condition.

nonlinear differential equations (the Lorenz equations), and for a representative one-dimensional partial differential equation whose uniform state undergoes a type I-s instability (the Swift–Hohenberg equation, see Section 5.1). In the final sections of this appendix, we also apply the method to the derivation of phase equations, and more briefly to other situations. In all these cases, we will only work out the lowest-order nontrivial approximation. The multiple scales method can be carried out to higher order but the mathematics becomes cumbersome and it is often difficult to extract scientifically useful insights from the higher-order equations because of the many terms that appear.

The multiple scales method differs in one important way from perhaps more familiar forms of perturbation theory such as the perturbation theory that is used in quantum mechanics to approximate the effect of making a small change to the energy potential in the Schrödinger equation (which we assume can be solved exactly for zero perturbation). The difference arises because, in the present case, the expansion is about a solution to the linear approximation which has arbitrary size since the equation is linear. Thus we are expanding about a solution with free parameters (the magnitude of the linear solution) which must somehow be fixed within the perturbation scheme.

A2.1 Multiple scales

To give an overall sense of what is involved with a multiple scale perturbation analysis for a spatiotemporal pattern-forming system, we consider a rotationally invariant system for which the extended directions $\mathbf{x}_\perp$ are infinite and that has a base state $\mathbf{u}_b(\mathbf{x}_\parallel)$ that becomes unstable to a type-I-s instability. An arbitrary perturbation $\mathbf{u}_p(\mathbf{x}, t) = \mathbf{u}(\mathbf{x}, t) - \mathbf{u}_b(\mathbf{x}_\parallel)$ about the solution $u_b(\mathbf{x}_\parallel)$ will satisfy a set of nonlinear pdes[2] of the form

$$\partial_t \mathbf{u}_p = \hat{\mathbf{L}} \mathbf{u}_p + \hat{\mathbf{N}}[\mathbf{u}_p]. \tag{A2.1}$$

Here $\hat{\mathbf{L}}$ is a linear differential-matrix operator that depends on the control parameter p, and $\hat{\mathbf{N}}$ is an operator that collects all the terms that are nonlinear in $\mathbf{u}_p$. Since a type-I-s instability has a purely real eigenvalue near onset, the linear instability of the uniform state is signaled by a zero eigenvalue of the operator $\hat{\mathbf{L}}$ for a perturbation at the critical wave vector $\mathbf{q}_c$. The rotational invariance of the system allows us to choose our coordinate system with x aligned along $\mathbf{q}_c$ and then the onset of

---

[2] Note that we are using $\hat{\mathbf{N}}$ here to denote just the nonlinear terms, rather than the whole of the right side as in Chapter 2.

instability is defined by

$$\hat{L}_0\left[e^{iq_c x}\bar{\mathbf{u}}_0(\mathbf{x}_\|)\right] = 0, \qquad \hat{L}_0 = \hat{L}\Big|_{p_c}, \qquad (A2.2)$$

with $\hat{L}_0$ equal to $\hat{L}$ evaluated at $p = p_c$, and we choose some convenient normalization convention for $\bar{\mathbf{u}}_0(\mathbf{x}_\|)$.[3]

Near threshold, we expect that the solution might saturate at small magnitude. Furthermore, we expect the solution to resemble the linear solution, i.e. to have largely the same combination of fields u_j and the same spatial structure. Thus we look for solutions

$$\mathbf{u}_p = \varepsilon^{s_1}\mathbf{u}_0 + \varepsilon^{s_2}\mathbf{u}_1 + \cdots, \qquad (A2.3)$$

with s_i some increasing set of powers and $\mathbf{u}_0$ given by a slow space and time modulation of the critical solution[4]

$$\mathbf{u}_0 = A_0(X, Y, T)e^{iq_c x}\bar{\mathbf{u}}_0(\mathbf{x}_\|) + \text{c.c.} \qquad (A2.4)$$

It is here that we introduce the multiple scales. Our intuition that A_0 describes slow modulations of the pattern is introduced at the outset through slow space and time variables, which are traditionally written as upper case letters X, Y, and T. These are scaled versions of x, y, and t so that an $O(1)$ change of X, Y, T corresponds to a large change in the physical variables

$$X = \varepsilon^{s_x} x, \qquad Y = \varepsilon^{s_y} y, \qquad T = \varepsilon^{s_t} t, \qquad (A2.5)$$

with the positive powers s_x, s_y, and s_t to be chosen appropriately.

The powers of ε introduced in Eqs. (A2.3)–(A2.5) can be found in one of two ways. The first is to leave them as unknowns, proceed with the expansion, and find what values are needed to get the various terms to balance. The second is to use phenomenological arguments to fix the powers at the outset, with of course the consistency of the formal procedure that follows as a check. We will follow the second approach since it leads to a more transparent development.

Phenomenological arguments suggest the scaling

$$X = \varepsilon^{1/2} x, \qquad (A2.6a)$$

$$Y = \varepsilon^{1/4} y, \qquad (A2.6b)$$

$$T = \varepsilon t. \qquad (A2.6c)$$

---

[3] We are using the notation $\bar{\mathbf{u}}_0$ for the linear onset solution $\bar{\mathbf{u}}_c$ of early chapters to emphasize its role as the zeroth-order term in the perturbation expansion.

[4] We use the symbol A_0 here, rather than $\bar{A}$, because the higher-order terms in Eq. (A2.3) contain terms analogous to Eq. (A2.4) but with amplitudes A_i that are determined by their own amplitude equations. These amplitudes give additional terms in the amplitude that are higher order in $\varepsilon^{1/2}$.

The scaling of the space variables is motivated by the $O(\varepsilon^{1/2})$ width of the wave number band near threshold. As described in Section 6.2.1, the different scalings in the x and y directions are motivated by the different dependence of the growth rate on wave vector perturbations in the x and y directions for stripes in a rotationally invariant system. For $\mathbf{q} = q_c \hat{x} + (k_x, k_y)$, the change in growth rate is quadratic in k_x (hence the $\varepsilon^{1/2}$ scaling of X) but fourth order in k_y (hence the $\varepsilon^{1/4}$ scaling of Y). Furthermore, the amplitude of saturation is expected to be $O(\varepsilon^{1/2})$, and so we write $\mathbf{u}_p$ as an expansion in powers of $\varepsilon^{1/2}$

$$\mathbf{u}_p = \varepsilon^{1/2} \mathbf{u}_0 + \varepsilon^1 \mathbf{u}_1 + \cdots, \quad (A2.7)$$

with $\mathbf{u}_0$ as in Eq. (A2.4) introducing the amplitude A_0.

The strategy is now to substitute Eq. (A2.7) into Eq. (A2.1) and to collect terms at each order in $\varepsilon^{1/2}$. To do this, we need to evaluate the derivatives in $\hat{\mathbf{L}}$ that act on $\mathbf{u}_p$, which is expressed in terms of the multiple scales Eq. (A2.6). To evaluate the derivatives, we use a general rule of differentiation, which is that if we have a dependent variable $y(x)$ and a function $f(x, y)$ that depends on x and y, then the derivative of f with respect to x is

$$\frac{df}{dx} = \frac{\partial f}{\partial x} + \frac{dy}{dx} \frac{\partial f}{\partial y}. \quad (A2.8)$$

This is the situation we have in Eq. (A2.4), with the dependent variable $X(x) = \varepsilon^{1/2} x$. It follows that a spatial derivative acting on $\mathbf{u}_p$ can be written

$$\left(\frac{\partial \mathbf{u}_p}{\partial x}\right)_y = \left(\frac{\partial \mathbf{u}_p}{\partial x}\right)_{X,y} + \varepsilon^{1/2} \left(\frac{\partial \mathbf{u}_p}{\partial X}\right)_{x,y}, \quad (A2.9)$$

or in short

$$\partial_x \to \partial_x + \varepsilon^{1/2} \partial_X, \quad (A2.10)$$

where on the right side the ∂_x will operate on the $e^{\pm i q_c x}$ dependence and the ∂_X will act on the X dependence of the A_i. Extending this scheme we have

$$\partial_y \to \partial_y + \varepsilon^{1/4} \partial_Y. \quad (A2.11)$$

Higher-order derivatives are readily evaluated, e.g.

$$\nabla^2 = \partial_x^2 + \partial_y^2 + \partial_z^2 \to \partial_x^2 + \partial_z^2 + \varepsilon^{1/2}(2 \partial_x \partial_X + \partial_Y^2) + \varepsilon \partial_X^2. \quad (A2.12)$$

The component at order $\varepsilon^{1/2}$ of Eq. (A2.12), which gives $2 i q_c \partial_X + \partial_Y^2$ when acting on a term in $e^{i q_c x}$, will become familiar as the representation at a particular order in ε of the rotationally invariant Laplacian. Similarly the time derivative becomes

$$\partial_t \to \varepsilon\,\partial_T, \qquad (A2.13)$$

since there is no fast time dependence in the base state $e^{iq_c x}$.[5]

A2.2 Solvability conditions

The equations of motion are now formally expanded in the small parameter ε, and the terms at each order collected and equated to zero. We first rewrite the general equations of motion as

$$\hat{\mathbf{L}}\mathbf{u}_p = \partial_t \mathbf{u}_p - \hat{\mathbf{N}}[\mathbf{u}_p]. \qquad (A2.14)$$

The linear part of the evolution equation is then expanded in ε using Eqs. (A2.9)–(A2.13), and in the proximity of the control parameter to onset $p = p_c(1+\varepsilon)$ to give

$$\hat{\mathbf{L}} = \hat{\mathbf{L}}_0 + \varepsilon^{1/2}\hat{\mathbf{L}}_1 + \cdots. \qquad (A2.15)$$

In particular, $\hat{\mathbf{L}}_0$ is the linearization of the equations of motion about the uniform solution $\mathbf{u} = \mathbf{u}_0$ evaluated at $p = p_c$, as in Eq. (A2.2). Note that the $\hat{\mathbf{L}}_i$ for $i > 0$ will typically contain both fast and slow derivatives (e.g. ∂_x and ∂_X).

Equations (A2.3), (A2.4), (A2.9), and (A2.13) are substituted into Eq. (A2.14) and terms at each order in ε are collected. At order $\varepsilon^{(n+1)/2}$, this will generate an equation for the higher-order unknown $\mathbf{u}_n$

$$\hat{\mathbf{L}}_0 \mathbf{u}_n = \text{rhs}. \qquad (A2.16)$$

The symbol rhs (right-hand side) denotes terms evaluated from lower-order calculations that depend on $\mathbf{u}_m$, and therefore depend on the amplitudes A_m, for $m < n$. Note that since there are only slow time derivatives $\partial_t \to \varepsilon\,\partial_T$, the time derivative only appears in rhs as slow time derivatives of the lower-order amplitudes.

As well as leading to the solution for $\mathbf{u}_n$, Eq. (A2.16) actually generates constraints on the rhs that are the solvability conditions, and it is such a solvability condition that leads to the important equation for the as yet unspecified amplitude A_0, and that leads to equations for its higher-order companions A_i when the expansion is continued further. The solvability conditions arise because $\hat{\mathbf{L}}_0$ has a non-empty null space, with at least one eigenvector with zero eigenvalue. We know this is true because the unstable mode has zero growth rate precisely at onset, and the growth rate is the eigenvalue of $\hat{\mathbf{L}}_0$; this is the content of Eq. (A2.2). The

---

[5] If we were to construct the amplitude equation for an oscillatory instability, this would not be true. In that case, we need to include the fast dependence ∂_t in the linear operator $\mathbf{L}$ since this is an essential part of constructing the linear solution.

constraints arise from the fact that Eq. (A2.16) only has finite solutions for $\mathbf{u}_n$ if the expression rhs has no components in this null space. Technically, this can be expressed by the condition that "the rhs is orthogonal to the zero-eigenvalue eigenvector of the adjoint operator $\hat{\mathbf{L}}_0^\dagger$." This is certainly a mouthful, and you may need to remind yourself from a linear algebra text about the mathematics of resolving a vector along basis vectors, the formal definition of the adjoint operator, and why this statement tells us that rhs then has no components along the zero eigenvector of $\hat{\mathbf{L}}_0$. Even after understanding the formal content of the expression, since $\hat{\mathbf{L}}_0$ is in general a matrix differential operator, finding the adjoint and its zero modes is often a difficult calculation in practice. The explicit implementation of the solvability condition in the examples below should help to clarify the concepts.[6]

With the solvability condition satisfied, we can formally invert Eq. (A2.16) to give

$$\mathbf{u}_n = \hat{\mathbf{L}}_0^{-1}(\text{rhs}) + \left[A_n(X, Y, T) e^{iq_c x} \bar{\mathbf{u}}_0(\mathbf{x}_\|) + \text{c.c.} \right], \tag{A2.17}$$

where the second term is the complementary function for the operator $\hat{\mathbf{L}}_0$ (the solution to the homogeneous equation given by setting rhs to zero) and introduces an unknown higher-order amplitude function A_n. The solvability conditions at higher orders eventually lead to equations for these new amplitudes.

This is actually a familiar scheme to those who know secular perturbation theory. We are perturbing about the zeroth-order solution $\mathbf{u}_0$, which however contains a free complex amplitude that corresponds to the arbitrary magnitude of the solution to the linear problem and to the arbitrary position of the stripes. A naive perturbation expansion will lead to corrections to this zeroth-order solution that grow without bound as time increases. Secular perturbation theory eliminates these problem terms by placing constraints on the zeroth-order solution via the solvability condition. We need to choose the "right" zeroth-order solution $\mathbf{u}_0$ in Eq. (A2.7) so that the "correction terms" expressed by the higher-order terms are indeed small.

The actual implementation of the scheme for realistic systems is quite involved, even when the zeroth-order solution is known analytically. We will demonstrate different aspects of the procedure in two examples. First we present an elementary

---

[6] As a simpler matrix case, consider a symmetric real $N \times N$ matrix $\mathbf{A}$ which has eigenvalue and eigenvector pairs $(\lambda_i, \mathbf{e}_i)$ where we think of the eigenvectors $\mathbf{e}_i$ as $N \times 1$ column vectors. Because $\mathbf{A}$ is a symmetric matrix, the eigenvectors can be chosen to be orthonormal, $\mathbf{e}_i^T \mathbf{e}_j = \delta_{ij}$, where $\mathbf{e}_i^T$ is the $1 \times N$ row vector obtained by the transpose operator ^{T}. Then you should be able to verify that the $N \times N$ matrix defined by $\sum_i \lambda_i \mathbf{e}_i \mathbf{e}_i^T$ applied to an arbitrary vector $\mathbf{v} = \sum_j v_j \mathbf{e}_j$ (expressed in terms of the eigenvectors) produces exactly the same result as $\mathbf{A}$ itself and indeed must equal $\mathbf{A}$ if all the eigenvalues are nonzero. Further, if the eigenvalues are all nonzero, then you can verify that $\mathbf{A}^{-1} = \sum_i \lambda_i^{-1} \mathbf{e}_i \mathbf{e}_i^T$. If $\mathbf{A}$ is singular with some zero eigenvalues, we can still define a kind of matrix inverse $\mathbf{B} = \sum_i' \lambda_i^{-1} \mathbf{e}_i \mathbf{e}_i^T$ by excluding the zero eigenvalues from the sum. We can solve the matrix problem $\mathbf{A}\mathbf{x} = \mathbf{y}$ formally by writing $\mathbf{x} = \mathbf{B}\mathbf{y}$ provided that the vector $\mathbf{x}$ lies inside the span of the eigenvectors with nonzero eigenvalues. But then $\mathbf{y} = \mathbf{A}\mathbf{x}$ must also lie in the span of such eigenvectors and this is equivalent to requiring that $\mathbf{y}$ be orthogonal to all the zero eigenvectors. This is the equivalent of the solvability conditions in this context.

introduction using the Lorenz equations. These are three coupled nonlinear odes that played an important role in the development of chaos theory. This first example illustrates the approach in a simple context, the technique of introducing the slow time scale T, and also promotes an understanding of solvability conditions in the context of matrix equations. However since these equations are ordinary differential equations, they do not illustrate the introduction of the slow spatial dependence. The second example on the Swift–Hohenberg equation illustrates the spatial aspects.

A2.3 Amplitude equations

A2.3.1 Lorenz model

As a simple illustration of how to derive amplitude equations by the method of multiple scales, we implement the approach for the Lorenz model. (These are discussed in most nonlinear dynamics texts, e.g. Strogatz [99].) These ordinary differential equations are the starting point for the subsequent analysis – you do not need to understand their derivation to benefit from the following discussion.

The Lorenz equations are three coupled evolution equations for the components of the vector $\mathbf{u}(t) = (X(t), Y(t), Z(t))$

$$d_t X = -\sigma(X - Y), \qquad (A2.18a)$$

$$d_t Y = rX - Y - XZ, \qquad (A2.18b)$$

$$d_t Z = -bZ + XY, \qquad (A2.18c)$$

where $d_t = d/dt$ denotes a time derivative. Briefly, the physical content of the equations is the following. In a severely truncated Galerkin approximation for two-dimensional Rayleigh–Bénard convection, X represents the circulation velocity in the convection, Y the spatially periodic temperature perturbation, and Z the heat transport due to the convection. The control parameter r is the reduced Rayleigh number $r = R/R_c$, σ is the Prandtl number, and $b = 8/3$ is a numerical constant.

A linear stability analysis of Eqs. (A2.18) shows that the simple solution $X = Y = Z = 0$ (which corresponds to the spatially uniform motionless non-convecting state) undergoes a bifurcation at the critical value $r_c = 1$, where two nonzero fixed points $(X, Y, Z) = (\pm\sqrt{b(r-1)}, \pm\sqrt{b(r-1)}, r-1)$ develop. (These two states correspond to time-independent convecting rolls that circulate clockwise or counterclockwise.) This is the topic of Exercise 2.4. We wish to derive an amplitude equation that describes the time evolution to the nonlinear state that develops for r slightly larger than $r_c = 0$. To develop the perturbation expansion, we write the bifurcation parameter $r = r_c(1 + \varepsilon) = 1 + \varepsilon$, and then ε is the small parameter.

We write Eqs. (A2.18) in the form of Eq. (A2.14)

$$\hat{\mathbf{L}}\mathbf{u}_p = \frac{d\mathbf{u}_p}{dt} - \hat{\mathbf{N}}[\mathbf{u}_p]. \qquad (A2.19)$$

Here $\mathbf{u}_p$ is the perturbation from the base state $\mathbf{u} = 0$ and $\hat{\mathbf{L}}$ is the evolution operator acting on $\mathbf{u}_p$ and linearized about $X = Y = Z = 0$

$$\hat{\mathbf{L}} = \begin{bmatrix} -\sigma & \sigma & 0 \\ 1+\varepsilon & -1 & 0 \\ 0 & 0 & -b \end{bmatrix}. \tag{A2.20}$$

The nonlinear term $\hat{\mathbf{N}}$ is given by

$$\hat{\mathbf{N}}[u] = \begin{bmatrix} 0 \\ -XZ \\ XY \end{bmatrix}. \tag{A2.21}$$

If we expand the linear operator $\hat{\mathbf{L}}$ in powers of $\varepsilon^{1/2}$,

$$\hat{\mathbf{L}} = \hat{\mathbf{L}}_0 + \varepsilon^{1/2}\hat{\mathbf{L}}_1 + \varepsilon\hat{\mathbf{L}}_2 + \cdots, \tag{A2.22}$$

we find that

$$\hat{\mathbf{L}}_0 = \begin{bmatrix} -\sigma & \sigma & 0 \\ 1 & -1 & 0 \\ 0 & 0 & -b \end{bmatrix}, \quad \hat{\mathbf{L}}_1 = \begin{bmatrix} 0 & 0 & 0 \\ 0 & 0 & 0 \\ 0 & 0 & 0 \end{bmatrix}, \quad \hat{\mathbf{L}}_2 = \begin{bmatrix} 0 & 0 & 0 \\ 1 & 0 & 0 \\ 0 & 0 & 0 \end{bmatrix}. \tag{A2.23}$$

The eigenvalues of the matrix $\hat{\mathbf{L}}_0$ are 0, $-(\sigma+1)$, and $-b$, and corresponding eigenvectors are $\mathbf{e}_0 = (1,1,0)$, $\mathbf{e}_1 = (-\sigma,1,0)$, and $\mathbf{e}_2 = (0,0,1)$. The zero eigenvalue corresponds to the onset of the linear instability. Also we expand $\mathbf{u}_p = (X, Y, Z)$ in powers of $\varepsilon^{1/2}$

$$\mathbf{u}_p = \varepsilon^{1/2}\mathbf{u}_0(T) + \varepsilon\mathbf{u}_1(T) + \cdots, \tag{A2.24}$$

where we introduce the slow time variable $T = \varepsilon t$ since we are looking for solutions corresponding to the growth of the weakly unstable mode to saturation. The term $d\mathbf{u}_p/dt$ first contributes at $O(\varepsilon^{3/2})$

$$\frac{d\mathbf{u}_p}{dt} = \varepsilon^{3/2}\frac{d\mathbf{u}_0}{dT} + \varepsilon^2\frac{d\mathbf{u}_1}{dT} + \cdots. \tag{A2.25}$$

These expressions are to be substituted into Eq. (A2.19) and then we require that the equation is true at each order in $\varepsilon^{1/2}$.

At $O(\varepsilon^{1/2})$, Eq. (A2.19) reduces to

$$\hat{\mathbf{L}}_0 \mathbf{u}_0 = 0.$$

This shows us that $\mathbf{u}_0$ is simply some amplitude of the zero-eigenvalue mode

$$\mathbf{u}_0 = A_0(T) \begin{bmatrix} 1 \\ 1 \\ 0 \end{bmatrix}, \quad (A2.26)$$

which introduces the amplitude function A_0. In this case, A_0 is the amplitude of a real vector, and can be taken as a real function.

At $O(\varepsilon)$, we get (since $\hat{\mathbf{L}}\mathbf{u} \to \hat{\mathbf{L}}_0\mathbf{u}_1 + \hat{\mathbf{L}}_1\mathbf{u}_0$ and $\hat{\mathbf{L}}_1$ is zero)

$$\hat{\mathbf{L}}_0 \mathbf{u}_1 = \begin{bmatrix} 0 \\ 0 \\ A_0^2 \end{bmatrix}, \quad (A2.27)$$

where the term in A_0^2 is the $O(\varepsilon)$ nonlinear term, namely bXY evaluated at the solution $\mathbf{u}_0$. The solution of this type of algebraic equation may cause problems because the matrix $\hat{\mathbf{L}}_0$ is singular, i.e. it has an eigenvalue that is zero so that the inverse $\hat{\mathbf{L}}_0^{-1}$ may not be formed. In the particular case of Eq. (A2.27), we see by inspection that the right side has no component along the zero-eigenvalue eigenvector $\mathbf{e}_0 = (1, 1, 0)$ of $\hat{\mathbf{L}}_0$. In this case, Eq. (A2.27) can be solved. We find

$$\mathbf{u}_1 = \begin{bmatrix} 0 \\ 0 \\ -b^{-1}A_0^2 \end{bmatrix} + A_1 \begin{bmatrix} 1 \\ 1 \\ 0 \end{bmatrix}, \quad (A2.28)$$

where the second term introduces the next order correction $A_1(T)$ to the amplitude that may be determined at a higher order of the expansion. You can check that Eq. (A2.27) is satisfied by Eq. (A2.28) for any value of A_1.

At $O(\varepsilon^{3/2})$, we find

$$\hat{\mathbf{L}}_0 \mathbf{u}_2 = \begin{bmatrix} \partial_T A_0 \\ -\varepsilon A_0 + b^{-1}A_0^3 + \partial_T A_0 \\ A_0 A_1 \end{bmatrix}, \quad (A2.29)$$

where the right side gets contributions from $-\hat{\mathbf{L}}_2\mathbf{u}_0$, $-\hat{N}$, and $d\mathbf{u}_p/dt$. Now to solve for $\mathbf{u}_2$, we must explicitly require that the right side of Eq. (A2.29) has vanishing component along the zero-eigenvalue eigenvector $(1, 1, 0)$. This is equivalent to the statement that the right side must be orthogonal to the zero-eigenvalue eigenvector of the adjoint operator $\hat{\mathbf{L}}_0^\dagger$. For a real matrix, the adjoint is just the transpose, so that

$$\hat{\mathbf{L}}_0^\dagger = \begin{bmatrix} -\sigma & 1 & 0 \\ \sigma & -1 & 0 \\ 0 & 0 & -b \end{bmatrix}. \quad (A2.30)$$

You can check that $\hat{\mathbf{L}}_0^\dagger$ does indeed have a zero eigenvalue, for which the corresponding eigenvector is $(1, \sigma, 0)$. The condition that the right side of Eq. (A2.29) is orthogonal to this vector yields the amplitude equation for A_0

$$\frac{1+\sigma}{\sigma}\partial_T A_0 = \varepsilon A_0 - b^{-1}A_0^3. \tag{A2.31}$$

In this way, the solvability condition for the existence of the solution $\mathbf{u}_2$ at $O(\varepsilon^{3/2})$ imposes constraints (in the form of a dynamical equation in the slow time dependence) for the amplitude A_0 introduced in the solution $\mathbf{u}_0$. Similarly, extending the procedure to $O(\varepsilon^2)$ would yield a dynamical equation for the next order correction to the amplitude A_1, as well as introducing a further correction A_2, and so on.

Equation (A2.31) describes the slow growth and nonlinear saturation of the amplitude of the unstable mode near threshold. Note that sufficiently close to onset, when $r - 1$ is tiny and positive, the two nonzero fixed points of the Lorenz equations, $(\pm\sqrt{b(r-1)}, \pm\sqrt{b(r-1)}, r-1)$, become approximately proportional to the unstable eigenvector $(1, 1, 0)$ in Eq. (A2.26). So close enough to onset, the amplitude equation Eq. (A2.31) with Eq. (A2.26) indeed describes the nonlinear motion of an orbit that starts near the unstable fixed point $(0, 0, 0)$ and that approaches one of the nonzero fixed points.

This example of the derivation of an amplitude equation introduces many of the features of the full calculation although there are some simplifications that may not occur in general. For example, in the present case $\hat{\mathbf{L}}_0$ is real and, for a real matrix, the adjoint is the transpose. In more general examples, the operator $\hat{\mathbf{L}}_0$ will be a complex matrix-differential operator and we must expand our notions of vector spaces, adjoint operators, etc., in the usual way to function spaces.

A2.3.2 One-dimensional Swift–Hohenberg equation

The one-dimensional Swift–Hohenberg equation

$$\partial_t u(x,t) = ru - (\partial_x^2 + 1)^2 u - u^3. \tag{A2.32}$$

was introduced in Section 2.2.2 as a simple mathematical model displaying the phenomenon of pattern formation. In that section, the linear stability analysis was used to show that there is a type-I-s instability from the uniform state $u = 0$ at the critical value of the control parameter $r = r_c = 0$. The critical wave vector is $q_c = 1$, so that the onset mode is $e^{\pm ix}$. We now show how to derive the lowest-order amplitude equation for this model. The new feature that goes beyond the previous example is the spatial dependence in the evolution equation and in the resulting amplitude equation.

The evolution equation (A2.32) can be written in the general form, Eq. (A2.14), with the linear operator $\hat{L}$ given by

$$\hat{L} u_p = (r-1) u_p - 2 \partial_x^2 u_p - \partial_x^4 u_p, \tag{A2.33}$$

and the nonlinear operator $\hat{N}$ by

$$\hat{N}[u_p] = -u_p^3. \tag{A2.34}$$

Following the general procedure, we introduce the small parameter ε as the distance of the control parameter r from its critical value (here $r_c = 0$), $r = \varepsilon$, and then expand the field $u(x,t)$ and the evolution equation in powers of $\varepsilon^{1/2}$. We expand $u(x,t)$ as

$$u = \varepsilon^{1/2} u_0 + \varepsilon u_1 + \text{h.o.t.}, \tag{A2.35}$$

with u_0 given as some slowly varying amplitude $A_0(X,T)$ of the critical mode

$$u_0 = A_0(X,T) e^{ix} + \text{c.c.} \tag{A2.36}$$

Here X and T are the slow space and time scales, $X = \varepsilon^{1/2} x$ and $T = \varepsilon t$. In the present case, unlike the previous example, the onset mode is complex and so the amplitude is a complex function.

In the expansion of the linear operator $\hat{L}$ in powers of ε, we introduce the multiple scales by the substitution $\partial_x \to \partial_x + \varepsilon^{1/2} \partial_X$. This leads to the replacements

$$\partial_x^2 \to \partial_x^2 + 2\varepsilon^{1/2} \partial_x \partial_X + \varepsilon \partial_X^2, \tag{A2.37a}$$

$$\partial_x^4 \to \partial_x^4 + 4\varepsilon^{1/2} \partial_x^3 \partial_X + 6\varepsilon \partial_x^2 \partial_X^2 + \cdots. \tag{A2.37b}$$

Thus we find at successive orders in $\varepsilon^{1/2}$

$$\hat{L}_0 = -1 - 2\partial_x^2 - \partial_x^4, \tag{A2.38a}$$

$$\hat{L}_1 = -4(\partial_x^2 + 1)\partial_X, \tag{A2.38b}$$

$$\hat{L}_2 = -2\partial_X^2 - 6\partial_x^2 \partial_X^2. \tag{A2.38c}$$

The nonlinear term, and the time derivative $\partial_t u \to \varepsilon \partial_T u$, contribute at $O(\varepsilon^{3/2})$ and higher.

Now collect terms at each order in $\varepsilon^{1/2}$ in the expansion of Eq. (A2.32). At $O(\varepsilon^{1/2})$, we find

$$\hat{L}_0 u_0 = 0, \tag{A2.39}$$

which is automatically satisfied by the expression Eq. (A2.36). At $O(\varepsilon)$, we have

$$\hat{L}_0 u_1 = -\hat{L}_1 u_0 = 0, \tag{A2.40}$$

since the operator $(\partial_x^2 + 1)$ in $\hat{L}_1$ gives zero when acting on $e^{\pm ix}$. Thus we simply have

$$u_1 = A_1(X, T)e^{ix} + \text{c.c.}, \tag{A2.41}$$

which introduces the next-order correction A_1 to the amplitude. At $O(\varepsilon^{3/2})$ and after some effort, we find the equation

$$\hat{L}_0 u_2 = \left[-(1 + 4\partial_X^2)A_0 + \partial_T A_0 + 3|A_0|^2 A_0\right]e^{ix} + A_0^3 e^{3ix} + \text{c.c.} \tag{A2.42}$$

The amplitude equation for A_0 arises as the solvability condition for this equation. The solvability condition arises because the functions $e^{\pm ix}$ satisfy the homogeneous equation

$$\hat{L}_0 e^{\pm ix} = 0, \tag{A2.43}$$

(i.e. they are zero-eigenvalue eigenvectors for $\hat{L}_0$). Thus the coefficient of the $e^{\pm ix}$ dependence on the right-hand side of Eq. (A2.42) must be set to zero. This yields the amplitude equation

$$\partial_T A_0 = 1 + 4\partial_X^2 A_0 - 3|A_0|^2 A_0. \tag{A2.44}$$

Returning to the unscaled variables, and writing at lowest order $A = \varepsilon^{1/2} A_0$, yields

$$\partial_t A_0 = \varepsilon A_0 + 4\partial_x^2 A_0 - 3|A_0|^2 A_0, \tag{A2.45}$$

which has the form of the general type-I-s amplitude equation Eq. (6.9), reduced to one spatial dimension, with values of the parameters

$$\tau_0 = 1, \quad \xi_0 = 2, \quad g_0 = 3. \tag{A2.46a}$$

We can easily verify that after the solvability constraint is satisfied since Eq. (A2.42) can indeed be solved. The terms remaining in Eq. (A2.42) are

$$\hat{L}_0 u_2 = A_0^3 e^{3ix} + \text{c.c.}, \tag{A2.47}$$

which can be solved by inspection to give

$$u_2 = -\frac{1}{64} A_0^3 e^{3ix} + A_1 e^{ix} + \text{c.c.}, \tag{A2.48}$$

with A_1 not determined at this order. Equation (A2.48) tells us the magnitude of the third spatial harmonic (proportional to $\varepsilon^{3/2}$) and can be used to extend the expansion to higher order.

A2.4 Phase equations

A2.4.1 General method

The method of multiple scales is not restricted to expansions in weak nonlinearity near threshold. Another application is to phase equations, which describe the slow modulation of stripe patterns away from threshold. Here we use the Goldstone-type argument of Chapter 9 to argue that a modulation of a regular pattern on length scales large compared to the basic periodicity should evolve slowly. This slow space-time variation can be extracted from the basic equations using the method of multiple scales.

This application differs somewhat from the derivation of amplitude equations in that the slow scale is not determined by an independent parameter such as ε, but itself defines the small parameter. The small parameter is essentially the reciprocal of the length scale of the spatial variation in units of the periodicity of the pattern.

The starting point for the derivation of the phase equation is the definition of the phase variable ϕ, Eq. (9.1), in terms of the wave vector field $\mathbf{q}$

$$\nabla \phi(\mathbf{x}_\perp, t) = \mathbf{q}(\mathbf{X}, t), \tag{A2.49}$$

or

$$\phi = \int \mathbf{q}(\mathbf{X}) \cdot d\mathbf{x}_\perp. \tag{A2.50}$$

In these equations, a slow space variable $\mathbf{X}$ is introduced by defining

$$\mathbf{X} = \eta \mathbf{x}_\perp, \tag{A2.51}$$

where η is the small parameter such that the slow spatial variations of interest in the pattern occur over a length scale of order unity in the $\mathbf{X}$ variable. Since the wave vector defines the orientation and local periodicity of the pattern, this variable varies on the long length scale and is a function of $\mathbf{X}$. Note that Eq. (A2.49) applies in regions of smooth variation of the pattern, away from defects and disordered regions.

The expressions Eqs. (A2.49) and (A2.50) are not easy to work with because they mix the fast and slow coordinates $\mathbf{x}_\perp$ and $\mathbf{X}$ in an inconvenient way. To develop the systematic perturbation expansion, it is useful to introduce a scaled phase variable $\Phi(\mathbf{X}, T)$ by defining

$$\Phi = \eta \phi, \tag{A2.52}$$

so that the derivatives of Φ with respect to $\mathbf{X}$ are $O(1)$ (the first derivative is just the wave vector). In terms of the scaled phase, we have

$$\mathbf{q}(\mathbf{X}) = \nabla_\mathbf{X} \Phi(X), \quad \Phi(\mathbf{X}) = \int \mathbf{q}(\mathbf{X}) \cdot d\mathbf{X}. \tag{A2.53}$$

This clever trick allows the inclusion in the same formal expansion scheme of both the first derivative of ϕ, which is $O(1)$ and gives the local wave vector, and of higher derivatives of ϕ, which are $O(\eta)$ and give the slow spatial variation.

With the definitions Eq. (A2.53), the derivation of the phase dynamics follows quite closely the multiple scales derivation of the amplitude equation. In the present case, we expand the evolution equations for the fields $\mathbf{u}(\mathbf{x}, t)$ in powers of η, corresponding to the slow spatial variation of $\mathbf{q}$.

The zeroth-order solution for $\mathbf{u}$ (i.e. no effect of the spatial variation of $\mathbf{q}$) is the fully nonlinear, spatially periodic solution $\mathbf{u_q}(\mathbf{x}_\perp, z)$, that corresponds to the ideal stripe state with wave vector $\mathbf{q}$. Since $\mathbf{u_q}$ is periodic in $\mathbf{x}_\perp$ with period $2\pi q^{-1}$ in the $\hat{\mathbf{q}}$ direction, we redefine the spatially periodic function in terms of the phase

$$\mathbf{u_q}(\mathbf{x}_\perp, z) = \bar{\mathbf{u}}_\mathbf{q}(\phi, z), \quad \phi = \mathbf{q} \cdot \mathbf{x}_\perp. \tag{A2.54}$$

The expansion in powers of η is then

$$\mathbf{u}(\mathbf{x}_\perp, z, t) = \mathbf{u}^{(0)}(\phi, z; \mathbf{X}, T) + \eta \mathbf{u}^{(1)} + \cdots, \tag{A2.55}$$

where the dependence of $\mathbf{u}^{(i)}$ on the slow variables $\mathbf{X}, T$ arises through the implicit dependence on $\mathbf{q}(\mathbf{X}, T)$. In particular we have for the zeroth-order term

$$u^{(0)}(\phi, z; \mathbf{X}, T) = \bar{\mathbf{u}}_{\mathbf{q}(\mathbf{X},T)}(\phi, z). \tag{A2.56}$$

Equation (A2.55) is substituted into the evolution equations for the system, and terms at each order in η are collected. For the phase equation, Eq. (9.4), we need only go up to terms that are first order in η. These terms arise from slow spatial derivatives, slow time dependence, and also the term $\eta \mathbf{u}^{(1)}$ in Eq. (A2.55). For example, a spatial derivative acting on $\mathbf{u}^{(i)}$ gives

$$\nabla \mathbf{u}^{(i)} \to \mathbf{q} \, \partial_\phi \mathbf{u}^{(i)} + \eta \, \nabla_\mathbf{X} \mathbf{u}^{(i)}. \tag{A2.57}$$

Higher-order derivatives may also be needed, for example

$$\nabla^2 \mathbf{u}^{(i)} \to q^2 \, \partial_\phi^2 \mathbf{u}^{(i)} + \eta D \, \partial_\phi \mathbf{u}^{(i)} + O(\eta^2), \tag{A2.58}$$

with the operator D defined by

$$D = 2\mathbf{q} \cdot \nabla_\mathbf{X} + (\nabla_\mathbf{X} \cdot \mathbf{q}). \tag{A2.59}$$

Also, the time derivative gives

$$\partial_t \mathbf{u}^{(i)}(\phi, z; \mathbf{X}, T) = \eta^2 \, \partial_T \phi \, \partial_\phi \mathbf{u}^{(i)} + \eta^2 \, \partial_T \mathbf{u}^{(i)} = \eta \, \partial_T \Phi \, \partial_\phi \mathbf{u}^{(i)} + O(\eta^2). \tag{A2.60}$$

At $O(\eta)$, there are also terms $\eta \hat{\mathbf{L}} \mathbf{u}^{(1)}$, with $\hat{\mathbf{L}}$ the linear operator given by linearizing the equations of motion about $\mathbf{u}^{(0)}$. We know from physical arguments that $\hat{\mathbf{L}}$ has an eigenvector with zero eigenvalue, and so the phase equation appears as the solvability condition that the equation for $\mathbf{u}^{(1)}$ has a finite solution. Here we see the close similarity with the derivation of the amplitude equation. The zero mode in the present case corresponds to a translation of the solution, and so takes the form $\nabla u^{(0)}$.

This procedure is illustrated for the simple example of the Swift–Hohenberg equation in the following section.

A2.4.2 Phase equation for the Swift–Hohenberg model

Here we will use the Swift–Hohenberg equation in two space dimensions, when it can be written in the form

$$\partial_t u(\mathbf{x}, t) = ru - (\nabla^2 + 1)^2 u - u^3, \tag{A2.61}$$

with $\mathbf{x} = (x, y)$, and $\nabla^2 = \partial_x^2 + \partial_y^2$. As in Eq. (A2.55), we expand u as an expansion in powers of η, to give

$$u(\mathbf{x}, t) = u^{(0)}(\phi, z; \mathbf{X}, T) + \eta u^{(1)} + \text{h.o.t.}, \tag{A2.62}$$

with $\mathbf{X}$ and T the slow space and time variables and $u^{(0)}$ the zeroth-order solution

$$u^{(0)}(\phi, z; \mathbf{X}, T) = \bar{u}_{q(\mathbf{X},T)}(\phi), \tag{A2.63}$$

where $\bar{u}_q(\phi = \mathbf{q} \cdot \mathbf{x})$ is the nonlinear, spatially periodic, time-independent solution for straight stripes which satisfies[7]

$$r\bar{u}_q(\phi) - (q^2 \partial_\phi^2 + 1)^2 \bar{u}_q(\phi) - \bar{u}_q^3(\phi) = 0. \tag{A2.64}$$

The h.o.t. in Eq. (A2.62) denotes terms that are of second and higher order in η.

We now substitute Eq. (A2.62) into the evolution equation, Eq. (A2.61). We will need the rather complicated operator involving up to fourth-order derivatives

$$(\nabla^2 + 1)^2 \to \left[(q^2 \partial_\phi^2 + 1) + \eta D \partial_\phi\right]\left[(q^2 \partial_\phi^2 - 1) + \eta D \partial_\phi\right] + \text{h.o.t.} \tag{A2.65}$$

$$= (q^2 \partial_\phi^2 + 1)^2 + \eta\left\{2\partial_\phi(q^2 \partial_\phi^2 + 1)D + \left[2\mathbf{q} \cdot \nabla_X (q^2)\right]\partial_\phi^3\right\} + \text{h.o.t.} \tag{A2.66}$$

The other terms in Eq. (A2.61) are easy to evaluate up to first order in η

$$\partial_t u(\mathbf{x}, t) \to \eta(\partial_T \Phi)\partial_\phi \bar{u}_q(\phi) + \text{h.o.t.}, \tag{A2.67}$$

$$ru - u^3 \to r\bar{u}_q - \bar{u}_q^3 + \eta\left[r - 3\bar{u}_q^2\right]u^{(1)} + \text{h.o.t.} \tag{A2.68}$$

---

[7] For the scalar field u the function $\bar{u}_\mathbf{q}$ cannot depend on the direction of $\mathbf{q}$, as so we write it as $\bar{u}_q$.

Now collecting terms at $O(\eta)$, we find the equation

$$\left[r - (q^2 \partial_\phi^2 + 1)^2 - 3\bar{u}_q^2\right]u^{(1)} = (\partial_T \Phi)\partial_\phi \bar{u}_q(\phi)$$
$$+ \left\{2\partial_\phi (q^2 \partial_\phi^2 + 1)D + \left[2\mathbf{q}\cdot\nabla_X(q^2)\right]\partial_\phi^3\right\}\bar{u}_q(\phi). \quad (A2.69)$$

It is straightforward to check that $\partial_\phi \bar{u}_q$ is a zero-eigenvalue eigenvector of the operator on the left-hand side

$$\left[r - (q^2 \partial_\phi^2 - 1)^2 - 3\bar{u}_q^2\right]\partial_\phi \bar{u}_q = 0, \quad (A2.70)$$

as is expected from the translational symmetry. The operator acting on $u^{(1)}$ in Eq. (A2.69) is self-adjoint, and so the solvability condition, that the right-hand side have no component along the zero-eigenvalue eigenvector of the adjoint operator, reduces to the orthogonality condition for the right-hand side with $\partial_\phi \bar{u}_q$

$$(\partial_T \Phi)\int_0^{2\pi} d\phi (\partial_\phi \bar{u}_q)^2 + \int_0^{2\pi} d\phi (\partial_\phi \bar{u}_q)\left\{2\partial_\phi(q^2\partial_\phi^2 + 1)D \right. \quad (A2.71)$$
$$\left. + \left[2\mathbf{q}\cdot\nabla_X(q^2)\right]\partial_\phi^3\right\}\bar{u}_q = 0.$$

After integrating by parts with respect to ϕ some terms in the second integral and rearranging, this reduces to

$$(\partial_T \Phi)\int_0^{2\pi} d\phi(\partial_\phi \bar{u}_q)^2 = \nabla_X \cdot \left\{\mathbf{q}\int_0^{2\pi} d\phi\left[q^2(\partial_\phi^2 \bar{u}_q)^2 - (\partial_\phi \bar{u}_q)^2\right]\right\}. \quad (A2.72)$$

Equation (A2.72) is in the form of Eq. (9.4) with

$$\tau(q) = \frac{1}{\pi}\int_0^{2\pi} d\phi(\partial_\phi \bar{u}_q)^2, \quad (A2.73a)$$

$$B(q) = \frac{1}{\pi}\int_0^{2\pi} d\phi\left[q^2(\partial_\phi^2 \bar{u}_q)^2 - (\partial_\phi \bar{u}_q)^2\right]. \quad (A2.73b)$$

(Since we can multiply τ and B by the same arbitrary constant without changing the physics, we have included a normalization constant $1/\pi$ in these expressions for later convenience.)

These integral expressions depend on knowing the full nonlinear, but spatially periodic, stripe solutions to some satisfactory level of approximation. We have obtained expressions for the saturated nonlinear stripe solution to the Swift–Hohenberg model in Section 4.1.3 using the lowest-order Galerkin method. The calculation there was done for the critical wave number $q = 1$, but the calculation is easily generalized to any q. Truncating at lowest order in the Galerkin expansion gives $\bar{u}_q \simeq a_q \cos\phi$, $\phi = qx$, with

$$a_q^2 = \frac{4}{3}\left[r - (q^2 - 1)^2\right]. \tag{A2.74}$$

Then we find

$$\tau(q) = a_q^2 \quad \text{and} \quad B(q) = (q^2 - 1)a_q^2. \tag{A2.75}$$

The function a_q^2 is positive everywhere between the neutral stability curve of the uniform state, and goes to zero on the neutral stability curve. The function $B(q)$ changes sign at $q = 1$. It is useful to plot $qB(q)$, since the slope of this curve is needed to calculate the parallel diffusion constant via Eq. (9.7). As discussed in Chapter 9, the signs of B and $(qB)'$ with the prime denoting the derivative with respect to q, are important in determining the stability of the stripe state against long-wavelength perturbations. The dependence of qB on the wave number q for the Swift–Hohenberg model at $r = 0.25$ is shown in Fig. 9.1 of Chapter 9.

A2.5 Other applications of the solvability condition

The solvability condition also arises in other situations not arising from a multiple scales expansion. The key ingredient that leads to solvability conditions is the need to invert a linear operator with a zero eigenvalue. In a perturbation context, a zero eigenvalue often arises from a symmetry. For example in a translationally invariant system, the spatial derivative of a stationary localized solution $\mathbf{u}_0(\mathbf{x})$ to

$$\partial_t \mathbf{u} = \hat{O}\mathbf{u}(\mathbf{x}, t) \tag{A2.76}$$

satisfies

$$\hat{L}\nabla u_0 = 0, \tag{A2.77}$$

where $\hat{L}$ is the linear operator given by expanding the operator $\hat{O}$ about $\mathbf{u}_0$. This approach can be used to calculate the climb of dislocations for example, where we are seeking the dynamics through symmetry related translations along the stripes.

A2.6 Further reading

(i) *Advanced Mathematical Methods for Scientists and Engineers* by Bender and Orszag [11].

Glossary

Absolute instability: A spatially dependent instability that grows exponentially in magnitude at all points within a system. See also the entry for "convective instability."

Ansatz: A mathematical expression that represents a plausible solution for some problem.

Aspect ratio Γ: The dimensionless ratio of the largest lateral width of a pattern-forming medium to its smallest transverse direction. For a cylindrical Rayleigh–Bénard convection cell, the aspect ratio is defined to be the ratio of the cell radius R to the fluid depth d, $\Gamma = R/d$. The larger the aspect ratio, the weaker the influence of the lateral boundary conditions on the pattern.

Base state: A nonlinear state about which a linear stability analysis is carried out.

Bifurcation: A discrete qualitative change in some property of a dynamical system as some parameter is varied. The value at which the discrete change occurs is called the bifurcation point. Many scientists use a casual meaning of bifurcation, often motivated by experimentally obvious changes in some system such as the appearance of convection rolls from a featureless conducting state, or some change in symmetry. Mathematically and more rigorously, a bifurcation corresponds to a change, as some system parameter is varied, in the topological structure of the vector field in phase space guiding the dynamics and so can be difficult to define or quantify.

Boussinesq equations: The five coupled nonlinear partial differential equations that determine how the state of an incompressible convecting fluid – given by the three components v_x, v_y, and v_z of the velocity field $\mathbf{v}(x,y,z,t)$, the temperature field $T(x,y,z,t)$, and the pressure field $p(x,y,z,t)$ – evolves in time. The Boussinesq equations are fundamental in the sense that they correspond to laws of mass conservation, momentum conservation, and energy conservation, and are believed to describe convection quantitatively over a large range of fluid parameters. These equations are difficult to study mathematically and the properties of their solutions are known mainly from laboratory experiments and from numerical simulations.

Broken symmetry: Situation where a state (equilibrium or nonequilibrium) has a different symmetry than the equations that govern the evolution of the state. For example, a periodic stationary stripe state of convection rolls of wave vector $\mathbf{q}$ in an infinite domain represents a broken symmetry solution of the Boussinesq equations. These equations are invariant under the continuous symmetry of a horizontal translation

$\mathbf{x}_\perp \to \mathbf{x}_\perp + \mathbf{d}_0$ in an arbitrary direction by an arbitrary amount, while the stripe state is invariant only under a discrete symmetry of horizontal translations $\mathbf{x}_\perp \to \mathbf{x}_\perp + m\lambda(\mathbf{q}/q)$ by integer multiples m of the wavelength $\lambda = 2\pi/q$. Broken symmetries often arise from bifurcations.

Chaos: Deterministic dynamics that is nontransient, bounded, and nonperiodic in time (neither periodic nor quasiperiodic). Scientists often use the criteria of a broad-band power spectrum $P(\omega)$ and of a positive largest Lyapunov exponent $\lambda_1 > 0$ to argue the existence of chaos in experimental or computational data.

Characteristic length scale: The most significant length of some spatially varying field $f(\mathbf{x})$. If the field f can be represented as a superposition of Fourier modes $\exp(i\mathbf{q} \cdot \mathbf{x})$ such that the wave numbers $q = \sqrt{\mathbf{q} \cdot \mathbf{q}}$ are closely clustered around their mean $\langle q \rangle$, then the characteristic length is defined to be $2\pi/\langle q \rangle$. The characteristic length scale of a cellular pattern near the onset of a type-I supercritical bifurcation of a uniform state is usually determined by the critical wave number q_c.

Characteristic time scale: The most significant time in some time-varying observable $u(t)$. If the observable u can be written as a superposition of Fourier modes $\exp(i\omega t)$ such that the frequencies ω are closely clustered around their mean $\langle \omega \rangle$, then the characteristic time is defined to be $2\pi/\langle \omega \rangle$ (and the characteristic frequency is $\langle \omega \rangle$).

Convective instability: A spatially dependent instability that propagates as it grows in such a way that, at any fixed point in space, only asymptotic decay is observed. See also the entry for "absolute instability."

Correlation function: A real-valued function $C(\mathbf{x}_1, t_1; \mathbf{x}_2, t_2, \ldots)$ that measures the extent to which the fields at position $\mathbf{x}_1$ at time t_1, position $\mathbf{x}_2$ at time t_2, etc., are correlated. Often the two-point correlation function with just two space-time points is used. If the notation $\langle A \rangle$ denotes an ensemble average of an observable A (the average over all possible realizations of the field u consistent with imposed constraints), then the two-point function C of the signal u at spatial points $\mathbf{x}_1$ and $\mathbf{x}_2$ at times t_1 and t_2 is defined to be

$$C(\mathbf{x}_1, t_1; \mathbf{x}_2, t_2) = \langle (u(\mathbf{x}_1, t_1) - \langle u \rangle)(u(\mathbf{x}_2, t_2) - \langle u \rangle) \rangle.$$

Different observables can lead to different conclusions about the amount of correlation. For a system that is homogeneous in space and time, the correlation function becomes a function of only the distances and times between the observation points, e.g. $C(\mathbf{x}_1, t_1; \mathbf{x}_2, t_2) = C(\|\mathbf{x}_2 - \mathbf{x}_1\|, |t_2 - t_1|)$.

Critical exponent: The exponent of a power law that determines how some particular quantity varies near a continuous transition. More specifically, if a quantity A varies as $(p - p_c)^\alpha$ as some parameter p approaches a critical value p_c of a continuous transition, then the exponent α is defined to be the critical exponent of the quantity A. The concept is useful for equilibrium and nonequilibrium systems.

Defect: A place in a pattern where the local pattern cannot be defined; a disruption in the periodicity of some periodic pattern. Defects can be localized to a finite region (a point defect such as a dislocation) or continuously extend along some line (e.g. a grain boundary).

Degree of freedom: For dissipative systems, one of the variables used in a mathematical model that describes the system. For energy-conserving Hamiltonian systems, a degree of freedom has a slightly different historical meaning, namely a pair of conjugate variables (e.g. position and momentum) that appear in the Hamiltonian equations of the system.

Disclination: A point defect in a locally periodic pattern such that the wave vector rotates by an integer multiple of π on any closed loop containing the disclination. A focus singularity is an example of a disclination. Disclinations are associated with the rotational symmetry of a pattern.

Dislocation: A point defect in a locally periodic pattern such that one or more wave vectors of the periodic field integrate to a nonzero multiple of 2π on any closed loop containing the dislocation. A convection roll that terminates abruptly in the middle of a cell is an example of a dislocation. Dislocations are associated with the discrete translational symmetry of a pattern.

Excitable dynamics: A dynamical system for which there is a stable fixed point with the property that a small perturbation of the fixed point decays but a sufficiently large perturbation of the fixed point grows in magnitude and then decays back to the fixed point. Neurons and heart muscle are examples of excitable dynamics.

Extensive chaos: A chaotic state of a homogeneous spatially extended nonequilibrium system which has the property that the fractal dimension D grows linearly with the system's volume V. Such systems have to reach some minimum size L before an extensive behavior of D can be observed.

First-order phase transition: A thermodynamic equilibrium transition between phases in which some thermodynamic quantity changes discontinuously across the transition. An example is the melting of ice to form water, for which the density changes discontinuously between the two phases. A subcritical nonequilibrium bifurcation is also sometimes called a first-order transition.

Front: A line in a two-dimensional pattern separating a region of one pattern from a region of another kind of pattern or no pattern at all. One example is the propagating boundary separating a stripe state and an unstable spatially uniform state. Another example is a grain boundary, where a region of stripes of one orientation meets a region of stripes with a different orientation.

Galerkin method: A numerical method for solving differential and partial differential equations, in which the unknown variables are written as a finite linear superposition of known basis functions. The coefficients of the superposition are then determined as a function of time. For certain basis functions such as Fourier modes or Chebyshev polynomials and for appropriate boundary conditions, Galerkin methods can require many fewer numerical degrees of freedom than finite difference methods.

Grain boundary: A one-dimensional or two-dimensional extended defect that separates patterns of two different orientations.

Odes: An abbreviation for "ordinary differential equations," pronounced "oh-dee-ees."

Onset: A nonequilibrium system is said to be "just above onset" if a control parameter p has been increased just beyond some critical value p_c at which a bifurcation occurs. More precisely, a system is just above onset if the reduced parameter $\epsilon = (p - p_c)/p_c$ is small, so that $0 < \epsilon \ll 1$.

Operator: A "function of functions," i.e. some mapping of input functions to output functions. A simple operator is $\hat{N}[f] = d^2f/dx^2 + 2f$, which associates with each input function $f(x)$ the output function $f''(x) + 2f(x)$.

Order parameter: A quantity that distinguishes one phase from another in an equilibrium phase transition, or one state from another for two states related by a bifurcation.

Pattern: A field $f(\mathbf{x}, t)$ is said to be a pattern if it varies periodically or nearly periodically in space, or has spatial structure that repeats in some other way, or consists of patches that themselves can be regarded as patterns. The concept includes time-varying fields.

Pattern formation: The formation of a pattern by some change in the parameters of a sustained nonequilibrium system.

Pdes: An abbreviation for "partial differential equations", pronounced "pee-dee-ees."

Phase space: The space of dynamical variables whose values define the state of a dynamical system. (See the glossary entry for "State of a dynamical system.") Also sometimes called the "state space" of a dynamical system.

Phase transition: A distinct change in the character of a system that is always in thermodynamic equilibrium, as some thermodynamic variable such as the temperature or pressure is slowly varied. For a second-order (also called continuous) phase transition, the features that distinguish one phase from another change continuously through the transition. An example is the continuous decrease to zero of the magnetization of a ferromagnet as the temperature is increased near the magnet's critical point. A discontinuous (also called first-order) phase transition involves a finite jump in the value of some quantity characterizing the phase. An example would be the discontinuous change in density as ice melts to form liquid water.

Power spectrum: The power spectrum $P(\omega)$ of some time-varying observable $u(t)$ is defined to be the magnitude-squared of the Fourier coefficient with frequency ω:

$$P(\omega) = \left| \frac{1}{2\pi} \int_{-\infty}^{\infty} u(t) e^{i\omega t} \, dt \right|^2 .$$

Similarly, the power spectrum $P(\mathbf{q})$ of some spatially varying field $u(\mathbf{x})$ is defined to be the magnitude-squared of the Fourier coefficient with wave vector $\mathbf{q}$. Power spectra are often one of the first quantities computed to analyze complicated temporal or spatial structure. For a statistically stationary signal u, the power spectrum can be shown to be the Fourier transform of the two-point correlation function of u, i.e. the information in a power spectrum and the information in the two-point correlation function are equivalent.

Primary bifurcation: The first instability of a spatially uniform time-independent nonequilibrium state. For example, the onset of convection is the primary bifurcation of the fluid from its featureless conducting state. See also the entry for "secondary bifurcation."

Quasicrystalline: An infinitely extended two- or three-dimensional pattern that is not periodic and yet is highly ordered in that the pattern's power spectrum $P(\mathbf{q})$ consists of sharp peaks (δ functions). Such patterns were first discovered in equilibrium crystals and later discovered in nonequilibrium fluid experiments.

Quasiperiodic: A function $f(t)$ that can be written in the form

$$f(t) = \sum_{m=-\infty}^{\infty} \sum_{n=-\infty}^{\infty} f_{mn} e^{i(m\omega_1 + n\omega_2)t},$$

for frequencies ω_1 and ω_2 such that their ratio ω_1/ω_2 is irrational, or the generalization to three and more frequencies; a superposition of sinusoids and their harmonics involving two (or more) incommensurate frequencies.

Secondary bifurcation: Any bifurcation of a nonlinear state that itself arose directly from the bifurcation of a spatially uniform time-independent state; any instability that occurs after a primary bifurcation. See the entry for "primary bifurcation."

Second-order phase transition: A thermodynamic equilibrium transition between two phases such that physical quantities vary continuously through the transition. Some quantity (the order parameter) that is zero in one of the phases continuously through the transition is often used to distinguish the two phases. Two examples are the continuous loss of magnetism as an iron ferromagnet is heated through its Curie point, and the continuous decrease of resistivity to zero of a metal like lead as it is cooled below its superconducting transition point. Sometimes used to describe a supercritical bifurcation of a nonequilibrium system.

Spatial disorder: A field $f(\mathbf{x}, t)$ defined over some spatial region of size L is said to be "spatially disordered" at time t if one or more correlation functions of that field decay significantly over a length scale that is small compared to L. A commonly used easily-computed measure of spatial disorder is the two-point correlation function $C(x)$, which sometimes decays asymptotically for large $|x|$ as an exponential $\exp(-|x|/\xi_2)$ with a length scale ξ_2 (the two-point correlation length). In this case, the field is spatially disordered if $\xi_2 \ll L$.

Spatially extended system: A nonequilibrium system that is large compared to some characteristic wavelength such as the size of a cellular structure (say a convection roll) or the wavelength of a propagating wave (say the arm of a rotating spiral). It is not always easy to identify an appropriate length scale with respect to which the size of a system can be measured, e.g. strongly driven turbulent flows for which there may be no cellular structure.

Spatiotemporal chaos: Informally, a spatially extended dynamical system that is chaotic and such that at least one field associated with the system is spatially disordered. Experts have not yet reached agreement about how to define spatiotemporal chaos since there are many ways to quantify spatial disorder and these definitions are not all consistent. A commonly used empirical criterion for the presence of spatiotemporal chaos is that the dynamics is chaotic (see the glossary entry for "chaos") and that the two-point correlation length ξ_2 for some field is substantially smaller than the lateral extent of the system.

State of a dynamical system: A set of numbers (possibly infinitely many) that are sufficient and necessary to determine a unique solution of the evolution equations that describe the system. The evolution equations must be known for a state to be defined. For Newton's law of motion $m\ddot{\mathbf{x}} = \mathbf{f}$ for a single point particle of mass m moving in three spatial dimensions according to a specified force $\mathbf{f}$, the state of the system would be the vector of six numbers defined by the three components of the position $\mathbf{x}$ and the three components of the velocity $d\mathbf{x}/dt = \mathbf{x}'$ at a given time. For a pattern-forming system like Rayleigh–Bénard convection, the state of the system is, at a given time, the infinitely many values of the fields (T, p, v_x, v_y, v_z) everywhere in the domain of interest.

Subcritical bifurcation: A bifurcation such that the lowest nonlinear terms near onset enhance, rather than saturate, instability. An important example in pattern-forming systems is the formation of a hexagonal lattice from a uniform state in reaction–diffusion systems. An important implication of a subcritical bifurcation is hysteresis: for a given set of parameter values, at least two distinct states can be observed, and which particular state is found depends on the history of how parameters are varied to reach a particular point in parameter space.

Truncation error: The difference between some mathematical expression and a discrete approximation of that mathematical expression, e.g. the difference between the derivative $f'(x)$ of some function and some finite-difference approximation of that derivative. The truncation error typically goes to zero in the limit that the spatial and temporal resolutions, Δx and Δt, go to zero. For most problems, the truncation error is large compared to the errors associated with round-off errors so the latter can be ignored.

Turbulence: Usually understood as meaning a temporally nonperiodic spatially disordered state of a fluid that has a large range of time and length scales. Some fluid dynamicists define turbulence as a complex fluid motion in which there is substantial generation of vorticity. The concept is often informally applied to non-fluid systems that are disordered in space and time, and is sometimes used interchangeably with spatiotemporal chaos, which is sometimes called "weak turbulence."

Two-point correlation function: A correlation function (see glossary entry) with two space-time points. Static correlations in a statistically time independent situation are given by setting the two times equal to give a function of just the two space points.

Uniform state: A system whose fields are translationally invariant with respect to at least one coordinate. One can have translational invariance along an infinite line or plane, and also translational invariance in a finite domain like a cylinder that is periodic with respect to one or more variables.

Wave number spectrum: The normalized wave number spectrum $P(q)$ of some spatial field $f(\mathbf{x})$ is the fraction of Fourier modes $f(\mathbf{q})$ with wave number q in the interval $[q, q + dq]$ out of all possible observed wave numbers.

References

[1] G. Ahlers and R. P. Behringer. The Rayleigh–Bènard instability and the evolution of turbulence. *Prog. Theor. Phys. Suppl.*, **64**:186–201, 1978.

[2] A. Aitta, G. Ahlers, and D. S. Cannell. Tricritical phenomena in rotating Couette–Taylor flow. *Phys. Rev. Lett.*, **54**(7):673, 1985.

[3] C. David Andereck, S. S. Liu, and H. L. Swinney. Flow regimes in a circular Couette system with independently rotating cylinders. *J. Fluid Mech.*, **164**:155–83, 1986.

[4] E. Anderson, Z. Bai, C. Bischof, L. S. Blackford, and J. Demmel. *LAPACK User's Guide*. Philadelphia, *SIAM*, 3rd edition, 2000.

[5] I. S. Aranson, L. Aranson, L. Kramer, and A. Weber. Stability limits of spirals and traveling waves in nonequilibria media. *Phys. Rev. A*, **46**(6):R2992–R2995, 1992.

[6] I. S. Aranson and L. Kramer. The world of the complex Ginzburg–Landau equation. *Rev. Mod. Phys.*, **74**:99–143, 2002.

[7] N. W. Ashcroft and N. D. Mermin. *Solid State Physics*. New York, Holt, Rinehart, and Winston, 1976.

[8] K. L. Babcock, G. Ahlers, and D. S. Cannell. Noise-sustained structure in Taylor–Couette flow with through-flow. *Phys. Rev. Lett.*, **67**(248):3388–91, 1991.

[9] P. Ball. *The Self-Made Tapestry: Pattern Formation in Nature*. New York, Oxford University Press, 1999.

[10] D. Barkley. A model for fast computer simulation of waves in excitable media. *Physica D*, **49**:61–70, 1991.

[11] C. M. Bender and S. A. Orszag. *Advanced Mathematical Methods for Scientists and Engineers*. International Series in Pure and Applied Mathematics. New York, McGraw-Hill, 1978.

[12] E. Bodenschatz, W. Pesch, and G. Ahlers. Recent developments in Rayleigh–Bènard convection. *Ann. Rev. Fluid Mech.*, **32**:709, 2000.

[13] E. Bodenschatz, W. Pesch, and L. Kramer. Structure and dynamics of dislocations in anisotropic pattern-forming systems. *Physica D*, **32**:135–45, 1988.

[14] C. Bowman and A. C. Newell. Natural patterns and wavelets. *Rev. Mod. Phys.*, **70**(1):289–301, 1998.

[15] F. H. Busse. Nonlinear properties of convection. *Rep. Prog. Phys.*, **41**:1929–67, 1978.

[16] F. H. Busse and J. A. Whitehead. Oscillatory and collective instabilities in large Prandtl number convection. *J. Fluid Mech.*, **66**:67–79, 1974.

[17] P. M. Chaikin and T. C. Lubensky. *Principles of Condensed Matter Physics*. Cambridge, Cambridge University Press, 1995.

[18] S. Chandrasekhar. *Hydrodynamic and Hydromagnetic Stability*. Oxford, Clarendon Press, 1968.

[19] H. Chaté and P. Manneville. Phase diagram of the two-dimensional complex Ginzburg–Landau equation. *Physica A*, **224**:348–68, 1996.
[20] J. Chazottes and B. Fernandez. *Dynamics of Coupled Map Lattices and of Related Spatially Extended Systems*. New York, Springer, 2005.
[21] J. M. Chomaz, P. Huerre, and L. G. Redekopp. Bifurcations to local and global modes in spatially developing flows. *Phys. Rev. Lett.*, **60**(1):25–8, 1988.
[22] S. Ciliberto, P. Coullet, J. Lega, E. Pampaloni, and C. Perezgarcia. Defects in roll-hexagon competition. *Phys. Rev. Lett.*, **65**(19):2370–3, 1990.
[23] J. D. Crawford. Introduction to bifurcation theory. *Rev. Mod. Phys.*, **63**:991–1038, 1991.
[24] M. C. Cross. Traveling and standing waves in binary-fluid convection in finite geometries. *Phys. Rev. Lett*, **57**:2935–8, 1986.
[25] M. C. Cross and P. C. Hohenberg. Pattern formation outside of equilibrium. *Rev. Mod. Phys.*, **65**(3):851–1112, 1993.
[26] M. C. Cross and P. C. Hohenberg. Spatiotemporal chaos. *Science*, **263**:1569–70, 1994.
[27] M. C. Cross, D. Meiron, and Y. Tu. Chaotic domains: a numerical investigation. *Chaos*, **4**(4):607–19, 1994.
[28] B. N. Datta. *Numerical Linear Algebra and Applications*. Pacific Grove, Brooks/Cole, 1995.
[29] P. Dayan and L. F. Abbott. *Theoretical Neuroscience: Computational and Mathematical Modeling of Neural Systems*. Cambridge, MA, MIT Press, 2001.
[30] J. R. deBruyn, E. Bodenschatz, S. Morris, S. Trainoff, Y.-C. Hu, D. S. Cannell, and G. Ahlers. Apparatus for the study of Rayleigh–Bènard convection in gases under pressure. *Rev. Sci. Instrum.*, **67**:2043–67, 1996.
[31] W. Decker and W. Pesch. Order parameter and amplitude equations for the Rayleigh–Bènard convection. *J. Phys. II France*, **4**:419–38, 1994.
[32] P. G. Drazin and W. H. Reid. *Hydrodynamic Stability*. Cambridge, Cambridge University Press, second edition, 2004.
[33] K. R. Elder, J. Viñals, and M. Grant. Dynamic scaling and quasi-ordered states in the 2-dimensional Swift–Hohenberg equation. *Phys. Rev. A*, **46**(12):7618–29, 1992.
[34] I. R. Epstein and J. A. Pojman. *An Introduction to Chemical Dynamics: Oscillations, Waves, Patterns, and Chaos*. New York, Oxford University Press, 1998.
[35] P. R. Fenstermacher, H. L. Swinney, and J. P. Gollub. Dynamical instabilities and the transition to chaotic Taylor vortex flow. *J. Fluid. Mech.*, **94**(1):103–28, 1979.
[36] J. M. Flesselles, A. J. Simon, and A. J. Libchaber. Dynamics of one-dimensional interfaces – an experimentalists view. *Advances in Physics*, **40**(1):1–51, 1991.
[37] A. V. Getling. *Rayleigh–Bénard Convection: Structures and Dynamics*. New Jersey, World Scientific, 1997.
[38] J. P. Gollub and S. V. Benson. Many routes to turbulent convection. *J. Fluid Mech.*, **100**:449–70, 1980.
[39] M. Golubitsky and I. Stewart. *The Symmetry Perspective: from Equilibrium to Chaos in Phase Space and Physical Space*. Basel, Birkhäuser, 2002.
[40] M. Golubitsky, I. Stewart, and D. G. Schaeffer. *Singularities and Groups in Bifurcation Theory: Volume 2*. New York, Springer, 1988.
[41] M. Gorman, P. J. Widmann, and K. A. Robbins. Nonlinear dynamics of a convection loop: A quantitative comparison of experiment with theory. *Physica D*, **19**:255–67, 1986.
[42] H. S. Greenside, W. M. Coughran, and N. L. Schryer. Nonlinear pattern formation near the onset of Rayleigh–Bénard convection. *Phys. Rev. Lett.*, **49**:726–9, 1982.
[43] H. S. Greenside and W. M. Coughran, Jr. Nonlinear pattern formation near the onset of Rayleigh–Bénard convection. *Phys. Rev. A*, **30**:398–428, 1984.
[44] P. S. Hagan and M. S. Cohen. Spiral waves in reaction diffusion equations. *SIAM J. Appl. Math*, **42**:762–86, 1982.

[45] A. Hagberg and E. Meron. From labyrinthine patterns to spiral turbulence. *Phys. Rev. Lett.*, **72**(15):2494–7, 1994.

[46] A. L. Hodgkin and A. F. Huxley. A quantitative description of membrane current and its application to conduction and excitation in nerve. *J. Physiology, London*, **117**(4):500–44, 1952.

[47] R. Hoyle. *Pattern Formation: An Introduction to Methods.* New York, Cambridge University Press, 2006.

[48] Y. Hu, R. Ecke, and G. Ahlers. Convection near threshold for Prandtl numbers near 1. *Phys. Rev. E*, **48**(6):4399–413, 1993.

[49] P. Coullet G. Iooss. Instabilities of one-dimensional cellular patterns. *Phys. Rev. Lett.*, **64**(8):866–9, 1990.

[50] J. P. Keener. Waves in excitable media. *SIAM J. Appl. Math.*, **39**(3):528–48, 1980.

[51] S. L. Judd and M. Silber. Simple and superlattice Turing patterns in reaction–diffusion systems: bifurcation, bistability, and parameter collapse. *Physica D*, **136**(1-2):45–65, 2000.

[52] K. Kaneko. *Theory and Applications of Coupled Map Lattices.* Chichester, Wiley, 1993.

[53] A. Karma. Scaling regime of spiral wave-propagation in single-diffusive media. *Phys. Rev. Lett.*, **68**(3):397–400, 1992.

[54] J. P. Keener. A geometrical-theory for spiral waves in excitable media. *SIAM J. Appl. Math.*, **46**(6):1039–56, 1986.

[55] D. Kincaid and E. W. Cheney. *Numerical Analysis: Mathematics of Scientific Computing.* Pacific Grove, CA, Brooks/Cole, third edition, 2002.

[56] E. Knobloch and J. De Luca. Amplitude equations for travelling wave convection. *Nonlinearity*, **3**:575–80, 1990.

[57] P. Kolodner, C. M. Surko, and H. L. Williams. Dynamics of traveling waves near the onset of convection in binary fluid mixtures. *Physica D*, **37**:319–33, 1989.

[58] A. Kudrolli and J. P. Gollub. Localized spatiotemporal chaos in surface waves. *Phys. Rev. E*, **54**(2):R1052–R1054, 1996.

[59] A. Kudrolli, B. Pier, and J. P. Gollub. Superlattice patterns in surface waves. *Physica D*, **123**:99–111, 1998.

[60] Y. Kuramoto. *Chemical Oscillations, Waves, and Turbulence.* New York, Springer-Verlag, 1984.

[61] S. D. Landy. Mapping the universe. *Scientific American*, **280**(6):38–45, 1999. A web version of this article is available at the URL www.sciam.com/1999/0699issue/0699landy.html.

[62] I. Lengyel, G. Rábai, and I. Epstein. Batch oscillation in the reaction of chlorine dioxide with iodine and malonic acid. *J. Am. Chem. Soc.*, **112**:4606–7, 1990.

[63] I. Lengyel, G. Rábai, and I. Epstein. Experimental and modeling study of oscillations in the chlorine dioxide–iodine–malonic acid reaction. *J. Am. Chem. Soc.*, **112**:9104–10, 1990.

[64] K. Libbrecht. *The Art of the Snowflake: A Photographic Album.* New York, Voyageur Press, 2007.

[65] E. M. Lifshitz and L. P. Pitaevskii. *Statistical Physics, Part 2*, volume 9 of *Course of Theoretical Physics*. Amsterdam, Elsevier, 1980.

[66] R. Lifshitz and D. M. Petrich. Theoretical model for Faraday waves with multiple-frequency forcing. *Phys. Rev. Lett.*, **79**(7):1261–4, 1997.

[67] S. S. Mao, J. R. deBruyn, and S. W. Morris. Electroconvection patterns in smectic films at and above onset. *Physica A*, **239**:189–203, 1997.

[68] M. Marder and J. Fineberg. How things break. *Physics Today*, **49**(9):24, 1996.

[69] C. Martel and J. M. Vega. Finite size effects near the onset of the oscillatory instability. *Nonlinearity*, **9**:1129–71, 1996.

[70] C. Martel and J. M. Vega. Dynamics of a hyperbolic system that applies at the onset of the oscillatory instability. *Nonlinearity*, **11**:105–42, 1998.

[71] R. M. May and W. J. Leonard. Nonlinear aspects of competition between three species. *SIAM J. Appl. Math*, **29**:243, 1975.

[72] D. Meiron and A. C. Newell. The shape of stationary dislocations. *Phys. Lett. A*, **113**:289–92, 1985.

[73] F. Melo, P. B. Umbanhowar, and H. L. Swinney. Hexagons, kinks, and disorder in oscillated granular layers. *Phys. Rev. Lett.*, **75**(21):3838–41, 1995.

[74] N. D. Mermin. The topological theory of defects in ordered media. *Rev. Mod. Phys*, **51**:591–648, 1979.

[75] S. Morris, E. Bodenschatz, D. Cannell, and G. Ahlers. The spatiotemporal structure of spiral-defect chaos. *Physica D*, **97**:164–79, 1996.

[76] S. W. Morris, J. R. deBruyn, and A. D. May. Patterns at the onset of electroconvection in freely suspended smectic films. *J. Stat. Phys.*, **64**:1025–43, 1997.

[77] S. W. Morris, E. Bodenschatz, D. S. Cannell, and G. Ahlers. Spiral defect chaos in large-aspect-ratio Rayleigh–Bènard convection. *Phys. Rev. Lett.*, **71**(13):2026–9, 1993.

[78] C. B. Muratov and V. V. Osipov. General theory of instabilities for patterns with sharp interfaces in reaction-diffusion systems. *Physical Review E*, **53**(4):3101–16, 1996.

[79] J. D. Murray. *Mathematical Biology I*. Berlin, Springer-Verlag, third edition, 2002.

[80] A. C. Newell and J. A. Whitehead. Finite bandwidth, finite amplitude convection. *J. Fluid Mech.*, **38**:279–303, 1969.

[81] A. C. Newell, T. Passot, and M. Souli. The phase diffusion and mean drift equations for convection at finite Rayleigh numbers in large containers. *J. Fluid Mech.*, **220**:187–252, 1990.

[82] L. Ning, G. Ahlers, and D. S. Cannell. Wave-number selection and traveling vortex waves in spatially ramped Taylor–Couette flow. *Phys. Rev. Lett.*, **64**(11):1235–8, 1990.

[83] L. Ning, Y. Hu, R. E. Ecke, and G. Ahlers. Spatial and temporal averages in chaotic patterns. *Phys. Rev. Lett.*, **71**(4):2216–19, 1993.

[84] Q. Ouyang and H. L. Swinney. Transition from a uniform state to hexagonal and striped Turing patterns. *Nature*, **352**:610–12, 1991.

[85] Q. Ouyang and H. L. Swinney. Spatial bistability of 2-dimensional Turing patterns in a reaction-diffusion system. *J. Phys. Chem.*, **96**(16):6773–6, 1992.

[86] Q. Ouyang and H. L. Swinney. Transition to chemical turbulence. *Chaos*, **1**(4):411–20, 1991.

[87] B. B. Plapp, D. A. Egolf, E. Bodenschatz, and W. Pesch. Dynamics and selection of giant spirals in Rayleigh–Bénard convection. *Phys. Rev. Lett.*, **81**(24):5334–7, 1998.

[88] A. La Porta and C. M. Surko. Quantitative characterization of 2d traveling-wave patterns. *Physica D*, **123**(1-4):21–35, 1998.

[89] W. H. Press, S. A. Teukolsky, W. T. Vetterling, and B. P. Flannery. *Numerical Recipes in C, Third Edition: The Art of Scientific Computing*. New York, Cambridge University Press, 2007.

[90] Z. Qu and A. Garfinkel. Nonlinear dynamics of excitation and propagation in cardiac muscle. In D. P. Zipes and J. Jalife, editors, *Cardiac Electrophysiology, From Cell to Bedside (4th Edition)*, Philadelphia, Saunders, 2004.

[91] Lord Rayleigh. On convection currents in a horizontal layer of fluid when the higher temperature in on the underside. *Phil. Mag. (Series 6)*, **32**(192):529–46, 1916.

[92] A. Schlüter, D. Lortz, and F. Busse. On the stability of steady finite amplitude convection. *J. Fluid Mech.*, **23**:129–44, 1965.

[93] L. A. Segel. Distant side-walls cause slow amplitude modulation of cellular convection. *J. Fluid Mech.*, **38**:203–24, 1969.

[94] L. A. Segel and J. L. Jackson. Dissipative structure: An explanation and an ecological example. *J. Theor. Biol.*, **37**:545–59, 1972.

[95] S. Setayeshgar and M. C. Cross. Turing instability in a boundary-fed system. *Phys. Rev. E*, **58**(4):4485–500, 1998.

[96] E. D. Siggia and A. Zippelius. Dynamics of defects in Rayleigh–Bènard convection. *Phys. Rev. A*, **24**:1036–49, 1981.

[97] J. Stoer, R. Bulirsch, R. Bartels, W. Gautschi, and C. Witzgall. *Introduction to Numerical Analysis*, volume 12 of *Texts in Applied Mathematics*. New York, Springer-Verlag, second edition, 1997.

[98] J. C. Strikwerda. *Finite Difference Schemes and Partial Differential Equations*. Pacific Grove, CA, Wadsworth & Brookes/Cole, 1989.

[99] S. H. Strogatz. *Nonlinear Dynamics and Chaos*. Reading, MA, Addison-Wesley, 1994.

[100] S. Tajima and H. Greenside. Microextensive chaos of a spatially extended system. *Phys. Rev. E*, **66**(1), 2002.

[101] W. Y. Tam, W. Horsthemke, Z. Noszticzius, and H. L. Swinney. Sustained spiral waves in a continuously fed unstirred chemical reactor. *J. Chem. Phys.*, **88**:3395–6, 1988.

[102] G. I. Taylor. Stability of a viscous liquid contained between two rotating cylinders. *Roy. Soc. Proc. A*, **CCXXIII**(612):289–343, 1923.

[103] G. Tesauro and M. C. Cross. Climbing of dislocations in nonequilibrium patterns. *Phys. Rev. A*, **34**:1363–79, 1986.

[104] O. Thual and S. Fauve. Localized structures generated by subcritical instabilities. *J. Phys. (Paris)*, **49**(11):1829–33, 1988.

[105] L. N. Trefethen, A. E. Trefethen, S. C. Reddy, and T. A. Driscoll. Hydrodynamic stability without eigenvalues. *Science*, **261**:578–84, 1993.

[106] A. Turing. The chemical basis of morphogenesis. *Phil. Trans. Roy. Soc. London*, **237**:37–72, 1952.

[107] P. B. Umbanhowar, F. Melo, and H. L. Swinney. Localized excitations in a vertically excited granular layer. *Nature*, **382**:793–6, 1996.

[108] W. van Saarloos. Front propagation into unstable states 2. Linear versus nonlinear marginal stability and rate of convergence. *Phys. Rev. A*, **39**(12):6367–90, 1989.

[109] W. van Saarloos. Front propagation into unstable states. *Phys. Reports*, **396**:29, 2003.

[110] I. Waller and R. Kapral. Spatial and temporal structure in systems of coupled nonlinear oscillators. *Phys. Rev. A*, **30**(4):2047–55, 1984.

[111] J. E. Wesfried and V. Croquette. Forced phase diffusion in Rayleigh–Bénard convection. *Phys. Rev. Lett.*, **45**:634–7, 1980.

[112] A. T. Winfree and S. H. Strogatz. Singular filaments organize chemical waves in 3 dimensions. 1. Geometrically simple waves. *Physica D*, **8**:35–49, 1983.

[113] A. T. Winfree. Varieties of spiral wave behavior: An experimentalist's approach to the theory of excitable media. *Chaos*, **1**(3):303–34, 1991.

[114] F. X. Witkowski, L. J. Leon, P. A. Penkoske, W. R. Giles, M. L. Spano, W. L. Ditto, and A. T. Winfree. Spatiotemporal evolution of ventricular fibrillation. *Nature*, **392**:78–82, March 5 1998.

Index

A^*, 130
$G(\theta)$, 257
$\Psi(p)$, 71
$\delta_{m,k}$, 454
λ_1, 474
$\mathbf{q}_c$, 62
$\mathbf{x}_\parallel$, 59
$\mathbf{x}_\perp$, 59
$\nabla_\parallel$, 59
$\nabla_\perp$, 60
ω_c, 62
$\sigma_\mathbf{q}$, 61
τ_0, 78
ε, 28, 132
ξ_0, 78
d_t, 89
q_N, 68
q_b, 282
u_b, 64

action potential, 406
activator, 104
adaptive system, 49
adiabatic elimination, 234
adjoint operator, 508
aeolian tune, 55
Airy function, 387
alternans, 438
amplitude, 130
amplitude equation, 133
antidromic stimulation, 408
aspect ratio, 25
ATP, 405
axial vector, 189

background wave number, 282
backward bifurcation, 130, 135, 498
backward-Euler method, 466
banded matrix, 470
Barkley model, 443
base state, 64
basis, 449

beating, 211
Belousov–Zhabotinsky reaction, 32, 50, 105, 108, 122, 180
Benjamin–Feir instability, 198, 364, 370, 371, 378, 380, 397, 438
bidomain heart model, 412
bifurcation, 23, 62, 496
    backward, 130, 135, 498
    codimension-n, 501
    codimension-two, 87, 133
    degenerate, 221, 500
    forward, 129, 135, 498
    Hopf, 23, 62, 358, 500
    imperfect, 72, 220, 501
    parity breaking, 330
    pitchfork, 135, 169, 498
    primary, 62
    reduced parameter, 77
    saddle-node, 266, 299, 498
    secondary, 62, 330, 360, 437
    simple, 500
    subcritical, 71, 130, 498
    supercritical, 71, 129, 498
    transcritical, 136, 170, 187, 265, 497
    transverse assumption, 78
biharmonic operator, 177, 492
binary fluid convection, 112, 360, 392, 393
blinking state, 358, 392
Bloch analysis, 140, 147, 231, 254, 330, 397, 438
BoenHoeffer–van der Pohl model, 414
boundary condition, 69
    amplitude, 219, 248
    enhancing, 219
    infinite, 65
    Neumann, 124
    no-slip, 93
    periodic, 60, 65
    suppressing, 219
Boussinesq equations, 8, 174, 204, 346
broken symmetry, 39, 318
Brusselator, 105, 121, 200, 240, 276, 403, 416, 472
Burgers equation, 373
Busse balloon, 127, 147, 474

BVDP model, 414

c.c., 130
CCD, 25
CDIMA chemical reaction, 114
CGLE, 196, 367
chaos, 44, 92
　domain, 28, 190, 261, 348
　Lorenz, 40, 89
　low-dimensional, 40
　spatiotemporal, 28, 30, 31, 34, 37, 198, 345, 349, 356, 365
　spiral defect, 27, 190, 195, 347
characteristic length scale, 41, 62, 80, 81
charge-coupled device, 25
chemotaxis, 18
chiral symmetry, 175, 257
climb, 164, 280, 282, 333
CML, 91, 356
coarsening, 49, 185, 332
codimension-two bifurcation, 87, 133
coefficient of thermal expansion, 8
coherence length, 54, 78, 227
Cole–Hopf transformation, 373, 397
complex Ginzburg–Landau equation, 196, 367
computer word, 447
condition number, 448, 478, 488, 489
confined coordinate, 58
conservation equation, 112
continuation method, 478
continuous transition, 129
continuously fed stirred tank reactor (CSTR), 109
control parameter, 59
control parameter ramp, 334
coordinate
　confined, 58
　extended, 58
correction, 478
correlation length, 42
coupled map lattice, 91, 356
Crank–Nicolson, 466
critical
　exponent, 36
　frequency, 62
　parameter value, 62
　Rayleigh number, 8
　slowing down, 73, 221
　wave number, 62
　wave vector, 62
cross-stripe instability, 145, 270, 277
CSTR, 109

defect, 129, 160, 279
　core, 162, 279, 282–284, 290, 338
　line, 163
　point, 163
defibrillator, 412
degeneracy, 83
degenerate bifurcation, 221, 500
delay-differential equation, 47
demodulation, 392, 394

depolarization, 406
diblock copolymer, 183
Dictyostelium discoideum, 17
diffusion coefficient, 112
diffusion length, 104
diffusion matrix, 98
director, 84, 162
disclination, 160, 162, 165
discontinuous transition, 130
discretization, 447
dislocation, 27, 160, 162
　climb, 164, 280, 282, 310, 333
　glide, 165, 283, 288, 310
　selected wave number, 164, 282, 333
domain chaos, 28, 190, 261, 316, 348
domain wall, 163
driven out of equilibrium, 5
driven-dissipative system, 6

Eckhaus instability, 142, 149, 171, 197, 233, 236, 242, 318, 325, 354, 371
eikonal approximation, 430
Ekman vortex, 220
electrocardiogram, 439
electroconvection, 84, 222
elementary reaction, 111
embedding, 39
equivariant bifurcation theory, 87, 501
excitable dynamics, 420
　bistability, 421
excitable medium, 33, 358, 402
　Barkley model, 443
explicit time integration method, 459
extended coordinate, 58
extended direction, 59
extensive variable, 350

fibrillation, 19, 402, 404
Fife scaling, 428, 432
filament, 429
finite wavelength instability, 141, 147
finite-difference approximation, 452
finite-difference method, 450
first-order transition, 130
FitzHugh–Nagumo model, 200, 403, 413, 415
floating-point number, 448
fluid turbulence, 23, 49, 316
focus, 160, 165, 333
　wave-number selection, 325
forward bifurcation, 129, 135, 498
forward-Euler method, 461
Fourier analysis, 61, 66
Fourier stability method, 465
fractional derivative, 177
front, 43, 280, 296, 342, 421, 436
　phase, 424
　pulled, 303
　pushed, 304
　trigger, 424
functional, 179

Galerkin method, 139, 147, 169, 171, 317, 449
gating variable, 409
Gauss's law, 203
generalized biharmonic equation, 492
ghost point, 464
Ginzburg–Landau equation, 238, 239
Goldstone mode, 319
gradient flow, 179
grain boundary, 160, 163
    amplitude, 290
    perpendicular, 293, 394
    phase, 290
    small angle, 290
    zipper, 394
granule, 13
Gross–Pitaevski equation, 196
group speed, 361
group theory, 86
    equivariant bifurcation theory, 87, 501
    representation, 87
growth factor, 491
growth rate, 61

healing length, 227
heart
    arrhythmia, 403
    bidomain model, 412
    monodomain model, 412
    vulnerable period, 440
Helmholtz equation, 468
heteroclinic orbit, 280, 301
Hodgkin–Huxley model, 403
homogeneous, 57
Hopf bifurcation, 23, 62, 358, 500
horizontal thermal diffusion time, 53
h.o.t., 212

I-o, 79
I-s, 77, 79
II-o, 80
II-s, 80
III-o, 81
ill-conditioned, 488
imperfect bifurcation, 72, 220, 501
implicit function, 415, 494
implicit time integration method, 459
incompressible flow, 9
inhibitor, 104
initial-boundary-value problem, 457
instability
    absolute, 43, 361, 380
    Benjamin–Feir, 198, 364, 370, 371, 378, 380, 397, 438
    convective, 43, 361, 380
    cross-stripe, 145, 270, 277
    Eckhaus, 142, 149, 171, 197, 233, 236, 242, 318, 325, 354, 371
    finite wavelength, 141, 147
    Kuppers–Lortz, 277
    long-wavelength, 147
    oscillatory, 62, 360
    short wavelength, 147
    skew-varicose, 150, 329, 355
    zigzag, 142, 145, 148, 250, 252, 274, 318, 325, 328, 329, 354
intensive variable, 350
ion channel, 405
ion pump, 405
Ising–Bloch transition, 343
iteration, 446, 476

kink, 107
Kolmogorov turbulence, 316
Kuppers–Lortz, 277
Kuramoto–Sivashinsky equation, 199, 492

laminar, 54
laminar state, 199
Landau theory, 158
largest Lyapunov exponent λ_1, 474
laser Doppler velocimetry, 71
lasing instability, 382
lateral boundary, 60
law of mass action, 111, 122, 200
leakage current, 410
leap-frog method, 492
Lewis number, 393
Lifshitz–Petrich equation, 192
linear marginal selection, 304
linear stability analysis, 8, 56, 63, 69
lipid bilayer, 405
liquid crystal, 222
    nematic, 84
    smectic, 180, 183
logistic map, 91, 356
long-wavelength instability, 147
Lorenz model, 39, 89, 509
Lyapunov functional, 179

machine precision, 457
marginal stability, 67, 305
matrix bandwidth, 470
mean-field approximation, 221
mean flow, 188, 327
mesh size, 451
meandering instability, 438
microstructure, 17
monodomain heart model, 412
morphogenesis, 35, 96, 174, 308
Mullin–Sekerka instability, 343
multiple root, 494
multiple scales, 218, 503
multistep explicit time integration method, 460

Navier–Stokes equations, 54
Nernst equation, 406
neuron
    channel, 405
    ion pump, 405
neutral stability curve, 67
Newell criterion, 370, 378

Newton method, 139, 446, 477
　Newton–Kantorovich, 477
　Newton–Raphson, 477
no-slip boundary condition, 93
Noble cardiac model, 413
noise sustained state, 384
non-Boussinesq convection, 187
non-normal matrix, 95
non-polynomial reaction rate, 123
nonadiabatic, 182, 283, 296, 340
nonlinear front selection, 306
nonlinear Schrödinger equation, 196
nonlinear steady state, 129
nonlinear WKB, 319
normal form, 496
Nozaki–Bekki hole, 375
null space, 507
nullcline, 99, 121, 404, 415, 494, 495
Nusselt number, 71

operator splitting, 447, 460, 470
order of a chemical reaction, 111
order parameter, 71
order-parameter equation, 193
Oregonator, 105, 122, 403, 442
orthogonal matrix, 82
oscillatory bifurcation, 62
oscillatory instability, 77, 149
oscillon, 341

paramecium, 402
parity symmetry, 359
pattern formation, 60
pattern selection, 84
pde, 41
phase, 127, 130, 159, 319
phase diffusion, 127, 160, 197, 234, 252, 319, 514
phase front, 424
phase portrait, 414
pinch point analysis, 304
pitchfork bifurcation, 135, 169, 498
potential density, 181
potential dynamics, 179, 224
power law, 36
power spectrum, 73
primary bifurcation, 62
pulse, 280, 340, 424
　width, 426
Purkinje fiber, 411

quadratic convergence, 478
quasicrystal, 31, 156, 192
quenched state, 49
quilt, 43

rate constant, 111
rate of reaction, 111
Rayleigh number, 8
　critical, 8
Rayleigh–Bénard convection, 3, 59

reaction diffusion, 59
reciprocal lattice, 156
reduced bifurcation parameter, 77
reduced control parameter, 132
reflection coefficient, 390
refractory phase, 408
relaxation oscillation, 414, 419
relaxational dynamics, 179
representation, 447
residual, 478
Reynolds number, 54
　axial, 386
　inner, 21, 386
　outer, 21
　rotational, 28
roll state, 147
rounding, 448
Routh–Hurwitz criterion, 102

saddle-node bifurcation, 266, 299, 498
saturated state, 129
Schrödinger equation, 206, 398, 504
scroll wave, 429
second-order transition, 129
secondary bifurcation, 62, 330, 360, 437
secular perturbation theory, 508
shadowgraphy, 24
shock, 364, 373, 378, 397, 398
sideband instability, 364, 371
siemens, 410
singularity, 160
sink, 364, 373, 394
sinoatrial node, 19
sinus node, 429
skew-varicose, 150, 329, 355
slime mold, 17
smectic, 180, 183
Snell's law, 442
solidification, 363
solidification front, 17
soliton, 196
solvability condition, 503, 507
Soret effect, 112
source, 364, 373, 394, 429
spatial resolution, 451
spatially extended, 41
spatiotemporal intermittency, 199
spiral, 364, 429
　meandering instability, 438
spiral defect chaos, 27, 190, 195, 347
stability balloon, 139, 194, 230, 250, 325, 370
stall speed, 423
starch indicator, 114
state of a dynamical system, 91
stationary bifurcation, 62
stationary instability, 77
stationary phase analysis, 304, 362
stoichiometric coefficient, 111
stripe coupling coefficient, 257
stripes, 26
structural stability, 305

structurally unstable, 238
subcritical bifurcation, 71, 130, 498
supercritical bifurcation, 71, 129, 498
superlattice, 30, 156, 276
supertransient, 205
Swift–Hohenberg model, 175, 492
   one-dimensional, 54, 63, 137, 504
   two-dimensional, 82

tachycardia, 404, 439
target, 160, 165, 333, 364, 377, 398, 429
Taylor number, 22
Taylor series, 446
Taylor–Couette flow, 20, 59, 74, 220, 306, 335, 356, 385
time-reversal symmetry, 216
time step, 458
topological defect, 27, 129, 160, 279, 375, 377
trace, 101
transcritical bifurcation, 136, 170, 187, 265, 266, 497
translationally invariant, 57
transverse assumption, 78
tridiagonal matrix, 469
trigger front, 424
Turing model, 97, 174
twist variable, 439

uniaxial medium, 84

uniform nonequilibrium state, 60
uniform state, 57
universal, 36, 81, 209, 221

Van der Pohl equation, 414
vertical thermal diffusion time, 53
vertical vorticity, 189, 327
voltage clamp, 411
von Kármán vortex street, 387
von Neumann stability analysis, 490

wave equation, 401
wave-number selection, 22, 165, 316, 331
   dislocation, 164, 282, 333
   focus, 325, 326, 333
   front, 333
   grain boundary, 291, 333
   ramp, 335
wave packet, 361
weakly nonlinear, 72
Wilson–Cohen model, 200
winding number, 161

zigzag instability, 142, 145, 148, 250, 252, 274, 318, 325, 329, 354
zone-boundary instability, 438

Printed in the United States
By Bookmasters